Lecture Notes in Mathematics

Edited by A. Dold and B. Eckmann

T0226012

1391

S. Cambanis A. Weron (Eds.)

Probability Theory on Vector Spaces IV

Proceedings of a Conference, held in Łańcut, Poland, June 10–17, 1987

Springer-Verlag

Berlin Heidelberg New York London Paris Tokyo Hong Kong

Editors

Stamatis Cambanis
Department of Statistics, University of North Carolina
Chapel Hill, NC 27599, USA

Aleksander Weron
Institute of Mathematics, Technical University of Wroclaw
50-370 Wroclaw, Poland

Mathematics Subject Classification: Primary 60 Bxx, 28 Cxx
Secondary 46 L 10, 60 E 07, 60 G 10, 60 G 15, 60 G 25, 60 G 46, 60 H 05, 81 C 20, 82 A 31

ISBN 3-540-51548-8 Springer-Verlag Berlin Heidelberg New York
ISBN 0-387-51548-8 Springer-Verlag New York Berlin Heidelberg

Printing and binding: Druckhaus Beltz, Hemsbach/Bergstr.
2146/3140-543210 – Printed on acid-free paper

Dedicated to the memory of

HUGO STEINHAUS (1887-1972)

on the occasion of his 100-th birthday

PREFACE

This volume consists of a selection of 34 contributions presented during the international conference PROBABILITY THEORY ON VECTOR SPACES IV held in Lancut (south-eastern Poland), June 10-17, 1987. The conference, the fourth of such meetings organized in Poland during the last decade, was aimed at bringing together a group of researchers to discuss their current work on functional analysis aspects of probability theory, to provide encouragement for further research through the lectures of invited scholars, and to introduce to younger colleagues the latest research techniques in infinite dimensional probability theory.

During the meeting the participants celebrated HUGO STEINHAUS's 100th birthday. The invited speakers on the special session were Z.Ciesielski and K. Urbanik. Professor Steinhaus had a world wide reputation. He was one of the founders of the internationally acknowledged "Lwow school" of functional analysis, and "Wroclaw school" of probability.

It has been the aim of the editors of these Proceedings to include new and significant results. We wish to express our deep gratitude to all the referees for their indispensable help. The selected papers tie together known results of the previous proceedings volumes: Springer-Verlag's Lecture Notes in Math. vol. 656 (1978), vol. 828 (1980), and vol. 1080 (1984), describe the underlying ideas, summarize the state-of-the-art, and state some open problems. Topics covered in this volume include: PROBABILITY MEASURES ON ALGEBRAIC AND TOPOLOGICAL STRUCTURES (A.Bartoszewicz, S.A.Chobanjan & G.J.Georgobiani, Ph.Feinsilver & R.Schott, W.Hazod, M.Lewandowski & W.Linde, A.Luczak, J.Misiewicz & C.Ryll-Nardzewski, and V.Tarieladze); LIMIT THEOREMS ON VECTOR SPACES (J.C.Alt, T. Byczkowski & P.Graczyk, E.Gine & J.Zinn, M.G.Hahn, J.Kuelbs & D.C.Weiner, E.Hensz, R.Jajte, and A.J.Rackauskas); INFINITELY DIVISIBLE (STABLE, GAUSSIAN) MEASURES AND PROCESSES (S.Csorgo, X.Fernique, R.LePage, A.Madrecki, M.Marques & S.Cambanis, G.Pap, M.Rybaczuk & K.Weron, W.Smolenski & R.Sztencel, and T.Zak); and STOCHASTIC INTEGRALS (J.Jurek, K.Podgorski, A.Russek, K.Samotij, Z.Suchanecki, and J.Wos).

The contribution of R.LePage includes, as an appendix, his very popular, never published earlier, Technical Report "Multidimensional infinitely divisible variables and processes: PartI: Stable case", Stanford Univ., 1980. Proceedings contain also two longer survey articles on recent developments in the integration theory of Banach space valued measures (P.Masani & H.Niemi), and in the prediction theory of infinite dimensional stationary sequences (A.Makagon & H.Salehi).

The conference took place in the Alfred Potocki palace in Lancut. The palace was built in an early Baroque style (1629-41) and set in a handsome park. It is ranked among the most superb stately homes to be seen in Eastern Europe. We wish to thank the authorities of the Lancut Museum for providing facilities which made it possible to hold our conference in so charming place. Cordial thanks are also due to M.Krol, T.Inglot, A.Makagon, and J.Misiewicz who helped in the organization of the meeting. The conference was made possible through the partial support of the Research Program CPBP 01.02. which is acknowledged with gratitude.

Aleksander Weron

CONTENTS

Probability Theory on Vector Spaces IV
Lancut, June'87, Springer's LNM 1391

UNE FORME GÉNÉRALE DE LA LOI FORTE DES GRANDS NOMBRES POUR DES VARIABLES ALÉATOIRES VECTORIELLES

Jean-Christian ALT
U.E.R de Mathématiques
7, rue René Descartes
67084 STRASBOURG Cédex

I. Introduction. — Cet article présente une forme générale de la loi forte des grands nombres (en abrégé : l. f. g. n) pour des variables aléatoires (v.a) à valeurs dans un espace de Banach. Etant données une suite $(X_n)_n$ de v.a. à valeurs dans un espace de Banach séparable $(B, \| \,\|)$, et une suite croissante de réels > 0 $(a_n)_n$ tendant vers l'infini, on cherche à caractériser la l.f.g.n., i.e. la convergence presque sûre (p.s) vers 0 de la suite des sommes partielles normalisées S_n/a_n (on a posé pour tout $n \geq 1 : S_n = X_1 + \cdots + X_n$). On se fixe un réel $M > 1$; on sait qu'il existe alors (cf [13], lemme 3.3) une suite strictement croissante d'entiers $(n_k)_{k\geq 1}$ telle que

$$\forall\, k \geq 1 \qquad M a_{n_k} \leq a_{n_{k+1}} \leq M^3 a_{(n_k)+1}.$$

Le principal objectif de ce travail est de montrer que si les variables $(X_n)_n$ sont convenablement bornées, il est possible de caractériser la l.f.g.n. au moyen de variances "faibles" similaires à celles déjà utilisées dans [2] et [3]. De façon précise, si les variables indépendantes et centrées $(X_n)_n$ vérifient l'hypothèse de bornitude

$$\forall\, n \geq 1 \qquad \|X_n\| \leq b_n \quad (p.s),$$

où $(b_n)_n$ est une suite croissante de réels > 0 telle que

$$\exists\, \gamma > 0 \quad \sum_{k\geq 1} \exp(-\gamma a_{n_k}/b_{n_k}) < \infty,$$

et si la loi faible des grands nombres est satisfaite, i.e. si S_n/a_n converge en probabilité vers 0, alors la l.f.g.n équivaut à la convergence pour tout $\epsilon > 0$ de la série $\Sigma_{k\geq 1} \exp(-\epsilon[a_{n_k}/\sigma_k]^2)$, où l'on a posé

$$\sigma_k^2 = \sup\Big\{ \sum_{n_{k-1} < i \leq n_k} E\langle X_i, x'\rangle^2 \ ; \quad \|x'\|_{B'} \leq 1 \Big\}.$$

Ce résultat tire son origine de la l.f.g.n de Prokhorov, dont il constitue une généralisation. Rappelons l'énoncé de ce théorème.

THÉORÈME 1 (YOU. V. PROKHOROV [9]). — *Soit $(X_n)_n$ une suite de variables aléatoires réelles (v.a.r) indépendantes et centrées satisfaisant la condition de bornitude*

$$\exists\, C > 0 \quad \forall\, n \in \mathbb{N} \quad |X_n| \leq Cn/\log_2 n \quad (p.s)$$

(avec pour tout $x \in \mathbb{R}_+ : \log_2 x = ln[ln(xVe^e)]$).

La suite $(S_n/n)_n$ converge alors p.s. vers 0 si et seulement si

$$(CP) \qquad \forall\, \epsilon > 0 \quad \sum_{n \geq 1} \exp(-\epsilon l_n^{-1}) < \infty$$

où l'on a posé

$$l_n = 2^{-2n} \sum_{2^n < i \leq 2^{n+1}} EX_i^2.$$

Prokhorov a de plus montré que cet énoncé est optimal au sens suivant : étant donnée une suite $(C_n)_n$ de réels tendant vers l'infini, il existe 2 suites $(X_n)_n$ et $(Y_n)_n$ de v.a.r. indépendantes et symétriques vérifiant

i) $\forall\, n \in \mathbb{N} \quad |X_n| \leq C_n n/\log_2 n \quad$ et $\quad |Y_n| \leq C_n n/\log_2 n$
ii) $\forall\, n \in \mathbb{N} \quad EX_n^2 = EY_n^2$
iii) $\forall\, \epsilon > 0 \quad \sum_{n \geq 1} \exp(-\epsilon l_n^{-1}) < \infty \quad$ (où $l_n = 2^{-2^n} \Sigma_{2^n < i \leq 2^{n+1}} EX_i^2$),

et telles que $(X_1 + \cdots + X_n/n)_n$ converge p.s. vers 0 et $(Y_1 + \cdots + Y_n/n)_n$ diverge p.s. Autrement dit l'hypothèse de bornitude du théorème 1 est la moins restrictive permettant de caractériser la l.f.g.n par la convergence d'une série exprimée à l'aide des moments d'ordre 2 des v.a.

Il est intéressant de comparer la l.f.g.n de Prokhorov avec celle de Kolmogorov qui concerne des v.a.r. indépendantes et identiquement distribuées (i.i.d) et qui a reçu très tôt l'extension suivante aux espaces de Banach.

THÉORÈME 2 (E. MOURIER [8]). — *Soit $(X_n)_n$ une suite de v.a. i.i.d. à valeurs dans un espace de Banach séparable B. La suite $(S_n/n)_n$ converge p.s. vers un élément de B si et seulement si $E\|X_1\| < \infty$, la limite p.s s'identifiant dans ce cas à EX_1.*

On constate que le passage du cas réel au cas banachique s'effectue simplement en remplaçant les valeurs absolues par des normes.

Il était donc naturel d'adopter une démarche similaire pour la généralisation du théorème de Prokhorov aux espaces de Banach en remplaçant dans la condition (CP) du théorème 1 les quantités l_n par

$$\Lambda_n = 2^{-2n} \sum_{2^n < i \leq 2^{n-1}} E\|X_i\|^2.$$

Cette démarche a permis à Kuelbs et Zinn de démontrer le théorème suivant.

THÉORÈME 3 (J. KUELBS ET J. ZINN [5], THÉORÈME 2). — *Soit* $(X_n)_n$ *une suite de v.a. indépendantes à valeurs dans un espace de Banach séparable, vérifiant les conditions*

1) $\exists\, C > 0 \quad \forall\, n \geq 1 \quad \|X_n\| \leq Cn/\log_2 n \quad (p.s)$
2) $\forall\, \epsilon > 0 \quad \sum_{n\geq 1} \exp\bigl(-\epsilon\Lambda_n^{-1}\bigr) < \infty$

On a alors l'équivalence suivante

$$\Bigl(S_n/n \xrightarrow{P} 0\Bigr) \Longleftrightarrow \Bigl(S_n/n \xrightarrow{p.s.} 0\Bigr)$$

Cet énoncé a l'avantage d'être vrai dans tout espace de Banach, mais nécessite une hypothèse 2) très restrictive. En effet, si dans le cas de v.a. équidistribuées les normes se comportent comme des équivalents naturels des valeurs absolues, leur utilisation pour des v.a non équidistribuées se traduit par contre par une perte d'information. Une seconde démarche consiste donc à rechercher des moments plus adéquats, permettant d'obtenir une condition (CP) moins forte, et, en contrepartie, à imposer des restrictions sur la nature de l'espace de Banach considéré. Cette démarche a été suivie par B. Heinkel qui a étendu le théorème de Prokhorov aux espaces de Banach p-uniformément lisses en employant une condition s'exprimant à l'aide des quantités

$$\tilde{\lambda}_n = 2^{-2n} \sum_{2^n < i \leq 2^{n+1}} \sup\Bigl\{E\langle X_i, x'\rangle^2;\quad \|x'\|_{B'} \leq 1\Bigr\}.$$

Rappelons qu'un espace de Banach B est dit p-uniformément lisse (pour $p \in]1,2]$) s'il existe une constante $K > 0$ telle que pour tout couple (x, y) d'éléments de B on ait

$$\|x+y\|^p + \|x-y\|^p \leq 2\|x\|^p + K\|y\|^p.$$

Voici l'un des théorèmes obtenus, ayant pour cadre les espace 2-uniformément lisses.

THÉORÈME 4 (B. HEINKEL [4], THÉORÈME 2). — *Soit* $(X_n)_n$ *une suite de v.a. indépendantes et centrées à valeurs dans un espace de Banach 2-uniformément lisse. Supposons vérifiées les 3 conditions :*

1) $\exists\, C > 0 \quad \forall\, n \in \mathbb{N} \quad \|X_n\| \leq Cn/\log_2 n \quad (p.s)$
2) $\lim_{n\to\infty} n^{-2} \sum_{i=1}^{n} E\|X_i\|^2 = 0$
3) $\forall\, \epsilon > 0 \quad \sum_{n\geq 1} \exp\bigl(-\epsilon\tilde{\lambda}_n^{-1}\bigr) < \infty$.

Sous ces hypothèses la suite $(S_n/n)_n$ *converge p.s. vers* 0.

La condition 3) du théorème précédent n'est pas encore pleinement satisfaisante car on vérifie sans difficulté qu'elle n'est pas nécessaire. En fait nous avons démontré dans [2] que la condition de Prokhorov naturelle qui est nécessaire dans n'importe quel espace de Banach pour la l.f.g.n sous l'hypothèse 1) du théorème 4 s'exprime à l'aide des quantités

$$\lambda_n = 2^{-2n} \sup\Big\{ \sum_{2^n < i \leq 2^{n+1}} E\langle X_i, x'\rangle^2 \ ; \ \|x'\|_{B'} \leq 1 \Big\},$$

la condition de Prokhorov qui en résulte s'écrivant

$$(CP) \qquad \forall\, \epsilon > 0 \quad \sum_{n \geq 1} \exp(-\epsilon \lambda_n^{-1}) < \infty.$$

La suffisance de cette condition a, dans un premier temps, été prouvée pour des v.a. à valeurs dans un espace de Banach de type $p \in [1, 2]$. En adoptant une méthode de randomisation gaussienne mise au point par M. Ledoux pour l'étude de la loi du logarithme itéré équidistribuée dans les espaces de type 2 ([6]) nous avons pu démontrer le théorème suivant.

THÉORÈME 5 ([2], THÉORÈME 5). — *Soit $(X_n)_n$ une suite de v.a. indépendantes et centrées à valeurs dans un espace de Banach de type $p \in]1, 2]$. Supposons vérifiées les deux conditions :*

1) *il existe une suite $(C_n)_n$ de réels tendant vers 0 telle que*

$$\forall\, n \in \mathbb{N} \qquad \|X_n\| < C_n\, n/\log_2 n \quad (p.s)$$

2) $$\lim_{n\to\infty} \qquad n^{-p} \textstyle\sum_{i=1}^n E\|X_i\|^p = 0.$$

Sous ces hypothèses on a l'équivalence

$$\Big(S_n/n \xrightarrow{p.s} 0\Big) \iff \Big(\forall\, \epsilon > 0 \quad \sum_{n \geq 1} \exp(-\epsilon \lambda_n^{-1}) < \infty\Big).$$

Finalement M. Ledoux et M. Talagrand ont réussi à s'affranchir des hypothèses restrictives de l'énoncé précédent, et à montrer que le théorème initial de Prokhorov est valable sans limitation dans tout espace de Banach. Leur énoncé est le suivant.

THÉORÈME 6 (M. LEDOUX, M. TALAGRAND [7], THÉORÈME 13). — *Soit $(X_n)_n$ une suite de v.a. indépendantes et centrées à valeurs dans un espace de Banach séparable. On suppose vérifiées les deux conditions :*

1) $\exists\, C > 0 \quad \forall\, n \in \mathbb{N} \quad \|X_n\| \leq Cn/\log_2 n \quad (p.s)$

2) $S_n/n \xrightarrow{P} 0.$

Sous ces hypothèses on a l'équivalence

$$\Big(S_n/n \xrightarrow{p.s} 0\Big) \iff \Big(\forall\, \epsilon > 0 \sum_{n \geq 1} \exp(-\epsilon \lambda_n^{-1}{}_n) < \infty\Big).$$

M. Ledoux et M. Talagrand ont énoncé le théorème précédent dans le langage des processus empiriques. Ils ont imaginé pour sa démonstration une méthode de randomisation par des variables de Rademacher. En utilisant cette méthode, nous avons été à même de prouver le résultat exposé au début de cette introduction; ce résultat constitue la forme définitive d'un énoncé qui figurait déjà dans [3] ([3], théorème 7). Rappelons en le cadre, qui est le plus général possible autorisant une caractérisation de la l.f.g.n au moyen des variances faibles des v.a. : on se donne 2 suites croissantes $(a_n)_n$ et $(b_n)_n$ de réels > 0 et une suite strictement croissante d'entiers $(n_k)_k$ satisfaisant les hypothèses :

$$H \begin{cases} 1)\exists\ M > 1 \quad \forall\ k \geq 1 \quad Ma_{n_k} \leq a_{n_{k+1}} \leq M^3 a_{(n_k)+1} \\ 2)\exists\ \gamma > 0 \quad \sum_{k\geq 1} \exp(-\gamma a_{n_k}/b_{n_k}) < \infty\ . \end{cases}$$

Il est évident de constater que le théorème 6 correspond au cas particulier $a_n = n$, $b_n = Cn/\log_2 n$ et $n_k = 2^k$. Etant donnée une suite $(X_n)_n$ de v.a. à valeurs dans B on posera :

$$I_k =]n_{k-1}, n_k] \qquad \text{et} \qquad \sigma_k^2 = \sup\Big\{\sum_{i\in I_k} E\langle X_i, x'\rangle^2\ ; \quad \|x'\|_{B'} \leq 1\Big\}\ .$$

Avec ces notations voici l'énoncé précis du résultat annoncé.

THÉORÈME 7. — *Soit $(X_n)_n$ une suite de v.a. indépendantes et centrées à valeurs dans un espace de Banach séparable $(B, \|\ \|)$; et soient $(a_n)_n$ et $(b_n)_n$ deux suites croissantes de réels > 0, et $(n_k)_k$ une suite strictement croissante d'entiers vérifiant les hypothèses H. On suppose satisfaites les conditions :*

1) $\forall\ n \geq 1 \quad \|X_n\| \leq b_n \quad (p.s)$
2) $S_n/a_n \xrightarrow{P} 0$.

Sous ces hypothèses $(S_n/a_n)_n$ converge p.s. vers 0 si et seulement si

$$\forall\ \epsilon > 0 \quad \sum_{k\geq 1} \exp\Big(-\epsilon[a_{n_k}/\sigma_k]^2\Big) < \infty.$$

Remarque 1. — Le théorème 7 admet le corollaire suivant. On se donne d'une part une suite $(X_n)_n$ de v.a. indépendantes et symétriques à valeurs dans un espace de Banach séparable B; et d'autre part une suite croissante de réels > 0 $(a_n)_n$ et une suite strictement croissante d'entiers $(n_k)_k$ telles que l'hypothèse $H1)$ soit satisfaite. On suppose vérifiées les trois conditions suivantes, dont les deux premières sont nécessaires pour la l.f.g.n.:

1) $\forall\ \epsilon > 0 \quad \sum_{n\geq 1} P\{\|X_n\| \geq \epsilon a_n\} < \infty$
2) $S_n/a_n \xrightarrow{P} 0$
3) $\forall\ \epsilon > 0 \quad \sum_{k\geq 1} \exp(-\epsilon[a_{n_k}/\sigma_k]^2) < \infty$

où l'on a posé

$$\sigma_k^2 = \sup\Big\{ \sum_{n_{k-1}<i\leq n_k} E\langle X_i, x'\rangle^2 \ ; \quad \|x'\|_{B'} \leq 1\Big\} .$$

Pour toute suite croissante $(b_n)_n$ de réels > 0 vérifiant $H2)$ on a alors l'équivalence :

$$\Big(S_n/a_n \xrightarrow{p.s.} 0\Big) \iff \Big(a_n^{-1} \sum_{i=1}^{n} X_i 1_{\{b_i\leq\|X_i\|\leq a_i\}} \xrightarrow{p.s.} 0\Big) .$$

Remarque 2. — Il est connu que la loi du logarithme itéré (l.l.i.) apparaît sous bien des aspects comme une forme limite de la l.f.g.n. Il n'est donc pas surprenant que le théorème 7 admette une "version l.l.i.". Cette version s'obtient en remplaçant l'hypothèse 2) par

2′) $\qquad (S_n/a_n)_n$ est bornée en probabilité,

la conclusion se transformant alors en :

$$((S_n/a_n)_n \text{ est p.s. bornée}) \iff \Big(\exists\, \epsilon > 0 \quad \sum_{k\geq 1} \exp(-\epsilon[a_{n_k}/\sigma_k]^2) < \infty\Big).$$

Nous donnerons en cas particulier de cette version l.l.i. une généralisation banachique de la l.l.i. de Kolmogorov, qui, comme le théorème 7, était déjà énoncée sous une forme partielle dans [3], ([3] théorème 6).

THÉORÈME 8. — *Soit $(X_n)_n$ une suite de v.a. indépendantes et centrées à valeurs dans un espace de Banach séparable B. On pose pour tout entier n*

$$\sigma_n^2 = \sup\big\{E\langle S_n, x'\rangle^2 \ ; \quad \|x'\|_{B'} \leq 1\big\}.$$

Soit $(s_n)_n$ une suite croissante de réels > 0 tendant vers $+\infty$. On suppose vérifiées les 3 conditions :

1) $\forall\, n \in \mathbb{N} \quad \|X_n\| \leq s_n(\log_2 s_n)^{-1/2}$ (p.s)
2) *la suite $\big(S_n/s_n(\log_2 s_n)^{1/2}\big)_n$ est bornée en probabilité*
3) $\exists\, c > 0 \quad \forall\, n \geq 1 \quad \sigma_n \leq c s_n$.

La suite $\big(S_n/s_n(\log_n s_n)^{1/2}\big)_n$ est alors p.s. bornée.

II. Démonstration du théorème 7.

A. Nécessité. Comme pour le théorème de Prokhorov dans le cas réel, elle repose sur le lemme suivant (cf. [10], théorème 5.2.2).

LEMME 9. — *Il existe 2 constantes strictement positives c_1 et c_2 telles que pour toute suite $Y_1, \ldots, Y_n$ de v.a.r indépendantes et centrées vérifiant*

$$\exists\, c > 0 \quad \forall\, i = 1, \ldots, n \quad |Y_i| \leq c s_n \quad (p.s)\ ,$$

où

$$s_n = \sum_{i=1}^{n} EY_i^2\ ,$$

on ait pour tout réel $\epsilon > 0$ *satisfaisant les inégalités*

$$\epsilon \geq c_1 \quad \text{et} \quad \epsilon c \leq c_2$$

la minoration

$$P\{Y_1 + \cdots + Y_n \geq \epsilon s_n\} \geq \exp(-\epsilon^2).$$

On vérifie aisément que, sans perte de généralité, les variables $(X_n)_n$ peuvent être supposées symétriques.
Des hypothèses H résulte l'égalité

$$\lim_{n\to\infty} b_n/a_n = 0. \tag{2.1}$$

Les conditions 1) et 2) du théorème 7 permettent donc d'affirmer, en vertu d'un théorème de De Acosta ([1], lemme 3.1) :

$$\lim_{n\to\infty} E\big\|S_n/a_n\big\|^2 = 0.$$

Posons pour tout $k \geq 1$: $T_k = \sum_{i\in I_k} X_i$; on a aussi

$$\lim_{k\to\infty} E\big\|T_k/a_{n_k}\big\|^2 = 0. \tag{2.2}$$

Fixons nous un réel $\epsilon > 0$; et choisissons un réel $\delta > 0$ tel que

$$2\gamma\delta \leq c_2 \qquad \text{et} \qquad 2\epsilon\delta^2 \leq 1. \tag{2.3}$$

La suite de v.a indépendantes $(T_k/a_{n_k})_k$ convergeant p.s. vers 0 d'après les hypothèses, on a

$$\sum_{k\geq 1} P\Big\{\|T_k/a_{n_k}\| \geq \epsilon\delta\Big\} < \infty. \tag{2.4}$$

Par définition de σ_k, il existe un élément f_k de B' tel que $\|f_k\| \leq 1$ et

$$s_k^2 = E\langle T_k/a_{n_k}, f_k\rangle^2 \leq \sigma_k^2/a_{n_k}^2 \leq 2s_k^2 \tag{2.5}$$

D'où, en conséquence de (2.2) :

$$\lim_{k\to\infty} s_k^2 = 0. \tag{2.6}$$

Définissons

$$A = \left\{k \in \mathbf{N} \mid \epsilon a_{n_k} b_{n_k} \le \gamma \sigma_k^2\right\}.$$

Pour tout élément k de A on a

$$\begin{aligned} P\Big\{\|T_k/a_{n_k}\| \ge \epsilon\delta\Big\} &\ge P\Big\{f_k(T_k/a_{n_k}) \ge \epsilon\delta\Big\} \\ &= P\Big\{f_k(T_k)/a_{n_k}s_k \ge \epsilon\delta/s_k\Big\}. \end{aligned} \tag{2.7}$$

En vue d'appliquer le lemme 9 posons pour tout $i \in I_k$

$$Y_i = f_k(X_i/a_{n_k}).$$

Par la condition 1) du théorème, ces variables vérifient :

$$\begin{aligned} |Y_i| \le \|X_i\|/a_{n_k} &\le b_{n_k}/a_{n_k} \qquad (p.s) \\ &= s_k(b_{n_k}/a_{n_k}s_k) \end{aligned} .$$

Posons $c_k = b_{n_k}/a_{n_k}s_k$; remarquons que par (2.6) on a pour k assez grand $\epsilon\delta/s_k \ge c_1$; d'autre part :

$$\begin{aligned} (\epsilon\delta/s_k)c_k &\le \epsilon\delta b_{n_k}/a_{n_k}s_k^2 \\ &\le 2\epsilon\delta a_{n_k}b_{n_k}/\sigma_k^2 && \text{(par (2.5))} \\ &\le 2\gamma\delta && \text{(car } k \in A) \\ &\le c_2 && \text{(par (2.3))} \end{aligned}$$

Le lemme 9 s'applique par conséquent pour k assez grand et donne :

$$\begin{aligned} P\Big\{f_k(T_k)/a_{n_k}s_k \ge \epsilon\delta/s_k\Big\} &\ge \exp(-\epsilon^2\delta^2/s_k^2) \\ &\ge \exp(-2\epsilon^2\delta^2 a_{n_k}^2/\sigma_k^2) && \text{(par (2.5))} \\ &\ge \exp(-\epsilon[a_{n_k}/\sigma_k]^2) && \text{(par (2.3))} \end{aligned}$$

On déduit alors de (2.4) et (2.7) :

$$\sum_{k\in A} \exp\Big(-\epsilon[a_{n_k}/\sigma_k]^2\Big) < \infty.$$

Pour $k \notin A$ on a

$$a_{n_k}^2/\sigma_k^2 \ge \gamma a_{n_k}/\epsilon b_{n_k}.$$

D'où, grâce à l'hypothèse $H2$)

$$\sum_{k\notin A} \exp\Big(-\epsilon[a_{n_k}/\sigma_k]^2\Big) < \infty.$$

ce qui achève la démonstration de la nécessité.

B. Suffisance. Supposant satisfaite la condition

$$\forall\, \epsilon > 0 \quad \sum_{k\geq 1} \exp\Big(-\epsilon[a_{n_k}/\sigma_k]^2\Big) < \infty\ ,$$

nous voulons montrer que la suite $(S_n/a_n)_n$ est p.s. convergente vers 0.

La condition 2) du théorème 7 permet d'utiliser un argument usuel de symétrisation (cf [5], lemme 2.1); les v.a. $(X_n)_n$ seront donc supposées symétriques dans toute la suite de la démonstration.

Tenant compte de l'hypothèse $H1$), il est connu (cf. [5] lemme 2.2) que la convergence p.s. vers 0 de $(S_n/a_n)_n$ équivaut à la convergence p.s. vers 0 de la suite $(T_k/a_{n_k})_k$. Les variables $(T_k/a_{n_k})_k$ étant indépendantes, cette dernière convergence équivaut à

$$\forall\, \eta > 0 \quad \sum_{k\geq 1} P\Big\{\|T_k\|/a_{n_k}\| \geq \eta\Big\} < \infty.$$

Pour toute suite $(\epsilon_n)_n$ de variables de Rademacher indépendantes, indépendante de la suite $(X_n)_n$, l'affirmation précédente équivaut à la propriété suivante, que l'on se propose de démontrer :

$$\text{(2.8)} \qquad \forall\, \eta > 0 \quad \sum_{k\geq 1} P\Big\{\|\tilde{T}_k\| \geq \eta a_{n_k}\Big\} < \infty\,,$$

où l'on a posé pour tout entier k

$$\tilde{T}_k = \sum_{i\in I_k} \epsilon_i X_i.$$

Fixons un réel $\eta > 0$; et soit $\epsilon > 0$ un réel qui sera précisé ultérieurement. Remarquons que d'après (2.1) et la condition 1) du théorème 7, on a pour n assez grand

$$\|X_n\| \leq a_n \quad (p.s).$$

Comme de plus $(S_n/a_n)_n$ converge en probabilité vers 0 par hypothèse, un théorème de De Acosta ([1], lemme 3.1) permet d'affirmer l'existence d'un entier k_0 tel que

$$\text{(2.9)} \qquad \forall\, k \geq k_0 \quad Ea_{n_k}^{-1}\|T_k\| \leq \epsilon.$$

Définissons pour k entier fixé plus grand que k_0

$$A_1 = \left\{ (x_i)_i \in B^{I_k} \mid E\Big\|\sum_{i\in I_k} \epsilon_i x_i\Big\| \le 4\epsilon a_{n_k} \right\}$$

et

$$A_2 = \left\{ (x_i)_i \in B^{I_k} \mid \sup_{\|x'\|\le 1} \Big\{ \sum_{i\in I_k} \langle x_i, x'\rangle^2 \Big\} \le 4(\sigma_k^2 + 2\epsilon\alpha a_{n_k} b_{n_k}) \right\},$$

où α est la constante apparaissant dans le lemme 10 cité plus bas. Désignons par P_X^k la loi de la suite $(X_i)_{i\in I_k}$. L'inégalité de Tchebychev implique :

$$\begin{aligned} P_X^k(A_1^c) &= P\Big\{ E_\epsilon \|\sum_{i\in I_k} \epsilon_i X_i\| > 4\epsilon a_{n_k} \Big\} \\ &\le (4\epsilon a_{n_k})^{-1} E\Big\|\sum_{i\in I_k} \epsilon_i X_i\Big\| \\ &\le 1/4. \qquad\qquad \text{(par (2.9)} \end{aligned}$$

D'où

$$P_X^k(A_1) \ge 3/4.$$

Pour minorer $P_X^k(A_2)$ on commence par remarquer qu'en vertu de l'inégalité du triangle on a

$$E \sup_{\|x'\|\le 1} \Big\{ \sum_{i\in I_k} \langle X_i, x'\rangle^2 \Big\} \le \sigma_k^2 + E \sup_{\|x'\|\le 1} \Big\{ \Big| \sum_{i\in I_k} \langle X_i, x'\rangle^2 - E\langle X_i, x'\rangle^2 \Big| \Big\}.$$

Soit $(X_n')_n$ une copie indépendante de la suite $(X_n)_n$, également indépendante de la suite $(\epsilon_n)_n$. On a clairement les majorations

$$\begin{aligned} E \sup_{\|x'\|\le 1} &\Big\{ \Big| \sum_{i\in I_k} \langle X_i, x'\rangle^2 - E\langle X_i', x'\rangle^2 \Big| \Big\} \\ &\le E \sup_{\|x'\|\le 1} \Big\{ \Big| \sum_{i\in I_k} \langle X_i, x'\rangle^2 - \langle X_i', x'\rangle^2 \Big| \Big\} \\ &= E \sup_{\|x'\|\le 1} \Big\{ \Big| \sum_{i\in I_k} \epsilon_i (\langle X_i, x'\rangle^2 - \langle X_i', x'\rangle^2) \Big| \Big\} \\ &\le 2E \sup_{\|x'\|\le 1} \Big\{ \Big| \sum_{i\in I_k} \epsilon_i \langle X_i, x'\rangle^2 \Big| \Big\}. \end{aligned}$$

Le dernier terme se majore grâce au lemme suivant de M. Talagrand ([7] théorème 5).

LEMME 10. — *Il existe une constante $\alpha > 0$ telle que pour toute suite $(x_1, \ldots, x_n)$ à valeurs dans un espace de Banach $(B, \| \ \|)$ on ait*

$$E \sup_{\|x'\| \leq 1} \mid \sum_{i=1}^{n} \epsilon_i \langle x_i, x' \rangle^2 \mid \leq \alpha (\sup_{i=1}^{n} \|x_i\|) E \| \sum_{i=1}^{n} \epsilon_i x_i \|.$$

Une application de ce lemme donne finalement, tenant compte de (2.9) et de la condition 1) :

$$E \sup_{\|x'\| \leq 1} \left\{ \sum_{i \in I_k} \langle X_i, x' \rangle^2 \right\} \leq \sigma_k^2 + 2\epsilon\alpha a_{n_k} b_{n_k}.$$

Ceci entraîne $P_X^k(A_2) \geq 3/4$; l'ensemble $A = A_1 \cap A_2$ est donc de probabilité plus grande que $1/2$.

La suite de la démonstration repose essentiellement sur un lemme de M. Talagrand ([11], théorème 4) dont nous énonçons une version adaptée à notre situation.

LEMME 11. — *Il existe une constante K telle que pour tous entiers $m \geq q \geq 2$, et toute partie mesurable C de B^{I_k} vérifiant $P_X^k(C) \geq 1/2$ on ait*

$$P_* \left\{ (X_i)_{i \in I_k} \in H(C, m, q) \right\} \geq 1 - (K/q)^m$$

où l'on a posé

$$H(C, m, q) = \Big\{ (y_i)_{i \in I_k} \in B^{I_k} \mid \exists\, x^1, \ldots, x^q \in C \quad \textit{tels que} \quad \text{card}\,\{ i \in I_k \mid y_i \notin \{x_i^1, \ldots, x_i^q\}\} \leq m \Big\}$$

et où P_* désigne la mesure intérieure associée à P.

Nous appliquons ce lemme à l'ensemble A en prenant

$$m_k = [\gamma \delta a_{n_k} / b_{n_k}] \quad \text{et} \quad q = [e\delta'^{-1} K] + 2,$$

où $[x]$ représente la partie entière du réel x, et où δ et δ' sont des réels > 0 qui seront précisés plus loin. Pour k assez grand on a bien $m_k \geq q \geq 2$. Posons

$$H'(A, m_k, q) = \left\{ (X_i)_{i \in I_k} \in H(A, m_k, q) \right\}.$$

Quitte à approximer $P_*(H'(A, m_k, q))$ à l'aide d'une suite $(B_n)_n$ de parties mesurables incluses dans $H'(A, m_k, q)$, on supposera $H'(A, m_k, q)$ mesurable. On peut alors écrire

$$P \left\{ \| \sum_{i \in I_k} \epsilon_i X_i \| \geq \eta a_{n_k} \right\} \leq (K/q)^{m_k} + \int_{H'(A, m_k, q)} P_\epsilon \left\{ \| \sum_{i \in I_n} \epsilon_i X_i \| \geq \eta a_{n_k} \right\} dP_X. \tag{2.10}$$

Ces notations supposent les suites indépendantes $(X_n)_n$ et $(\epsilon_n)_n$ définies sur un espace produit $(\Omega = \Omega' \times \Omega',\ \mathcal{F} \times \mathcal{F}, P = P_X \otimes P_\epsilon)$ construit de façon que pour tout $n \quad X_n$ ne dépende que de la première variable et ϵ_n de la seconde.

Soit $\omega \in H'(A, m^k, q)$; alors par définition de $H'(A, m_k, q) \quad y = [(X_i)_{i \in I_k}](\omega)$ est un élément de $H(A, m_k, q)$. On pourra supposer $X = (X_i)_{i \in I_k}$ définie sur Ω^{I_k} et vérifiant

$$\forall\, \omega' = (\omega'_i)_{i \in I_k} \in \Omega^{I_k} \qquad X_i(\omega') = X_i(\omega'_i).$$

On a donc

$$y_i = X_i(\omega_i).$$

Par définition de $H(A, m_k, q)$, il existe $x^1, \ldots, x^q \in A$ tels que

$$\operatorname{card}\Big\{ i \in I_k \mid y_i \notin \{x_i^1, \ldots, x_i^q\} \Big\} \leq m_k.$$

Il existe par conséquent j éléments $i_1, \ldots, i_j$ de I_k avec $j \leq m_k$ et q parties disjointes $(I_k^l)_{l=1}^q$ de I_k vérifiant

$$\forall\, l = 1, \ldots, q \quad I_k^l \subset \{ i \in I_k \mid y_i = x_i^l \}$$

et

$$I_k = \{i_1, \ldots, i_j\} \cup \Big(\bigcup_{l=1}^q I_k^l \Big).$$

Les variables $(X_i)_{i \in I_k}$ étant p.s. majorées en norme par b_{n_k}, on peut supposer que

$$\forall\, i \in I_k \quad \|y_i\| = \|X_i(\omega_i)\| \leq b_{n_k}.$$

Il vient alors, en posant $J = \bigcup_{l=1}^q I_k^l$:

$$\begin{aligned} \Big\| \sum_{I_k} \epsilon_i y_i \Big\| &\leq m_k \sup_{i \in I_k} \|y_i\| + \Big\| \sum_{i \in J} \epsilon_i y_i \Big\| \\ &\leq m_k b_{n_k} + \Big\| \sum_{i \in J} \epsilon_i y_i \Big\|. \end{aligned}$$

Les ensembles I_k^l étant disjoints, on a d'après leur définition :

$$\begin{aligned} E \Big\| \sum_{i \in J} \epsilon_i j_i \Big\| &\leq \sum_{l=1}^q E \Big\| \sum_{i \in I_k^l} \epsilon_i x_i^l \Big\| \\ &\leq \sum_{l=1}^q E \Big\| \sum_{i \in I_k} \epsilon_i x_i^l \Big\| && \text{(inégalité de Jensen)} \\ &\leq 4 \epsilon a_{n_k} q\ . && (\text{car } x^l \in A_1) \end{aligned}$$

On déduit de ce qui précède :

$$P\Big\{\|\sum_{i\in I_k}\epsilon_i y_i\| \geq \eta a_{n_k}\Big\}$$
$$\leq P\Big\{\|\sum_{i\in J}\epsilon_i y_i\| \geq E\|\sum_{i\in J}\epsilon_i y_i\| + (\eta a_{n_k} - 4\epsilon q a_{n_k} - m_k b_{n_k})\Big\}\,.$$

La définition de m_k montre que si ϵ, δ et δ' ont été choisis assez petits, on a pour k suffisamment grand

(2.11) $$\eta a_{n_k} - 4\epsilon q a_{n_k} - m_k b_{n_k} \geq \eta a_{n_k}/2.$$

D'où pour k assez grand

$$P\Big\{\|\sum_{i\in I_k}\epsilon_i y_i\| \geq \eta a_{n_k}\Big\} \leq P\Big\{\|\sum_{i\in J}\epsilon_i y_i\| \geq E\|\sum_{i\in J}\epsilon_i y_i\| + \eta a_{n_k}/2\Big\}.$$

Le membre de droite de l'inégalité précédente se majore grâce au résultat suivant de M. Talagrand ([12]).

LEMME 12. — *Il existe une constante $\beta > 0$ telle que pour toute suite $x_1, \ldots, x_n$ d'éléments d'un espace de Banach $(B, \|\ \|)$ on ait, en posant $\sigma^2 = \sup\{\sum_{i=1}^q \langle x_i, x'\rangle^2\ ;\ \|x'\|_{B'} \leq 1\}$:*

$$\forall\, t > 0 \quad P\Big\{\Big\|\sum_{i=1}^n \epsilon_i x_i\Big\| \geq E\Big\|\sum_{i=1}^n \epsilon_i x_i\Big\| + t\Big\} \leq \exp(-\beta t^2/\sigma^2).$$

Ce lemme fournit immédiatement la majoration

(2.12) $$P\Big\{\|\sum_{i\in I_k}\epsilon_i y_i\| \geq \eta a_{n_k}\Big\} \leq \exp\Big\{(-\beta\eta^2 a_{n_k}^2/4\sigma_k'^2)\Big\},$$

où l'on a posé

$$\sigma_k'^2 = \sup\Big\{\sum_{i\in J}\langle y_i, x'\rangle^2\ ;\ \|x'\| \leq 1\Big\}.$$

σ_k' se majore en remarquant que $x^1, \ldots, x^q$ sont éléments de A_2 :

$$\sigma_k'^2 \leq \sum_{l=1}^q \sup\Big\{\sum_{i\in I_k^l}\langle x_i^l, x'\rangle^2\ ;\ \|x'\| \leq 1\Big\}$$

(2.13) $$\leq \sum_{l=1}^q \sup\Big\{\sum_{i\in I_k}\langle x_i^l, x'\rangle^2\ ;\ \|x'\| \leq 1\Big\}$$

$$\leq 4q(\sigma_k^2 + 2\epsilon\alpha a_{n_k} b_{n_k})\,.$$

Il résulte finalement de (2.10), (2.12) et (2.13) :

$$P\Big\{\Big\|\sum_{i\in I_k}\epsilon_i X_i\Big\|\geq \eta a_{n_k}\Big\}\leq (K/q)^{m_k}+\exp\Big(-\beta\eta a_{n_k}^2/16q[\sigma_k^2+2\epsilon\alpha a_{n_k}b_{n_k}]\Big).$$

Par choix de q et m_k on a, en supposant $\delta'\leq e$:

$$\begin{aligned}(K/q)^{m_k}\leq(\delta'/e)^{m_k}&\leq(\delta'/e)^{\gamma\delta a_{n_k}/b_{n_k}-1}\\ &<(e/\delta')(\delta'/e)^{\gamma\delta a_{n_k}/b_{n_k}}.\end{aligned}$$

Posons $\delta'/e=\exp(-\delta'')$ où δ'' est un réel >0 d'autant plus grand que δ' est petit ; on a

$$(K/q)^{m_k}\leq(e/\delta')\exp(-\gamma\delta\delta'' a_{n_k}/b_{n_k}).$$

L'hypothèse $H2$) montre qu'une fois le choix de δ effectué en vue d'obtenir l'inégalité (2.11), on peut trouver un réel δ' tel que la série $(K/q)^{m_k}$ soit convergente (le choix d'un réel ϵ assez petit venant finalement rendre possible (2.11)).

Utilisant l'inégalité

$$\forall\, x,y>0\quad \exp\Big(-\frac{1}{x+y}\Big)\leq\exp(-1/2x)+\exp(-1/2y),$$

on a d'autre part

$$\exp\Big(-\frac{\beta\eta^2a_{n_k}^2}{16q[\sigma_k^2+2\epsilon\alpha a_{n_k}b_{n_k}]}\Big)\leq\exp\Big(-\frac{\beta\eta^2a_{n_k}^2}{32q\sigma_k^2}\Big)+\exp\Big(-\frac{\beta\eta^2a_{n_k}}{64q\alpha\epsilon b_{n_k}}\Big).$$

En fin de compte si ϵ a été choisi de façon que

$$\beta\eta^2/64q\alpha\epsilon\geq\gamma,$$

les hypothèses assurent la convergence de la série $\sum_{k\geq1}P\{\|\sum_{i\in I_k}\epsilon_iX_i\|\geq\eta a_{n_k}\}$, ce qui achève la démonstration du théorème 7.

RÉFÉRENCES

[1] De Acosta A. : *Inequalities for B valued random vectors with applications to the strong law of large numbers*, Ann. Prob., 9 (1981) , 157-161.

[2] Alt J.C. : *La loi des grands nombres de Prokhorov dans les espaces de type p*, à paraître dans les Annales de l'Institut Henri Poincaré.

[3] Alt J.C. : *Sur le comportement asymptotique presque sûr des sommes de variables aléatoires à valeurs vectorielles*, Annales Scientifiques de l'Université de Clermont, 90 (1987), 3 - 24.

[4] Heinkel B. : *Une extension de la loi des grands nombres de Prokhorov*, Z. Wahr. verw. Gebiete, 67 (1984), 349-362.

[5] Kuelbs J. et Zinn J. : *Some stability results for vector valued random variables*, Ann. Prob. 7 (1979), 75-84.

[6] Ledoux M. : *Gaussian randomization and the law of the iterated logarithm in type 2 Banach spaces*, Preprint (1985).

[7] Ledoux M. et Talagrand M. : *Comparison theorems, random geometry and some limit theorems for empirical processes*, à paraître dans Annals of Probability.

[8] Mourier E. : *Eléments aléatoires dans un espace de Banach*, Ann. Inst. Poincaré, 13 (1953), 161-244.

[9] Prokhorov You.V. : *Some remarks on the strong law of large numbers*, Theor. Prob. Appl., 4 (1959), 204-208.

[10] Stout W.F. : *Almost sure convergence*, Academic Press, New-York, 1974.

[11] Talagrand M. : *Isoperimetry and integrability of the sum of independant Banach-spaces valued random variables*, à paraître dans Annals of Probability.

[12] Talagrand M. : *An isoperimetric theorem on the cube and the Khintchin-Kahane inequalities*, à paraître dans Proc. Amer. Math. Soc.

[13] Wittmann R. : *A general law of the iterated logarithm*, Z. Wahrs. verw. Gebiete, 68 (1985), 521-543.

Probability Theory on Vector Spaces IV
Lancut, June'87, Springer's LNM 1391

ON SOME SUFFICIENT CONDITIONS FOR THE EXISTENCE OF A CONSISTENT TEST IN NONCOMMUTATIVE STATISTICS

Artur Bartoszewicz (Łódź)

0. The approach to statistcics, based on decision theory, was generalized by Holevo in [5] to the von Neumann algebras context.

A noncommutative test is, by Holevo, a positive operator X between 0 and $\mathbb{1}$. In [2] we introduced the notions of consistency and uniform consistency of a sequential noncommutative test.

For the sake of clarity, recall some definitions. Let A be a W^*-algebra and $\{A_i,\ i = 1,2,\ldots\}$ a filtration of A, i.e. an increasing sequence of W^*-subalgebras of A, generating A. A sequence of positive operators X_n, $0 \leq X_n \leq \mathbb{1}$, belonging to A_n, respectivily, will be called a sequential test. Consider two families of normal normed states on A, say $\{\mu_\theta\}_{\theta\in\Theta}$ and $\{\nu_\lambda\}_{\lambda\in\Lambda}$, where Θ and Λ are arbitrary sets of parameters.

A noncommutative sequential test is called consistent if, for all $\theta \in \Theta$, $\mu_\theta(X_n)$ converges to zero, and, for all $\lambda \in \Lambda$, $\nu_\lambda(X_n)$ converges to one. Such a test is uniformly consistent if the above convergence is uniform on Λ and Θ.

In [2] we gave and proved some necessary and sufficient conditions for the existence of consistent and uniformly consistent tests, which were noncommutative analogues of the conditions given by Kraft in [7]. The aim of the present paper is to consider two particular cases. In the first case we assume the parameter spaces can be broken up into disjoint subsets and study how the existence of consistent or uniformly consistent tests between these subsets will imply the orthogonality or existence of test for all the sets of parameters. Some of the results of this part of the paper are obtained under the assumption that the states considered are tracial. The second case will be connected with the orthogonality of product and assymptotically product states on the infinite tensor product of W*-algebras.

1. PROPOSITION 1.1. If

(i) all the states μ_θ, $\theta \in \Theta$, and ν_λ, $\lambda \in \Lambda$, are tracial,

(ii) for every n, the sets Θ and Λ can be broken up into the finite sum of the same number of disjoint subsets, say $\Theta = \bigcup_{i=1}^{k_n} \Theta_i$ and $\Lambda = \bigcup_{j=1}^{k_n} \Lambda_j$, and, for any $i,j = 1,\ldots,k_n$, there exists a test $X_n^{ij} \in A_n$ such that $\nu_\lambda(X_n^{ij}) < \varepsilon_n$ for $\lambda \in \Lambda_j$ and $\mu_\theta(\mathbb{1} - X_n^{ij}) < \varepsilon_n$ for $\theta \in \Theta_i$,

(iii) $k_n \varepsilon_n^{\frac{1}{2}} \to 0$, then there exists a uniformly consistent test for $\{\mu_\theta\}_{\theta\in\Theta}$ against $\{\nu_\lambda\}_{\lambda\in\Lambda}$.

P r o o f. Denote by E_n^{ij} the spectral projection of X_n^{ij} corresponding to the interval $(\varepsilon^{\frac{1}{2}},1]$. Evidently, we have the following inequality

$$\varepsilon_n^{\frac{1}{2}} E_n^{ij} \leq X_n^{ij} \leq \varepsilon_n^{\frac{1}{2}} \mathbb{1} + E_n^{ij}$$

Put

$$E_n = \bigvee_{i=1}^{k_n} \bigwedge_{j=1}^{k_n} E_n^{ij}.$$

We have

$$\nu_\lambda(E_n) = \nu_\lambda(\bigvee_i \bigwedge_j E_n^{ij}) \le \sum_i \nu_\lambda(\bigwedge_j E_n^{ij}) \le \sum \nu_\lambda(E_n^{ij_o}) \le$$

$$\le \varepsilon_n^{-\frac{1}{2}} \sum_i \nu_\lambda \ (X_n^{ij_o}) \le \varepsilon^{-\frac{1}{2}} k_n \varepsilon_n = k_n \varepsilon^{\frac{1}{2}}.$$

On the other hand,

$$\mu_\theta(\mathbb{1} - E_n) = \mu_\theta(\mathbb{1} - \bigvee_i \bigwedge_j E_n^{ij}) \le \mu_\theta(\bigvee_j (\mathbb{1} - E_n^{i_oj})) \le$$

$$\le \sum_j \mu_\theta(\mathbb{1} - E_n^{i_oj}) \le \sum_j (\mu_\theta(\mathbb{1} - X_n^{i_oj}) + \varepsilon_n^{\frac{1}{2}}) \le$$

$$\le k_n(\varepsilon_n + \varepsilon_n^{\frac{1}{2}}).$$

So, E_n is the desired test. □

In the classical case, the analogous theorem was formulated by Kraft in [7] with the condition $\varepsilon_n k_n \to 0$. This weaker condition is sufficient if all the tests X_n^{ij} are not randomized, i.e. X_n^{ij} are projections.

Suppose now that the parameter spaces Θ and Λ are divided into countably many classes.

PROPOSITION 1.2. If the states μ_θ, $\theta \in \Theta$ and ν_λ, $\lambda \in \Lambda$ are tracial, $\Theta = \bigcup_{i=1}^{\infty} \Theta_i$, $\Lambda = \bigcup_{j=1}^{\infty} \Lambda_j$ and there exists a uniformly consistent test $\{X_n^{ij}\}$ for Θ_i against Λ_j for any i,j, then there exists a consistent test for Θ against Λ.

P r o o f. Evidently, there is a sequence ε_n tending to zero such that

$$\mu_\theta(\mathbb{1} - X_n^{ij}) \le \varepsilon_n$$

and

$$\nu_\lambda(X_n^{ij}) \le \varepsilon_n$$

for $\theta \in \Theta_i$, $\lambda \in \Lambda_j$.

For every positive integer k, we can choose a number n_k so

$$\mu_\theta(\mathbb{1} - X_n^{ij}) \le k^{-4}$$

and

$$\nu_\lambda(X_n^{ij}) \le k^{-4}$$

for $n \ge n_k$, $i,j = 1,2,\ldots,k$.

Denote by E_n^{ij} the spectral projection of X_n^{ij} corresponding to the interval $(k^{-2},1]$. Similarly as in Proposition 1 we have

$$k^{-2}E_n^{ij} \le X_n^{ij} \le k^{-2}\cdot\mathbb{1} + E_n^{ij}.$$

Put

$$E_{n_k} = \bigvee_{i=1}^{k} \bigwedge_{j=1}^{k} E_{n_k}^{ij} .$$

For $\lambda \in \Lambda_j$, $\theta \in \Theta_i$ and $i,j \le k$, we can conclude that

$$\nu_\lambda(E_{n_k}) \le k^{-1}$$

and

$$\mu_\theta(\mathbb{1} - E_{n_k}) \le \mu_\theta(\bigvee_j (\mathbb{1} - E_{n_k}^{i_oj})) \le$$

$$\le \sum_j [\mu_\theta(\mathbb{1} - X_n^{i_oj}) + k^{-2}] \le k(k^{-4} + k^{-2}).$$

Putting $E_n = E_{n_k}$ for n between n_k and n_{k+1}, we obtain the consistent test for Θ against Λ. □

Consider now the case when the partition of Λ into $\bigcup_{j=1}^{\infty} \Lambda_j$ depends upon the partition of Θ. Now, the conclusion will be weaker but we do not need to assume that the states considered are tracial.

PROPOSITION 1.3. If $\Theta = \bigcup_{i=1}^{\infty} \Theta_i$ and, for each i, there exists a partition of Λ into $\bigcup_{j=1}^{\infty} \Lambda_{ij}$ so that there is a uniformly consistent test for Θ_i against Λ_{ij}, then $\{\mu_\theta\}$ is orthogonal to $\{\nu_\lambda\}$.

P r o o f. Let (i,j) be an arbitrary pair of indices $\{X_n\}$ a test for which

$$\sup_{\theta\in\Theta_i} (X_n) \to 0$$

and

$$\inf_{\lambda\in\Lambda_{ij}} (X_n) \to 1.$$

Since the set of operators X_n is relatively compact in the ultraweak topology in A, there is a test $X \in A$ such that $\mu_\theta(X) = 0$ and $\nu_\lambda(X) = 1$ for any $\theta \in \Theta_i$ and $\lambda \in \Lambda_{ij}$. Denote by E_{ij} the spectral projection of X corresponding to the one-point set $\{0\}$. Evidently, $\mu_\theta(E_{ij}) = 1$ and $\nu_\lambda(E_{ij}) = 0$ for $\theta \in \Theta_i$ $\lambda \in \Lambda_{ij}$. It is easy to see that, for $E = \bigvee_{i=1}^{\infty} \bigwedge_{j=1}^{\infty} E_{ij}$, $\mu_\theta(E) = 1$ and $\nu_\lambda(E) = 0$ for all θ and λ, so the sets $\{\mu_\theta\}$ and $\{\nu_\lambda\}$ are orthogonal. □

2. Consider now one of the most important examples of a W*-algebra with the countable filtration - the infinite tensor product of von Neumann algebras (for the definitions and basic properties, see [3]).

Let A be a Γ-tensor product of W*-algebras A_i. The filtration of A is the sequence of finite tensor products of A_i (more precisely, the sequence of subalgebras of A canonically isomorphic to finite tensor products of A_i). Let μ and ν be asymptotically product states on A, i.e. the limits (in $\sigma(A_*,A)$ or in norm in A_*, see [1]) of the states $\mu_n = \mu^n \bullet \bigotimes_{i>n} \alpha_i$ and $\nu_n = \nu^n \bullet \bigotimes_{i>n} \beta_i$, respectively. μ^n and ν^n are here the restrictions of μ and ν to $\bigodot_{i=1}^{n} A_i$ and $\{\alpha_i\}$, $\{\beta_i\}$ are the families of states on A_i such that there exist product states $\bigotimes_{i=1}^{\infty} \alpha_i$ and $\bigotimes_{i=1}^{\infty} \beta_i$ on A.

PROPOSITION 2.1. If $\{\alpha_i\}$ and $\{\beta_i\}$ are not equivalent in the sense of Bures, i.e. if the product state $\bigotimes_{i=1}^{\infty} \beta_i$ does not exist on $\bigotimes_{i=1}^{\infty} (A_i,\alpha_i)$, then there exists a consistent test for μ against ν.

P r o o f. By the same considerations as in [3], we get that the states μ_n and ν_n have the supports contained in projections on local tensor products $\bigotimes_{i=1}^{\infty} (A_i,\alpha_i)$ and $\bigotimes_{i=1}^{\infty} (A_i,\beta_i)$, respectively. If $\{\alpha_i\}$ and $\{\beta_i\}$ are not equivalent, these projections are orthogonal. Hence, as in Proposition 1 of [2], we can construct a consistent test for μ against ν (making use of the theorem of Kosaki [6] and the lemma of Bures [4]). □

COROLLARY 2.2. If all the states α_i and β_i are faithful, then there exists a consistent test for $\bigotimes_{i=1}^{\infty} \alpha_i$ against $\bigotimes_{i=1}^{\infty} \beta_i$ if and only if the families $\{\alpha_i\}$ and $\{\beta_i\}$ are not equivalent.

REFERENCES

1. A. Bartoszewicz, The Projective Limit of Consistent Family of States on Tensor Products of W*-Algebras, Bull. Pol. Acad. Sci. Math. 31, No. 5-8 (1983), 273-279.

2. A. Bartoszewicz, Remarks on the Consistency of Tests in Noncommutative Statistics, to appear in Prob. and Math. Stat.

3. D. Bures, Tensor product of W*-algebras, Pac. J. Math. 27 (1968), 13-37.

4. D. Bures, An extension of Kakutani´s theorem on infinite product measures to the tensor product of semifinite W*-algebras, Trans. Amer. Math. Soc. 135 (1969), 199-212.

5. A.S. Holevo, The Analogue of Statistical Decision Theory in the Noncommutative Probability Theory, Proc. Moscow Math. Soc. 26 (1972), 133-149.

6. H. Kosaki, On the Bures distance and the Uhlmann transition probability of states on a von Neumann algebra, preprint.

7. C. Kraft, Some Conditions for Consistency and Uniform Consistency of Statistical Procedures, Univ. Calif. Pub. Stat. 2, No. 6 (1955), 125-141.

Institute of Mathematics
Łódź University
ul. Banacha 22
90-238 Łódź, POLAND

Probability Theory on Vector Spaces IV
Lancut, June'87, Springer's LNM 1391

LÉVY-KCHINCHINE FORMULA ON VECTOR SPACES

By

T. Byczkowski and P. Graczyk

Institute of Mathematics, Wrocław Technical University

1. INTRODUCTION

The purpose of this note is to prove the Lévy-Kchinchine formula for semigroups of probability measures on a class of complete separable metric vector spaces which expands essentially the class of locally convex spaces. They contain, in particular, the spaces $L_p[0,1]$, $0 < p < 1$. We employ there the convolution semigroups technique. Our paper is a generalization of [2], where the Lévy-Kchinchine formula was proved for symetric semigroups.

2. PRELIMINARIES

Throughout the whole paper E will stand for a complete separable metric vector space over the real numbers. D will denote Q^+ or $\mathbb{R}^+$. A family $(\mu_t)_{t\in D}$ of probability measures on E is called a convolution semigroup if

$$\mu_t * \mu_s = \mu_{t+s}$$

for all $t,s \in D$. It is called continuous if μ_t converges weakly to δ_0 as $t \to 0+$, $t \in D$. $\mathcal{W}(\mu)$ will denote the root set of μ, i.e. $\mathcal{W}(\mu) = \bigcup_{n\in\mathbb{N}} \{\upsilon^m : \upsilon^n = \mu,\ 1 \le m \le n\}$.

In the sequel we will need the following lemma taken from [8] :

LEMMA 1. Let $(\mu_r)_{r\in Q^+}$ be a convolution semigroup such that for some $a > 0$ the family $(\mu_r)_{r\in Q\cap(0,a)}$ is conditionally compact. Then there exist a unique continuous convolution semigroup $(\nu_t)_{t\in\mathbb{R}^+}$ such that $\nu_r = \mu_r$ for every $r \in Q^+$.

Now we will give a topological criterion of conditional compactness of families of measures like $(\mu_r)_{r\in Q\cap[0,1]}$ where $(\mu_r)_{r\in Q}$ is a semigroup. The criterion is taken from Böge ([1]).

DEFINITION. A topological vector space E is strongly root compact if for every compact set $C \subset E$ there exists a compact set $C_o \subset E$ such that for every $n \in \mathbb{N}$ all finite sequences $\{x_1,\dots,x_n\}$, $x_n = 0$, fulfilling the condition

$$(C + x_i + C + x_j) \cap (C + x_{i+j}) \neq \emptyset \quad \text{for} \quad i+j \le n, \quad i,j = 1,\dots,n$$

have their terms contained in C_o.

LEMMA 2. If E is strongly root compact and $M = \{\mu_i\}_{i\in I}$ is a uniformly tight family of Borel measures on E then the family

$$\mathcal{W}(M) = \bigcup_{i\in I} \mathcal{W}(\mu_i)$$

is conditionally compact.

REMARK. Böge proved in [1] that C is strongly root compact. Using the technique of Böge, Siebert showed in [7] that all locally convex spaces have the property of strong root compactness. Considering p-convex sets instead of convex ones one can prove similarily that the spaces $L_p[0,1]$, $0 < p < 1$ are strongly root compact. The problem of conditional compactness of roots of probability measures on the space $L_o[0,1]$ remains unsolved.

3. LÉVY-KCHINCHINE FORMULA

In this section we prove the main result of this note, namely the Lévy-Kchinchine formula for continuous semigroups of probability measures. The proof is partially based on the method used in [2] for symmetric semigroups.

THEOREM 1. Let E be a strongly root compact Polish vector space and $(\mu_t)_{t>0}$ a continuous convolution semigroup on E. Then there exists a nonnegative measure ν such, that for every open neighbourhood U of 0 the measure $\nu|_{U^c}$ is finite and $(1/t)\mu_t|_{U^c} \Rightarrow \nu|_{U^c}$ as $t \to 0+$ whenever $\nu(\partial U) = 0$. Moreover, the following decomposition holds:

$$\mu_t = \gamma_t * \kappa_t \,, \quad t > 0 \tag{1}$$

with γ_t, κ_t satysfying:

(i) $(\gamma_t)_{t>0}$, $(\kappa_t)_{t>0}$ are continuous convolution semigroups,

(ii) for every increasing sequence of Borel sets $\{F_n\}_{n\in\mathbb{N}}$ such that $\bigcap_{n=1}^{\infty} F_n^c = \{0\}$ and $\nu(F_n) < \infty$ for every $n \in \mathbb{N}$, there exists a sequence $\{x_n\}_{n\in\mathbb{N}}$ of elements of E such that

$$\lim_n \exp(t\nu|_{F_n}) * \delta_{tx_n} = \gamma_t \,, \quad t > 0 \,.$$

(iii) $\lim_{t\to 0+} (1/t)\gamma_t|_{U^c} = \nu|_{U^c}$ for every open neighbourhood U of 0 such that $\nu(\partial U) = 0$.

(iv) $\lim_{t\to 0+} (1/t)\kappa_t(U^c) = 0$ for every open neighbourhood of 0.

If the semigroup $(\mu_t)_{t>0}$ is symmetric then the assumption of strong root compactness of E may be omitted. In this case the semigroups $(\gamma_t)_{t>0}$, $(\kappa_t)_{t>0}$ are symmetric and if the sets F_n in (ii) are symmetric, $(F_n = -F_n)$, then (ii) holds for $x_n = 0$, $n \in \mathbb{N}$.

Proof. We divide the proof into several steps, dealing simultaneously with the non-symmetric and symmetric case.

Step 1. We define the measure ν and show that $(1/t)\mu_t|_{U^c} \Rightarrow \nu|_{U^c}$ for open neighbourhoods U of 0 such that $\nu(\partial U) = 0$.

Let q be a pseudonorm generating the topology of E ([6]). Let us choose a sequence (η_n) of positive numbers decreasing to 0. Then for $s,t > 0$ and $n \in \mathbb{N}$

$$(2) \qquad \exp(s/t)\mu_t = \exp((s/t)\mu_t|_{\{q>\eta_n\}}) * \exp((s/t)\mu_t|_{\{q\le\eta_n\}}).$$

Using elementary facts from the abstract semigroup theory (see, e.g., [4], ch.X) we obtain that

$$\exp((s/t)\mu_t) \Rightarrow \mu_s \ , \quad \text{as} \quad t \to 0+$$

for every $s > 0$. Since the measures $(1/t)\mu_t|_{\{q>\eta\}}$ are uniformly bounded for every $\eta > 0$ (see [3]), standard compactness arguments ([5]) imply the existence of a sequence t' decreasing to 0 such that

$$(3) \qquad (1/t')\mu_{t'}|_{\{q>\eta_n\}} \Rightarrow \nu^{(n)} \quad \text{as} \quad t' \to 0+$$

for some finite measures $\nu^{(n)}$, $n \in \mathbb{N}$.

$\{\nu^{(n)}\}_{n=1}^{\infty}$ is an increasing sequence of measures and

$$\nu^{(n)}|_{\{q>\eta_n\}} = \nu^{(n+1)}|_{\{q>\eta_n\}} .$$

Define

$$\nu = \lim_n \nu^{(n)}$$

ν is a σ-additive, in general infinite measure on E. Let $\eta > 0$ and $\nu\{q = \eta\} = 0$. Let n be such that $\eta_n < \eta$. Clearly $\nu^{(n)}\{q=\eta\} = \nu\{q=\eta\} = 0$. Suppose that for a sequence t'' decreasing to 0 $(1/t'')\mu_{t''}|_{\{q>\eta\}} \Rightarrow \nu_\eta$, for a measure ν_η. Then $\nu_\eta\{q=\eta\} = 0$. Finally

$$\lim_{t\to 0^+}(1/t)\mu_t\big|_{\{q>\eta\}} = \nu_\eta = \nu^{(n)}\big|_{\{q>\eta\}} = \nu\big|_{\{q>\eta\}} .$$

Step 2. We will prove the decomposition (1).

From (2) and (3) we infer that for every $n \in \mathbb{N}$ there exists a semigroup $(\kappa_s^{(n)})_{s\in\mathbb{R}^+}$ such that

$$\mu_s = \exp(s\nu^{(n)}) * \kappa_s^{(n)}, \quad s \in \mathbb{R}^+, \quad n \in \mathbb{N} .$$

Since $\nu^{(n)} \precsim \nu^{(n+1)}$ we can choose a sequence $\{x_n\}_{n=1}^{\infty}$ of elements of E such that the sequence $(\exp(\nu^{(n)}) * \delta_{x_n})_{n\in\mathbb{N}}$ converges as $n \to \infty$. For a fixed $m \in \mathbb{N}$ we consider the sequence $(\exp((1/m)\nu^{(n)}) * \delta_{(1/m)x_n})_{n\in\mathbb{N}}$. It is conditionally compact and its points of accumulation differ by translations. Since their m-th convolution powers are equal, the sequence is convergent. Finally, the sequence $(\exp(s\nu^{(n)}) * \delta_{sx_n})_{n\in\mathbb{N}}$ converges for every $s \in Q_+$.

In the symmetric case we obtain immediately the convergence of the sequences $(\exp s\nu^{(n)})_{n\in\mathbb{N}}$, $s \in \mathbb{R}^+$.

Denoting $\gamma_s = \lim_{n\to\infty} \exp(s\nu^{(n)}) * \delta_{sx_n}$ we get

$$\mu_s = \gamma_s * \kappa_s , \quad s \in Q^+$$

for a convolution semigroup $(\kappa_s)_{s\in Q^+}$.
E is strongly root compact, so by virtue of lemmas 1 and 2, $(\gamma_s)_{s\in Q^+}$ and $(\kappa_s)_{s\in Q^+}$ can be extended uniquely to continuous semigroups $(\gamma_t)_{t>0}$ and $(\kappa_t)_{t>0}$ and

$$\mu_t = \gamma_t * \kappa_t \quad \text{for} \quad t \in \mathbb{R}^+.$$

If $(\mu_t)_{t>0}$ are symmetric then $(\gamma_s)_{s\in Q^+}$ and $(\kappa_s)_{s\in Q^+}$ are symmetric semigroups, so they are continuous and the assumption of the strong root compactness of E is superfluous.

Step 3. We prove the properties (iii) and (iv) for semigroups

$(\gamma_t)_{t>0}$ and $(\kappa_t)_{t>0}$ obtained in Step 2. The proof is the same as in the symmetric case (see [2]).

Step 4. We show that

$$(4) \qquad \lim_{n\to\infty} \exp(t\nu^{(n)}) * \delta_{tx_n} = \gamma_t$$

for every $t \in \mathbb{R}^+$. We know from Step 2 that (4) holds for $t \in Q^+$. In the symmetric case, uniqueness of the extension of a symmetric rational semigroup to a continuous real semigroup and continuity of symmetric semigroups imply immediately (4).

In the non-symmetric case denote

$$M = \{\exp(s\nu^{(n)}) * \delta_{sx_n} : s \in Q \cap [0,1],\ n \in \mathbb{N}\} .$$

Then

$$M \subseteq \bigcup_n \mathcal{W}(\exp(\nu^{(n)}) * \delta_{x_n}) .$$

Since the sequence $\{\exp(\nu^{(n)}) * \delta_{x_n}\}_{n\in\mathbb{N}}$ is convergent it follows from Lemma 2 that $\overline{M}$ is compact in the topology of weak convergence and, in particular, the family $\{\exp(t\nu^{(n)}) * \delta_{tx_n}\}$ is conditionally compact for every $t \in (0,1)$.

Using the fact that for $s \in Q^+ \cap [0,1]$ $\exp(s\nu^{(n)}) * \delta_{sx_n} \quad \gamma_s$ we obtain for $n \in \mathbb{N}$ continuous semigroups $(\lambda_t^{(n)})_{t>0}$ such that

$$\gamma_t = \exp(t\nu^{(n)}) * \delta_{tx_n} * \lambda_t^{(n)} , \quad t > 0 .$$

Let $t \in (0,1)$ be fixed. Suppose that for a sequence $\{n'\}$ $\exp(t\nu^{(n')}) * \delta_{tx_{n'}} \Rightarrow \tilde{\gamma}$ as $n' \to \infty$. Then we have

$$\gamma_t = \tilde{\gamma} * \tilde{\lambda}$$

for a measure $\tilde{\lambda}$ which is a point of accumulation of the sequence $\{\lambda_t^{(n)}\}$. Since $\lambda_t^{(n)} * \lambda_{1-t}^{(n)} = \lambda_1^{(n)} \Rightarrow \delta_0$, we get $\tilde{\lambda} \prec \delta_0$. Hence points of accumulation $\{\exp(t\nu^{(n)}) * \delta_{tx_n}\}_{n\in\mathbb{N}}$ are translations of γ_t . Suppose that

(5) $$\exp(t\nu^{(n')}) * \delta_{tx_{n'}} \Rightarrow \gamma_t * \delta_x \quad \text{as} \quad n' \to \infty$$

for some $x \in E$.

Put $t_1 = t$. Choose $r_1 \in Q$ such that $0 < r_1 - t_1 < t_1/2$. Suppose that for a subsequence $\{n''\}$ of the sequence $\{n'\}$ we have

(6) $$\exp((r_1 - t_1)\nu^{(n'')}) * \delta_{(r_1-t_1)x_{n''}} \Rightarrow \gamma_{r_1-t_1} * \delta_y$$

for some $y \in E$. Then, by the formula (4) used for $r \in Q$ and by (5) and (6) we obtain $y = -x$ and we can replace $\{n''\}$ in (6) by $\{n'\}$. Denoting $t_2 = 2t_1 - r_1$ we have then

$$\delta_{t_2x_{n'}} * \exp(t_2\nu^{(n')}) \Rightarrow \gamma_{t_2} * \delta_{2x} .$$

Repeating the above construction by induction we obtain a sequence $\{t_m\}_{m=1}^{\infty}$, $t_m = 2t_{m-1} - r_{m-1}$, $0 < r_m - t_m < t_1/2^m$ such that for every $m \in \mathbb{N}$

$$\exp(t_m\nu^{(n')}) * \delta_{t_mx_{n'}} \Rightarrow \gamma_{t_m} * \delta_{2^{m-1}x} \quad \text{as} \quad n' \to \infty .$$

Thus, the sequence $\{\gamma_{t_m} * \delta_{2^{m-1}x}\}_{m\in\mathbb{N}}$ is conditionally compact. On the other hand, $\lim_{m\to\infty} t_m = t_o$ for some $t_o > 0$ and $\gamma_{t_m} \Rightarrow \gamma_{t_o}$ as $m \to \infty$. Finally, we have $x = 0$ and it completes the proof of (4).

Step 5. Using Tortrat's technique ([9]) and the method presented in Step 4 we prove (ii) what completes the proof of the theorem.

4. PROPERTIES OF LÉVY MEASURES

The measure ν defined in Theorem 1 will be called the Lévy measure of the convolution semigroup $(\mu_t)_{t>0}$. The Lévy measure is defined uniquely, and as we have seen in Step 1 of the proof, ν may be defined for a continuous semigroup of measures on a complete separable vector metric space.

A measure ν on E will be called a Lévy measure if there exists a continuous convolution semigroup $(\mu_t)_{t>0}$ such that ν is its Lévy measure. By Theorem 1, if ν is a Lévy measure on E then ν is the Lévy measure of the semigroup $(\exp(t\nu))_{t>0}$ defined up to semigroups of the form $(\delta_{tx})_{t>0}$, $x \in E$.

On Hilbert spaces, Lévy measures have a full chracterisation in the language of measure and integral theory. For Banach spaces this is already imposible because of diversified geometrical structure of these spaces. Nevertheless, some properties of Lévy measures may be formulated. As before, in this section E is a complete separable vector metric space, strongly root compact.

THEOREM 2.

(i) If ν_1, ν_2 are Lévy measures of $(\mu_t^{(1)})_{t>0}$, $(\mu_t^{(2)})_{t>0}$ then $\nu_1 + \nu_2$ is the Lévy measure of the semigroup $(\mu_t^{(1)} * \mu_t^{(2)})_{t>0}$.

(ii) If ν_1, ν_2 are as in (i) and $(\mu_t^{(1)})_{t>0} \prec (\mu_t^{(2)})_{t>0}$, then $\nu_1 \le \nu_2$.

(iii) If ν is a Lévy measure and $\tilde{\nu} \le \nu$, then $\tilde{\nu}$ is also a Lévy measure.

Proof. Part (i) of the theorem we prove estimating separately $\liminf_{t\to 0^+}(1/t)(\mu_t^{(1)} * \mu_t^{(2)})\big|_{\{q>\eta\}}(A)$ and $\limsup_{t\to 0^+}(1/t)(\mu_t^{(1)} * \mu_t^{(2)})\{q>\eta\}$ for $\eta > 0$ and A being open sets. (ii) follows immediately from (i). In (iii), let $\gamma_t = \lim_{n\to\infty} \exp(t\nu^{(n)}) * \delta_{tx_n}$, $t > 0$, be a convolution semigroup of the Lévy measure ν. Let $\{y_n\}_{n=1}^{\infty}$ be such a sequence of elements of E that the sequence $(\exp(\tilde{\nu}^{(n)}) * \delta_{y_n})_n$ converges, weakly.

In the same way as in the proof of Theorem 1 we show that there exists a continuous semigroup $(\tilde{\gamma}_t)_{t>0}$ such that

$$\tilde{\gamma}_t = \lim_{n\to\infty} \exp(t\tilde{\nu}^{(n)}) * \hat{\delta}_{ty_n}, \quad t > 0 .$$

Then, for a fixed $m \in \mathbb{N}$ we have:

$$\gamma_t = \exp(t\nu^{(m)}) * \lambda_t^{(m)} , \quad t > 0$$

$$\tilde{\gamma}_t = \exp(t\tilde{\nu}^{(m)}) * \tilde{\lambda}_t^{(m)} , \quad t > 0$$

for some continuous semigroups $(\lambda_t^{(m)})_{t>0}$, $(\tilde{\lambda}_t^{(m)})_{t>0}$ such that $\tilde{\lambda}_t^{(m)} \prec \lambda_t^{(m)}$. Hence

$$\lim_{t\to 0^+} (1/t)\tilde{\lambda}_t^{(m)}\{q > \eta\} = 0 \quad \text{for} \quad \eta > \eta_m$$

and as in Step 3 of the proof of Thm. 1 we prove that

$$\lim_{t\to 0^+} (1/t)\tilde{\gamma}_t\Big|_{\{q>\eta\}} = \tilde{\nu}\Big|_{\{q>\eta\}}$$

for $\eta > 0$ such that $\tilde{\nu}\{q = \eta\} = 0$, so $\tilde{\nu}$ is the Lévy measure of the semigroup $(\tilde{\gamma}_t)_{t>0}$.

<u>COROLLARY</u>. A Borel measure ν on E is a Lévy measure if and only if $\nu(U^c) < \infty$ for every open neighbourhood U of 0 and the sequence $\{\exp \nu^{(n)}\}_{n\in\mathbb{N}}$ is conditionally compact after suitable translations.

<u>Proof</u>. Necessarity of both conditions follows from Thm. 1. If the sequence $(\exp(\nu^{(n)}))$ is conditionally compact after translations then there exists a sequence $\{x_n\}_{n\in\mathbb{N}}$ of elements of E such that the sequence $\{\exp(\nu^{(n)}) * \delta_{x_n}\}_{n\in\mathbb{N}}$ is convergent.

As in the proof of Theorem 1 we construct a continuous semigroup $(\gamma_t)_{t>0}$ such that $\gamma_t = \lim_{n\to\infty} \exp(t\nu^{(n)}) * \delta_{tx_n}$, $t > 0$.

Then (see [2]) for $\eta > 0$ and A open:

$$\liminf_{t\to 0^+} (1/t)\gamma_t\Big|_{\{q>\eta\}}(A) \geq \nu\Big|_{\{q>\eta\}}(A) .$$

Denoting the Lévy measure of the semigroup $(\gamma_t)_{t>0}$ by $\tilde{\nu}$, we get $\tilde{\nu} \geq \nu$ and thus ν is a Lévy measure and $\nu = \tilde{\nu}$.

REFERENCES

[1] Böge, W. (1964). Zür Charakterisierung sukzessiv unendlich teilbarer Wahrscheinlichkeitsverteilungen auf lokalkompakten Gruppen. Z. Wahr. verw. Gebiete 2, 380-394.

[2] Byczkowski, T. and Samotij, K. (1986). Absolute continuity of stable seminorms. Ann. Probability 14, 299-312.

[3] Byczkowski, T. and Żak, T. (1981). Asymptotic properties of semigroups of measures on vector spaces. Ann. Probability 9, 211-220.

[4] Dunford, N. and Schwartz, J. (1958). Linear Operators, Part I. Interscience Publishers, New York, London.

[5] Parthasarathy, K.R. (1967). Probability Measures on Metric Spaces. Academic Press, New York.

[6] Rolewicz, S. (1972). Metric Linear Spaces. Monografie Matematyczne 56, PWN Warszawa.

[7] Siebert, E. (1974). Einbettung unedlich teilbarer Wahrscheinlichkeitsmasse auf topologischen Gruppen. Z. Wahr. Verw. Gebiete 28, 227-247.

[8] Siebert, E. (1976). Convergence and convolutions of probability measures on a topological group. Ann. Probability 4, 433-443.

[9] Tortrat, A. (1967). Structure des lois indéfiniment divisibles dans un espace vectoriel topologique (separé) X. Lect. Notes in Math. 31, 299-328.

Probability Theory on Vector Spaces IV
Lancut, June '87, Springer's LNM 1391

A PROBLEM ON REARRANGEMENTS OF SUMMANDS IN NORMED SPACES AND RADEMACHER SUMS

BY

S.A. CHOBANYAN AND G.J. GEORGOBIANI
Acedemy of Sciences of Georgian SSR
Muskchelishvili Institute of Computer Mathematics
Tbilisi 380093 USSR

1. INTRODUCTION. The main aim of this paper is solution of the following two problems. Let $A = (a_1,\ldots,a_n)$ be a finite collection of elements of a normed space X (real or complex), $a_1 + \ldots + a_n = 0$, and let $\sigma : [1,\ldots,n] \to [1,\ldots,n]$ be a permutation. Denote $A_\sigma = (a_{\sigma(1)},\ldots,a_{\sigma(n)})$ and

$$|A_\sigma| = \max_{1\le k\le n} \|a_{\sigma(1)} + \ldots + a_{\sigma(k)}\| \ . \tag{1}$$

PROBLEM 1. Estimate $\min_\sigma |A_\sigma|$.

PROBLEM 2. Estimate $\frac{1}{n!} \sum_\sigma |A_\sigma|$.

Though probability does not participate in formulations of the problems, both the method of proof and estimating expressions do use the probability theory on vector spaces.

For applications upper estimates are important in both problems. Obviously, any upper estimate in Problem 2 provides an estimate for Problem 1. However, as we shall see, the best upper estimate in Problem 2 does not give the best upper estimate for Problem 1.

The first upper estimate for $\min_\sigma |A_\sigma|$ has been obtained by

M.I. Kadec [6] for $L_p(\Omega,\mu)$ spaces, $1 \le p < \infty$. He proved that

$$\min_\sigma |A_\sigma| \le c\Big(\sum_{k=1}^n \|a_k\|^d\Big)^{1/d} , \tag{2}$$

where $d = \min(p,2)$, and c is a constant which does not depend on A.

In [16] Nikishin has improved this estimate for $1 \le p \le 2$ by showing that

$$\min_\sigma |A_\sigma| \le c\Big(\int_\Omega \Big(\sum_{k=1}^n |a_k|^2\Big)^{p/2} d\mu\Big)^{1/p} . \tag{3}$$

The first author in [1,2] has proved the validity of the following estimate for general normed spaces

$$\min_\sigma |A_\sigma| \le 8E\Big\|\sum_{i=1}^n a_i r_i\Big\| , \tag{4}$$

where (r_i) is the sequence of Rademacher functions. In particular, (4) shows that (3) holds for any p, $1 \le p < \infty$; (4) covers also other results for concrete Banach spaces (cf. Corollary of Theorem 4).

In the present paper it is found a result more exact than (4). Namely, we show that there exists a permutation σ such that

$$|A_\sigma| = \min_\theta |A_\sigma\theta| , \tag{5}$$

where θ is a collection of sings, $\theta = (\theta_1,\dots,\theta_n)$, while

$$|A_\sigma\theta| = \max_{1\le k\le n} \|a_{\sigma(1)}\theta_i + \dots + a_{\sigma(k)}\theta_k\| .$$

Inspite of simplicity, of both formulation and proof of (5), it implies all known by now results concerning Problem 1. For example if we note (having in mind the Lévy inequality) that

$$\min_\theta |A_\sigma\theta| \le E\max_k \|a_{\sigma(1)}r_{\sigma(1)} + \dots + a_{\sigma(k)}r_{\sigma(k)}\| \le 2E\Big\|\sum_{i=1}^n a_i r_i\Big\| ,$$

then (5) implies (4) with the better constant:

$$\min_\sigma |A_\sigma| \le 2E\Big\|\sum_{i=1}^n a_i r_i\Big\| .$$

Investigation of Problem 2 has been initiated by Garsia [4,5] for real $a_1,\dots,a_n$. In [5] he has showh that

$$\frac{1}{n!}\sum_{\sigma}|A_\sigma| \le c\left(\sum_{i=1}^{n} a_i^2\right)^{1/2},$$

where c is an absolute constant.

Here, for the case of a normed space, we prove that it takes place the following two-sided inequality

$$\frac{1}{2}E\left\|\sum_{i=1}^{n} a_i r_i\right\| \le \frac{1}{n!}\sum_{\sigma}|A_\sigma| \le 2E\left\|\sum_{i=1}^{n} a_i r_i\right\| . \qquad (6)$$

We did not think of the left inequality up to the Łańcut Conference. Just during the conference Professor Kwapień kindly informed us that it also is true. As we found out later, the left inequality in (6) had been obtained first by V.M. Kadec. As for the right inequality, it was proved first by Maurey and Pisier [14] (apparently with a greater constant). It is evident that this inequality implies (4) with other constant.

Our proofs in both problems are simple. Actually, we use only the Lévy inequality.

In Section 1 we give the solution of Problem 1, Section 2 deals with Problem 2. In Section 3, (5) is applied to the problem on structure of the set of sums of a conditionally convergent series in a normed space. We note that our inequalities are applicable also to the problem on existence of a permutation of a functional series under which the series is almost everywhere convergent (cf. [4,5,16]).

2. SOLUTION OF PROBLEM 1

Theorem 1. Let X be a normed space, $a_1,\dots,a_n \in X$ and let $\sum_{i=1}^{n} a_i = 0$. Then there exists a permutation σ such that

$$\max_k \|a_{\sigma(1)} + \dots + a_{\sigma(k)}\| = \min_\theta \max_k \|a_{\sigma(1)}\theta_1 + \dots + a_{\sigma(k)}\theta_k\| .$$

<u>Proof</u>. For an ordered collection $b = (b_1,\dots,b_n) \in X^n$ denote $|b| = \max_k \|b_1 +\dots+ b_k\|$. It is obvious that $|b|$ is a norm in the space of all sequences of elements of X of finite leght. Further for a collection of signs $\theta = (\theta_1,\dots,\theta_n)$ denote $A_\pi\theta =$ $= (a_{\pi(1)}\theta_1,\dots,a_{\pi(n)}\theta_n)$ and $|A_\pi\theta| = \max_k \|a_{\pi(1)}\theta_1 +\dots+ a_{\pi(k)}\theta_k\|$.

For any permutation π and any collection of sings θ we have

$$|A_\pi\theta| = |A_\pi\theta| + |A_\pi| - |A_\pi| \geq 2\max(|A_\pi^+|,|A_\pi^-|) - |A_\pi|. \qquad (7)$$

Here A_π^+ denotes the ordered collection which is obtained from $a_{\pi(1)},\dots,a_{\pi(n)}$ by writing out successively of elements $a_{\pi(i)}$ corresponding to $\theta_i = +1$. A_π^- is defined similarly. We note that

$$\max(|A_\pi^+|,|A_\pi^-|) = |A_{\pi^*(\pi,\theta)}| , \qquad (8)$$

where the permutation $\pi^* : [1,\dots,n] \to [1,\dots,n]$ is defined by virtue of π and θ as follows. Beforehand one writes out those numbers $\pi_{i_1},\dots,\pi_{i_m}$ for which $\theta_{i_1},\dots,\theta_{i_m} = +1$ and then in the inverse order those numbers for which $\theta_i = -1$. (The condition $\sum_{i=1}^n a_i = 0$ is important to prove (8).) Now (7) implies

$$|A_\pi\theta| \geq 2|A_{\pi^*}| - |A_\pi| . \qquad (9)$$

If we now put $\pi = \sigma$, where σ is such that $|A_\sigma| = \min_\pi |A_\pi|$, then $|A_{\sigma^*(\sigma,\theta)}| \geq |A_\sigma|$ for any θ , and hence

$$|A_\sigma\theta| \geq 2|A_\sigma| - |A_\sigma| .$$

Thus for such σ $\min_\theta |A_\sigma\theta| \geq |A_\sigma|$. Then inverse inequality is trivial. Theorem is proved.

<u>Remark</u>. It is clear that Theorem 1 remains valid if we replace the norm $\|\cdot\|$ by any seminorm on a vector space X . One might also con-

sider a quasi-norm g on a vector space X with the property $g(2x) \geq cg(x)$, $x \in X$, where $c > 1$. Then the method used above gives the existence of σ such that

$$\max_k g(a_{\sigma(1)} + \ldots + a_{\sigma(k)}) \leq \frac{1}{c-1} \min_\theta \max_k g(a_{\sigma(1)}\theta_1 + \ldots + a_{\sigma(k)}\theta_k).$$

Theorem 1 implies the following

Corollary 1. Let X be a normed space, $a_1, \ldots, a_n \in X$ and let $\sum_{i=1}^n a_i = 0$.

(i) There exists such a permutation σ that

$$\max_k \|a_{\sigma(1)} + \ldots + a_{\sigma(k)}\| \leq 2E\|\sum_{i=1}^n a_i r_i\| ,$$

where (r_i) is the sequence of Rademacher functions.

(ii) If X is a Banach lattice of some cotype q, $2 \leq q < \infty$, then there exists a permutation σ such that

$$\max_k \|a_{\sigma(1)} + \ldots + a_{\sigma(k)}\| \leq c\|(\sum_{i=1}^n |a_i|^2)^{1/2}\| ,$$

where c is a constant depending only on the lattice X.

(iii) If $X = L_p(\Omega, A, \mu)$, $1 \leq p < \infty$ then there exists a permutation σ such that

$$\max_k \|a_{\sigma(1)} + \ldots + a_{\sigma(k)}\| \leq c_p(\int_\Omega (\sum_{i=1}^n |a_i|^2)^{p/2} d\mu)^{1/p} .$$

Proof. (i). According to Theorem 1 there exists a permutation σ such that

$$\begin{aligned}\max_k \|a_{\sigma(1)} + \ldots + a_{\sigma(k)}\| &\leq \min_\theta \max_k \|a_{\sigma(1)}\theta_1 + \ldots + a_{\sigma(k)}\theta_k\| \\ &\leq E \max_k \|a_{\sigma(1)} r_{\sigma(1)} + \ldots + a_{\sigma(k)} r_{\sigma(k)}\| \\ &\leq 2E\|\sum_{i=1}^n a_i r_i\| .\end{aligned}$$

The latter inequality is a consequence of the Lévy inequality (cf. [22],

Ch. V). Assertion (ii) is a consequence of (i) since for any Banach lattice of cotype q, $2 \le q < \infty$ it takes place the Maurey inequality [13]

$$E\|\sum_{i=1}^{n} a_i r_i\| \le c\|(\sum_{i=1}^{n} |a_i|^2)^{1/2}\| ,$$

where c is a constant depending only on X. (iii) is a direct corollary of (ii).

The main part (i) of Corollary 1 is the direct consequence of the Mourier and Pisier result [14] (cf. also [1,2]). The assertion (iii) for $1 \le p \le 2$ has been proved earlier in a paper by Nikishin [17], which became starting point for us.

Now, we will give an estimate for $\min_{\sigma} |A_{\sigma}|$ in the case when $\sum_{i=1}^{n} a_i$ is not necessarily zero. We add to the collection $A = (a_1,\dots,a_n)$ one more element $a_{n+1} = -\sum_{i=1}^{n} a_i$. Applying to this new collection Theorem 1 and using the triangular inequality we get the following

Corollary 2. Let X be a normed space, $a_1,\dots,a_n \in X$. There exists a permutation $\sigma : [1,\dots,n] \to [1,\dots,n]$ such that

$$\max_k \|a_{\sigma(1)} +\dots+ a_{\sigma(k)}\| \le \min_{\theta} \max_k \|a_{\sigma(1)}\theta_1 +\dots+ a_{\sigma(k)}\theta_k\| + 2\|a_1 +\dots+ a_n\| .$$

In conclusion we mention an analogous inequality for normed space valued functions defined on a measure space.

Theorem 1'. Let X be a normed space, let (Ω,A,μ) be a measure space and let $A = (a_1,\dots,a_n)$ be a collection of elements of $L_1(\Omega,A,\mu;X)$. Then there exists a permutation σ such that

$$\int_{\Omega} \max_k \|a_{\sigma(1)} +\dots+ a_{\sigma(k)}\| d\mu \le \min_{\theta} \int_{\Omega} \max_k \|a_{\sigma(1)}\theta_1 +\dots+ a_{\sigma(k)}\theta_k\| d\mu + 2\int_{\Omega} \|a_1 +\dots+ a_n\| d\mu .$$

Proof of Theorem 1' can be carried out completely in the same way as that of Theorem 1. One is to consider first the case $\sum_{i=1}^{n} a_i = 0$ and to estimate $\int_\Omega |A_\sigma \theta| d\mu$ in the same manner as it was done for $|A_\sigma \theta|$ in Theorem 1.

Theorem 1 may be interpreted as a particular case of Theorem 1' (when μ is concentrated at a point and $\sum_{i=1}^{n} a_i = 0$).

Remark. In Theorem 1' one could take $L_\phi(X)$ instead of $L_1(X)$ for an increasing ϕ, $\phi(0) = 0$. Then the inequality would have the following form

$$\int_\Omega \phi(|A_\sigma| - \|\sum_{i=1}^{n} a_i\|)d\mu \le \inf_\theta \int_\Omega \phi(|A_\sigma \theta| + \|\sum_{i=1}^{n} a_i\|)d\mu .$$

The space of values of a_i's also can be chosen more general. Namely, the inequality remains valid if the normed space X is replaced by a quasi-normed space with a quasi-norm g satisfying the condition $g(2x) \ge cg(x)$, $x \in X$, where c is a constant greater than 1.

3. SOLUTION OF PROBLEM 2

Theorem 2. Let X be a normed space, and let $a_1,\dots,a_n \in X$, $\sum_{i=1}^{n} a_i = 0$. Then for any convex increasing function $f : R^+ \to R^+$

$$Ef(\tfrac{1}{2}\|\sum_{i=1}^{n} a_i r_i\|) \le \frac{1}{n!}\sum_\sigma \max_k f(\|a_{\sigma(1)} + \dots + a_{\sigma(k)}\|)$$
$$\le 2E\, f(\|\sum_{i=1}^{n} a_i r_i\|) ,$$

where (r_i) is the sequence of Rademacher functions.

Proof. As in the proof of Theorem 1' denote

$$A = (a_1,\dots,a_n), \quad |A| = \max_k \|a_1 + \dots + a_k\| ,$$

$$A_\pi = (a_{\pi(1)},\dots,a_{\pi(n)}) , \quad A_\pi\theta = (a_{\pi(1)}\theta_1,\dots,a_{\pi(n)}\theta_n) .$$

Let us prove first the right inequality. Using the Lévy inequality and convexity of f we get

$$2E\, f(\| \sum_{i=1}^{n} a_i r_i \|) \ge \frac{1}{n!} \sum_{\sigma} E f(\max_k \| \sum_{i=1}^{k} a_{\sigma(i)} r_{\sigma(i)} \|)$$
$$= \frac{1}{2^n n!} \sum_{\sigma,\theta} f(|A_\sigma \theta|) =$$
$$= \frac{1}{2^n n!} \sum_{\sigma,\theta} f(|A_\sigma \theta| + |A_\sigma| - |A_\sigma|)$$
$$\ge \frac{1}{2^n n!} \sum_{\sigma,\theta} f(2|A_{\sigma^*(\sigma,\theta)}| - |A_\sigma|)$$
$$\ge \frac{2}{n!} \sum_{\sigma^*} f(|A_{\sigma^*}|) - \frac{1}{n!} \sum_{\sigma} f(|A_\sigma|) = \frac{1}{n!} \sum_{\sigma} f(|A_\sigma|) ,$$

where $\sigma^*(\sigma,\theta)$ is a permutation which is constructed by means of σ and θ according the rule described in the proof of Theorem 1. Thus, the right inequality is shown. Now prove the left one. We have (A_σ^+ and A_σ^- are defined in the proof of Theorem 1):

$$E f(\tfrac{1}{2} \| \sum_{i=1}^{n} a_i r_i \|) \le \frac{1}{n!} \sum_{\sigma} E f(\tfrac{1}{2} \max_k \| \sum_{i=1}^{n} a_{\sigma(i)} r_{\sigma(i)} \|)$$
$$= \frac{1}{2^n n!} \sum_{\sigma,\theta} f(\tfrac{1}{2}|A_\sigma \theta|) \le \frac{1}{2^n n!} \sum_{\sigma,\theta} f(\tfrac{1}{2}(|A_\sigma^+| + |A_\sigma^-|))$$
$$\le \frac{1}{2^n n!} \cdot \frac{1}{2} (\sum_{\sigma,\theta} f(|A_\sigma^+|) + \sum_{\sigma,\theta} f(|A_\sigma^-|)) \le \frac{1}{n!} \sum_{\sigma} f(|A_\sigma|) .$$

Theorem is proved.

Remark. 1. As the proof shows the left inequality holds without the assumption $\sum_{i=1}^{n} a_i = 0$. The only fact we have used is the triangular inequality for the norm $|\cdot|$ and convexity of f .

2. Theorem 2, of course, implies Corollary 1 of Theorem 1, i.e., it yields an upper estimate for $\min_\sigma |A_\sigma|$. However, this estimate as Theorem 1 shows is not exact. Summing up Theorems 1 and 2 we can say that the best estimate in Problem 1 is effected by the expression $\min_\theta |A_\sigma \theta|$ for some permutation σ, while the best estimate in Problem 2 is realized by the expression $E\| \sum a_i r_i \|$.

3. The right inequality in Theorem 2 for $f(t) = t^p$, $1 \le p < \infty$ has been proved (with a greater constant) by Maurey and Pisier [14]; the case $X = R^1$ also for $f(t) = t^p$ has been treated earlier by

Garsia [4,5]. The left inequality for $f(t) = t$, as it was mentioned in Introduction, has been found first by V.M. Kadec (personal communication). In the monograph [10], by Kashin and Saakian, it is proved the following result for the scalar case: if $a_i \in R^1$, $a_1 +\ldots+ a_n = 0$, then for each $y > 0$

$$\frac{1}{n!}\operatorname{card}\{\sigma : |A_\sigma| > y\} \le c_1 P[|\sum_{i=1}^{n} a_i r_i| > c_2 y] ,$$

where c_1 and c_2 are some absolute constants. We can state as a conjecture that the analogous assertion takes place in the general case of a normed space.

4. We could state Theorem 2 for the case of a quasi-normed space. Namely, the following is true. Let q be a quasi-norm on a linear space X satisfying the condition $q(2x) \ge cq(x)$, where c is a constant greater than 1. Then if $a_1,\ldots,a_n \in X$ and $\sum_{i=1}^{n} a_i = 0$, then

$$\frac{1}{n!}\sum_{\sigma}\max_{k} q(\sum_{i=1}^{k} a_{\sigma(i)}) \le \frac{1}{c-1} Eq(2\sum_{i=1}^{n} a_i r_i).$$

5. Let us mention also the case of $f(t) = t^p$, $0 < p < 1$, which is not covered by Theorem 2. The same proof with the application of the application of the inequality $a^p + b^p \ge (a+b)^p$, $a > 0$, $b > 0$, yields that if X is a normed space, $a_1,\ldots,a_n \in X$ and $\sum_{i=1}^{n} a_i = 0$, then

$$\frac{1}{n!}\sum_{\sigma}\max_{k} \|a_1 +\ldots+ a_k\|^p \le \frac{2}{2^p-1} E\|\sum_{i=1}^{n} a_i r_i\|^p .$$

4. APPLICATION TO A PROBLEM ON SUMS OF A CONDITIONALLY CONVERGENT SERIES

Here we use Theorem 1 to investigate the set of sums of a conditionally convergent series in a normed space. The problem goes back to the well-known Riemann theorem [19] asserting that the set of sums of a conditionally convergent series $\sum a_k$, $a_k \in R^1$ under all its convergent rearrangements fills out the real line. For a series $\sum a_k$ with members from the Euclidean space R^n P. Lévy [11] and Steinitz [20] have established that the set $\tilde{\sigma}_{(a_k)}$ of all sums of the series $\sum a_k$, $a_k \in R^n$ under all its convergent rearrangements is linear, i.e., if $s_1, s_2 \in \sigma_{(a_k)}$, then $\alpha_1 s_1 + \alpha_2 s_2 \in \tilde{\sigma}_{(a_k)}$ for any real α_1 and α_2, $\alpha_1 + \alpha_2 = 1$.

S. Banach posed the question on linearity of $\tilde{\sigma}_{(a_k)}$ for a series $\sum a_k$ with members from a normed space (Problem 106 form the "Scotish book"). An unknown author (it is believed that it was Marcinkiewicz) has provided the Banach's question in "Scotish book" with a note answering the question in negative. His counterexample constructed for L_p spaces shows that $\tilde{\sigma}_{(a_k)}$ may be even nonconvex. Now it is known (V.M. Kadec [8]) that for any infinite dimensional Banach space X there exists a series $\sum a_k$, $a_k \in X$ such that $\tilde{\sigma}_{(a_k)}$ is non-convex.

The main question considered in this section is: what are additional conditions on a series $\sum a_k$ with members in a normed space X under which $\tilde{\sigma}_{(a_k)}$ is linear and closed? The first results answering this question were those by M.I. Kadec [7], Nikishin [16,17] and Pecherski [18]. (These results as well as those by other authors for concrete Banach spaces are cited at the end of this section.)

In [1,2] the first authors has obtained the following result for a general normed space. Throughout this section it is assumed that X is a normed space over the field of reals.

<u>Theorem 3</u>. Let $\sum a_k$ be a conditionally convergent series in X. If the series $\sum a_k r_k$ is almost everywhere convergent in X ((r_k) is the sequence of Rademacher functions), then $\tilde{\sigma}_{(a_k)}$ is linear and closed.

Theorem 1 allows to refine this result. Namely, the following statement is true.

Theorem 4. Let $\sum a_k$ be a conditionally convergent series in X. If for any permutation $\pi : N \to N$ there exists a sequence of signs $\theta = (\theta_1, \theta_2, \ldots)$ such that $\sum a_{\pi(k)} \theta_k$ is convergent, then $\tilde{\sigma}_{(a_k)}$ is linear and closed.

Theorem 4 contains all results on linearity of $\tilde{\sigma}_{(a_k)}$ we know for the case of a normed space. In particular, Theorem 3, as it can easily be seen, also is a consequence of Theorem 4. Indeed, the a.e. convergence of $\sum a_k r_k$ implies the a.e. convergence of any rearrangement $\sum a_{\pi(k)} r_{\pi(k)}$ which guarantees with a considerable reserve fulfilment of the condition of Theorem 4. Note also that Theorem 4, in contrast to other results for infinite dimensional normed space cases, gives for the finite dimensional case the Lévy-Steinitz theorem (in other words the condition of Theorem 4 in the finite dimensional case is automatically fulfilled).

Proof of Theorem 4 is based on the following lemma which is a consequence of Theorem 1, as well as on the developed technique (cf. [7, 16, 17, 18]).

Lemma. Let $\sum a_k$ be a series in X such that a subsequence S_{n_k} of its partial sums converges to $s \in X$. If the series $\sum a_k$ satisfies the condition of Theorem 4, i.e., for any permutation $\pi : N \to N$ there exists a sequence of signs $\theta = (\theta_1, \theta_2, \ldots)$ such that the series $\sum a_{\pi(k)} \theta_k$ is convergent, then there exists a rearrangement $\sum a_{\sigma(k)}$ of the series $\sum a_k$ such that $\sum a_{\sigma(k)} = s$.

In the following corollary there are collected results on conditions for $\tilde{\sigma}_{(a_k)}$ to be linear and closed for concrete Banach spaces.

Corollary 3. Any of the listed below conditions on X and on $\sum a_k$ is sufficient for $\tilde{\sigma}_{(a_k)}$ to be linear and closed.

(i) X is a Banach lattice of cotype q, $2 \le q < \infty$, and there exists

$$\lim_{n\to\infty} \left(\sum_{i=1}^{n} |a_i|^2 \right)^{1/2} \quad \text{in } X ; \tag{10}$$

(ii) X is an Orlicz space L_ϕ with the Δ_2 condition, and

$$(\sum_{k=1}^{\infty} |a_k|^2)^{1/2} \in L_\phi \; ; \tag{11}$$

(iii) X is a Banach space, and

$$\sum_{k=1}^{\infty} \rho(\|a_k\|) < \infty , \tag{12}$$

where ρ is the modulus of smoothness of X ;

(iv) X is a space of type p, $1 \le p \le$, and

$$\sum_{k=1}^{\infty} \|a_k\|^p < \infty \; ; \tag{13}$$

(v) X is a space of infratype p, $1 \le p \le 2$, and

$$\sum_{k=1}^{\infty} \|a_k\|^p < \infty .$$

Proof. Sufficiency of (i) follows from the fact that for a Banach space of some cotype the a.e. convergence of $\sum a_k r_k$ is equivalent to existence of the limit (10) (cf. [13]). Sufficiency of (i) implies sufficiency of (ii), since an Orlicz space with Δ_2 condition is Banach lattice with some cotype. It is well-known that (12) implies a.e. convergence of $\sum a_k r_k$. This implies sufficiency of (iii). By the definition of type p spaces, (13) also implies a.e. convergence of $\sum a_k r_k$. Hence, condition (iv) is sufficient.

Sufficiency of (v). We say that a Banach space X is of infratype p, $1 \le p \le 2$, if $a_k \in X$, $k \in N$ and $\sum \|a_k\|^p < \infty$ imply existence of a sequence of signs θ_k such that the series $\sum a_k \theta_k$ is convergent in X. It is clear that for such a space the condition $\sum \|a_k\|^p < \infty$ implies the condition of Theorem 4. Corollary is proved.

Sufficiency of the condition (i) of Corollary 3 has been proved by Maleev [12] for the case of lattices of cotype 2, for general X it was proved in [1,2]. Sufficiency of (ii) has been independetly shown by Megrabian [15]. Sufficiency of (ii) goes back to Nikishin's paper [17] where the case of L_p , $1 \le p \le 2$ has been treated. Sufficiency of (iii) has been proved by S. Troyanski [20] (cf. also paper by Fonf [3]). M.I. Kadec in [7] has first proved sufficiency of the condition (iv) for the case of L_p, $1 \le p < \infty$ spaces; it was the first condition of linearity of $\tilde{\sigma}_{(a_k)}$ for infinite dimensional spaces. Sufficiency of (v) has been proved by V.M. Kadec [10].

Remark. Theorem 4 remains valid for quasi-normed X, if the quasi-norm q satisfies the inequality $q(2x) \ge cq(x)$, $x \in X$, where c is a constant greater than 1. This is a consequence of the fact that Theorem 1 holds for such spaces (cf. Remark after Theorem 1).

In conclusion the authors express their gratitude to Professor S. Kwapień for useful discussions.

REFERENCES

[1] S.A.Chobanyan, The structure of the set of sums of a conditionally convergent series in a Banach space, Soviet Math. Dokl., v. 30 (1984), p. 438-441.

[2] S.A.Chobanyan, The structure of the set of sums of a conditionally convergent series in a normed space, Mat. Sb., v. 128 (170) (1985), p. 50-65; English transl. in Math. USSR Sb., v. 56 (1987).

[3] V.P.Fonf, On conditionally convergent series in a uniformly smooth Banach space, Mat. Zametki, v. 11 (1972), p. 209-214; English transl. in Math. Notes, v. 11 (1972).

[4] A.Garsia, Rearrangements for Fourier series, Ann. Math., v. 79 (1964), p. 623-629.

[5] A.Garsia, Topics in Almost Everywhere Convergence, Markham, Chicago 1970.

[6] M.I.Kadec, On a property of vector broken lines in the n-dimensional space, Uspekhi Mat. Nauk, 8 (1953), p. 139-143. (Russian)

[7] M.I.Kadec, On conditionally convergent series in L_p spaces, Uspekhi Mat. Nauk, v. 9 (1954), p. 107-109. (Russian)

[8] V.M.Kadec, On a problem of S.Banach (Problem 106 from "Scotish book"), Functional Anal. Appl., v. 20 (1986), 4, p. 74-75.

[9] V.M.Kadec, B-convexity and a Steinitzś lemma, Proc. of N. Cauc. Sci. Center of Higher School, Nat. Sci., v. 4 (1984), p. 27-29. (Russian)

[10] B.S.Kashin and A.A.Saakian, Orthogonal Series, Nauka, Moscow 1984. (Russian)

[11] P.Lévy, Sur les séries semi-convergents, Nouv. Ann. Math., v. 5, (1905), p. 506-511.

[12] R.Maleev, On conditionally convergent series in some Banach lattices, C.R. Acad. Bulgare Sci., v. 32 (1979), p. 1015-1018. (Russian)

[13] B.Maurey, Type et cotype dans les espaces munis de structures locales inconditionneles, Sém. Maurey-Schwartz 1973/1974, Exposés 24, 25, Centre Math. École Polytech., Paris, 1974.

[14] B.Maurey and G.Pisier, Remarques sur l'expose d'Assouad, Sém. Maurey-Schwartz 1974/1975, Annexe 1, Centre Math. École Polytech., Paris, 1975.

[15] R.M.Megrabian, On the set of sums of functional series in L_ϕ spaces, Teor. veroyatmost. i Primenen., v. 30 (1985), p. 511-523; English transl. in Theory Probab. Appl., v. 30 (1985).

[16] E.M.Nikishin, On convergent rearrangements of functional series, Mat. Zametki, v. 1 (1967), p. 129-136. (Russian)

[17] E.M.Nikishin, On rearrangements of series in L_p, Mat. Zametki, v. 14 (1973), p. 31-38; English transl. in Math. Notes 14 (1973).

[18] D.V.Pecherski, A theorem on projections of rearranged series with members in L_p, Izv. Akad. Nauk SSSR, Ser. Math., v. 41 (1977), p. 203-214; English transl. in Math. USSR Izv., v. 11 (1977).

[19] B.Riemann, Über die Darstellbarkeit einer Funktion durch eine trigonometrische Reihe, Abhandl. d. Königl. Gessellsch. der Wiss., v. 13 (1867).

[20] E.Steinitz, Bedingt konvergente Reihen und konvexe Systeme, J. reine und angew. Math., v. 143 (1913), p. 128-175; v. 144 (1913), p. 1-49; v. 146 (1916), p. 68-111.

[21] S.Troyanski, On conditionally convergent series and certain F-spaces, Teor. Funktsii Funktsional. Anal. i Prolozhen., v. 5 (1967), p. 102-107. (Russian)

[22] N.N.Vakhania, V.I.Tarieladze and S.A.Chobanyan, Probability Distributions on Banach Spaces, D. Reidel, Dordrecht 1987.

Probability Theory on Vector Spaces IV

Lancut, June'87, Springer's LNM 1391

AN EXTREME-SUM APPROXIMATION TO INFINITELY DIVISIBLE LAWS WITHOUT A NORMAL COMPONENT

SÁNDOR CSÖRGŐ*

1. The result and illustrative examples

Consider a non-degenerate infinitely divisible distribution on the real line without a normal component. With $-\infty < t < \infty$, it has the characteristic function

$$\varphi_{\theta,\Psi_1,\Psi_2}(t) = \exp\left\{it\theta + \int_{-\infty}^{0}\left(e^{itx} - 1 - \frac{itx}{1+x^2}\right)dL(x) + \int_{0}^{\infty}\left(e^{itx} - 1 - \frac{itx}{1+x^2}\right)dR(x)\right\},$$

where $\theta \in (-\infty,\infty)$ is a uniquely determined constant and L and R are uniquely determined left-continuous and right-continuous Lévy measures on $(-\infty, 0)$ and $(0,\infty)$, respectively, that is, $L(\cdot)$ and $R(\cdot)$ are non-decreasing functions such that $L(-\infty) = R(\infty) = 0$ and

$$\int_{-\varepsilon}^{0} x^2 dL(x) + \int_{0}^{\varepsilon} x^2 dR(x) < \infty \quad \text{for any} \quad \varepsilon > 0.$$

The reason for the subscripts Ψ_1 and Ψ_2 is that we will prefer to work with the inverse functions

$$\Psi_1(u) = \inf\{x < 0 : L(x) > u\}, \quad 0 < u < \infty,$$

and

$$\Psi_2(u) = \inf\{x < 0 : -R(-x) > u\}, \quad 0 < u < \infty,$$

* Supported by the Hungarian National Foundation for Scientific Research, Grant No. 1808.

where the infimum of the empty set is taken to be zero. These are non-decreasing, non-positive, right-continuous functions on $(0,\infty)$ such that

$$\int_\varepsilon^\infty \Psi_1^2(u)du + \int_\varepsilon^\infty \Psi_2^2(u)du < \infty \quad \text{for all} \quad \varepsilon > 0. \tag{1}$$

Let $F_{\theta,\Psi_1,\Psi_2}(\cdot)$ denote the distribution function determined by the characteristic function $\varphi_{\theta,\Psi_1,\Psi_2}$. Let $Y_1, Y_2, \ldots$ be independent exponentially distributed random variables with mean 1, and put $S_n = Y_1 + \ldots + Y_n, n \geq 1$, for the corresponding partial sums. Let $\{S_n^{(1)}\}$ and $\{S_n^{(2)}\}$ be two independent copies of the sequence $\{S_n\}$. In this note, we consider the problem of approximation of F_{θ,Ψ_1,Ψ_2} by the distribution functions

$$P\left\{\sum_{j=1}^m \Psi_1\left(S_j^{(1)}\right) - \sum_{j=1}^k \Psi_2\left(S_j^{(2)}\right) - \left(\int_1^m \Psi_1(u)du - \int_1^k \Psi_2(u)du\right) + \theta - \theta_1 - \theta_2 < x\right\}$$

which, for any $x \in (-\infty,\infty)$, we denote by $F^{m,k}_{\theta,\Psi_1,\Psi_2}(x); m,k = 1,2,\ldots$; where the constants θ_1 and θ_2 are defined by

$$\theta_l = (-1)^{l+1}\left\{\int_0^1 \frac{\Psi_l(u)}{1+\Psi_l^2(u)}du - \int_1^\infty \frac{\Psi_l^3(u)}{1+\Psi_l^2(u)}du\right\}, \quad l = 1,2.$$

The motivation for this "extreme-sum" approximation will be explained in the next section.

We consider the Lévy distances

$$L_{m,k}(\Psi_1,\Psi_2) =$$
$$\inf\left\{\varepsilon > 0 : F^{m,k}_{\theta,\Psi_1,\Psi_2}(x-\varepsilon) - \varepsilon \leq F_{\theta,\Psi_1,\Psi_2}(x) \leq F^{m,k}_{\theta,\Psi_1,\Psi_2}(x+\varepsilon) + \varepsilon \quad \text{for all} \quad x\right\}$$

and for $a,\delta,\rho > 0$ and $n = 1,2,\ldots$, set

$$r_n^{(l)}(a,\delta,\rho) = \min\left(\delta\left(P\{S_{n+1} \leq a\}\right)^{-1}, \left(\int_a^\infty \sqrt{u}\,d\Psi_l(u)\right)^{-\rho}\right), \quad l = 1,2.$$

THEOREM. *For any pair of* Ψ_1 *and* Ψ_2,

$$\lim_{m,k\to\infty} L_{m,k}(\Psi_1,\Psi_2) = 0. \tag{2}$$

Furthermore, if

$$\int_1^\infty \sqrt{u}\,d\Psi_1(u) + \int_1^\infty \sqrt{u}\,d\Psi_2(u) < \infty, \tag{3}$$

then for any three sequences $\left\{a_n^{(1)}\right\}$, $\left\{a_n^{(2)}\right\}$ *and* $\{\delta_n\}$ *such that*

$$a_n^{(l)} \to \infty \text{ and } \limsup_{n\to\infty} \frac{a_n^{(l)}}{n} < 1, \; l = 1,2, \text{ and } \delta_n \to 0 \text{ as } n \to \infty,$$

and any positive number $\rho < 1$,

$$\lim_{m,k\to\infty} \min\left(r_m^{(1)}(a_m^{(1)},\delta_m,\rho), r_k^{(2)}(a_k^{(2)},\delta_k,\rho)\right) L_{m,k}(\Psi_1,\Psi_2) = 0. \tag{4}$$

It would be difficult to prove a theorem like this by the traditional characteristic function technique. On the other hand, we shall see in Section 3 that the result follows from a recent representation of an arbitrary infinitely divisible random variable, given by S. Csörgő, Haeusler and Mason [3], in a rather straightforward fashion. The main aim of the present note is, therefore, to expose and popularize that representation.

The rate of convergence above depends on a trade-off between the two terms in $r_n^{(l)}$ for both $l = 1$ and $l = 2$. The probability

$$F_{n+1,1}(a_n) = P\{S_{n+1} \le a_n\} = \frac{1}{n!}\int_0^{a_n} e^{-t}t^n dt$$

goes to zero fast if $a_n \to \infty$ slowly, while the integral terms

$$\left(\int_{a_n}^{\infty} \sqrt{u}\, d\Psi_l(u)\right)^{\rho}, \quad l = 1,2,$$

go to zero fast if the convergence $a_n \to \infty$ is fast. By Stirling's formula and the well-known asymptotics of the incomplete gamma function [5; section 9.5], for slow sequences we have the asymptotic equality

$$P\{S_{n+1} \le a_n\} \sim \left(\frac{a_n}{n}\right)^{n+1} \frac{1}{\sqrt{2\pi n}} \frac{e^n}{e^{a_n}} \quad \text{whenever } \frac{a_n}{n} \to 0 \text{ as } n \to \infty.$$

On the other hand, for the very fast sequence $a_n = \tau(n+1)$ Lemma 3.1 of Devroye [4] gives

$$P\{S_{n+1} \le \tau(n+1)\} \le \exp\left(-\frac{(1-\tau)^2}{2} n\right) \text{ whenever } 0 < \tau < 1.$$

A result of the type of the theorem, though slightly different in nature, was first proved by Hall [6] for the approximation of stable laws. (For further comments see the end of the next section.) We note that in the broader context when F_{θ,Ψ_1,Ψ_2} is absolutely continuous with a bounded density, the Lévy distance $L_{m,k}(\Psi_1,\Psi_2)$ may be replaced by the uniform distance

$$K_{m,k}(\Psi_1,\Psi_2) = \sup_{-\infty<x<\infty} |F_{\theta,\Psi_1,\Psi_2}^{m,k}(x) - F_{\theta,\Psi_1,\Psi_2}(x)|$$

in the theorem. This follows from Lemma 2 in [7]. Applying the theorem to a stable distribution with exponent $0 < \alpha < 2$, obtained with the functions $\Psi_l(u) = \Psi_l^{(\alpha)}(u) = -c_l u^{-1/\alpha}, u > 0, l = 1,2$, where the constants $c_1 \geq 0, c_2 \geq 0, c_1 + c_2 > 0$, are determined by the other parameters of the distribution, the choice $a_n^{(1)} = a_n^{(2)} = \tau n, 0 < \tau < 1$, gives the corollary that for any $\rho < 1$,

$$K_{m,k}\left(\Psi_1^{(\alpha)}, \Psi_2^{(\alpha)}\right) = o\left(\max\left(c_1 m^{-\rho\left(\frac{1}{\alpha}-\frac{1}{2}\right)}, c_2 k^{-\rho\left(\frac{1}{\alpha}-\frac{1}{2}\right)}\right)\right) \quad \text{as } m,k \to \infty.$$

This is exactly Theorem 2.2 in [2], proved by a very similar but somewhat more direct technique, based on the representation of an arbitrary stable random variable from [1]. Just as the representation of an arbitrary infinitely divisible random variable in [3] is an extension of that of a stable random variable in [1] (cf. the remark following Corollary 3 in [3] with some corrections for [1], also noted in [2]), the theorem above is an extension of Theorem 2.2 in [2].

We give one more example of the absolutely continuous case. Consider the gamma distribution function of order $\alpha > 0$ and parameter $\lambda > 0$, given by

$$F_{\alpha,\lambda}(x) = \frac{\lambda^\alpha}{\Gamma(\alpha)} \int_0^x e^{-\lambda t} t^{\alpha-1} dt, \quad x \geq 0,$$

and zero otherwise. This is a textbook example of an infinitely divisible distribution without a normal component, produced by the Lévy measures $L(x) = 0, x < 0$, and

$$R(x) = -\alpha \int_{\lambda x}^{\infty} e^{-t} t^{-1} dt, \quad x > 0,$$

and parameter

$$\theta = \theta(\alpha, \lambda) = \alpha \int_0^\infty \frac{e^{-\lambda t}}{1+t^2} dt.$$

Now for $u > 0$, let $h_{\alpha,\lambda}(u) > 0$ be the unique solution $x > 0$ of the equation

$$\int_{\lambda x}^{\infty} e^{-t} t^{-1} dt = \frac{u}{\alpha},$$

and set

$$C_{\alpha,\lambda} = \alpha \left\{ \int_0^\infty \frac{e^{-\lambda t}}{1+t^2} dt - \int_{-\infty}^{-h_{\alpha,\lambda}(1)} \frac{e^t}{1+t^2} dt + \int_{-h_{\alpha,\lambda}(1)}^{0} \frac{t^2 e^t}{1+t^2} dt \right\}.$$

Then the theorem and elementary computations give the following, rather strange limit theorem: as $k \to \infty$,

$$\sup_{-\infty < x < \infty} \left| P\left\{ \sum_{j=1}^{k} h_{\alpha,\lambda}(S_j) - \alpha\left(e^{-h_{\alpha,\lambda}(k)} - e^{-h_{\alpha,\lambda}(1)}\right) + C_{\alpha,\lambda} \leq x \right\} - F_{\alpha,\lambda}(x) \right|$$
$$= O(e^{-\eta k})$$

for any $\eta < 1/2$.

While the rate of convergence in (2) may presumably be arbitrarily slow if a stronger integrability condition such as (3) is not satisfied, i.e. if (1) is just barely satisfied, there are rates faster than the above algebraic, or even the exponential rate. The simplest case when these obtain is when the functions Ψ_1 and Ψ_2 vanish beyond some positive value. In this case, the proof allows us the interpretation

$$\left(\int_{a_n}^{\infty} \sqrt{u}\, d\Psi_l(u)\right)^{-\rho} = \infty, \quad l = 1,2,$$

for all sufficiently large n if $a_n \to \infty$ as $n \to \infty$. As a concrete example of this situation, we consider the Poisson distribution with parameter $\lambda > 0$, given by the distribution function

$$F_\lambda(x) = \sum_{n=0}^{[x]} \frac{\lambda^n}{n!} e^{-\lambda}, \quad -\infty < x < \infty.$$

The corresponding characteristic function is obtained by taking $L(x) = 0, x < 0$, and $R(x) = -\lambda$, if $0 < x < 1$, and $R(x) = 0$, if $x \geq 1$, and $\theta = \lambda/2$. Then, if $L_k(\lambda)$ denotes the Lévy distance between F_λ and the distribution function $P\left\{\sum_{j=1}^{k} I(S_j < \lambda) \leq x\right\}, -\infty < x < \infty$, where I is the indicator function, the theorem gives

$$L_k(\lambda) = O\left(\left(\frac{ea_k}{k}\right)^{k+1} \frac{1}{\sqrt{k}} \frac{1}{e^{a_k}}\right) \text{ as } k \to \infty$$

for any sequence $\{a_k\}$ such that $a_k \to \infty$ and $a_k/k \to 0$. Exactly the same obtains for the Lévy distance $L_k(p), 0 < p < 1$, between the geometric distribution given by

$$G_p(x) = \sum_{n=0}^{[x]} (1-p)p^n, \quad -\infty < x < \infty,$$

and the distribution function defined for $x \in (-\infty, \infty)$ by

$$P\left\{\sum_{j=1}^{k} \sum_{n=1}^{\infty} nI\left(-\log(1-p) - \sum_{l=1}^{n} \frac{p^l}{l} \leq S_j < -\log(1-p) - \sum_{l=1}^{n-1} \frac{p^l}{l}\right) \leq x\right\}.$$

2. Representation and motivation

Pertaining to the independent jump-point sequences $\{S_n^{(l)}\}, l = 1,2$, of the preceding section, consider the independent standard left-continuous Poisson processes $N_l(t) = \sum_{j=1}^{\infty} I(S_j^{(l)} < t), 0 \leq t < \infty$, and for $l = 1,2$, set

$$V_0^{(l)} = V_0^{(l)}(\Psi_l) = (-1)^{l+1} \left\{\int_{S_1^{(l)}}^{\infty} (u - N_l(u))\, d\Psi_l(u) + \int_1^{S_1^{(l)}} u\, d\Psi_l(u) + \Psi_l(1)\right\}$$

and $V_{0,0} = V_{0,0}(\theta, \Psi_1, \Psi_2) = V_0^{(1)} + V_0^{(2)} + \theta - \theta_1 - \theta_2$. Then, by Theorem 3 in [3],

$$Ee^{itV_{0,0}(\theta,\Psi_1,\Psi_2)} = \varphi_{\theta,\Psi_1,\Psi_2}(t), \quad -\infty < t < \infty, \tag{5}$$

or, in other words, the random variable $V_0^{(1)} + V_0^{(2)} + \theta - \theta_1 - \theta_2$ represents in distribution the arbitrarily given infinitely divisible random variable with no normal component. (Adding to it a normal random variable $N(0, \sigma^2)$ with mean 0 and variance $\sigma^2 \geq 0$, independent of (N_1, N_2), we obtain a completely arbitrary infinitely divisible random variable with characteristic function $\exp(-\sigma^2 t^2/2)\varphi_{\theta,\Psi_1,\Psi_2}(t), -\infty < t < \infty$. In the present note, we assume that $\sigma = 0$.) This representation is basic for our method of proof.

We note that in case of the Poisson distribution with parameter $\lambda > 0$, this representation reduces to saying that $N_2(\lambda)$ has the distribution function F_λ in the preceding section. This is, of course, trivial. However, in the case of the gamma (α, λ) example of the preceding section, it says that the distribution function of the random variable

$$\sum_{j=1}^{\infty} h_{\alpha,\lambda}(S_j) - \alpha\left(1 - e^{h_{\alpha,\lambda}(1)}\right) + C_{\alpha,\lambda}$$

is $F_{\alpha,\lambda}$, while in the geometric (p) example it says that the distribution function of

$$\sum_{j=1}^{\infty}\sum_{n=1}^{\infty} nI\left(-\log(1-p) - \sum_{l=1}^{n}\frac{p^l}{l} \leq S_j < -\log(1-p) - \sum_{l=1}^{n-1}\frac{p^l}{l}\right)$$

is G_p. As the proof in the next section shows, these series forms of the representation of gamma and geometric variables are obtained from the above integral form by integrating out between the jump-points of N_2. Further non-trivial examples can be derived in a similar fashion. Concerning the special case of stable random variables, we have already referred to [1], [2] and [3].

The rest of the present section is not going to be used in the proof section, it is only for motivation.

Following [3], let $X_1, X_2, \ldots$ be independent random variables with a common distribution function F, assumed to be non-degenerate, and for each integer $n \geq 1$, let $X_{1,n} \leq \cdots \leq X_{n,n}$ denote the order statistics based on the sample $X_1, \ldots, X_n$. Introduce the inverse or quantile function Q of F as $Q(s) = \inf\{x : F(x) \geq s\}, 0 < s \leq 1, Q(0) = Q(0+)$, consider the truncated variance function

$$\sigma^2(s) = \int_s^{1-s}\int_s^{1-s} (\min(u,v) - uv)dQ(u)dQ(v), \quad 0 < s < 1,$$

and the truncated mean function

$$\mu(s,t) = \int_s^t Q(u+)du.$$

Choose a sequence $\{\alpha_n\}$ of positive constants such that $\alpha_n \downarrow 0$ and $n\alpha_n \to 0$, as $n \to \infty$, and define two right-continuous, non-decreasing functions of $s \in (0,\infty)$ by putting

$$q_1(n,s) = \begin{cases} Q((s/n)+) & , \text{if } 0 < s \le n - n\alpha_n, \\ Q((1-\alpha_n)+) & , \text{if } n - n\alpha_n < s < \infty, \end{cases}$$

and

$$q_2(n,s) = \begin{cases} -Q(1-s/n) & , \text{if } 0 < s \le n - n\alpha_n, \\ -Q(\alpha_n) & , \text{if } n - n\alpha_n < s < \infty. \end{cases}$$

All these quantities make sense if n is large enough. Now assume that there exists a subsequence $\{n'\}$ of the positive integers such that for two (necessarily) non-decreasing, non-positive, right-continuous functions Ψ_1 and Ψ_2 defined on $(0,\infty)$, for a sequence of positive constants A_n, and for a non-negative constant δ we have

$$\frac{\sqrt{n'}\sigma(1/n')}{A_{n'}} \to \delta < \infty \quad \text{and} \quad \frac{q_l(n',s)}{A_{n'}} \to \Psi_l(s), \quad \text{as } n' \to \infty, \tag{6}$$

at every continuity point $s \in (0,\infty)$ of $\Psi_l, l = 1,2$. Then it is shown in [3] (cf. Lemmas 2.1, 2.5, Theorems 1, 2 and 4, and the whole structure of the proof of Theorems 1 and 2) that (1) is already satisfied and there exists a sequence $\{r_{n'}\}$ of positive integers such that $r_{n'} \to \infty$ and $r_{n'}/n' \to 0$ as $n' \to \infty$, and for each quadruplet (m_1, k_1, m_2, k_2) of fixed integers $m_1, k_1 \ge 0$ and $m_2, k_2 \ge 1$,

$$\begin{aligned} &\frac{1}{A_{n'}}\left(V_{n'}(m_1,k_1,r_{n'}), X_{1,n'},\ldots,X_{m_2,n'}, X_{n'-k_2+1,n'},\ldots,X_{n',n'}\right) \\ &\xrightarrow{D} \left(V_{m_1}^{(1)} + V_{k_1}^{(2)}, \Psi_1(S_1^{(1)}),\ldots,\Psi_1(S_{m_2}^{(1)}), -\Psi_2(S_{k_2}^{(2)}),\ldots,-\Psi_2(S_1^{(2)})\right) \end{aligned} \tag{7}$$

as $n' \to \infty$, where

$$\begin{aligned} V_{n'}(m_1,k_1,r_{n'}) = &\sum_{j=m_1+1}^{r_{n'}} X_{j,n'} + \sum_{j=n'-r_{n'}+1}^{n'-k_1} X_{j,n'} \\ &- n'\left\{\mu\left(\frac{m_1+1}{n'}, \frac{r_{n'}+1}{n'}\right) + \mu\left(1 - \frac{r_{n'}+1}{n'}, 1 - \frac{k_1+1}{n'}\right)\right\} \end{aligned}$$

and for $m \ge 0$,

$$\begin{aligned} V_m^{(l)} = (-1)^{l+1}\Bigg\{ &\int_{S_{m+1}^{(l)}}^{\infty} (u - N_l(u))\,d\Psi_l(u) + \int_1^{S_{m+1}^{(l)}} u\,d\Psi_l(u) - m\Psi_l(S_{m+1}^{(l)}) \\ &+ \int_1^{m+1} \Psi_l(u)\,du + \Psi_l(1)\Bigg\}, \quad l = 1,2, \end{aligned}$$

and where $\xrightarrow{D}$ denotes convergence in distribution. Furthermore, if $\delta = 0$ in (6) then necessarily $\Psi_l(s) = 0$ for all $1 \le s < \infty, l = 1,2$, and $V_{n'}(m_1, k_1, r_{n'})$ may be replaced in (7) by

$$V_{n'}(m_1,k_1) = \sum_{j=m_1+1}^{n'-k_1} X_{j,n'} - n'\mu\left(\frac{m_1+1}{n'}, 1 - \frac{k_1+1}{n'}\right).$$

Also, since $0 \le \sigma((r_{n'}+1)/n')/\sigma(1/n') \le 1$, this ratio must converge along a further subsequence $\{n''\}$ of $\{n'\}$ to some number $\sigma \in [0,1]$. If this happens and $\delta > 0$ in (6) then (7) holds true along $\{n''\}$ with $V_{n''}(m_1, k_1, r_{n''})$ replaced by $V_{n''}(m_1, k_1)$ and $V_{m_1}^{(1)} + V_{k_1}^{(2)}$ replaced by $V_{m_1}^{(1)} + N(0,\sigma^2) + V_{k_2}^{(2)}$ in the limit, where the three terms of this last sum are independent. Finally, we note that by Theorem 5 in [3] the condition (6) is optimal for all these conclusions (and more refined ones presented in [3]) in the sense that if $m \ge 0$ and $k \ge 0$ are fixed integers and there exist two sequences of constants $A_n > 0$ and B_n and a sequence $\{n'\}$ of positive integers such that

$$\frac{1}{A_{n'}}\left\{\sum_{j=m+1}^{n'-k} X_{j,n'} - B_{n'}\right\}$$

converges in distribution to a non-degenerate limit, then we must have (6) along a subsequence $\{n''\}$ of $\{n'\}$.

Now we see that the sum

$$\sum_{j=1}^{m} \Psi_1(S_j^{(1)}) - \sum_{j=1}^{k} \Psi_2(S_j^{(2)}) \tag{9}$$

may be looked upon as the asymptotic contribution of the extreme-sums

$$\sum_{j=1}^{m} X_{j,n''} + \sum_{j=n''-k+1}^{n''} X_{j,n''}$$

of the smallest m and largest k elements of the sample $X_1, \dots, X_{n''}$ to the two independent non-normal components of the limiting infinitely divisible distribution of the whole sum $\sum_{j=1}^{n''} X_j$. It is, therefore, reasonable to expect that these limiting extreme sums in (9) themselves will approximate in distribution the corresponding infinitely divisible random variable without a normal component, at least after a suitable centring. This is exactly what is stated in the theorem in the preceding section, and this is the underlying motivation for calling this approximation an "extreme-sum approximation".

The approximation of Hall [5] to a non-normal stable law is in terms of the distributions of the suitably centred asymptotic contributions of a fixed number of extreme elements of a sample from the domain of attraction of that stable law, where these extreme elements are ***largest in absolute value***. Since the contributions of the smallest and largest sample elements in the natural order are asymptotically independent, it is just as reasonable to consider them separately. This is what is done in [2] in the stable case and presently in the greatest possible generality. It would be interesting to see how Hall's approximation generalizes for an arbitrary infinitely divisible law without a normal component.

3. Proof of the theorem

Let $Z_{m,k} = Z_{m,k}(\theta, \Psi_1, \Psi_2)$ denote the random variable with distribution function $F^{m,k}_{\theta,\Psi_1,\Psi_2}$. According to (5), we have

$$F_{\theta,\Psi_1,\Psi_2}(x) = P\{V_{0,0}(\theta, \Psi_1, \Psi_2) \leq x\}, \quad -\infty < x < \infty.$$

Since $L_{m,k}(\Psi_1, \Psi_2) \leq 2\varepsilon$ for an $\varepsilon > 0$ if

$$\begin{aligned} P\{|V_{0,0} - Z_{m,k}| \geq 2\varepsilon\} &\leq P\left\{\left|V_0^{(1)} - \sum_{j=1}^{m} \Psi_1(S_j^{(1)}) + \int_1^m \Psi_1(u)du\right| \geq \varepsilon\right\} \\ &\quad + P\left\{\left|V_0^{(2)} + \sum_{j=1}^{k} \Psi_2(S_j^{(2)}) - \int_1^k \Psi_2(u)du\right| \geq \varepsilon\right\} \\ &\leq 2\varepsilon, \end{aligned}$$

to prove (2) it is sufficient to show that

$$W_n = \left|\int_{S_1}^{\infty} (u - N(u))d\Psi(u) + \int_1^{S_1} ud\Psi(u) + \Psi(1) - \sum_{j=1}^{n} \Psi(S_j) + \int_1^n \Psi(u)du\right|$$

converges to zero in probability as $n \to \infty$, where Ψ is any one of Ψ_1 and Ψ_2 and $N(u) = \sum_{j=1}^{\infty} I(S_j < u)$ is a standard Poisson process. Since the jump-points hit the possible discontinuity points of Ψ with probability zero, we almost surely have

$$\begin{aligned} &\int_{S_1}^{\infty} (u - N(u))d\Psi(u) + \int_1^{S_1} ud\Psi(u) + \Psi(1) \\ &\quad = \int_{S_{n+1}}^{\infty} (u - N(u))d\Psi(u) + \sum_{j=1}^{n}\left\{\int_{S_j}^{S_{j+1}} ud\Psi(u) - j(\Psi(S_{j+1}) - \Psi(S_j))\right\} \\ &\qquad + \int_1^{S_1} ud\Psi(u) + \Psi(1) \\ &\quad = \int_{S_{n+1}}^{\infty} (u - N(u))d\Psi(u) + \sum_{j=1}^{n} \Psi(S_j) + \int_1^{S_{n+1}} ud\Psi(u) - n\Psi(S_{n+1}) + \Psi(1) \\ &\quad = \int_{S_{n+1}}^{\infty} (u - N(u))d\Psi(u) + \sum_{j=1}^{n} \Psi(S_j) - \int_1^{S_{n+1}} \Psi(u)du + \Psi(S_{n+1})(S_{n+1} - n). \end{aligned}$$

Hence

$$W_n \le \left|\int_{S_{n+1}}^{\infty}(u - N(u))d\Psi(u)\right| + \left|\int_{n}^{S_{n+1}}\Psi(u)du\right| + |\Psi(S_{n+1})|\,|S_{n+1} - n|$$
$$\le \left|\int_{S_{n+1}}^{\infty}(u - N(u))d\Psi(u)\right| + 2\max(|\Psi(S_{n+1})|, |\Psi(n)|)|S_{n+1} - n|.$$

Since by (1) the random improper Riemann-Stieltjes integral

$$\int_{S_1}^{\infty}(u - N(u))d\Psi(u)$$

is finite with probability one, and since $S_{n+1} \to \infty$ almost surely as $n \to \infty$, the first term of this bound of W_n converges to zero almost surely. Also, (1) implies that

$$\lim_{u\to\infty}\sqrt{u}|\Psi(u)| = 0, \tag{10}$$

and since $|S_{n+1} - n|/\sqrt{n}$ or $|S_{n+1} - n|/\sqrt{S_{n+1}}$ converge in distribution, the second term of the bound goes to zero in probability. Thus we have proved (2).

In order to show (4) under (3), we put

$$w_n = \max\left(P\{S_{n+1} \le a_n\}, \varepsilon_n\left(\int_{a_n}^{\infty}\sqrt{u}d\Psi(u)\right)^{\rho}\right),$$

where $a_n \equiv a_n^{(l)}$ if $\Psi = \Psi_l, l = 1, 2$. Glancing through the above proof of (2), we see that it is sufficient to show that

$$P\{W_n \ge 8w_n\} \le 5w_n, \text{ for all large enough } n, \tag{11}$$

for a sequence $\{\varepsilon_n\}$ of positive numbers converging to zero sufficiently slowly.

For the given $\rho < 1$, choose the integer $l > 0$ so large that $l > (l+1)\rho$. Then by the Markov inequality

$$P\left\{\left|\int_{S_{n+1}}^{\infty}(u - N(u))d\Psi(u)\right| \ge 2w_n\right\}$$
$$\le P\{S_{n+1} \le a_n\} + P\left\{\int_{a_n}^{\infty}|u - N(u)|d\Psi(u) \ge 2\varepsilon_n\left(\int_{a_n}^{\infty}\sqrt{u}d\Psi(u)\right)^{\rho}\right\}$$
$$\le w_n + (2\varepsilon_n)^{-l}\left(\int_{a_n}^{\infty}\sqrt{u}d\Psi(u)\right)^{-l\rho} C_1(l)\left(\int_{a_n}^{\infty}\sqrt{u}d\Psi(u)\right)^{l}$$
$$= w_n + \varepsilon_n\left(\int_{a_n}^{\infty}\sqrt{u}d\Psi(u)\right)^{\rho} C_1(l)2^{-l}D_n(l)$$
$$\le 2w_n,$$

where

$$C_1(l) = \sup_{u_1,\dots,u_l\geq 1} E|N(u_1)-u_1|\cdots|N(u_l)-u_l|u_1^{-1/2}\cdots u_l^{-1/2}$$
$$\leq \left(\sup_{u\geq 1} E(|N(u)-u|u^{-1/2})^l\right)^{1/l}$$

is finite by the moment convergence theorem combined with elementary considerations, and where the last inequlity is for all sufficiently large n, provided $\varepsilon_n \to 0$ so slowly that

$$D_n(l) = \varepsilon_n^{-l-1}\left(\int_{a_n}^{\infty} \sqrt{u}d\Psi(u)\right)^{l-(l+1)\rho} \to 0, \tag{12}$$

which is possible on account of the choice of l.

Next, with the same choice of l,

$$P\{2\max(|\Psi(S_{n+1})|,|\Psi(n)|)\,|S_{n+1}-n| \geq 6w_n\}$$
$$\leq P\{|\Psi(n)||S_{n+1}-n| \geq w_n\} + P\{|\Psi(S_{n+1})||S_{n+1}-n| \geq 2w_n\}$$

where (10) and integration by parts shows that the first term is not greater than

$$P\left\{\int_n^{\infty} \sqrt{u}d\Psi(u)|S_{n+1}-n|n^{-1/2} \geq \varepsilon_n\left(\int_{a_n}^{\infty} ud\Psi(u)\right)^{\rho}\right\}$$
$$\leq \varepsilon_n\left(\int_{a_n}^{\infty} \sqrt{u}d\Psi(u)\right)^{\rho} C_2(l)D_n(l)$$
$$\leq w_n$$

for all large enough n, on account of the choice of ε_n in (12), since

$$C_2(l) = \sup_{n\geq 1} E\left(|S_{n+1}-n|/n^{1/2}\right)^l < \infty,$$

again by the moment convergence theorem. Similarly, the second term is less than or equal to

$$P\left\{\int_{S_{n+1}}^{\infty} \sqrt{u}d\Psi(u)|S_{n+1}-n|S_{n+1}^{-1/2} \geq 2w_n\right\}$$
$$\leq P\{S_{n+1}\leq a_n\} + P\left\{\int_{a_n}^{\infty} \sqrt{u}d\Psi(u)|S_{n+1}-n|S_{n+1}^{-1/2} \geq 2w_n\right\}$$
$$\leq w_n + \varepsilon_n\left(\int_{a_n}^{\infty} \sqrt{u}d\Psi(u)\right)^{\rho} C_3(l)2^{-l}D_n(l)$$
$$\leq 2w_n$$

by (12), for all large enough n, since

$$\begin{aligned} C_3(l) &= \sup_{n\geq l} E\left(|S_{n+1}-n|/S_{n+1}^{1/2}\right)^l \\ &= \sup_{n\geq l} E\left(nS_{n+1}^{-1}\right)^{l/2}\left(|S_{n+1}-n|/n^{1/2}\right)^l \\ &\leq \sup_{n\geq l}\left(E\left(nS_{n+1}^{-1}\right)^l E\left(|S_{n+1}-n|/n^{1/2}\right)^{2l}\right)^{1/2} \\ &\leq (C_2(2l))^{1/2}\left(\sup_{n\geq l} n^l\frac{(n-l)!}{n!}\right)^{1/2} \\ &< \infty. \end{aligned}$$

Collecting together these estimates we obtain (11). Hence the theorem is completely proven.

References

[1] M.CSÖRGŐ, S.CSÖRGŐ, L.HORVÁTH, and D.M.MASON: Normal and stable convergence of integral functions of the empirical distribution function, *Ann. Probab.*, 14(1986), 86-118.

[2] S.CSÖRGŐ: Notes on extreme and self-normalised sums from the domain of attraction of a stable law, *J. London Math. Soc.*, to appear.

[3] S.CSÖRGŐ, E.HAEUSLER, and D.M.MASON: A probabilistic approach to the asymptotic distribution of sums of independent, identically distributed random variables, *Adv. in Appl. Math.*, 9(1988), to appear.

[4] L.DEVROYE: Laws of the iterated logarithm for order statistics of uniform spacings, *Ann. Probab.*, 9(1981), 860-867.

[5] A.ERDÉLYI, W.MAGNUS, F.OBERHETTINGER, and F.G.TRICOMI: *Higher transcendental functions*, Vol. II (McGraw-Hill, New York, 1953).

[6] P.HALL: On the extreme terms of a sample from the domain of attraction of a stable law, *J. London Math. Soc.* (2), 18(1978), 181-191.

[7] V.M.ZOLOTAREV: A generalization of the Lindeberg-Feller theorem, *Theory Probab. Appl.*, 12(1967), 608-618.

Bolyai Institute
University of Szeged
Aradi vértanúk tere 1
H-6720 Szeged
Hungary

Probability Theory on Vector Spaces IV
Lancut, June '87, Springer's LNM 1391

AN OPERATOR APPROACH TO PROCESSES ON LIE GROUPS

by

Philip Feinsilver
Southern Illinois University
Carbondale, Illinois

René Schott
Université de Nancy I
Nancy, France

I. Introduction

This work is based on an observation and a question. The "observation," to be presented in more detail below, is that processes on the real line can (should, perhaps) be thought of as coming from processes on the Heisenberg group. The "question" is: What is a process on a Lie group? -- in the sense of how to get concrete presentations of such processes. Here we focus on processes with stationary independent increments. Also, the naive construction presented here works only on exponential groups (where one has global exponential coordinates). Both of these restrictions can be overcome by various extensions of the basic method presented here -- e.g. by martingale techniques [F1]. In any case, the processes considered here can be thought of as the "building blocks" for more general Markov processes (e.g. by forming stochastic integrals).

The observation referred to above is the following [F2]. Let L(D) denote the generator of a process with stationary independent increments on the real line, D denoting d/dx, and L having the familiar Lévy-Khinchine form. Then one typically considers the flow (semigroup) $T_t f(x) = e^{tL} f(x)$, say for continuous bounded functions f. One is thinking of the operators acting on functions of x. The idea is to think of both D and x, x acting as a multiplication operator, as operators in the same operator algebra and the action $e^{tL}f(x)$ as coming from a representation with cyclic vector Ω. Thus, $e^{tL}f(x) = e^{tL}f(x)e^{-tL}\Omega$, where $L\Omega = 0$, is seen as arising from an automorphism, conjugation by $\exp(tL)$. In fact, except for $\sqrt{-1}$, this action is the same as the Hamiltonian flow that gives the Heisenberg representation in quantum mechanics. So actually we are taking the viewpoint of "quantum probability" as it is called nowadays. I.e. one takes an operator algebra to start and looks at various representations of (and functionals on) it. In the Lie group case it is natural to take the enveloping algebra of the corresponding Lie algebra to start.

We are left, though, with the question of what is a process on a Lie group? One can, of course, make definitions as for the line. Random elements and processes can be intrinsically abstractly defined as measures on the appropriate spaces. But still you want to know what they "look like." One way is to look locally at an exponential coordinate patch and say that the coordinates -- coefficients of a Lie algebra basis

-- are given by real-valued stochastic processes. For an exponential group, in fact, one can map processes on Euclidean spaces to processes on the group by the exponential map. We want to show explicitly how this works. The case of Brownian motion on the Heisenberg group is presented in detail.

One should look at [McK] and [S] for product integral constructions which are one way of proving that the constructions given below work. Basic work is found in [H] which shows that, put briefly, if $L(D)$ is a Lévy generator on $\mathbb{R}^n$, $D = (D_1,\dots,D_n)$, $D_j = d/dx_j$, then $L(\xi)$ is a Lévy generator on a Lie group G with $\xi = (\xi_1,\dots,\xi_n)$, where the ξ_j form a basis for the Lie algebra at the identity. See [H] and [F] for precise details. The work of [HL] and [E] apply in these considerations as well. Here we are saying that the ideas and constructions -- the operator formulation of stochastic processes -- employed in [F2] for the Heisenberg case to deal with the real line do, in fact, extend to more general Lie algebras.

II. Brownian Motion and the Heisenberg Group

One can think of the approach to be indicated below as based on a well-known trick of operational calculus [Y]. Namely, if one wants to construct a functional calculus for the operator A^2 from A, take a standard Gaussian random variable X and look at $\langle e^{AX}\rangle$, $\langle\ \rangle$ denoting expected value. The formal identity

$$\langle e^{AX}\rangle = \exp(A^2/2)$$

can be made rigorous under appropriate conditions and, in fact, the same idea is used to construct fractional powers of A by using symmetric stable random variables [Y].

Now let $B_1(t)$, $B_2(t)$ be standard Brownian motions on the line, and let dB_1, dB_2 denote increments over an interval of length dt, i.e. think of $dB_1 = B_1(kt/N)-B_1((k-1)t/N)$, $dt = 1/N$, $1 \le k \le N$, and similarly for dB_2. (In the following, a similar shorthand will be employed when the meaning is clear from the context.) For real x,y:

$$\langle e^{dB_1 x}\, e^{dB_2 y}\rangle = \exp(x^2 dt/2)\cdot\exp(y^2 dt/2)$$

If x and y are bounded, possibly non-commuting, operators then one can multiply the above relation for $k = 1,2,\dots,N$, i.e. one can "multiply over dt," and take the limit as $N \to \infty$. By independence of the increments on the left-hand-side (LHS) we can take the $\langle\ \rangle$ outside. Then apply the Trotter product formula to both sides to yield (in the limit as $N \to \infty$)

$$\langle :e^{B_1(t)x}\, e^{B_2(t)y}:\rangle = \exp[t(x^2+y^2)/2].$$

(The (Wick-ordering) dots on the left-hand side refer to the fact that if x and y do not commute, the result involves commutators of x and y. This is discussed in detail below.) By use of appropriate probability limit theorems we can extend this approach to unbounded operators x,y as long as they generate a finite-dimensional Lie algebra. Furthermore, replacing $B_1(t)$, $B_2(t)$ by different real-valued processes, we obtain the corresponding processes on Lie groups. As mentioned above, for non-exponential groups one needs techniques to patch together the processes across different coordinate patches (look at [HL] e.g.).

The Heisenberg group may be defined as $\mathbb{R}^3$ with the multiplication $(x,y,z)\cdot(x',y',z') = (x+x',y+y',z+z'+(x'y-xy')/2)$. Alternatively, we can specify elements via the coordinates (a,b,c) in the representation

$$g(a,b,c) = \exp(ax)\exp(bd)\exp(ch)$$

where x,d,h are a basis for the Lie algebra at the identity. They satisfy the commutation relations $[d,x] = h$, $[d,h] = [x,h] = 0$. If one uses the vector fields derived from the group law given above on $\mathbb{R}^3$ one finds $x_1 = \partial_x+(y/2)\partial_z$, $x_2 = \partial_y-(x/2)\partial_z$, $x_3 = \partial_z$, the ∂'s denoting derivatives. One can say "Brownian motion on the Heisenberg group is the diffusion process with generator $(x_1^2+x_2^2)/2$ on $\mathbb{R}^3$." This Laplacian is the "Kohn Laplacian." (Our approach is independent of any particular realization of the group. We use only the commutation relations on the Lie algebra.) The components of the process are $(B_1(t),B_2(t),B_3(t))$ where B_1 and B_2 are the standard real Brownian motions and $B_3(t) = (1/2)\int_0^t (B_1dB_2 - B_2dB_1)$ -- Lévy's stochastic area process.

We now present our approach. In the coordinates mentioned above we have $g(a,b,c)\cdot g(a',b',c') = g(a+a',b+b',c+c'+a'b)$.

<u>Proposition</u>. Let $B_1(t),B_2(t)$ be standard Brownian motions, and $B_3(t) = \int_0^t B_1dB_2$. Then

$$\langle g(B_1(t),B_2(t),B_3(t))\rangle = \exp[(\tan ht)x^2/2h](\sec ht)^{xd/h+1/2}\exp[(\tan ht)d^2/2h]. \quad (*)$$

<u>Remarks</u>. 1. If $h \to 0$, on the right (RHS) we find $\exp(t(x_1^2+x_2^2)/2)$ with $[x_2,x_1] = 0$ -- i.e. the abelian case is recovered.

2. The correspondence $g(a,b,c) = \exp[ax+bd+(c-ab/2)h]$ gives us the components (B_1,B_2,B_3) mentioned previously -- i.e. with B_3 as the Lévy area.

3. The formula holds for $t \geq 0$ up until the first zero of cos ht. This is due to the fact that the RHS involves the Lie algebra sl(2) -- and we lose global coordinates.

<u>Proof</u>: Let $\Delta = d^2/2h$, $R = x^2/2h$, $\rho = [\Delta,R] = xd/h + 1/2$. Then $[\Delta,\rho] = 2\Delta$, $[\rho,R] = 2R$. I.e. Δ,R,ρ is a representation of sl(2). We use the method indicated above and apply the Trotter product formula. Namely

$$\langle g(dB_1,dB_2,0)\rangle = \exp(dt\ x^2/2)\exp(dt\ d^2/2) = \exp(dt\ hR)\exp(dt\ h\Delta).$$

Referring to the group law (stated just before the proposition) we see that "multiplying over dt" yields for the LHS $\langle g(B_1,B_2,B_3)\rangle$ as stated in the proposition. On the RHS, via the Trotter formula, we find (in the limit) $\exp[ht(R+\Delta)]$. To find the result as stated, use local coordinates at the identity: $g(A,B,C) = \exp(AR)\cdot\exp(b\rho)\exp(C\Delta)$ where $b = \log B$. Using the correspondence with 2×2 matrices given by $\Delta = \begin{bmatrix} 0 & 0 \\ -1 & 0 \end{bmatrix}$, $R = \begin{bmatrix} 0 & 1 \\ 0 & 0 \end{bmatrix}$, $\rho = \begin{bmatrix} 1 & 0 \\ 0 & -1 \end{bmatrix}$, one can do the calculations explicitly.

We can derive some further formulae involving the joint distribution of the components B_1,B_2,B_3. We now take the specific representation $x \to x$, $d \to z\partial_x$, $h \to z$, and apply both sides of (*) to $\eta = \exp(xy/z)$. We find, cancelling η on both sides (note that x,y,z commute here):

$$\langle \exp(xB_1)\exp(yB_2)\exp(zB_3)\rangle = (\sec zt)^{1/2}\exp[(\tan zt)(x^2+y^2)/2z+(\sec zt)xy/z].$$

(We use the operational formula $\sigma^{xD}f(x) = f(x\sigma)$, with $\sigma = \sec zt$.)

Next we switch notation for clarity. Denote the process (B_1,B_2,B_3) by $w = (w_1,w_2,w_3)$, let $\xi = (x,d,h)$, and $\alpha = (A,B,C)$ so that $g(\alpha) = \exp(Ax)\exp(Bd)\exp(Ch)$ $= \prod_{j=1}^{3}\exp(\alpha_j\xi_j)$, ordered product. We reduce further to the standard representation $x \to x$, $d \to \partial_x$, $h \to 1$. Let $L = (x^2+d^2)/2$. Note that in this representation L is simultaneously the generator for Brownian motion on the Heisenberg group and the harmonic oscillator Hamiltonian.

We need a "ground state." Take $\Omega(x) = \exp(ix^2/2)$ which satisfies $L\Omega = (i/2)\Omega$. We find $\exp(\lambda d)\Omega = \exp[i(x+\lambda)^2/2] = \exp(ix\lambda+i\lambda^2/2)\Omega$. Now consider $\langle g(w_1,w_2,w_3)\rangle g(\alpha)\Omega$. We write this as:

$$\langle g(w_1+A,w_2+B,w_3+C+w_2A)\rangle\Omega = e^{tL}g(\alpha)\Omega.$$

Using $\lambda = w_2+B$ in the calculation of $\exp(\lambda d)\Omega$, we can eliminate d from the LHS. On the RHS we have $e^{tL}g(\alpha)\Omega = e^{it/2}e^{tL}g(\alpha)e^{-tL}\Omega$. This latter is just the Hamiltonian flow with Hamiltonian L, the harmonic oscillator. Splitting $g(\alpha) = \exp(Ax)\exp(Bd)e^C$, using the fact that the quantum and classical flows agree in phase space, and using the group law one finds for the RHS:

$$e^{it/2}\, e^P\cdot e^Q\cdot e^R\,\Omega$$

where $P = x(A\cos t - B\sin t)$,

$Q = C + (1/2)(A^2-B^2)\sin t\cos t - AB\sin^2 t$,

$R = ix(A\sin t + B\cos t) + (i/2)(A\sin t + B\cos t)^2$.

This is thus the expected value

$$\langle e^{U(x,w_1,w_2,w_3,A,B,C)}\rangle$$

with $U = x(w_1+iw_2+A+iB) + (i/2)(w_2^2+B^2) + w_3 + C + w_2(A+iB)$ which is derived by eliminating d on the LHS as mentioned above.

__Remarks on the nilpotent case__: As seen in the above example, we start with a subset of the basis $\xi_1, \xi_2, \ldots$, that generates the Lie algebra (as a Lie algebra). With $w_1(t), w_2(t), \ldots$ denoting the corresponding components the remaining components arise as multiple stochastic integrals with respect to these "basic components." The group law thus determines the nature of the higher components. The method presented here is directly applicable to the nilpotent case (since these are exponential groups).

III. General Approach

Consider a Lie group G with Lie algebra $\mathfrak{g}$ with generators $\xi = (\xi_1, \xi_2, \ldots, \xi_d)$, $d \leq n < \infty$, where $\{\xi_1, \ldots, \xi_n\}$ is a linear basis of $\mathfrak{g}$.

A. To find the distribution of a process w(t) with generator $L(\xi)$ we follow the procedure mentioned previously. Namely, start with the expectations (ordered products):

$$\prod_{j=1}^{d} \langle \exp(\xi_j dw_j) \rangle = \prod_{j=1}^{d} \exp[L(\xi_j)dt].$$

In these products the terms are conventionally ordered with increasing subscripts. Now "multiply over dt":

$$\prod_{dw} \prod_{j} \langle \exp(\xi_j dw_j) \rangle = \prod_{dt} \prod_{j} \exp[L(\xi_j)dt].$$

Using the group law and independence of the increments the LHS can be expressed in the form (in the limit $dt \to 0$):

$$\langle \prod_{j=1}^{n} \exp(w_j \xi_j) \rangle \quad \text{or} \quad \langle \exp(W \cdot \xi) \rangle$$

which defines, via the group structure, the components of the process $W(t) = (W_1(t), \ldots, W_n(t))$, where $W \cdot \xi = \sum_j W_j \xi_j$.

The major theoretical questions arise exactly at this point. Since for $dt = N^{-1}$ there is no problem, one has to deal with the limit $N \to \infty$. One can deal with operator-theoretic problems by taking finite-dimensional or unitary representations (and using spectral theory) or else using abstract theorems for Lie groups (see [F1] and [S]). Here are two types of limit theorems that one can formulate:

(1) If the LHS converges by general theory, then for every representation of the RHS, we have convergence. (This is the "easy side" of Lévy continuity.)

(2) If the RHS converges by general theory, then for every representation of the LHS we can construct a corresponding process. See [B] for Lévy continuity

theorems for Lie groups. For example, a Lévy theorem holds for amenable groups -- which include nilpotent groups and groups of rigid type (i.e. in the adjoint representation all of the eigenvalues have absolute value one).

The RHS becomes, using the Trotter product formula, $\exp[t\,\mathcal{L}(\xi)]$, with $\mathcal{L}(\xi) = \sum_{j=1}^{d} L(\xi_j)$. Notice that a probability limit theorem for the LHS will *imply* that $\mathcal{L}$ generates a process, without the necessity of invoking the Trotter formula. Finally, observe that one could take different L's for different components.

B. Consider a representation of G acting on functions $f(x)$. Here and in the following L will denote the generator of the process on G (i.e. denoted by $\mathcal{L}$ above). Denote the action: $\exp(W\cdot\xi)f(x) = f(x\odot W)$, "translation by W." Then we have

$$\langle \exp(W\cdot\xi)\rangle f(x) = \langle f(x\odot W)\rangle = e^{tL}f(x).$$

If Ω is a "vacuum state" -- i.e. $L\Omega = 0$ -- then

$$e^{tL}f(x)\Omega = e^{tL}f(x)e^{-tL}\Omega = f(x(t))\Omega$$

which can be computed as a Hamiltonian flow. This gives an operator formulation and operator calculus as presented for real-valued processes in [F2]. Notice that the operator path $x(t)$ is not "random." If $L\Omega = 0$ and Ω never vanishes, then one can compute as follows:

$$e^{tL}f(x) = e^{tL}(f(x)/\Omega(x))\Omega(x) = e^{tL}(f(x)/\Omega(x))e^{-tL}\Omega(x).$$

C. One can also consider the action on functions of ξ (defined in the enveloping algebra or by operational calculus):

$$\langle \exp(W\cdot\xi)f(\xi)\rangle = e^{tL}f(\xi).$$

And with $L\Omega = 0$:

$$\langle \exp(W\cdot\xi)f(\xi)\Omega\rangle = e^{tL}f(\xi)e^{-tL}\Omega = f(\xi(t))\Omega.$$

Here we are working directly with operators defining the representation " ξ ".

D. Just as in the passage from the Heisenberg group to the real line, we can consider certain induced representations. Let $\mathfrak{h}$ be a subalgebra of $\mathfrak{g}$ and let $\mathfrak{z}$ denote the center of $\mathfrak{g}$. Let Ω be a cyclic vector, "vacuum state," such that $\mathfrak{h}$ annihilates Ω and elements of $\mathfrak{z}$ act as scalars on Ω. Denote the basis elements of $\mathfrak{h}$ by δ. Let η denote a complementary basis (to δ) in $\mathfrak{g}$. Then

$$\langle \exp(W\cdot\delta)f(\eta)\Omega\rangle = e^{tL(\delta)}f(\eta)\Omega$$

where we take the generator L to depend only on elements of δ. If $L(\delta)\Omega = 0$, we have

$$\langle \exp(W\cdot\delta)f(\eta)\Omega\rangle = e^{tL}f(\eta)e^{-tL}\Omega = f(\eta(t))\Omega.$$

Again the operator path $\eta(t)$ provides a deterministic operator description of the process. Since $\delta\Omega = 0$, we can write

$$\langle \exp(W\cdot\delta)f(\eta)\exp(-W\cdot\delta)\Omega\rangle = e^{tL(\delta)}f(\eta)e^{-tL(\delta)}\Omega,$$

an exact transference of the adjoint action.

E. Cases B and C can be compared by choosing, say $f(\xi) = \exp(\alpha\cdot\xi)$ which can equally be thought of as $f(\alpha)$. The action $\exp(W\cdot\xi)\exp(\alpha\cdot\xi) = \exp((W\odot\alpha)\cdot\xi)$ is just the group law. And

$$\langle \exp[(W\odot\alpha)\cdot\xi]\rangle = e^{tL(\xi)}e^{\alpha\cdot\xi}.$$

Let ξ^* denote a representation of ξ as right vector fields, say, acting on functions of α. Then

$$\langle \exp(W\cdot\xi^*)\exp(\alpha\cdot\xi)\rangle = \langle \exp[(\alpha\odot W)\cdot\xi]\rangle = e^{tL(\xi^*)}e^{\alpha\cdot\xi}.$$

These formulas and constructions, illustrated in A through E, are analogs (in fact, extensions) of characteristic functions and Fourier methods used in the study of real-valued processes.

Acknowledgement

The first author would like to thank the University of Nancy and the C.N.R.S. for their support. This joint work was sponsored under NATO grant #86/0321.

References

[B] Ph. Bougerol, Extension du théorème de continuité de Paul Lévy aux groupes moyennables, Springer Lecture Notes, v. 1064, 1983, 10-22.

[E] M. Emery, Stabilité des solutions des équations differentielles stochastiques, application aux intégrales multiplicative stochastiques, Z. Wahr. 41, 1978, 241-262.

[F1] Ph. Feinsilver, Processes with independent increments on a Lie group, Trans. A.M.S. 242, 1978, 73-121.

[F2] Ph. Feinsilver, Special functions, probability semigroups, and Hamiltonian flows, Springer Lecture Notes, v. 696, 1978.

[HL] M. Hakim-Dowek and D. Lepingle, L'exponentielle stochastique des groupes de Lie, Springer Lecture Notes, v. 1204, 1986, 352-374.

[H] G. Hunt, Semigroups of measures on Lie groups, Trans. A.M.S. 81, 1956, 264-293.

[McK] H. P. McKean, Stochastic Integrals, Academic Press, 1969.

[S] E. Siebert, Continuous hemigroups of probability measures on a Lie group, Springer Lecture Notes, v. 928, 1982, 362-402.

[Y] K. Yosida, Functional Analysis, Springer, 1980.

Probability Theory on Vector Spaces IV
Lancut, June'87, Springer's LNM 1391

RÉGULARITÉ DE FONCTIONS ALÉATOIRES GAUSSIENNES STATIONNAIRES À VALEURS VECTORIELLES.

Xavier FERNIQUE

Sommaire : On caractérise la régularité des trajectoires des fonctions aléatoires gaussiennes stationnaires X sur $\mathbf{R}^n$ à valeurs dans un espace de Banach séparable E à partir des seules lois des f.a. gaussiennes réelles $\langle X, y\rangle$ où y parcourt le dual de E. On utilise dans la preuve l'outil des mesures majorantes.

0. Introduction.

0.1. J'ai établi en 1985 un critère simple pour la régularité des trajectoires des f.a. gaussiennes stationnaires sur $\mathbf{R}$ ou $\mathbf{R}^n$ et à valeurs vectorielles (Théorème 3.3 ci-dessous). Ce résultat a déjà été utilisé ((5), (6)), mais n'a jamais été publié. Je répare ici cette omission et je donne la preuve.

0.2. Notations. — On note $(\Omega, \mathcal{A}, \mathbf{P})$ un espace d'épreuves *complet*, E un espace de Banach séparable muni de sa tribu canonique; E' est son dual, on note (E'_1, w) la boule unité de ce dual munie d'une distance définissant la topologie faible; T est un ensemble et $X = \{X(t), t \in T\} = \{X(\omega, t), \omega \in \Omega, t \in T\}$ est un ensemble de vecteurs aléatoires sur $(\Omega, \mathcal{A}, \mathbf{P})$ à valeurs dans E indexés par T. La fonction aléatoire X sur T à valeurs vectorielles définit aussi une fonction aléatoire X sur $T \times E'_1$ à valeurs réelles. Soient X et Y deux f.a. sur T à valeurs dans E, on dit que Y est une modification vectorielle de X si :

$$\forall t \in T\,,\ \mathbf{P}\{\omega : X(\omega) = Y(\omega)\} = 1.$$

Dans le cas où $T = \mathbf{R}^n$, nous noterons V l'ensemble $(-1/2, +1/2)^n$ et λ sera la mesure de Lebesgue normalisée sur V.

1. Modification régulières de f.a. à valeurs vectorielles.

L'existence des modifications régulières des f.a. à valeurs vectorielles a été présentée pour l'essentiel dans (5); nous rappelons sans démonstration les résultats utilisés ici en les complétant sur certains points.

1.1 *Modifications séparables ((5),1.4)* : en général, une f.a. à valeurs vectorielles n'a pas de modification vectorielle séparable même si $T = (T, \delta)$ est un espace métrique séparable; par contre dans ce cas et puisque (E'_1, w) est aussi un espace métrique séparable, X a des modifications réelles séparables sur $(T \times E'_1; \delta, w)$, mais ces modifications ne sont pas nécessairement associables à des modifications vectorielles de X.

Supposant toujours (T,δ) séparable et de plus X continue en probabilité sur (T,δ), on peut pourtant construire de bonnes modifications vectorielles de X de la façon suivante : Soit S une partie dénombrable et dense de T , à tout élément t de T, on peut associer une suite $(s_n(t))$ convergeant vers t dans (T,δ) telle que de plus $(X(s_n(t))$ converge p.s. vers $X(t)$; on définit alors la modification Y de X en posant :

$$Y(t) = \lim X(s_n(t) \text{ si cette limite existe}, Y(t) = 0 \text{ sinon ;}$$

on vérifie alors que pour toute partie convexe fermée C de E *contenant l'origine* et pour toute partie ouverte U de T, on a:

$$\{\forall t \in U,\ Y(t) \in C\} \supseteq \{\forall t \in S \cap U, X(t) \in C\} ;$$

en particulier si Xest p.s. à trajectoires bornées sur S, on constate, puisque $(\Omega, \mathcal{A}, \mathbf{P})$ est complet, que Y a la même propriété sur T. C'est ce procédé de construction que nous utiliserons quand nous parlerons de *modification très régulière* de X.

1.2. *Modifications mesurables ((5),1.3).* Ici la situation est plus simple, en effet :

THÉORÈME 1.2. — *On suppose (T,δ) séparable, alors les deux propriétés suivantes sont équivalentes :*
1.2.1 $\forall y \in E_1', t \to \langle X(t), y\rangle$ a une modification réelle mesurable sur $\Omega \times T$.
1.2.2 $t \to X(t)$ a une modification vectorielle mesurable sur $\Omega \times T$.

Les propriétés classiques des f.a. réelles permettent de déduire de ce théorème :

COROLLAIRE 1.2.3. — *On suppose (T,δ) métrique séparable, on suppose aussi que X vérifie la propriété suivante :*
1.2.4 $\forall y \in E_1', t \to \langle X(t), y\rangle$ est continue en probabilité sur T.
Sous ces deux hypothèses, X a une modification vectorielle mesurable sur $\Omega \times T$.

1.3. *Modifications à trajectoires continues ((5),1.2).*

THÉORÈME 1.3. — *On suppose (T,δ) métrique séparable, alors les deux propriétés suivantes sont équivalentes :*
1.3.1 Il existe une modification réelle de X sur $(T \times E_1'; \delta, w)$ ayant p.s. des trajectoires réelles continues.
1.3.2 Il existe une modification vectorielle de X sur (T,δ) ayant p.s. des trajectoires vectorielles continues.

1.4. *Modifications à trajectoires bornées* : nous détaillons ce point qui n'est pas exposé dans (5).

THÉORÈME 1.4. — *On suppose (T,δ) métrique séparable et X continue en probabilité; alors les deux propriétés suivantes sont équivalentes :*

1.4.1 Il existe une modification réelle Z de X sur $T\times E_1'$ ayant p.s. des trajectoires localement bornées au sens suivant : il existe une application U de $\Omega \times T \times E_1'$ dans l'ensemble des parties ouvertes de $T \times E_1'$ telle que pour tout $(\omega, t, y) \in \Omega \times T \times E_1', (t,y)$ appartienne à $U(\omega,t,y)$ et que de plus l'ensemble $\{\forall (t,y) \in T \times E_1', \sup_{U(t,y)} |Z| < \infty\}$ soit mesurable et de probabilité 1.
1.4.2 Il existe une modification vectorielle Y de X sur T ayant p.s. des trajectoires localement bornées dans E.

Démonstration. — Démontrons seulement que 1.4.1 implique 1.4.2. Soient F et S des parties dénombrables et denses de E_1' et T; sous l'hypothèse 1.4.1, la w-compacité de E_1' montre qu'il existe une application V de $\Omega \times T$ dans l'ensemble des parties ouvertes de T telle que pour tout $(\omega, t) \in \Omega \times T$, t appartienne à $V(\omega,t)$ et que l'ensemble $\Omega_0 = \{\forall t \in T, \sup_{V(t)\cap S} \sup_F |Z| < \infty\}$ soit mesurable et de probabilité 1. On note alors Y une modification *très régulière* de X; l'ensemble $\Omega_1 = \{\forall t \in T, \sup_{V(t)} \|Y\| < \infty\}$ contient l'ensemble $\{\forall t \in T, \sup_{V(t)\cap S} \sup_F |X| < \infty\}$ qui ne diffère de Ω_0 que par un ensemble négligeable; Ω_1 est donc aussi mesurable et de probabilité 1 de sorte que la propriété 1.4.2 est vérifiée.

1.5. *Sur la continuité en probabilité ((5),1.5).* Le corollaire 1.2.3. et le théorème 1.4. utilisent deux notions différentes de continuité en probabilité; dans certains cas, ces notions sont équivalentes :

THÉORÈME 1.5. — *Soit X une f.a. vectorielle sur un espace topologique $(T,\mathcal{T})$; on suppose que X est tendu au sens suivant :*
1.5.1 pour tout $\epsilon > 0$, il existe une partie compacte K de E telle que:

$$\forall t \in T, \mathbf{P}\{X(t) \notin K\} < \epsilon \;;$$

dans ces conditions, X est continue en probabilité si et seulement si la propriété 1.2.4 est satisfaite.

COROLLAIRE 1.5.2. — *Soit X une f.a. vectorielle stationnaire sur $\mathbf{R}^n$ et vérifiant la propriété 1.2.4; alors X est continue en probabilité.*

2. Rappels : Critère de Dudley, Fernique, Talagrand pour la régularité des f.a. gaussiennes à valeurs réelles.

Pour étudier la régularité des trajectoires d'une f.a. gaussienne X sur T à valeurs réelles, on dispose de critères liés à l'existence de mesures majorantes; ils s'expriment à partir de la pseudo-distance d_X sur T associée à X et définie par :

2.0.1 $$\forall (s,t) \in T \times T,\ d_X(s,t)^2 = \mathbf{E}\,|X(s) - X(t)|^2$$

et des boules fermées $B_X(t,u)$ de centre $t \in T$ et de rayon $u > 0$ pour cette distance.

THÉORÈME 2.1. ((3),(7)). — *Soient T un ensemble et X une f.a. réelle gaussienne séparable sur (T, d_X); soit de plus S une partie de T. On suppose qu'il existe une probabilité μ sur (T, d_X) telle que :*

2.1.1 $$\sup_{s \in S} \int (\log \frac{1}{\mu\{B_X(s,u)\}})^{1/2} du < \infty,$$

alors X est p.s. à trajectoires bornées sur S.

Inversement supposons que X soit p.s. à trajectoires bornées sur S, il existe alors une probabilité μ sur (S, d_X) vérifiant 2.1.1.

De même pour que X soit p.s. à trajectoires continues sur (S, d_X), il faut et il suffit qu'il existe une probabilité μ sur (T, d_X) telle que :

2.1.2 $$\lim_{\epsilon \to 0} \sup_{s \in S} \int_0^{\epsilon} (\log \frac{1}{\mu\{B_X(s,u)\}})^{1/2} du = 0.$$

Le critère ci-dessus se simplifie si X est une f.a. gaussienne sur $\mathbf{R}^n$ stationnaire; V et λ ayant les significations 0.2., on a :

THÉORÈME 2.2. ((3),(4)). — *Soit X une f.a. gaussienne réelle stationnaire et séparable sur $\mathbf{R}^n$ alors X a p.s. ses trajectoires continues si et seulement si :*

2.2.1 $$\int (\log \frac{1}{\lambda\{V \cap B_X(0,u)\}})^{1/2} du < \infty.$$

De plus, il existe des constantes $c > 0$ et $c_n < \infty$ telles que pour tout $\epsilon > 0$ on ait :

$$c\,\mathbf{E} \sup_{\epsilon V} X \leq \sup_{s \in \epsilon V} d_X(0,s) + \int (\log \frac{\epsilon^n}{\lambda\{\epsilon V \cap B_X(0,u)\}})^{1/2} du \leq c_n \mathbf{E} \sup_{\epsilon V} X.$$

3. Le résultat principal, énoncé et démonstration.

Dans la suite de ce travail, on se propose de caractériser la régularité des trajectoires d'une f.a. gaussienne stationnaire X sur $\mathbf{R}^n$ à valeurs vectorielles. Pour qu'il existe une modification vectorielle de X ayant p.s. des trajectoires localement bornées ou continues, il faut que pour tout $y \in E'_1$, $< X, y >$ ait la même propriété. Ceci impose que X vérifie la propriété 1.2.4.; le corollaire 1.5.2 montre alors que X est continue en probabilité; nous supposerons donc cette condition réalisée dans la suite.

Les théorèmes 1.3 et 1.4 ramènent la construction de modifications vectorielles à trajectoires régulières à celle de modifications réelles à trajectoires régulières pour la f.a. gaussienne réelle X associée à X sur $\mathbf{R}^n \times E'_1$. Le théorème 2.1. permet de caractériser ces régularités à partir de probabilités sur $\mathbf{R}^n \times E'_1$; les conditions ainsi obtenues sont trop lourdes pour être directement utilisables; on cherche donc des conditions plus maniables.

Antérieurement aux résultats de Talagrand, la régularité des trajectoires de certaines f.a. gaussiennes à valeurs vectorielles pouvait s'étudier à partir du lemme suivant ((1),(2)) basé sur les inégalités de Slépian ((3)) :

LEMME 3.0. — *Soit X une f.a. gaussienne à valeurs réelles sur un ensemble produit $T \times F$; on suppose qu'il existe des f.a. gaussiennes G et H à trajectoires p.s. bornées sur T et F (resp. p.s. à trajectoires continues sur (T, d_G) et (F, d_H)) telles que :*

3.0.1 $$\forall (s,t) \in T \times T,\ \forall (y,z) \in F \times F,\ d_X(s,y;t,z) \leq d_G(s,t) + d_H(y,z);$$

alors X admet une modification X' à trajectoires bornées sur $T \times F$ (resp. à trajectoires continues sur $(T \times F, d_X)$) et on a :

3.0.2 $$\mathbf{E} \sup_{T \times F} X' \leq 2^{1/2}\{\mathbf{E} \sup_T G + \mathbf{E} \sup_F H\}.$$

En fait ce lemme a un champ d'application trop limité; même si X est stationnaire, ses hypothèses sont trop fortes; nous l'illustrons par l'exemple suivant :

Exemple 3.1. : Nous supposons $E = c_0$; nous notons (a_n) une suite positive décroissante et (g_n) une suite gaussienne normale, nous posons :

3.1.0 $$\forall t \in \mathbf{R},\ X(t) = \{a_n(g_{2n} \cos 2^n t + g_{2n+1} \sin 2^n t)\}_{n \in \mathbf{N}},$$

c'est une f.a. gaussienne stationnaire sur $\mathbf{R}$ et elle est p.s. à valeurs dans E si et seulement si $\lim a_n (\log n)^{1/2} = 0$; on constate que sous cette seule condition, la f.a. gaussienne réelle X sur $\mathbf{R} \times E_1'$ associée à X est p.s. à trajectoires bornées.

Pourtant cette condition est insuffisante pour que X vérifie les hypothèses du lemme 3.0 : elles seront vérifiées si et seulement si (a_n) vérifie de plus la propriété suivante : en posant $\Phi^2(t) = \sup_n a_n^2(1 - \cos 2^n t)$, l'intégrale

$$\int (\log \frac{1}{\lambda\{\Phi(t) \leq u\}})^{1/2} du$$

est finie. Or cette condition n'est pas vérifiée si par exemple $a_n = n^{-\alpha}, \alpha \leq 1/2$.

3.2. *Conditions nécessaires de régularité.* Pour que la f.a. gaussienne $X = \{X(t), t \in \mathbf{R}^n\}$, stationnaire ou non, à valeurs dans E, soit p.s. à trajectoires localement bornées, il faut qu'en utilisant les notations 0.2., on ait:

3.2.1 $$\sup_{y \in E_1'} \mathbf{E} \sup_{t \in V} \langle X(t), y \rangle < \infty ;$$

de même pour que X soit p.s. à trajectoires continues, il faut que :

3.2.2 $$\lim_{\epsilon\to 0}\sup_{y\in E_1'} \mathbf{E} \sup_{t\in\epsilon V}\langle X(t),y\rangle = 0 ;$$

si de plus X est stationnaire, ces conditions impliquent respectivement (th. 2.2) :

3.2.1′ $$\sup_{y\in E_1'}\{\sup_{s\in V} d_{\langle X,Y\rangle}(0,s) + \int(\log \frac{1}{\lambda\{V\cap B_{\langle X,y\rangle}(0,u)\}})^{1/2}du\} < \infty,$$

3.2.2′ $$\lim_{\epsilon\to 0}\sup_{y\in E_1'}\{\sup_{s\in\epsilon V} d_{\langle X,y\rangle}(0,s) + \int(\log \frac{\epsilon^n}{\lambda\{\epsilon V\cap B_{\langle X,y\rangle}(0,u)\}})^{1/2}du\} = 0.$$

Le fait remarquable s'énonce alors :

THÉORÈME 3.3. — *On suppose que X est stationnaire sur $\mathbf{R}^n$, alors on a les équivalences suivantes :*

X a une modification X' à trajectoires localement bornées si et seulement si 3.2.1' est satisfaite, de même X a une modification X' à trajectoires continues si et seulement si 3.2.2' est satisfaite.

Démonstration. — (a) Supposons 3.2.1' vérifiée; pour tout couple $(s,y),(t,z)$ d'éléments de $T\times E_1'$, la stationnarité montre qu'on a :

3.3.1 $$d_X(s,y;t,z) \le d_{\langle X,y\rangle}(s,t) + d_{X(0)}(y,z);$$

c'est "presque" la relation 3.0.1 : $X(0)$ borné sur E_1' peut jouer le rôle de H, mais $\langle X,y\rangle$ bien que borné sur V ne peut jouer le rôle de G puisqu'il dépend de y, si bien que les inégalités de Slépian sont inaptes ici à exploiter 3.3.1; au contraire, le théorème 2.1. permet d'opérer : notons en effet μ une probabilité sur E_1' telle que :

3.3.2 $$\sup_{y\in E_1'}\int(\log \frac{1}{\mu\{B_{X(0)}(y,u)\}})^{1/2}du < \infty,$$

posons de plus sur $2V\times E_1', \pi = 2^{-n}\lambda\times\mu$; pour tout $u>0$, la relation 3.3.1 fournit :

$$\pi\{(2V\times E_1')\cap B_X(s,y;u)\} \le 2^{-n}\lambda\{(2V)\cap B_{\langle X,y\rangle}(s,u/2)\cdot\mu\{B_{X(0)}(y,u/2)\} ;$$

et en intégrant et utilisant les propriétés algébriques de $(log)^{1/2}$, ceci implique pour tout couple (s,y) de $V\times E_1'$:

3.3.3
$$\int (\log \frac{1}{\pi\{(2V \times E_1') \cap B_X(s,y;u)\}})^{1/2} du \leq$$

$$2\int (\log \frac{1}{\mu\{B_{X(0)}(y,u)\}})^{1/2} du + 2\int (\log \frac{2^n}{\lambda\{2V \cap B_{\langle X,y\rangle}(0,u)\}})^{1/2} du.$$

Puisque s appartient à V, alors $2V \cap B_{\langle X,y\rangle}(s,u)$ contient le translaté $\tau_s\{V \cap B_{\langle X,y\rangle}(0,u)\}$; de plus l'intégrand du dernier terme est nul si u est supérieur à $2\sup_{t\in 2V} d_{\langle X,y\rangle}(0,t)$; dans ces conditions, le deuxième membre de 3.3.3 se majore indépendamment de s et y par :

$$2 \sup_{y\in E_1'} \int (\log \frac{1}{\mu\{B_{X(0)}(y,u)\}})^{1/2} du + 4 \sup_{t\in 2V} d_{\langle X,y\rangle}(0,t) \cdot (n\log 2)^{1/2} +$$

$$+ \int (\log \frac{1}{\lambda\{V \cap B_{\langle X,y\rangle}(0,u)\}}^{1/2} du,$$

de sorte que sous l'hypothèse 3.2.1', le théorème 2.1 et la propriété 3.3.2 permettent d'affirmer que la modification Z de X séparable sur $\mathbf{R}^n \times E_1'$ est p.s. à trajectoires bornées sur $V \times E_1'$; la stationnarité implique la même propriété sur $\mathbf{R}^n \times E_1'$. Le théorème 1.4 montre alors que X a une modification vectorielle Y p.s. à trajectoires localement bornées sur $\mathbf{R}^n$. Ceci justifie la première affirmation du théorème.

(b)Supposons maintenant 3.2.2' vérifiée; le même schéma de preuve utilisant une probabilité μ sur E_1' et la mesure de probabilité $\pi = 2^{-n}\lambda \times \mu$ telles que :

3.3.4
$$\lim_{\epsilon\to 0} \sup_{y\in E_1'} \int_0^{\epsilon} (\log \frac{1}{\mu\{B_{X(0)}(y,u)\}})^{1/2} du = 0$$

montre que pour tout $s \in V$, tout $y \in E_1'$ et tout $\epsilon \in (0,1)$, on a :

$$\int_0^{\epsilon} (\log \frac{1}{\pi\{(2V \times E_1') \cap B_X(s,y;u)\}})^{1/2} du \leq$$

$$2\int_0^{\epsilon/2} (\log \frac{1}{\mu\{B_{X(0)}(y,u)\}})^{1/2} du + 2\epsilon(n \log 2/\epsilon)^{1/2} +$$

$$+2 \int (\log \frac{\epsilon^n}{\lambda\{\epsilon V \cap B_{\langle X,y\rangle}(0,u)\}})^{1/2} du,$$

de sorte que sous l'hypothèse 3.2.2', le théorème 2.1 et la propriété 3.3.4 permettent d'affirmer que la modification Z de X séparable sur $\mathbf{R}^n \times E_1'$ est p.s. à trajectoires continues sur $V \times E_1'$; à partir du théorème 1.3 et en suivant le schéma (a), on construit alors une modification vectorielle Y de X dont toutes les trajectoires sont continues sur $\mathbf{R}^n$; ceci justifie donc la deuxième affirmation du théorème.

3.4. *Remarque* : La démonstration (a) montre qu'il existe une constante finie C_n telle que toute modification vectorielle *très régulière* Y de$X$ vérifie :

$$\mathbf{E} \sup_{t \in V} \|Y(t)\| \leq C_n\{\mathbf{E}\|Y(0)\| + \sup_{y \in E_1'} \mathbf{E} \sup_{t \in V} \langle Y(t), y\rangle\}.$$

On notera par ailleurs que le théorème 3.3 peut s'étendre aux f.a. gaussiennes à accroissements stationnaires ((5)).

En conclusion, le résultat ci-dessus montre que dans certains cas, l'étude de la régularité des trajectoires des f.a. à valeurs vectorielles est moins complexe qu'on ne s'y attend.

RÉFÉRENCES.

[1] R. CARMONA, *Tensor product of Gaussian measures*, Springer Lecture Notes in Math. 644, 1978, 96-124.

[2] S. CHEVET, *Un résultat sur les mesures gaussiennes*, C.R. Acad. Sci. Paris, A, 284, 1977, 441–444.

[3] X. FERNIQUE, *Régularité des trajectoires des f.a. gaussiennes*, Springer Lecture Notes in Math. 480, 1975, 1-96.

[4] X. FERNIQUE, *L'ordre de grandeur à l'infini de certaines fonctions aléatoires*, Colloques internationaux du C.N.R.S. 307, 1981, 271–302.

[5] X. FERNIQUE, *Fonctions aléatoires à valeurs vectorielles*, à paraître.

[6] M. LEDOUX, M. MARCUS, *Some remarks on the uniform convergence of Gaussian and Rademacher Fourier quadratic forms. Geometrical and statistical aspects of Probability in Banach spaces*, Springer Lecture Notes in Math. 1193, 1986, 53–72.

[7] M. TALAGRAND, *Regularity of Gaussian processes*, à paraître in Acta Math.

Institut de Recherche Mathématique Avancée
Unité associée n°1
7, rue René Descartes
67084 Strasbourg Cédex (France)

Probability Theory on Vector Spaces IV
Lancut, June '87, Springer's LNM 1391

L_p multipliers in the central limit theorem with p-stable limit*

Evarist Giné and Joel Zinn
Texas A&M University

Let B be a separable Banach space and let X be a B-valued random variable such that $X \epsilon CLT_p$, i.e. such that for i.i.d. X_i with the law of X,

$$\mathcal{L}\left(\sum_{i=1}^{n} X_i/n^{1/p}\right) \to_w \gamma \tag{1}$$

where γ is a p-stable measure on $B, 0 < p \leq 2$. The question of what real valued symmetric random variables ξ independent of X verify $\xi X \epsilon CLT_p$ has been partially answered as follows:

(i) for $1 < p \leq 2$, if $\xi \epsilon L_{p,1}$, i.e. if

$$\int_0^\infty (P\{|\xi| > t\})^{1/p} dt < \infty, \tag{2}$$

then

$$X \epsilon CLT_p \Rightarrow \xi X \epsilon CLT_p \tag{3}$$

and there are Banach spaces where this implication holds for all $X \epsilon CLT_p$ only if $\xi \epsilon L_{p,1}$ (Pisier - e.g. [2] - and Ledoux and Talagrand [5]; see also [3]);

(ii) for $0 < p < 1$, and for $1 \leq p < 2$ if B is of type p-stable, the implication (3) holds if and only if $\xi \epsilon L_p$ ([3]);

(iii) there is no definitive answer for $p = 1$, but $\xi \epsilon L_{1+\delta}$ fo some $\delta > 0$ suffices for (3) (same proof as for $p > 1$).

In this note we address the related question of determining those Banach spaces for which the implication (3) holds whenever $\xi \epsilon L_p, 1 \leq p \leq 2$. (Note that finite dimensional Banach spaces have this property for all p.) We show that this property for $p \neq 2$ characterizes type p stable Banach spaces, and that for $p = 2$ there are some nice Banach

* This research has been partially supported by grants from the National Science Foundation, U.S.A.

spaces (in particular, type 2 and cotype $2+\varepsilon$ for every $\varepsilon > 0$) for which the "L_2-multiplier property" does not hold. In view of (ii) we only have to consider $1 \leq p \leq 2$.

We begin with some notation. Let X be a B-valued random variable (tight if B is not separable); then $X_i, i \in \mathbb{N}$, denote independent copies of X, $S_n(X) = \Sigma_{i=1}^n X_i$, $i \in \mathbb{N}$, and

$$CLT_p(X) := sup_n E\|S_n(X)/n^{1/p}\|, \qquad 1 < p \leq 2, \tag{4}$$

$$CLT_p(X) := sup_n (E\|S_n(X)/n\|^{p/2})^{2/p}, \qquad p \leq 1. \tag{5}$$

We say that X satisfies the bounded $CLT_p, X \in BCLT_p$, if the sequence $\{\|S_n(X)/n^{1/p}\|\}_{n=1}^{\infty}$ is stochastically bounded. In order to avoid centering considerations, whenever we write $X \in CLT_1$ or $X \in BCLT_1$ we automatically mean that X is ***symmetric;*** for $p > 1, X \in BCLT_p$ implies already that X is centered. In what follows, ξ will always denote a real valued symmetric random variable ***independent*** of the B-valued random variable X. Finally $\cong$ and $\lesssim, \gtrsim$ denote two-sided and one-sided inequalities, up to constants independent of the variables appearing in them.

Theorem 1. Let B be a Banach space and let $1 \leq p < 2$. Then the following properties for B are equivalent:

(i) B is of type p-stable;

(ii) $X \in B, X \in CLT_p \Rightarrow \xi X \in CLT_p$ for all symmetric real ξ independent of X for which $E|\xi|^p < \infty$;

(iii) $X \in B, X \in BCLT_p \Rightarrow \xi X \in BCLT_p$ for all ξ as in (ii);

(iv) for all $X \in B$ centered if $p > 1$, symmetric if $p = 1$, and all ξ as in (ii),

$$CLT_p(\xi X) \lesssim \|\xi\|_p CLT_p(X); \tag{6}$$

(v) (iv) holds under the additional condition that X takes on only finitely many values.

Proof: (i) $\Rightarrow$ (ii) is proved in [3], Theorem 4.1.
(ii) $\Rightarrow$ (v): Let CLT_p^0 be the closure for the CLT_p norm of the simple random variables. Note that $CLT_p^0 \subset CLT_p$ and that for $X \in CLT_p^0$, the p-stable limit is zero. For each ξ as in (ii), the map $M_\xi : CLT_p^0 \to CLT_p$, $M_\xi(X) = \xi X$, is well defined (by (ii)) and, by independence, it has a closed graph. So, M_ξ is continuous. Similarly, the map $T_X : L_p \to CLT_p$, $T_X(\xi) = \xi X$, for $X \in CLT_p^0$ is also well defined and has a closed graph, hence it is

continuous. Then the principle of uniform boundedness applied to $\{M_\xi : \|\xi\|_p \le 1\}$ gives (v).

An analogous proof gives (iii) $\Rightarrow$ (iv). Trivially (iv) $\Rightarrow$ (iii).

(v) $\Rightarrow$ (iv): The cases $p = 1$ and $p > 1$ are considered separately. Let $p = 1$ and let $X^\lambda = XI(\|X\| \le \lambda)$, let $\mathcal{F}_n, n\epsilon\mathrm{N}$, be increasing finite σ-algebras such that $\overline{\vee\mathcal{F}_n} = \sigma(X)$, and let $Y_n = E(X^\lambda|\mathcal{F}_n, \sigma(\xi))$. We then have that for each r,

$$E\|S_r(\xi X^\lambda)/r\|^{1/2} = lim_n\ E\|S_r(\xi Y_n)/r\|^{1/2}.$$

Also

$$sup_r\ E\|S_r(X^\lambda - Y_n)/r\|^{1/2} \le (E\|X^\lambda - Y_n\|)^{1/2} \to 0 \text{ as } n \to \infty,$$

and therefore,

$$CLT_1(X^\lambda) = \lim_n\ CLT_1(Y_n).$$

These two observations and (v) give that for all r and λ,

$$\begin{aligned}(E\|S_r(\xi X^\lambda)/r\|^{1/2})^2 &\le lim\ inf_n CLT_1(\xi Y_n) \le \|\xi\|_1 lim\ inf_n CLT_1(Y_n) = \|\xi\|_1 CLT_1(X^\lambda)\\ &\le 2\|\xi\|_1 CLT_1(X).\end{aligned}$$

Since $lim_{\lambda\to\infty}\ E\|S_r(\xi X^\lambda)\|^{1/2} = E\|S_r(\xi X)\|^{1/2}$, (iv) follows. Let now $p > 1$. For finite $\mathcal{F}_n \nearrow$ with $\overline{\vee\mathcal{F}_n} = \sigma(X)$, define $Y_n = E(X|\mathcal{F}_n)$ (in this case $E\|X\| < \infty$). Then by convexity,

$$E\|S_r(Y_n)\| \le E\|S_r(X)\|,$$

so that

$$CLT_p(Y_n) \le CLT_p(X). \tag{7}$$

Also, for all r,

$$lim_{n\to\infty} E\|S_r(\xi Y_n)\| = E\|S_r(\xi X)\|. \tag{8}$$

Hence, since by (3), for all r,

$$E\|S_r(\xi Y_n)/r^{1/p}\| \le CLT_p(\xi Y_n) \le \|\xi\|_p CLT_p(Y_n)$$

we obtain (iv) from (7) and (8) by taking limits as $n \to \infty$ and then sup on r.

Finally, it remains to prove that (iv) $\Rightarrow$ (i). We begin by showing that the constant in inequality (6) for $B = \ell_p^m$ blows up as $n \to \infty$. Let $\{e_\alpha\}_{\alpha=1}^\infty$ be the canonical basis of ℓ_p. Let η be a real random variable such that $t^p P\{|\eta| > t\} = 1, t \geq 1$, and let ε be a symmetric Bernoulli random variable independent of η. Let

$$X = \varepsilon \sum_{\alpha=1}^{\infty} 2^{\alpha/p} I\left(2^{(\alpha-1)/p} < |\eta| \leq 2^{\alpha/p}\right) e_\alpha \tag{9}$$

and

$$X(m) = \varepsilon \sum_{\alpha=1}^{m} 2^{\alpha/p} I\left(2^{(\alpha-1)/p} < |\eta| \leq 2^{\alpha/p}\right) e_\alpha . \tag{10}$$

We show that

$$sup_m \; CLT_p(X(m)) < \infty \tag{11}$$

For this, it is obviously enough to prove that the random variable X satisfies the bounded CLT_p in ℓ_p. To see that $X \epsilon BCLT_p$, we apply the obvious bounded versions of Theorems 6.9 and 5.3 in [1]: $X \epsilon \ell_p$ satisfies the $BCLT_p$ (i.e. $CLT_p(X) < \infty$) if and only if:

(a) $sup_n n P\{\|X\| > n^{1/p}\} < \infty$,
(b) $sup_n n^{-1} \Sigma_\alpha \delta_{n,\alpha}^p < \infty$,
(c) $sup_n \Sigma_\alpha \int_{\delta_{n,\alpha}}^\infty t^{p-1} P\{|X_\alpha^n| > t\} dt < \infty$,
(d) $sup_n n^{p/2-1} \Sigma_\alpha \left(\int_0^{\delta_{n,\alpha}} t P\{|X_\alpha^n| > t\} dt\right)^{p/2} dt < \infty$,

where $X_\alpha^n = X_\alpha I(\|X\| \leq n^{1/p})$, X_α being the α-th coordinate of X, and where

$$\delta_{n,\alpha} = inf[t : P\{|X_\alpha^n| > t\} \leq 1/8 \cdot 3^p \cdot n\}].$$

Now, for X as defined in (9),

$$\|X\| \leq 2|\eta|$$

and

$$P\{|X_\alpha^n| > t\} = I(t^p < 2^\alpha \leq n)/2^\alpha, \;\; \alpha \epsilon \mathrm{N}.$$

Hence,

$$\delta_{n,\alpha} = 0 \quad \text{if} \quad 2^\alpha > n,$$
$$= 2^{\alpha/p} \quad \text{if} \quad 2^\alpha \leq n,$$

and therefore the quantities in (a)-(d) are bounded by: (a) ≤ 2, (b) $= sup_n n^{-1}\Sigma_{\{\alpha:2^\alpha\leq n\}}2^\alpha \leq 2$, (c) $= 0$ and (d) $= sup_n n^{p/2-1}\Sigma_{\{\alpha:2^\alpha\leq n\}}\left(\int_0^{2^{\alpha/p}} 2^{-\alpha}tI(t^p < 2^\alpha \leq n)dt\right)^{p/2} = sup_n n^{p/2-1}\Sigma_{\{\alpha:2^\alpha\leq n\}}\left(2^{(2/p-1)\alpha}/2\right)^{p/2} \leq 2^{1-p/2}/(2-2^{p/2})$. So, $X\epsilon BCLT_p$ and (10) holds. Let now ξ be a real random variable independent of X, with distribution

$$P\{|\xi| > t\} \simeq 1/\{t^p Lt(L_2t)^2\}, \quad t \geq 1,$$

where $Lt = log(t \vee e)$ and $L_2t = LLt$. We estimate from below the quantity (c) for the random variable ξX, which we denote by $(c)_\xi$. The new $\delta_{n,\alpha}$ for $n \leq m$ is now

$$\bar{\delta}_{n,\alpha} = inf[t : P\{(t^p/2^\alpha)^{1/p} < |\xi| \leq (n/2^\alpha)^{1/p}, 2^{(\alpha-1)/p} < |\eta| \leq 2^{\alpha/p}\} \leq 1/3^p \cdot 8 \cdot n].$$

The probability in this definition is bounded from above by

$$\frac{1}{t^p L(t/2^{\alpha/p})(L_2(t/2^{\alpha/p}))^2}.$$

Hence, there are numbers $1 \leq A, B < \infty$ such that for $n \geq A2^\alpha$,

$$\bar{\delta}_{n,\alpha} \leq \left(\frac{Bn}{L(n/2^\alpha)(L_2(n/2^\alpha))^2}\right)^{1/p}$$

There are numbers $1 \leq A', B' < \infty$ such that for $n > A't^p$ and $t^p \geq 2^\alpha$, the same probability is bounded from below by

$$\frac{1}{B't^p L(t^p/2^\alpha)\left(L_2(t^p 2^\alpha)\right)^2}.$$

Then, for some $A'', C > 0$,

$$\begin{aligned}(c)_\xi &\geq \frac{1}{B'}\sum_{\{\alpha:2^\alpha\leq n/A''\}} \frac{1}{L\left(n/2^\alpha A\right)\left(L_2\left(n/2^\alpha A\right)\right)^2}\int_{\frac{n^{1/p}}{L\left(\frac{n}{2^\alpha}\right)\left(L_2\left(\frac{n}{2^\alpha}\right)\right)^2}}^{(n/A')^{1/p}} dt/t \\ &\geq \frac{1}{B'}\sum_{\{\alpha:2^\alpha\leq n/A''\}} \frac{-L(A')/p + L_2\left(n/2^\alpha A\right)}{L\left(n/2^\alpha A\right)\left(L_2\left(n/2^\alpha A\right)\right)^2} \\ &\geq C\sum_{\{\alpha:2^\alpha\leq n/A''\}} \frac{1}{L\left(n/2^\alpha A\right)L_2\left(n/2^\alpha A\right)} \simeq CL_3\left(\frac{n}{A}\right).\end{aligned}$$

Therefore, for m large,

$$(c)_\xi \gtrsim L_3 m.$$

Now, using Theorem 3.3 and Lemma 3.2 in [1], it is easy to see that

$$CLT_p(\xi X(m)) \gtrsim ((c)_\xi)^{1/p}.$$

(For instance, if $p > 1$ and $Z = \xi \Sigma_{\alpha=1}^m X_\alpha$, we have, for $n \simeq m$,

$$\begin{aligned} L_3 m &\gtrsim \sum_{\alpha=1}^m E\, max_{i \leq m} |Z_{\alpha i}|^p I(\|Z_i\| \leq m^{1/p})/n \\ &\gtrsim \sum_{\alpha=1}^m E|\sum_{i=1}^n Z_{\alpha i} I(\|Z_i\| \leq n^{1/p})/n^{1/p}|^p = E\|\sum_{i=1}^n Z_i I(\|Z_i\| \leq n^{1/p})/n^{1/p}\|^p \\ &\gtrsim E\|\sum_{i=1}^n Z_i/n^{1/p}\|, \end{aligned}$$

assuming, as we may, that $E|\xi| < 1/3^p \cdot 8$.) Hence,

$$CLT_p(\xi X(m)) \gtrsim (L_3 m)^{1/p}. \tag{12}$$

If B is not of type p-stable, then ℓ_p is finitely representable in B ([6]). Hence there are in B random variables $\tilde{X}(m)$ such that

$$CLT_p(X(m)) \simeq CLT_p(\tilde{X}(m)), \; CLT_p(\xi X(m)) \simeq CLT_p(\xi \tilde{X}(m))$$

for all m. This, together with (11) and (12), contradicts inequality (6) in B. □

We do not know exactly for what Banach spaces inequality (6) with $p = 2$ holds. It certainly holds in all Banach spaces for which the conditions $t^2 P\{\|X\| > t\} \to 0$ and X preGaussian imply $X \epsilon CLT_2$ (the Λ-Ros (2) spaces which include in particular the cotype 2 spaces and the L_p spaces, $1 \leq p < \infty$ - [4], [8]). What we show next is that the inequality does not hold in all type 2 spaces, not even in all of these with a very good cotype (but different from 2). Ros(p) Banach spaces are used below; see [4] for the definition, and note that ℓ_p is Ros(p) for $p > 2$ and Ros(r) for all r if $p \leq 2$. In connection with the next result we note that for $p = 2$ condition (ii) to (v) of Theorem 1 for B are all equivalent: (v) ⇒ (ii) is a consequence of Theorem 3.1 in [7].

Proposition 2. If B is a Banach space which is Ros(p) for some $p > 2$ and is not of cotype 2, then there exists an $\ell_2(B)$-valued symmetric random variable X and a real valued variable ξ independent of X with $E\xi^2 < \infty$ such that $CLT_2(X) < \infty$ and $CLT_2(\xi X) = \infty$.

Poof. Since B is not of cotype 2 we can find B-valued symmetric preGaussian random variables $Y_\alpha, \alpha \in \mathbb{N}$, such that $\|Y_\alpha\| = 1$ and $\Sigma(E\|G_\alpha\|^p)^{2/p} < \infty$, where G_α are centered Gaussian variables with the covariance of Y_α ([1]). Take now a real variable η such that $P\{|\eta| > t\} = 1/t^2$, $t \geq 1$, and define

$$X = \sum_\alpha 2^{\alpha/2} Y_\alpha I(2^{\alpha-1} < \eta^2 \leq 2^\alpha) e_\alpha$$

where $\{e_\alpha\}$ is the canonical basis for ℓ_2. We then have

$$E\|\sum_{i=1}^n X_i/n^{1/2}\| \leq E\|\Sigma_{i=1}^n X_i I(\eta_i^2 > n)/n^{1/2}\| + E\|\sum_{i=1}^n X_i I(\eta_i^2 \leq n)/n^{1/2}\|$$

$$\leq 2 + \left(\sum_{\{\alpha: 2^{\alpha-1} \leq n\}} 2^\alpha E\|\sum_{i=1}^n Y_{\alpha i} I(2^{\alpha-1} < \eta_i^2 \leq 2^\alpha)/n^{1/2}\|^2 \right)^{1/2} = (I).$$

Now, here we could use the L_2-multiplier property of B and Rosenthal's inequality (e.g. [4]) on $\Sigma_{i=1}^n Y_{\alpha i}$, which may have some advantages for other spaces. But in this case the Rosenthal inequality for $\Sigma_{i=1}^n Y_{\alpha i} I(2^{\alpha-1} < \eta_i^2 \leq 2^\alpha)$ gives directly

$$(I) \leq 2 + c\left(\sum_{\{\alpha: 2^{\alpha-1} \leq n\}} 2^\alpha \left[n^{1-p/2} 2^{-\alpha} + 2^{-\alpha p/2} E\|G_\alpha\|^p \right]^{2/p} \right)^{1/2}$$

$$\leq 2 + c' \left[\sum_{\{\alpha: 2^{\alpha-1} \leq n\}} (2^\alpha/n)^{1-2/p} + \sum_\alpha (E\|G_\alpha\|^p)^{2/p} \right]^{1/2} < \infty$$

for some $c, c' < \infty$. So, $CLT_2(X) < \infty$. On the other hand, if ξ has the same distribution as in the proof of Theorem 1, and is independent of X, then $CLT_2(\xi X) = \infty$. In fact

$$(CLT_2(\xi X))^2 \gtrsim sup_n n^{-1} \sum_\alpha E\, max_{i \leq n} \|\xi_i Y_{\alpha i} I(\|\xi_i Y_{\alpha i}\| \leq n^{1/2})\|^2$$

$$= sup_n n^{-1} \sum_\alpha E\, max_{i \leq n} |\xi_i I\left(2^{\alpha-1} < \eta_i^2 \leq 2^\alpha, |\xi_i| \leq (n/2^\alpha)^{1/2}\right)|^2$$

$$\gtrsim sup_n L_3(n) = \infty,$$

where in the first inequality we use Theorem 3.3 in [1] and, in the last one, the computations form the last part of the proof of Theorem 1 with $p = 2$. □

In view of Proposition 2 and the comments preceding it, it would be interesting to further investigate the relationship between the L_2-multiplier property and the Λ-Ros(2) property (i.e. the best possible CLT_2) for B.

Acknowledgement. We thank Gilles Pisier for an interesting conversation on the subject of this note.

References.

1. Giné, E. and Zinn, J. (1983). Central limit theorems and weak laws of large numbers in certain Banach spaces. Zeits. Wahrs. v. Geb. 62, 323-354.
2. Giné, E. and Zinn, J. (1984). Some limit theorems for empirical processes. Ann. Probability 12, 929-989.
3. Giné, E.; Marcus, M.B.; and Zinn, J. (1987). On random multipliers in the central limit theorem with p-stable limit, $0 < p < 2$. Proceedings of the VI-th International Conference of Probability in Banach Spaces, Sondeborg 1986, to appear.
4. Ledoux, M. (1985). Sur une inegalité de H.P. Rosenthal et le théorème limite central dans les espaces de Banach. Israel J. Math. 50, 290-318.
5. Ledoux, M. and Talagrand, M. (1986). Conditions d'integrabilité pour les multiplicateurs dans le TLC banachique. Ann. Probability 13, 916-921.
6. Maurey B. and Pisier G. (1976). Séries de variables aléatoires vectorielles independantes et proprietés géométriques des espaces de Banach. Studia Math. 58, 45-90.
7. Pisier, G. (1976). Le théorème de la limite centrale et la loi du logarithme iteré dans les espaces de Banach. Séminaire Maurey-Schwartz 1975-76, Exposés III et IV. Ecole Polytechnique, Paris.
8. Pisier, G. and Zinn, J. (1977). On the limit theorems for random variables with values in the spaces $L_p (2 \leq p < \infty)$. Zeits. Wahrs. v. Geb. 41, 289-304.

Texas A&M University
Department of Mathematics
College Station, TX 77843
USA

Probability Theory on Vector Spaces IV
Lancut, June '87, Springer's LNM 1391

A UNIVERSAL LAW OF THE ITERATED LOGARITHM FOR TRIMMED AND CENSORED SUMS

M. G. Hahn*
Department of Mathematics
Tufts University
Medford, MA 02155 U.S.A.

J. Kuelbs**
Department of Mathematics
213 Van Vleck Hall
University of Wisconsin
Madison, WI 53706 U.S.A.

D. C. Weiner***
Department of Mathematics
Boston University
Boston, MA 02215 U.S.A.

Abstract

The conditionally trimmed (resp. censored) sums formed from an arbitrary i.i.d. sample are shown to satisfy a universal law of the iterated logarithm (LIL). The specific method of trimming (resp. censoring) attempts to retain as many summands as possible, and trims (resp. censors) only terms of sufficient magnitude.

* Supported in part by NSF grant DMS-87-02878
** Supported in part by NSF grant DMS-85-21586
*** Supported in part by NSF grant DMS-86-03188.

1. Introduction.

In [4], [5], and [9] universal asymptotic normality is obtained for suitably normalized conditionally trimmed (resp. censored) sums. Since it is well understood that the classical LIL is a consequence of the asymptotic normality of the suitably normalized sample sums and the regularity of the norming sequence which produces this normality, it is natural to question whether there is a universal LIL for conditionally trimmed (resp. censored) sums. Some related LIL results appear in [7], [2], [3], but each of these papers considers a special class of distributions, and [3] investigates the delicate problem of an LIL for unconditionally trimmed sums. Our goal here is to produce a universal result using conditionally trimmed (resp. censored) partial sums.

Let $X, X_1, X_2, \ldots$ be a sequence of i.i.d. real-valued random variables with the same distribution as X. Since the LIL always holds when $E(X^2) < \infty$, we will assume throughout that $E(X^2) = \infty$. However, with minor modifications both theorems presented can be formulated to apply to non-degenerate random variables with $E(X^2) < \infty$. Given $X, X_1, X_2, \ldots$, for each integer $n \geq 1$ arrange $X_1, \cdots, X_n$ in descending order of magnitude, breaking ties by priority of index, and denote the resulting order statistics by $\{X_n^{(j)}: 1 \leq j \leq n\}$. Thus

$$|X_n^{(1)}| \geq |X_n^{(2)}| \geq \cdots \geq |X_n^{(n)}| .$$

As usual let $S_n = X_1 + \cdots + X_n$ for $n \geq 1$.

Fix a real sequence $\{r_n\}$ satisfying

$$\begin{aligned} & r_n \uparrow \infty \\ (1.1) \qquad & n/r_n \uparrow \infty \\ & \underline{\lim}_{n}\, r_n/L_2 n = \eta > 0 \end{aligned}$$

where $L_2 x = L(Lx)$ and $Lx = \max(1, \log x)$. The sequence $\{r_n\}$ will have a companion sequence

$$t_n = n/r_n \qquad (n \geq 1) . \tag{1.2}$$

One should think of r_n as a bound on the expected number of terms of excessive magnitude to be modified via trimming or censoring. This is not completely accurate in all situations, but is a good guideline nevertheless.

To define the normalization required for the LIL for conditionally trimmed sums set

$$h(t) = t^{-2}E(X^2 I(|X| \leq t)) \qquad (t > 0) , \tag{1.3}$$

and let

$$b = \inf\{t \geq 1: h(t) > 0\} .$$

Then define for $s > 0$

$$d(s) = \inf\{t \geq b+1: sh(t) \leq 1\} . \tag{1.4}$$

It is then easy to show (see [4]) that

$$\lim_{s \to \infty} d(s) = \infty , \tag{1.5}$$

and for $s \geq b + 1$,

$$h(d(s)) = 1/s . \tag{1.6}$$

Given $\beta > 1$, let $n_k = [\beta^k]$ where $[\cdot]$ denotes the greatest integer function, and denote the kth block of integers by

$$I(k) = [n_k, n_{k+1}) . \tag{1.7}$$

Next define centering constants

$$\delta_n(\beta) = nE(XI(|X| \leq d(t_{n_k}))) \qquad (n \in I(k)) , \tag{1.8}$$

and norming constants

(1.9) $$\gamma_n = (r_n L_2 n)^{1/2} d(t_n)$$

where $\{t_n\}$ is as in (1.2). For each number r, $0 \le r \le n$, the conditionally trimmed sum is given by

(1.10) $$^{(r)}S_n(\beta) = S_n - \sum_{j=1}^{[r]} X_n^{(j)} I(|X_n^{(j)}| > d(t_{n_k})) \qquad (n \in I(k)).$$

Note that the trimming level is constant within each block, but may change from block to block. Further, each of the largest $[r]$ terms is deleted only if it exceeds $d(t_{n_k})$ in magnitude.

Theorem 1. Let X be a random variable with $E(X^2) = \infty$, and suppose $\{r_n\}$ is a sequence of positive numbers satisfying (1.1) with $\{t_n\}$ as in (1.2). Then, for every $\beta > 1$, $\eta > 0$ as above, there exists a sequence $\{\xi_n\}$ such that

(1.11) $$\lim_n \xi_n r_n / n = 0 ,$$

and

(1.12) $$\overline{\lim_n} \left| \frac{^{(\xi_n r_n)}S_n(\beta) - \delta_n(\beta)}{\gamma_n} \right| = C \qquad \text{a.s.}$$

where $0 < C < \infty$ is a constant. One particular choice of $\{\xi_n\}$ always satisfying (1.11) and (1.12) is

(1.13) $$\xi_n = \max\{e^2\beta^2 n_k P(|X| > d(t_{n_k})),\ 2L_2 n_k\}/r_n \qquad (n \in I(k)) .$$

Remark 1. Often the sequences $\{\xi_n\}$ arising in Theorem 1 are unbounded. However, if X is in the Feller class, i.e. if

$$\Lambda \equiv \overline{\lim_{t\to\infty}} \frac{P(|X|>t)}{h(t)} < \infty ,$$

then $\{\xi_n\}$ may remain bounded. To see this notice that the $\{\xi_n\}$ defined in (1.13) satisfy

$$\xi_n \le e^2\beta^2 \frac{n_k}{r_{n_k}} P(|X| > d(t_{n_k})) \vee 2L_2 n_k / r_n$$

$$\le 2(e^2\beta^2\Lambda \vee 2/\eta) < \infty .$$

In particular, any $\xi_n \to \infty$ for which (1.11) holds will work if X is in the Feller class. (cf. the related result in [7].)

In the conditionally censored companion result, to which we now turn, the boundedness of the analogue of ξ_n is <u>universal</u> for all X. The LIL for conditionally censored sums uses normalizations based on the function $a(t)$ defined by

$$tE((X^2 \wedge a^2(t))/a^2(t)) = 1 \tag{1.14}$$

when $t \ge 1/P(X \ne 0)$. Note $tP(|X| > a(t)) \le 1$ for all large $t > 0$, and $a(t)$ is strictly increasing and continuous for all large t. Given $\{r_n\}$ as in (1.1), we define norming constants

$$\Gamma_n = (r_n L_2 n)^{1/2} a(t_n) . \tag{1.15}$$

For $\beta > 1$ and $I(k)$ as above, we define centerings

$$\Delta_n(\beta) = nE((|X| \wedge a(t_{n_k}))\mathrm{sgn}(X)) \qquad (n \in I(k)). \tag{1.16}$$

For each number r, $0 \le r \le n$, define the conditionally censored sum by

$$S_n^{(r)}(\beta) = \sum_{j=[r]+1}^{n} X_n^{(j)} + \sum_{j=1}^{[r]} (|X_n^{(j)}| \wedge a(t_{n_k}))\mathrm{sgn}(X_n^{(j)}) \quad (n \in I(k)) . \tag{1.17}$$

Note that the censoring level is constant in each block, but may vary from block to block. Further, only the $[r]$ largest magnitudes are possibly altered.

<u>Theorem 2</u>. Let X be a random variable with $E(X^2) = \infty$, and suppose $\{r_n\}$ is a sequence of positive real numbers satisfying (1.1) with $\{t_n\}$ as in (1.2). Then, for every $\beta > 1$, $\eta > 0$ as above, whenever

$$(1.18) \qquad 0 < \lambda_n = o(t_n)$$

satisfies

$$(1.19) \qquad \lambda = \varliminf_n \lambda_n > (\beta e^2 \vee \frac{1}{\eta}) ,$$

then there exists a strictly positive finite constant C such that

$$(1.20) \qquad \varlimsup_n \left| \frac{S_n^{(\lambda_n r_n)}(\beta) - A_n(\beta)}{\Gamma_n} \right| = C \qquad \text{a.s.}$$

In particular, any $\lambda_n \longrightarrow \infty$ such that $\lambda_n = o(n/r_n)$ will yield (1.20) independent of β, η as above, except that C may depend on β.

2. Proofs.

The form of the proofs for both theorems is much the same, but they differ in detail. We will present the proof of Theorem 1, and then indicate the necessary modifications to obtain the proof of Theorem 2.

Proof of Theorem 1. Recall $\beta > 1$ and $n_k = [\beta^k]$ with $I(k) = [n_k, n_{k+1})$ for $k \geq 0$. Then define

$$u_j = u_j(k) = X_j \, I(|X_j| \leq d(t_{n_k}))$$
$$(1 \leq j < n_{k+1})$$
$$v_j = X_j - u_j ,$$

and note that these definitions depend on the blocks $I(k)$. For $n \in I(k)$ set

$$U_n = \sum_{j=1}^{n} u_j$$

$$V_n = \sum_{j=1}^{n} v_j .$$

Then

$$S_n = U_n + V_n ,$$

and

$$(2.1)\quad ({}^{(\xi_n r_n)}S_n(\beta) - \delta_n(\beta)) = (U_n - EU_n)$$

$$+ (V_n - \sum_{j=1}^{[\xi_n r_n]} X_n^{(j)} I(|X_n^{(j)}| > d(t_{n_k})))$$

since $\delta_n(\beta) = EU_n$ when $n \in I(k)$.

We will prove

$$(2.2)\qquad \lim_k \max_{n\in I(k)} |V_n - \sum_{j=1}^{[\xi_n r_n]} X_n^{(j)} I(|X_n^{(j)}| > d(t_{n_k}))| = 0 \qquad \text{a.s.}$$

and then

$$(2.3)\qquad \overline{\lim_k} \max_{n\in I(k)} |U_n - EU_n|/\gamma_n = C \qquad \text{a.s.}$$

where $0 < C < \infty$ is a constant. By (2.1), we then obtain (1.12). That $\{\xi_n\}$ as defined in (1.13) works is part of the proof.

To prove (2.2) observe that for any sequence $\{\xi_n\}$

$$J_k = P(\max_{n\in I(k)} |V_n - \sum_{j=1}^{[\xi_n r_n]} X_n^{(j)}\, I(|X_n^{(j)}| > d(t_{n_k}))| > 0)$$

$$\le P(\text{at least } (M_k+1) \text{ of the } X_j\text{'s } (1 \le j < n_{k+1}) \text{ satisfy } |X_j| > d(t_{n_k}))$$

where $M_k = \min_{n\in I(k)} [\xi_n r_n]$.

Choose $\{\xi_n\}$ such that for $n \in I(k)$

$$M_k = \min_{n\in I(k)} [\xi_n r_n] = \max\{e^2\beta^2 n_k p_{n_k}, 2L_2 n_k\}$$

where $p_n = P(|X| > d(t_n))$. That is, for each $n \in I(k)$ a suitable choice is to define

$$\xi_n = \max\{e^2\beta^2 n_k p_{n_k}, 2L_2 n_k\}/r_n .$$

Then

$$\lim_n \xi_n r_n/n = 0 ,$$

and for all sufficiently large k with $N_k = n_{k+1} - 1$

$$J_k \le \sum_{j=M_k+1}^{N_k} \binom{n}{j} p_{n_k}^j (1 - p_{n_k})^{N_k-j}$$

$$= N_k \binom{N_k-1}{M_k} \int_0^{p_{n_k}} t^{M_k}(1 - t)^{N_k-M_k-1} dt$$

by [1, p. 173]

$$\le 2(n_{k+1}\, p_{n_k} e/M_k)^{M_k}\, M_k^{-1/2}\, (N_k/(N_k - M_k))^{1/2}$$

by Stirling's formula

$$\le 2 \exp\{-2L_2 n_k\} .$$

Hence

$$\sum_k J_k < \infty$$

and (2.2) holds.

To prove (2.3), there are as usual two parts, the upper bound and the lower bound. The zero-one law will show that the a.s. positive finite variable C in (2.3) which will result must be a constant a.s..

To prove the upper bound observe that since γ_n is non-decreasing it suffices to show

$$(2.4) \qquad \sum_k P\Big(\max_{n\in I(k)} |U_n - EU_n| > 64\sqrt{e\beta}\ \gamma_{n_k}/\min(\eta,1)\Big) < \infty$$

where $\eta > 0$ is as in (1.1). It is easy to prove

$$(2.5) \qquad \lim_k \max_{n\in I(k)} E|U_n - EU_n|^2/\gamma_{n_k}^2 = 0 ,$$

using the definition (1.9) of $\{\gamma_n\}$ and the property (1.6). Now (2.4) follows from a standard application of Ottaviani's inequality if

$$(2.6) \qquad \sum_k P(|U_{N_k} - EU_{N_k}| > 32\sqrt{e\beta}\ \gamma_{n_k}/\min(\eta,1)) < \infty ,$$

recalling $N_k = n_{k+1} - 1$.

To prove (2.6) define

$$\text{(i)} \quad b_{n_k} = 8\sqrt{e\beta}\ \sqrt{r_{n_k}}\ d(t_{n_k})/\min(\eta,1)$$

$$(2.7) \qquad \text{(ii)} \quad \epsilon_{n_k} = 2\sqrt{L_2 n_k}$$

$$\text{(iii)} \quad c_{n_k} = \min(\eta,1)/(4\sqrt{r_{n_k}}) .$$

Hence for $1 \le j \le N_k$ and all large k,

$$\text{(i)} \quad |u_j - Eu_j| = |u_j(k) - Eu_j(k)| \le 2d(t_{n_k}) \le c_{n_k} b_{n_k}$$

$$\text{(2.8)} \quad \text{(ii)} \quad \epsilon_{n_k} c_{n_k} = \sqrt{\frac{L_2 n_k}{r_{n_k}}} \min(\eta,1)/2 \le 1$$

$$\text{(iii)} \quad 2\epsilon_{n_k} b_{n_k} = 32\sqrt{e\beta}\, \gamma_{n_k}/\min(\eta,1) ,$$

and by using the estimate $1 + x/3 + x^2/(4\cdot 3) + \cdots \le e^x$ the proof of Lemma 2.1 of [6] implies that

$$P(|U_{N_k} - EU_{n_k}| > 32\sqrt{e\beta}\, \gamma_{n_k}/\min(\eta,1))$$

$$\le \exp\{-\epsilon_{n_k}^2 [1 - N_k E(u_1(k) - Eu_1(k))^2 e^{\epsilon_{n_k} c_{n_k}}/(2b_{n_k}^2)$$

$$- E|U_{N_k} - EU_{N_k}|/(2\epsilon_{n_k} b_{n_k})]\} .$$

Since (2.5) holds and $\epsilon_{n_k} b_{n_k} = 16\sqrt{e\beta}\, \gamma_{n_k}/\min(\eta,1)$, it follows for all k sufficiently large that

$$P(|U_{N_k} - EU_{N_k}| > 32\sqrt{e\beta}\, \gamma_{n_k}/\min(\eta,1))$$

$$\le \exp\{-\epsilon_{n_k}^2 [3/4 - N_k E(u_1(k) - EU_1(k))^2 e^{\epsilon_{n_k} c_{n_k}}/(2b_{n_k}^2)]\}$$

$$\le \exp\{-\epsilon_{n_k}^2/2\}$$

$$= \exp\{-2L_2 n_k\} .$$

The second inequality follows from the definition of $d(n)$ and that $E(X^2) = \infty$ implies

$$nd^{-2}(n)\, E[XI(|X| \le d(n)) - E(XI(|X| \le d(n)))]^2 \sim 1$$

and hence

$$N_k E(u_1(k) - Eu_1(k))^2 e^{\epsilon_{n_k} c_{n_k}}/(2b_{n_k}^2) \le \frac{2\beta n_k e E(X^2 I(|X| \le d(t_{n_k})))\min(\eta^2,1)}{128\ e\beta\ r_{n_k} d^2(t_{n_k})}$$

$$\le \min(\eta^2,1)/32 .$$

Thus (2.6) holds, and the upper bound part of (2.3) follows.

For the lower bound part of (2.3) it suffices to prove that

$$(2.9) \qquad \overline{\lim_k}\ |U_{m_k} - EU_{m_k}|/\gamma_{m_k} > 0 \qquad \text{a.s.}$$

where $m_k = [\beta^{k^2}]$ and $U_{m_k} = \sum_{j=1}^{m_k} u_j(k^2)$. That is, for $n \in I(k^2) = [[\beta^{k^2}],[\beta^{k^2+1}])$

$$U_n = \sum_{j=1}^{n} (u_j(k^2) - Eu_j(k^2))$$

and hence (2.9) is sufficient.

To prove (2.9) it is useful to have what is known as Kolmogorov's lower bound exponential inequality. This appears in the following lemma which is part of Theorem 5.2.2 on p. 262 of [8].

Lemma 1. Let $Y_1, Y_2, \cdots, Y_n$ be independent mean zero random variables, and let

$$\sigma_n^2 = E(Y_1 + \cdots + Y_n)^2 .$$

Further, assume

$$|Y_j| \le c_n \sigma_n \quad \text{a.s.} \qquad (1 \le j \le n) .$$

Suppose $\epsilon_n > 0$, $\gamma > 0$ and $T_n = \sum_{j=1}^{n} Y_j$ for $n \ge 1$. Then there exist constants $\epsilon(\gamma)$ and $\pi(\gamma)$ such that for each $n \ge 1$ if $\epsilon_n \ge \epsilon(\gamma)$ and $\epsilon_n c_n \le \pi(\gamma)$, then

$$P(T_n/\sigma_n \ge \epsilon_n) \ge \exp\{-\epsilon_n^2(1+\gamma)/2\} .$$

Now for each $\alpha > 0$ and $k \geq 1$, define

$$E_k = \{|\sum_{j=m_{k-1}+1}^{m_k} (u_j(k^2) - Eu_j(k^2))| > 2\alpha\, \gamma_{m_k}\}$$

and

$$F_k = \{|\sum_{j=1}^{m_{k-1}} (u_j(k^2) - Eu_j(k^2))| \leq \alpha\, \gamma_{m_k}\} .$$

Then the $\{E_k\}$ are independent, and for all sufficiently large k it follows that

$$P(F_k^c) \leq \frac{m_{k-1}E[XI(|X|\leq d(t_{m_k}))-E(XI(|X|\leq d(t_{m_k})))]^2}{\alpha^2\gamma_{m_k}^2}$$

$$\leq \frac{2m_{k-1}}{\alpha^2(L_2m_k)m_k} .$$

Hence

$$\sum P(F_k^c) < \infty$$

for all $\alpha > 0$. Since

$$\{|\sum_{j=1}^{m_k} (u_j(k^2) - Eu_j(k^2))| > \alpha\gamma_{m_k}\} \supseteq E_k \cap F_k$$

it follows that for all $\alpha > 0$

$$\{|\sum_{j=1}^{m_k} (u_j(k^2) - Eu_j(k^2))| > \alpha\gamma_{m_k} \text{ i.o.}\}$$

(2.10) $$\supseteq \{E_k \cap F_k \text{ i.o.}\}$$

$$= \{E_k \text{ i.o.}\} .$$

Of course, the equality in (2.10) is modulo a set of probability zero.

Letting $\ell_k = m_k - m_{k-1}$, since $E(X^2) = \infty$ it follows from the definition of $d(n)$ that

$$\sigma_{m_k}^2 \equiv E\Big(\sum_{j=1}^{\ell_k} (u_j(k^2)-Eu_j(k^2))\Big)^2$$

(2.11)

$$= \ell_k\ E(XI(|X|\le d(t_{m_k}))-E(XI(|X|\le d(t_{m_k})))^2$$

$$\sim \ell_k \frac{d^2(t_{m_k})r_{m_k}}{m_k}$$

$$\sim d^2(t_{m_k})r_{m_k} .$$

Now for all large k,

$$|u_j(k^2) - Eu_j(k^2)| \le 2d(t_{m_k})$$

(2.12)

$$\le \sigma_{m_k} c_{m_k}$$

where $\sigma_{m_k}^2$ is as in (2.11) provided

$$c_{m_k} = 4/\sqrt{r_{m_k}} .$$

Fix $\gamma > 0$ and take

$$\epsilon_{n_k} = 4\alpha\sqrt{L_2 m_k} .$$

Then for any $\gamma > 0$ and $\pi(\gamma)$ as in Kolmogorov's lower bound, there is an $\alpha > 0$ sufficiently small so that $\epsilon_{n_k} c_{n_k} \le \pi(\gamma)$. Hence by stationarity

$$P(E_k) = P\Big(\sum_{j=1}^{\ell_k} (u_j(k^2) - Eu_j(k^2)) > 2\alpha\sqrt{L_2 m_k}\ \sqrt{r_{m_k}}\, d(t_{m_k})\Big)$$

$$\geq P\Big(\sum_{j=1}^{\ell_k} (u_j(k^2) - Eu_j(k^2)) > \epsilon_{n_k}\sigma_{n_k}\Big)$$

$$\geq \exp\{-8(1+\gamma)\alpha^2 L_2 m_k\} .$$

Thus for all $\alpha > 0$ small enough

$$\sum_k P(E_k) = \infty .$$

Since $\{E_k\}$ are independent, the Borel Cantelli lemma gives

$$P(E_k \ \text{ i.o.}) = 1 .$$

Thus the lower bound is strictly positive and the proof of Theorem 1 is complete. □

<u>Proof of Theorem 2</u>. Let $\Lambda_n = [\lambda_n r_n]$. For $n \in I(k)$, replacing r by Λ_n in (1.17) yields

$$S_n^{(\Lambda_n)}(\beta) = \sum_{j=1}^{n} \{(|X_j| \wedge a(t_{n_k}))\operatorname{sgn}(X_j)\}$$

$$+ \sum_{j=\Lambda_n+1}^{n} (X_n^{(j)} - \{(|X_n^{(j)}| \wedge a(t_{n_k}))\operatorname{sgn}(X_n^{(j)})\}) \tag{2.13}$$

$$\equiv U_n + V_n ,$$

say. Further, $\Lambda_n(\beta) = E(U_n)$. We will prove

$$\lim_k \max_{n\in I(k)} |V_n| = 0 \qquad \text{a.s.} \tag{2.14}$$

and then

$$\overline{\lim_{k\to\infty}} \max_{n\in I(k)} \frac{|U_n - EU_n|}{\Gamma_n} = C \qquad \text{a.s.} \tag{2.15}$$

for some constant $C \in (0,\infty)$. By (2.13), we will then have (1.20). For (2.14), note that

$$J_k \equiv P(\exists n \in I(k): V_n \neq 0)$$

$$\leq P(\exists n \in I(k): |X_n^{(\Lambda_n+1)}| > a(t_{n_k}))$$

$$\leq P(\text{at least } (M_k+1) \text{ of the } X_j\text{'s } (1\leq j\leq n_{k+1}) \text{ satisfy } |X_j| > a(t_{n_k}))$$

where $M_k = \min_{n\in I(k)} \Lambda_n$.

Letting $p_n = P(|X| > a(t_n))$, the calculation to prove (2.2) in Theorem 1 yields for large k that

$$J_k \leq 2(n_{k+1}p_{n_k}e/M_k)^{M_k} M_k^{-1/2}\left[\frac{N_k}{N_k - M_k}\right]^{1/2}$$

where $N_k = n_{k+1} - 1$. Now $p_n \leq r_n/n$ and $M_k \leq \Lambda_{n_k} = [\lambda_{n_k} r_{n_k}] = o(t_{n_k})r_{n_k} = o(n_k) = o(N_k)$, so $\frac{N_k}{N_k - M_k}$ remains bounded. Further, $M_k \to \infty$ since for large k $M_k = \min_{n\in I(k)} [\lambda_n r_n] > (\beta e^2 \vee 1/\eta) r_{n_k}$ and $r_n \to \infty$. Hence by (1.1)

$$\underline{\lim}_k\, M_k/L_2 n_k > \lambda\eta > 1$$

while

$$n_{k+1}p_{n_k}e/M_k \sim \beta e(n_k p_{n_k})/M_k \leq \beta e\, r_{n_k}/M_k$$

and

$$\overline{\lim_{n}} \frac{r_{n_k}}{M_k} < \frac{1}{\beta e^2} .$$

Thus there exists $\delta > 0$ such that for all large k,

$$J_k = O(e^{-(1+\delta)L_2 n_k}) .$$

Hence $\sum_k J_k < \infty$, and (2.14) holds.

To prove (2.15) it again suffices to prove an upper and lower bound result as in Theorem 1. Γ_n is nondecreasing, so the upper bound follows upon showing that for some constant $B > 0$

(2.16) $$\sum_k P(\max_{n\in I(k)} |U_n - EU_n| > B\Gamma_{n_k}) < \infty .$$

Now

$$\max_{n\in I(k)} E|U_n - EU_n|^2/\Gamma_{n_k}^2$$

(2.17) $$\leq n_{k+1}E(|X|^2 \wedge a^2(t_{n_k}))/((L_2 n_k) r_{n_k} a^2(t_{n_k}))$$

$$\leq O(\frac{n_k}{r_{n_k}} E\left[\frac{X^2 \wedge a^2(t_{n_k})}{a^2(t_{n_k})}\right]\frac{1}{L_2 n_k})$$

$$= O(\frac{1}{L_2 n_k})$$

since $n_k/r_{n_k} = t_{n_k}$ and $a(t)$ satisfies (1.14) for large t.

Thus by Ottaviani's inequality it suffices to prove for some $A > 0$,

(2.18) $$\sum_k P(|U_{n_k} - EU_{n_k}| > A\Gamma_{n_k}) < \infty .$$

Given k, define

$$(2.19)\qquad u_j = u_j(k) = (|X_j| \wedge a(t_{n_k}))\operatorname{sgn} X_j \qquad (j < n_{k+1}) .$$

Then $U_{n_k} = \sum_{j=1}^{n_k} u_j$ and $|u_j| \le a(t_{n_k})$. Verification of (2.18) now follows as in the argument in (2.7)-(2.8) including the paragraph ending prior to (2.9) upon replacing $d(t)$ by our $a(t)$ throughout, and noting that

$$n \operatorname{Var}((|X| \wedge a(n))\operatorname{sgn}(X))/a^2(n) \sim 1$$

since $E(X^2) = \infty$.

For the lower bound, let $m_k = [\beta^{k^2}]$ as in Theorem 1 and proceed exactly as indicated in the proof of Theorem 1 replacing $d(t)$ by $a(t)$ and using (2.19) throughout.

Thus (1.20) holds as in Theorem 1, and the final statement of Theorem 2 now follows easily.

References

[1] Feller, W. (1968). An introduction to probability theory and its applications: vol. I, third edition. John Wiley & Sons, New York.

[2] Griffin, P. S. (1985). Laws of the iterated logarithm for symmetric stable processes. Z. Wahrsch. verw. Gebiete, 68, 271-285.

[3] Griffin, P. S. (1985). The influence of extremes on the law of the iterated logarithm. Preprint.

[4] Hahn, M. G. and Kuelbs, J. (1987). Universal Asymptotic Normality for Trimmed Sums. Preprint.

[5] Hahn, M. G., Kuelbs, J. and Weiner, D. C. (1987), Joint estimation of some center and scale parameters via self-normalization for distributions with possibly infinite mean and variance, Preprint.

[6] Kuelbs, J. (1977). Kolmogorov's law of the iterated logarithm for Banach space valued random variables. Illinois J. 21, 784-800.

[7] Kuelbs, J. and Ledoux, M. (1985). Extreme values and LIL behavior. Preprint.

[8] Stout, W. F. (1974), Almost sure convergence, Academic Press, New York.

[9] Weiner, D. C. (1986), Centrality, scale, and asymptotic normality for sums of independent random variables, Preprint.

Probability Theory on Vector Spaces IV
Lancut, June '87, Springer's LNM 1391

On the decomposability group of a convolution semigroup

W. HAZOD
Universität Dortmund

The study of stable and semistable convolution semigroups of probability measures on a group G leads in a natural way to the investigation of groups of automorphisms of the group G, which decompose the given convolution semigroup. On the other hand, given a convolution semigroup $(\mu_t)_{t \geq o}$, the structure of the decomposability group leads to an intrinsic definition of stability resp. semistability.

Let $(\mu_t)_{t \geq o}$ be a continuous convolution semigroup. Let $\mathfrak{B}$ be a group of (admissible) automorphisms (see § 1 for precise definitions). $\mathfrak{J} := \mathfrak{J}((\mu_t)) := \{\tau \in \mathfrak{B} : \tau\mu_t = \mu_t,\ t \geq o\}$ is called the invariance group and $\mathfrak{z} := \mathfrak{z}((\mu_t)) := \{\tau \in \mathfrak{B}$: There exists $c \in \mathbb{R}_+^*$, such that $\tau\mu_s = \mu_{sc}, s \geq o\}$ is called the decomposability group. (μ_t) is stable iff there exists a continuous one - parameter subgroup $(\tau_t)_{t > o}$ in $\mathfrak{z}$, such that $(\tau_t) \cong \mathfrak{z}/\mathfrak{J}$, and (μ_t) is semistable iff $\mathfrak{z} \neq \mathfrak{J}$. For general definitions of stability resp. semistability the reader is referred to [24] resp. [2-6] .

We start with a general definition of convolution structures and automorphisms covering most of the examples of stability, which had been studied in the past. The main examples however are operator - semistable laws on finite dimensional vector spaces on the one hand and semi - stable laws on nilpotent Lie groups on the other hand. It turns out that the characterizations of $\mathfrak{z}$ and $\mathfrak{J}$, which had been obtained in the vector space case (see e.g. [20], [14] , [25])) also hold in the Lie group case.

§ 1 Definitions and elementary facts.

Let X be a completely metrizable second countable space and let $\mathcal{N} \subseteq M^1(X)$ be a subset of probability measures, closed with respect to the vague topology. $\mathcal{N}$ is endowed with a convolution structure $* : \mathcal{N} \times \mathcal{N} \to \mathcal{N}$, such that $\mathcal{N}$ becomes an affine topological semigroup. (See [6].)

A convolution semigroup $(\mu_t)_{t \geq o}$ is a continuous $*$ - homomorphism $\mathbb{R}_+ \ni t \mapsto \mu_t \in \mathcal{N}$.

An automorphism of $\mathcal{N}$ is a homeomorphism $\tau : X \to X$, such that the induced map on $M^1(X)$ - which is again denoted by τ - induces a $*$ - isomorphism of $\mathcal{N}$. I.e. $\tau(\mathcal{N}) = \mathcal{N}$ and $\tau(\mu * \nu) = \tau(\mu) * \tau(\nu)$ for $\mu, \nu \in \mathcal{N}$.

In the sequel we fix a group $\mathcal{B}$ of automorphisms of $\mathcal{N}$ endowed with a topology, such that $\mathcal{B}$ is a topological group which acts continuously on $\mathcal{N}$. The following § 2 will clearify the examples of $\mathcal{N}$, $*$, $\mathcal{B}$ we have in mind.

A one parameter group $T = (\tau_t)$ in $\mathcal{B}$ is a subset $\{\tau_t : t > o\} \subseteq \mathcal{B}$, such that $t \mapsto \tau_t x$ is continuous for $x \in X$ and such that $\tau_t \tau_s = \tau_{ts}$ for $t,s > o$.

A convolution semigroup $(\mu_t)_{t \geq o} \subseteq \mathcal{N}$ is stable w.r.t. $T := (\tau_t) \subseteq \mathcal{B}$

if $\tau_t(\mu_s) = \mu_{st}$ for $t,s > o$.

(μ_t) is called semistable w.r.t. $\tau \in \mathcal{B}$, $c \in \mathbb{R}_+^* \setminus \{1\}$

if $\tau(\mu_t) = \mu_{ct}$, $t > o$.

For the definitions see e.g. [24] [2] [6].

<u>1.1 Definition.</u> Let $\mathcal{N}$, $\mathcal{B}$, (μ_t) be given.
The invariance group of (μ_t) is defined by

$$\mathcal{J} = \mathcal{J}((\mu_t)) := \{\tau \in \mathcal{B} : \tau\mu_t = \mu_t, \quad t > o\}.$$

The decomposability group is defined by

$$\mathcal{Z} = \mathcal{Z}((\mu_t)) := \{\tau \in \mathcal{B} : \text{There exists } c \in \mathbb{R}_+^*, \text{ such that } \tau\mu_t = \mu_{ct}, \quad t > o\}.$$

Finally we define

<u>1.2 Definition.</u> $\tilde{\mathfrak{B}} := \mathfrak{B} \otimes \mathbb{R}_+^*$ endowed with the product topology, $\tilde{\mathfrak{g}} := \{(\tau,1) \in \tilde{\mathfrak{B}} : \tau \in \mathfrak{g}\}$ and $\tilde{\mathfrak{z}} := \{(\tau,c) \in \tilde{\mathfrak{B}} : \tau(\mu) = \mu_{ct},\ t > o\}$. Furthermore define $\tilde{\varphi} : \tilde{\mathfrak{z}} \to \mathbb{R}_+^*$ resp. $\varphi : \mathfrak{z} \to \mathbb{R}_+^*$ via $\tilde{\mathfrak{z}} \ni (\tau,c) \mapsto c$ resp. $\mathfrak{z} \ni \tau \mapsto c$.

<u>1.3 Proposition.</u> $\tilde{\varphi}$ resp. φ are homomorphisms from $\tilde{\mathfrak{z}}$ resp. $\mathfrak{z}$ to $\mathbb{R}_+^*$ with kernel $\tilde{\mathfrak{g}}$ resp. $\mathfrak{g}$.

[Let $(\sigma, d), (\tau, c) \in \tilde{\mathfrak{z}}$. Then by definition

$\tau(\mu_t) = \mu_{ct}, \quad \sigma(\mu_s) = \mu_{ds}, \quad t,s > o.$

Hence $(\tau\sigma)(\mu_t) = \mu_{cdt}$, $t > o$. I.e. $(\tau\sigma, cd) \in \tilde{\mathfrak{z}}$.]

<u>1.4 Proposition.</u> $\mathfrak{g}$, $\tilde{\mathfrak{g}}$ and $\tilde{\mathfrak{z}}$ are closed subgroups of $\mathfrak{B}$ resp. $\tilde{\mathfrak{B}}$. If $t \mapsto \mu_t$ is not constant, then also $\mathfrak{z}$ is closed in $\mathfrak{B}$. φ resp. $\tilde{\varphi}$ are continuous homomorphisms.

[Let $(\tau_\alpha)_{\alpha \in I}$ be a net in $\mathfrak{g}$ converging to some $\tau \in \mathfrak{B}$. Hence $(\tau_\alpha,1) \to (\tau,1)$. Then $\tau_\alpha\mu_t = \mu_t$, $t > o$. And since $\mathfrak{B}$ acts continuously on $\mathcal{N}$, we obtain $\tau\mu_t = \mu_t$, $t > o$. Hence $\tau \in \mathfrak{g}$, $(\tau,1) \in \tilde{\mathfrak{g}}$. Let (τ_α,c_α) be a net in $\tilde{\mathfrak{z}}$ converging to (τ,c) in $\tilde{\mathfrak{B}}$. Hence $\tau_\alpha\mu_t = \mu_{tc_\alpha}$. From $\tau_\alpha \to \tau$, $c_\alpha \to c$ we obtain $\tau\mu_t = \mu_{tc}$, hence $(\tau,c) \in \tilde{\mathfrak{z}}$.

Finally let $\tau_\alpha \in \mathfrak{z}$, $\varphi(\tau_\alpha) = c_\alpha \in \mathbb{R}_+^*$. Assume $\tau_\alpha \to \tau \in \mathfrak{B}$. If there is a subnet (α') such that $c_{\alpha'} \to c \in \mathbb{R}_+^*$, then as before $(\tau,c) \in \tilde{\mathfrak{z}}$, hence $\tau \in \mathfrak{z}$.

Assume that $c_{\alpha'} \to o$. Then $\tau_{\alpha'}\mu_t = \mu_{c_{\alpha'}t} \to \mu_o$, whence $\tau\mu_t = \mu_o$ for any $t > o$ follows. Since $t \mapsto \mu_t$ is not constant, this is a contradiction to τ being an automorphism.

If $c_{\alpha'} \to \infty$, consider $\tau_{\alpha'}^{-1} =: \sigma_{\alpha'}$, $\sigma_{\alpha'} \to \sigma := \tau^{-1}$, $d_{\alpha'} := 1/c_{\alpha'} \to o$. Since $\sigma_\alpha\mu_t = \mu_{d_\alpha t}$, $t > o$, we are led to a contradiction as before.

The continuity of φ resp. $\tilde{\varphi}$ is proved in the same way.]

Obviously the following holds.

1.5 Proposition. $(\mu_t)_{t \geq o}$ is semistable (with respect to some (τ,c)) iff $\mathfrak{Z}((\mu_t)) \neq \mathfrak{J}((\mu_t))$. $(\mu_t)_{t \geq o}$ is stable (with respect to some $(\tau_t)_{t > o}$) iff there is a one - parameter group (τ_t) in $\mathfrak{B}$ such that $(\tau_t, t) \in \tilde{\mathfrak{Z}}$, $t > o$.

Before we study the structure of $\mathfrak{Z}$ more profoundly (under further restrictions) we list some examples of convolution structures, by which this investigation was motivated.

§ 2 Examples.

In [6] we already mentioned several types of convolution structures and stability concepts, which are covered by the definition of stability given in the preceeding § 1. Therefore we will restrict here essentially to two types of examples which are closely connected.

2.1 Operator [semi-]-stable laws on finite dimensional vector spaces.

Operator stable laws were introduced by M. Sharpe [22], the concept of operator-semi-stability was introduced by R. Jajte [15] following a pioneer work of K. Urbanik [26] on operator-selfdecomposable measures. Recently a series of papers on this subject was published (see [6] for detailed hints), we only mention [10-14,16,17,19,20,25]. Since we are interested in non-commutative structures we always consider operator-stability and - semi-stability in the strict sense. See e.g. Sato [23] for a recent treatment. Let $X \cong \mathbb{R}^d$ be a finite - dimensional vector space. The convolution structure on $M^1(\mathbb{R}^d)$ is the usual convolution defined by addition.
Let $\mathfrak{B}$ be a closed subgroup of $Gl(\mathbb{R}^d)$ endowed with the induced topology of $Gl(\mathbb{R}^d)$. Then, if $\mathcal{N}$ is a vaguely closed affine $*$ - subsemigroup of $M^1(\mathbb{R}^d)$, invariant under the action of $\mathfrak{B}$, the assumptions of § 1 are fulfilled.
Let $\mathcal{N} = M^1(\mathbb{R}^d)$.
If $\mathfrak{B} = \{\alpha \cdot id : \alpha \in \mathbb{R}_+^*\} \cong \mathbb{R}_+^*$ is the group of scalars, then we obtain the classical concepts of (strictly) stable resp. semistable laws on $\mathbb{R}^d$.

If $\mathfrak{B} = Gl(\mathbb{R}^d)$ we obtain the (strictly) operator (-semi-) stable laws of Sharpe resp. Jajte mentioned above.
In this case the structures of the decomposability groups $\mathfrak{Z}$, $\tilde{\mathfrak{Z}}$ and the invariance groups $\mathfrak{J}$, $\tilde{\mathfrak{J}}$ are well known, at least for full measures,

see A. Łuczak [20], E. Siebert [25] , W.N. Hudson, Z.J. Jurek, J.A. Veeh [14], J.P. Holmes, W.N. Hudson, J.D. Mason [10].

Remark: Since a convolution semigroup (μ_t) on $\mathbb{R}^d$ is uniquely determined by a single measure μ_1, the decomposability group resp. invariance group of the semigroup (μ_t) is defined by μ_1.

There are different generalizations of operator-semi-stability for infinite dimensional vector spaces (see e.g. [6] and the literature cited there. For the validity of the convergence of type - theorem see W. Linde, G. Siegel [18]). However in this paper we will always assume vector spaces to be finite dimensional.

2.2 Semi-stable laws on simply - connected nilpotent Lie groups.

Let X = G be a locally compact group. Then $M^1(G)$, endowed with the vague topology, convolution defined by the group-operation, and any subgroup $\mathfrak{B} \subseteq \mathrm{Aut}(G)$, which acts continuously on G, fulfill the assumptions of § 1. It was pointed out in several papers (see e.g. [2 - 6, 7, 8]) that the following reduction is possible:

Let (τ_t) be a continuous one-parameter group resp. $\{\tau^k\}$ a discrete group in Aut(G). Let $(\mu_t)_{t \geq 0}$ be a convolution semigroup which is stable w.r.t. (τ_t) resp. semistable w.r.t. (τ, c). Then the measures are concentrated on the subgroup C of G on which (τ_t) resp. (τ^k) act contracting mod K. Here K is a compact subgroup of G such that $\mu_o = \omega_K$, the normalized Haar measure.

In the case of a continuous group (τ_t) the contractible part C is a closed subgroup of G, isomorphic to a semidirect product of a simply connected nilpotent Lie group N and K. The convolution semigroup (μ_t) on G is representable in a an unique manner by a convolution semigroup $(\nu_t)_{t \geq 0}$ on N, with $\nu_o = \varepsilon_e$, which is stable with respect to the restrictions of τ_t to N, and the measures ν_t are invariant under the action of K. See [7] .

In the case of a Lie group G a similar result is proved in [8] for semistable convolution semigroups (μ_t).

2.3 So we are well motivated to restrict considerations to the case where X = G is a simply connected nilpotent Lie group on which a

compact group $K \subseteq \mathrm{Aut}(G)$ acts. Let $\mathcal{N} \subseteq \{\mu \in M^1(G)$, such that $\rho\mu = \mu,\ \rho \in K\}$.
Furthermore let $\mathfrak{B} \subseteq \mathrm{Aut}(G)$ be a subgroup, which leaves $\mathcal{N}$ invariant. Then $G, \mathcal{N}, \mathfrak{B}$ fulfill the assumptions of § 1.

The concepts of semistability and stability in this context coincide with the concepts of operator stability described in 2.1:

<u>2.4</u> Let G be a nilpotent simply connected Lie group of dimension d, let $\mathfrak{g}$ be its Lie algebra. The exponential map $\exp : \mathfrak{g} \to G$ is a C^∞ - isomorphism. $\mathfrak{g}$ is - as a vector-space - isomorphic to $\mathbb{R}^d$, exp induces isomorphisms $f \mapsto \mathring{f}$ from $C_o(G)$ onto $C_o(\mathfrak{g})$ and $\mu \mapsto \mathring{\mu}$ from $M^1(G)$ onto $M^1(\mathfrak{g})$.

Let for $\tau \in \mathrm{Aut}(G)$ $d\tau =: \mathring{\tau}$ be the differential. Then $\tau \mapsto \mathring{\tau}$ is an (algebraic and topological) isomorphism from $\mathrm{Aut}(G)$ onto $\mathrm{Aut}(\mathfrak{g}) \subseteq \mathrm{Gl}(\mathbb{R}^d)$.

There is a 1 - 1 - correspondence between convolution semigroups $(\mu_t)_{t \geq o}$ on G and convolution semigroups $(\nu_t)_{t \geq o}$ on $\mathfrak{g}$.

<u>Remark.</u> If G is non-abelian, i.e. G is not a vector space, in general $(\mu *_G \lambda)^\circ \neq \mu^\circ *_{\mathfrak{g}} \lambda^\circ$. Hence, if the semigroup (μ_t) on G corresponds to (ν_t) on $\mathfrak{g}$, in general $\mathring{\mu}_t \neq \nu_t$. ($(\mathring{\mu}_t)$ need not to be a convolution semigroup !)
Furthermore, the semigroup $(\mu_t)_{t \geq o}$ on G is (τ, c) - semistable resp. $(\tau_t)_{t > o}$ - stable, iff the corresponding semigroup (ν_t) on $\mathfrak{g}$ is $(\mathring{\tau}, c)$ operator - semistable resp. $(\mathring{\tau}_t)_{t > o}$ operator stable (in the strict sense).

We obtain the following structure (see [2 - 6, 7, 8] for details):

<u>2.5 Theorem.</u> Let $\mathcal{N} = M^1(G)$ be endowed with the convolution induced by the group operation, and let $\mathfrak{B} := \mathrm{Aut}(G)$.
Let further $\mathring{\mathcal{N}} := M^1(\mathfrak{g}) = M^1(\mathbb{R}^d)$ endowed with the usual convolution, and let $\mathring{\mathfrak{B}} := \mathrm{Aut}(\mathfrak{g})$ (the Lie-algebra automorphisms form a closed subgroup of $\mathrm{Gl}(\mathbb{R}^d)$).

Then the correspondence of convolution semigroups (μ_t) on G with (ν_t) on $\mathfrak{g}$ transforms the concept of semi-stability of $(\mathcal{N}, \mathfrak{B})$ onto the concept of operator semistability of $(\mathring{\mathcal{N}}, \mathring{\mathfrak{B}})$.

Let moreover $K \subseteq \mathrm{Aut}(G)$ be a compact subgroup and $\mathcal{N}_K := \{\mu \in M^1(G): \tau\mu = \mu, \tau \in K\}$ and let $\mathcal{B}_K$ be a closed subgroup of Aut(G) with $\mathcal{B}_K \supseteq K$ and $\tau\mathcal{N}_K = \mathcal{N}_K$ for $\tau \in \mathcal{B}_K$. (This is the case if $K \triangleleft \mathcal{B}_K$).

Let $\mathring{K} := \{\mathring{\tau} : \tau \in K\}$, define $\mathring{\mathcal{N}}_K \subseteq M^1(\mathfrak{g})$ and $\mathring{\mathcal{B}}_K$ in an analogous manner. Then again semistability of $(\mathcal{N}_K, \mathcal{B}_K)$ is transformed into operator- semistability on $(\mathring{\mathcal{N}}_K, \mathring{\mathcal{B}}_K)$.

So the investigation of stability on locally compact groups resp. semistability on Lie groups leads to the investigation of strict operator stability resp. operator semistability on $\mathbb{R}^d$ with certain restrictions on the admissible operators.

Remark: The measures on the vector space are not supposed to be full in the sense of Sharpe!

§ 3 The structure of $\mathfrak{z}((\mu_t))$.

Let $X, \mathcal{N}, *, \mathcal{B}$ be defined as in § 1.
Let $(\mu_t)_{t \geq 0}$ be a continuous convolution semigroup.
Let $\mathfrak{z}, \tilde{\mathfrak{z}}, \mathfrak{y}, \tilde{\mathfrak{y}}, \varphi, \tilde{\varphi}$ be defined as in § 1:

$\mathfrak{z} = \{\tau \in \mathcal{B}:$ there exists $c = c(\tau) \in \mathbb{R}_+^*$, such that $\tau\mu_t = \mu_{ct}, t \geq 0\}$.

$\tilde{\mathfrak{z}} = \{(\tau, c) : \tau\mu_t = \mu_{ct}, t \geq 0\} \subseteq \mathcal{B} \otimes \mathbb{R}_+^*$,

$\mathfrak{y} = \{\tau \in \mathcal{B} : \tau\mu_t = \mu_t, t \geq 0\}$ and $\tilde{\mathfrak{y}} = \{(\tau, 1) : \tau \in \mathfrak{y}\}$,

$\tilde{\varphi} : \tilde{\mathfrak{z}} \ni (\tau, c) \mapsto c, \quad \varphi : \mathfrak{z} \ni \tau \mapsto c$.

For probabilities on $\mathbb{R}^d$ M. Sharpe introduced the concept of <u>full</u> measures, i.e. probability measures, which are not concentrated on hyperplanes of $\mathbb{R}^d$. If strict (semi-) stability is considered, it is sufficient only to suppose, that μ is not concentrated on a proper subspace. (See also W. Linde and G. Siegel [18]). This weaker fullness condition is, as easily seen, equivalent to the compactness of $\{\tau \in \mathrm{Gl}(\mathbb{R}^d) : \tau\mu = \mu\}$. Since for convolution semigroups (μ_t) on a vector space the invariance group of a single measure $\{\tau : \tau\mu_t = \mu_t\}$ is independent of $t > 0$, we obtain: $\mu_t, t > 0$, is not concentrated on a subspace iff $\mathfrak{y}((\mu_t))$ is compact.

<u>3.1 Definition.</u> (See also [6]). A <u>convolution semigroup</u> $(\mu_t)_{t \geq o} \subseteq \mathcal{N}$ is called <u>full</u> (w.r.t $\mathfrak{B}$) if $\mathcal{G}((\mu_t))$ is a compact subgroup of $\mathfrak{Z}$ (and hence of $\mathfrak{B}$).

In the sequel we will work out the structure of $\mathfrak{Z}$ under the assumption of fullness.

<u>3.2 Proposition.</u> Assume $\mathfrak{B}$ to be locally compact and second countable. If $S = \varphi(\mathfrak{Z}) = \mathbb{R}_+^*$ then there exists a continuous one-parameter group $T = (\tau_t)_{t > o}$ in $\mathfrak{Z}$, such that $(\tau_t, t) \in \tilde{\mathfrak{Z}}$, $t > o$.

I.e. $(\mu_s)_{s \geq o}$ is stable w.r.t. (τ_t).

<u>Proof:</u> $\mathfrak{Z}$ is a locally compact second countable group, hence $\mathfrak{Z}/\mathfrak{Z}_o$ - $\mathfrak{Z}_o$ denoting the identity component - is countable.
Therefore $\varphi : \mathfrak{Z}_o \to \mathbb{R}_+^*$ is a continuous surjective homomorphism of the connected locally compact group $\mathfrak{Z}_o$ onto the (connected Lie) group $\mathbb{R}_+^*$.
Therefore there exists a continuous one - parameter group $(\tau_t)_{t > o}$ in $\mathfrak{Z}_o$, such that $\varphi : (\tau_t) \to \mathbb{R}_+^*$ is bijective. □

<u>3.3 Definition.</u> Let $(\mu_t)_{t \geq o}$ be stable with respect to some group $(\tau_t)_{t > o}$. Then (τ_t) is called an "exponent" of $(\mu_t)_{t \geq o}$.

[In the case of operator - stable laws on $\mathbb{R}^d$ τ_t is represented infinitesimally, i.e. there exists a an uniquely determined operator B on $\mathbb{R}^d$, such that $\tau_t = t^B$ $(:= \exp(\log t \cdot B))$. B is called an "exponent of μ_t" then. See e.g. [22, 10-14, 19, 20, 25].

<u>3.4 Proposition.</u> Let $T = (\tau_t)_{t > o}$ be an exponent of the stable semigroup $(\mu_t)_{t > o}$. Let $(\sigma_t)_{t > o}$ be a continuous one - parameter group in $\mathfrak{B}$. Then we have:

(σ_t) is an exponent iff $\sigma_s^{-1}\tau_s \in \mathcal{G}((\mu_t))$ for any $s > o$.
(Obviously $\sigma_s^{-1}\tau_s \in \mathcal{G}_o$, then).

Especially we obtain:
$\mathfrak{Z}$ is a semidirect product of $T (\cong \mathbb{R}_+^*)$ and $\mathcal{G}$.

<u>Proof:</u> The first assertion follows obviously from

$\sigma_s\mu_t = \sigma_s(\sigma_s^{-1}\tau_s)(\mu_t)$:

If δ_s is an exponent, then $\delta_s^{-1}(\mu_{ts}) = \mu_t$, hence $\delta_s^{-1}\tau_s(\mu_t) = \mu_t$, $t > o$.
If $\delta_s^{-1}\tau_s \in \mathcal{Y}$, then $\delta_s\mu_t = \tau_s\mu_t$.
The second assertion follows from the first one, since $T \cap \mathcal{Y} = \{id\}$ by definition and since $\mathcal{Y} \lhd \mathfrak{Z}$. □

<u>3.5 Remark.</u> If $\mathcal{B}$ is locally compact, especially if $\mathcal{B}$ is a Lie group, then the "exponents" are identified with elements of the tangent space of $\mathfrak{Z}$ (as mentioned in 3.3 for $\mathbb{R}^d$).
In this case 3.4. reads as follows:

The tangent space of $\mathfrak{Z}$ is the semidirect product of $\mathbb{R}$ and $\dot{\mathcal{Y}}$, where $\dot{\mathcal{Y}}$ is the tangent space of the invariance group $\mathcal{Y}$.

[In the case of operator stable laws on $\mathbb{R}^d$ this result is well known, see [14]. See also [5,6] for the case of locally compact groups.]

<u>3.6 Theorem.</u> Let $(\mu_t)_{t \geq o}$ be a full convolution semigroup, which is stable with respect to some group $T = (\tau_t) \subseteq \mathcal{B}$. Then there exists an "exponent" $\overline{T} = (\overline{\tau}_t)$ of (μ_t) which commutes with $\mathcal{Y}$, i.e. $\overline{\tau}_t\rho = \rho\overline{\tau}_t$ for $t > o$, $\rho \in \mathcal{Y}$.

<u>Proof:</u> $\mathfrak{Z}$ is isomorphic to a semidirect product of $\mathbb{R}_+^*$ and the compact group $\mathcal{Y}$ hence $\mathfrak{Z}$ is locally compact. Following the splitting theorem of [9] Proposition 9.4, we obtain the existence of a one-parameter group $\overline{T} = (\overline{\tau}_t)$ such that $\mathfrak{Z}$, the semidirect product of T and $\mathcal{Y}$ is the direct product of $\overline{T}$ and $\mathcal{Y}$.
Hence the assertion. □

<u>3.7 Remark.</u> For operator stable laws on $\mathbb{R}^d$ an equivalent result is proved in a direct way by W.N. Hudson, Z.J. Jurek and J.A. Veeh [14] Theorem 2. See also [5,6].

<u>3.8 Remark.</u> The fact, that the automorphisms $\overline{\tau}_t$ commute with the action of K has strong impact on the structure of $\overline{\tau}_t$.
If K is large, the variety of "commuting exponents" becomes small.

In the case of operator stable laws on $\mathbb{R}^d$ we obtain: If K is similar to SO(d), i.e. if (μ_t) is elliptically symmetric, then by Schur's Lemma commuting exponents are of the form $\overline{\tau}_t = t^\alpha id$ for some $\alpha \in \mathbb{R}$.
(See e.g. [20], [10].)

Indeed, even the assumption that τ_t is in the normalizer of K, has impact on the structure of τ_t. For stable probabilities on certain nilpotent groups see [7] , § 3.

<u>3.9</u> Now let the full convolution semigroup $(\mu_t)_{t\geq o}$ be semistable, i.e. assume $\mathfrak{z}((\mu_t)) \neq \mathfrak{y}((\mu_t))$. Put $S := \varphi(\mathfrak{z}) \subseteq \mathbb{R}_+^*$.
If $\mathfrak{B}$ is locally compact and second countable, and if (μ_t) is not stable, then according to 3.2 S is a proper subgroup of $\mathbb{R}_+^*$.
Hence the connected component $\mathfrak{z}_o$ is contained in $\mathfrak{y}$.

If $\varphi : \mathfrak{z} \to \mathbb{R}_+^*$ resp. $\tilde{\varphi} : \tilde{\mathfrak{z}} \to \mathbb{R}_+^*$ are closed, S must be discrete and hence of the form $\{c^k : k \in \mathbb{Z}, c \neq 1\}$.
In the case of (full) operator semistable laws on $\mathbb{R}^d$, the compactness of $\mathfrak{y}$ implies the validity of Sharpe's compactness lemma resp. the convergence of type theorem [22] , whence the closedness of φ follows, (cf. also [19] [25]). For the validity of the compactness lemma for probabilities on (nilpotent) Lie groups see [0] [1] [21].

In the general situation, if we assume $\mathfrak{B}$ to be a Lie group, it can only be shown that S is finitely generated.

<u>3.10</u> Fix $c \neq 1$ in $S = \varphi(\mathfrak{z})$ and $\tau \in \mathfrak{z}$ with $\varphi(\tau) = c$.
Let $S_c := \{c^k : k \in \mathbb{Z}\}$ then $T = \{\tau^k\}$ is a discrete subgroup of $\mathfrak{z}$.
Let $\mathfrak{z}_c := \{\sigma \in \mathfrak{z} : \varphi(\sigma) \in S_c\}$.
We restrict our considerations to $\mathfrak{z}_c$:

<u>Proposition</u> $\mathfrak{z}_c$ is the semidirect product of T and $\mathfrak{y}$. Especially : $\varphi(\sigma) = c$ iff $\sigma = \tau\varkappa$ for some $\varkappa \in \mathfrak{y}$.

<u>Proof:</u> Let $\tau\mu_t = \mu_{tc}$, $t > o$. Then for $\varkappa \in \mathfrak{y}$ we have $\tau\varkappa\mu_t = \mu_{tc}$.
Conversely, let $\tau\mu_t = \mu_{tc}$, $\sigma\mu_t = \mu_{tc}$, $t \geq o$.
Then $\tau^{-1}\mu_s = \mu_{s/c}$, $s \geq o$ and therefore $\tau^{-1}\sigma\mu_t = \mu_t$, $t \geq o$,
i.e. $\tau^{-1}\sigma \in \mathfrak{y}$. Since $\mathfrak{y} \triangleleft \mathfrak{z}_c$ and $\mathfrak{y} \cap T = \{\tau^o\} = id$, $\mathfrak{z}_c$ is the semidirect product of $\mathfrak{y}$ and T. □

<u>Remark 1:</u> The compactness of $\mathfrak{y}$ is not used in the proof.

<u>Remark 2:</u> In contrast to the case of stable semigroups it is not possible to proof the existence of "commuting" automorphisms, i.e. of $\bar{\tau} \in \mathfrak{z}_c$, such that $\bar{\tau}\varkappa = \varkappa\bar{\tau}$ for all $\varkappa \in \mathfrak{y}$.

Only particular results are possible: E.g. let $\mathfrak{G}$ be a (compact) Lie group and $\tau \in \mathfrak{z}_c$. Since $\mathfrak{J} \triangleleft \mathfrak{z}_c$ τ acts on $\mathfrak{J}$.
Then there exist $n \in \mathbb{N}$, such that τ^n acts only on the center of $\mathfrak{G}$.
Especially, if $\mathfrak{G}$ is connected and semisimple, then - $[\mathrm{Aut}(\mathfrak{J}) : \mathrm{Int}(\mathfrak{J})] = n$ being finite - τ^n acts trivially on $\mathfrak{J}$.
Hence $\tau^n \varkappa = \varkappa \tau^n$ for any $\varkappa \in \mathfrak{G}$.

<u>Remark 3:</u> The set of all "commuting" automorphisms $\mathfrak{z}_c^{comm} := \{\sigma \in \mathfrak{z}_c : \sigma\varkappa = \varkappa\sigma$ for all $\varkappa \in \mathfrak{J}\}$ is easily described:
If there exists $\tau \in \mathfrak{z}_c^{comm}$ with $\varphi(\tau) = c$ then $\mathfrak{z}_c^{comm} = \tau \cdot Z(\mathfrak{J})$, $Z(\mathfrak{J})$ being the center of $\mathfrak{G}$. (For stable measures on $\mathbb{R}^d$ see [14]).
This follows immediately from Proposition 3.10.

References

[0] P. Baldi: Lois stables sur les déplacements de $\mathbb{R}^d$. In: Probability measures on groups. Proceedings Oberwolfach (1978). Lecture Notes in Math. 706, 1 - 9. Springer (1979).

[1] T. Drisch, L. Gallardo: Stable laws on the Heisenberg group. In: Probability measures on groups. Proceedings Oberwolfach (1983). Lecture Notes Math. 1064, 56 - 79 (1984).

[2] W. Hazod: Stable probabilities on locally compact groups. In: Probability measures on groups. Proceedings Oberwolfach (1981). Lecture Notes Math. 928, 183 - 211 (1982).

[3] W. Hazod: Remarks on [semi-] stable probabilities. In: Probability measures on groups. Proceedings Oberwolfach (1983). Lecture Notes Math. 1064, 182 - 203 (1984).

[4] W. Hazod: Stable and semistable probabilities on groups and vector spaces. In: Probability theory on vector spaces III. Proceedings Lublin (1983). Lecture Notes Math. 1080, 69 - 89 (1984).

[5] W. Hazod: Semigroupes de convolution [demi-] stable et autodécomposables sur les groupes localement compacts. In: Probabilitiés sur les structures géometriques. Actes des Journees Toulouse (1984). Publ. du Lab. Stat. et Prob. Université de Toulouse, 57 - 85 (1985).

[6] W. Hazod: Stable probability measures on groups and on vector spaces. A survey. In: Probability measures on Groups VIII. Proceedings, Oberwolfach (1985). Lecture Notes Math. 1210 304 - 352 (1986).

[7] W. Hazod, E. Siebert: Continuous automorphism groups on a locally compact group contracting modulo a compact subgroup and applications to stable convolution semigroups. Semigroup Forum 33, 111 - 143 (1986).

[8] W. Hazod, E. Siebert: Automorphisms on a Lie group contracting modulo a compact subgroup and applications to semistable convolution semigroups. To appear in: J. of Theoretical Probability.

[9] K.H. Hofmann, P. Mostert: Splitting in topological groups. Mem. Amer. Math. Soc. 43 (1963).

[10] J.P. Holmes, W.N. Hudson, J.D. Mason: Operator - stable laws: multiple exponents and elliptical symmetry. Ann. Probab. 10, 602 - 612 (1982).

[11] W.N. Hudson: Operator - stable distributions and stable marginals. J. Mult. Analysis 10, 26 - 37 (1980).

[12] W.N. Hudson, J.D. Mason: Exponents of operator stable laws. In: Probability in Banach spaces III. Proceedings Medford (1980). Lecture Notes in Math. 860, 291 - 298 (1981).

[13] W.N. Hudson, J.D. Mason: Operator stable laws. J. Mult. Analysis 11, 434 - 447 (1981).

[14] W.N. Hudson, Z.J. Jurek, J.A. Veeh: The symmetry group and exponents of operator stable probability measures. Ann. of Probability 14, 1014 - 1023 (1986).

[15] R. Jajte: Semistable probability measures on $\mathbb{R}^N$. Studia Math. 61, 29 - 39 (1977).

[16] Z.J. Jurek: On stability of probability measures in Euclidean spaces. In: Probability theory on vector spaces II. Proceedings Błażejewko (1979). Lecture Notes Math. 828, 129 - 145 (1980).

[17] Z.J. Jurek: Convergence of types, self-decomposibility and stability of measures on linear spaces. In: Probability in Banach spaces III. Proceedings Medford (1980). Lecture Notes in Math. 860, 257 - 284 (1981).

[18] W. Linde, G. Siegel: On the convergence of types for Radon probability measures in Banach spaces. To appear.

[19] A. Łuczak: Operator semi-stable probability measures on $\mathbb{R}^N$. Coll. Math. 45, 287 - 299 (1981).

[20] A. Łuczak: Elliptical symmetry and characterization of operator-stable and operator semi-stable measures. Ann. Probab. 12, 1217 - 1223 (1984).

[21] S. Nobel: Ph.D. Thesis. Univ. Dortmund. In preparation.

[22] M. Sharpe: Operator stable probability measures on vector groups. Trans. Amer. Math. Soc. 136, 51 - 65 (1969).

[23] K. Sato: Strictly operator-stable distributions. Nagoya Univ. Preprint 1985.

[24] E. Siebert: Semistable convolution semigroups on measurable and topological groups. Ann. Inst. H. Poincaré 20, 147 - 164 (1984).

[25] E. Siebert: Supplements to operator-stable and operator-semistable laws on Euclidean spaces. J. Mult. Analysis 19, 329 - 341 (1986).

[26] K. Urbanik: Lévy's probability measures on Euclidean space Studia Math. 44, 119 - 148 (1972).

Probability Theory on Vector Spaces IV
Lancut, June'87, Springer's LNM 1391

STRONG LAWS OF LARGE NUMBERS FOR ORTHOGONAL SEQUENCES IN VON NEUMANN ALGEBRAS

by

Ewa Hensz

1. There are many important theorems in probability theory concerning the almost sure convergence of series of pairwise orthogonal random variables in L_2. The fundamental result in this theory is the celebrated Rademacher-Menshov theorem. A simple consequence (via the Kronecker lemma) of this theorem is a strong law of large numbers; in other words, the theorem about the almost sure convergence of the first Cesàro means of random variables (c.f. [6]).

Recently, in [4] Moricz has shown the almost sure convergence of the second Cesàro means, obviously, under the weaker conditions.

This result can be evidently threated as a strong law of large numbers.

On the other hand, a great deal of work has been done lately to genralize various strong limit theorems to the von Neumann algebras context (c.f. [3]). In particular, in [2] the non-commutative version of the Rademacher-Manshov theorem and some of its consequences (among them the theorem about the convergence of the first means) have been proved.

The aim of this paper is twofold. Firstly, we give an extension of Moricz's law of large numbers to the case of L_2 over a von Neumann algebra with a faithful normal state. Secondly, we prove some strong laws of large numbers being other consequences of the Rademacher-Menshov theorem in this case. The classical versions of these theorems can be found in [5].

Our paper is closely connected with the papers [2,1] where the basic definitions have been introduced. However, for the sake of completeness, we begin with some notations and definitions.

2. Throughout the paper, M is a σ-finite von Neumann algebra with a faithful normal state Φ. Proj M will denote the lattice of all selfadjoint projections in M. We assume that M acts, in a standard way, on the Hilbert space $H = L_2(M,\Phi)$ which is the comletion of M under the norm $x \to \Phi(x^*x)^{1/2}$ with a cylic and separating vector ξ_o such that $\Phi(x) = (x\xi_o,\xi_o)$ for $x \in M$ (cf. [7]). We shall identify M with the subset $M\,\xi_o = \{x\,\xi_o : x \in M\}$ of H. The norm in H will be denoted by $\|\,.\,\|$, and the norm in M by $\|\,.\,\|_\infty$.

For a $\xi \in H$ and a $p \in$ Proj M we set

$$S_{\xi,p} = \{(x_k) \quad M : \sum_{k=1}^{\infty} x_k\,\xi_o = \xi \quad \text{in } H$$

$$\text{and } \sum_{k=1}^{\infty} x_k\, p \text{ converges in norm in } M\}$$

and

$$\|\,\xi\,\|_p = \inf\,\{\|\sum_{k=1}^{\infty} x_k p\|_\infty : (x_k) \in S_{\xi,p}\}$$

with $\inf \emptyset = \infty$.

Obviously, for all $\xi, \eta \in H$, we have

$$\| \xi + \eta \|_p \leq \| \xi \|_p + \| \eta \|_p$$

and for $x \in M$

$$\| x \xi_o \|_p \leq \| xp \|_\infty .$$

A sequence (ξ_n) in $H = L_2(M,\Phi)$ is said to be almost surely (a.s.) convergent to $\xi \in H$ if for every $\varepsilon > 0$, there exists a $p \in \text{Proj } M$ such that $\Phi(1-p) < \varepsilon$ and $\| \xi_n - \xi \|_p \to 0$ as $n \to \infty$.

It is easily seen that in the classical case of $M = L_\infty$ (over a probability space) the convergence just given coincides with the usual almost everywhere convergence (via Egorov theorem).

3. The following lemma is a basic tool in our consideration.

<u>LEMMA 1</u> ([2]). Let (D_n) be a sequence of positive operators from M, and (ε_n) be a sequence of positive numbers. Let (α_n) be a sequence in H. Then there exists a $p \in \text{Proj } M$ such that

$$\Phi(1-p) \leq 4 \sum_{n=1}^{\infty} \varepsilon_n^{-1} (\Phi(D_n) + \| \alpha_n \|^2)$$

and

$$\| {}_p D_n p \|_\infty \leq 2 \varepsilon_n$$

$$\| \alpha_n \|_p \leq 5 \varepsilon_n^{1/2}, \qquad n = 1,2,\ldots .$$

We also need following three small lemmas.

LEMMA 2 ([2]). Let $\alpha_1,\ldots,\alpha_N$ be complex numbers, and $x_1,\ldots,x_N \in M$. Then

$$\left| \sum_{i=1}^{N} \alpha_i x_i \right|^2 \leq \left(\sum_{i=1}^{N} |\alpha_i|^2 \right) \left(\sum_{i=1}^{N} |x_i|^2 \right)$$

where $|x| = (x^* x)^{1/2}$ for $x \in M$.

LEMMA 3 ([3]). Let $\alpha_1,\ldots,\alpha_N$ be complex numbers, and $\xi_1,\ldots,\xi_N \in H$ (H denotes here an arbitrary Hilbert space). Then

$$\left\| \sum_{i=1}^{N} \alpha_i \xi_i \right\|^2 \leq \left(\sum_{i=1}^{N} |\alpha_i|^2 \right) \left(\sum_{i=1}^{N} \|\xi_i\|^2 \right).$$

LEMMA 4. Let X be an arbitrary linear space, and $(\eta_j) \subset X$. For all $m > n$ we have

$$\frac{1}{m} \sum_{k=1}^{m} \left(1 - \frac{k-1}{m}\right) \eta_k - \frac{1}{n} \sum_{k=1}^{n} \left(1 - \frac{k-1}{n}\right) \eta_k$$

$$= \frac{m^2-n^2}{m^2 n^2} \sum_{k=1}^{n} (k-1)\eta_k - \frac{m-n}{mn} \sum_{k=1}^{n} \eta_k + \frac{1}{m^2} \sum_{k=n+1}^{m} (m-k+1)\eta_k .$$

We shall need the following version of Kronecker's lemma.

LEMMA 5. Let $b_n \uparrow \infty$ be real numbers and $(\xi_k) \subset H$ with $\sum_{k=1}^{n} \xi_k \to \eta$ a.s. Then $\frac{1}{b_n} \sum_{k=1}^{n} b_k \xi_k \to 0$ a.s.

P r o o f. It is analogous to the Kronecker lemma in [2].

Finally, for the sake of completeness it is worth reminding the extension of the Rademacher-Menshov theorem.

THEOREM 0 ([2]). Let (ξ_n) be an orthogonal sequence in $H = L_2(M,\Phi)$ such that

$$\sum_{n=1}^{\infty} (\log n)^2 \, \|\xi_n\|^2 < \infty.$$

Then $\sigma_n = \sum_{k=1}^{n} \xi_k$ converges almost surely to σ where σ is the sum of the series $\sum_{k=1}^{\infty} \xi_k$ in H.

4. Now, we extend Moricz's law of large numbers.

THEOREM 1. Let (ξ_n) be an orthogonal sequence in $H = L_2(M,\Phi)$ such that

$$\sum_{k=1}^{\infty} \frac{1}{k^2} \, \|\xi_k\|^2 < \infty.$$

Then the Cesàro means

$$\tau_n = \frac{1}{n} \sum_{k=1}^{n} \left(1 - \frac{k-1}{n}\right) \xi_k \to 0 \qquad \text{a.s.}$$

P r o o f. We are going to follow the general idea of Moricz, so we do it in five steps. The first two steps are the same like in the classical case.

(i) First, we show that

$$\sum_{n=0}^{\infty} \|\zeta_{2^n}\|^2 < \infty \tag{1}$$

where

$$\zeta_n = \frac{1}{n} \sum_{k=1}^{n} \xi_k, \qquad n = 1,2,\ldots .$$

In fact, by orthogonality, we have

$$\|\zeta_n\|^2 = \frac{1}{n^2} \sum_{k=1}^{n} \|\xi_k\|^2.$$

Thus,

$$\begin{aligned} \sum_{n=0}^{\infty} \|\zeta_{2^n}\|^2 &= \sum_{n=0}^{\infty} 2^{-2n} \sum_{k=1}^{2^n} \|\xi_k\|^2 \\ &= \sum_{k=1}^{\infty} \|\xi_k\|^2 \sum_{\{n:2^n \geq k\}} 2^{-2n} \\ &= O(1) \sum_{k=1}^{\infty} \frac{1}{k^2} \|\xi_k\|^2 < \infty. \end{aligned}$$

(ii) Now, we prove that

$$\sum_{n=0}^{\infty} \|\zeta_{2^n} - \tau_{2^n}\|^2 < \infty. \tag{2}$$

In fact, we have

$$\zeta_n - \tau_n = \frac{1}{n^2} \sum_{k=1}^{n} (k - 1)\, \xi_k .$$

Thus,

$$\sum_{n=0}^{\infty} \| \xi_{2^n} - \tau_{2^n} \|^2$$

$$= \sum_{n=0}^{\infty} 2^{-4n} \sum_{k=2}^{2^n} (k-1)^2 \| \xi_k \|^2$$

$$= \sum_{k=2}^{\infty} (k-1)^2 \| \xi_k \|^2 \sum_{\{n:2^n \geq k\}} 4^{-2n}$$

$$= 0(1) \sum_{k=1}^{\infty} \frac{1}{k^2} \| \xi_k \|^2 < \infty.$$

By (1) and (2) we obtain

$$(3) \qquad \sum_{n=0}^{\infty} \| \tau_{2^n} \|^2 < \infty.$$

(iii) Now, we approximate the vectors ξ_i $(i=1,2,\ldots)$ by the "almost orthogonal" operators $x_i \in M$. Namely, we choose $x_i \in M$ such that $\| \eta_i \| < \delta_i$ where $\eta_i = \xi_i - x_i \xi_o$.

The sequence (δ_i) denotes here a sequence of positive numbers with $\delta_i < 2^{-i}$ and

$$\sum_{k=2^n+1}^{2^{n+1}} (2^{n+1} - k + 1)\delta_k^2 = 0(1), \qquad n=1,2,\ldots .$$

Putting for $n=1,2,\ldots$, $t_n = \frac{1}{n} \sum_{k=1}^{n} (1 - \frac{k-1}{n})\, x_k$.

we shall estimate the operators $t_m - t_{2^n}$ for $2^n < m \leq 2^{n+1}$.

Writting $t_m - t_{2^n}$ on the form

$$t_m - t_{2^n} = \sum_{j=2^n+1}^{m} (t_j - t_{j-1})$$

by Lemma 2, we get

$$|t_m - t_{2^n}|^2 \leq (m - 2^n) \sum_{j=2^n+1}^{m} |t_j - t_{j-1}|^2$$
$$\leq 2^n \sum_{j=2^n+1}^{2^n} |t_j - t_{j-1}|^2.$$

Setting for $n = 0,1,\ldots$

$$D_n = 2^n \sum_{j=2^n+1}^{2^{n+1}} |t_j - t_{j-1}|^2$$

we obtain $D_n \in M$, $D_n \geq 0$ and

(4) $$|t_m - t_{2^n}|^2 \leq D_n \qquad \text{for } 2^n < m \leq 2^{n+1}.$$

We shall prove that

(5) $$\sum_{n=0}^{\infty} \Phi(D_n) < \infty.$$

By the representation

$$t_j - t_{j-1} = \sum_{k=1}^{j} \left(\frac{(k-1)(2j-1)}{j^2(j-1)^2} - \frac{1}{j(j-1)}\right) x_k, \quad j \geq 2$$

and, by Lemma 2, we obtain

$$|t_j - t_{j-1}|^2 \le j \sum_{k=1}^{j} \left(\frac{(k-1)^2(2j-1)^2}{j^4(j-1)^4} + \frac{1}{j^2(j-1)^2}\right) |x_k|^2$$

$$\le 5 \sum_{k=1}^{j} \frac{1}{j(j-1)^2} |x_k|^2.$$

Hence, we have

$$\Phi(D_n) = 2^n \sum_{j=2^n+1}^{2^{n+1}} \Phi(|t_j - t_{j-1}|^2)$$

$$\le 5 \cdot 2^n \sum_{j=2^n+1}^{2^{n+1}} \sum_{k=1}^{j} \frac{1}{j(j-1)^2} \Phi(|x_k|^2)$$

$$\le 5 \cdot 2^n \sum_{k=1}^{2^{n+1}} \Phi(|x_k|)^2 \sum_{j=2^n+1}^{2^{n+1}} \frac{1}{j(j-1)^2}$$

$$\le \frac{10}{(2^n+1)^2} \sum_{k=1}^{2^{n+1}} \Phi(|x_k|^2).$$

Finally, we get

$$\sum_{n=0}^{\infty} \Phi(D_n) \le 10 \sum_{n=0}^{\infty} \frac{1}{(2^n+1)^2} \sum_{k=1}^{2^{n+1}} \Phi(|x_k|^2)$$

$$= 10 \sum_{k=1}^{\infty} \Phi(|x_k|^2 \sum_{\{n: 2^{n+1} \ge k\}} \frac{1}{(2^n+1)^2}$$

$$= 0\ 1 \sum_{k=1}^{\infty} \frac{1}{k^2} \Phi(|x_k|^2).$$

But

$$\Phi(|x_k|^2) = \| x_k \xi_o \|^2 \le 2 \| \xi_k \|^2 + 2 \| \eta_k \|^2$$

$$\le 2 \| \xi_k \|^2 + 2 \delta_k^2.$$

Thus, taking into account the assumption and the rate of convergence of the sequence (δ_k) to zero, we get (5).

(iv) Now, we show that

$$(6) \qquad \sum_{m=1}^{\infty} \| \tau_m - t_m \xi_o - \tau_{2^n} + t_{2^n} \xi_o \|^2 < \infty .$$

(Here $n = n(m)$ is uniquely determined by m via the condition $2^n < m \leq 2^{n+1}$).

In fact, by Lemma 4, we obtain

$$\tau_m - t_m \xi_o - \tau_{2^n} + t_{2^m} \xi_o$$

$$= \frac{m^2 - 2^{2n}}{m^2\ 2^{2n}} \sum_{k=1}^{2^n} (k-1)\eta_k - \frac{m-2^n}{m\ 2^n} \sum_{k=1}^{2^n} \eta_k$$

$$+ \frac{1}{m^2} \sum_{k=2^n+1}^{m} (m-k+1)\eta_k \qquad A_m - B_m + C_m .$$

Estimating the summands one after the other using Lemma 3 we obtain

$$\| A_m \|^2 \leq \frac{9}{2^{4n}} \| \sum_{k=1}^{2^n} (k-1)\eta_k \|^2$$

$$\leq \frac{9}{2^{4n}} \sum_{k=1}^{2^n} \frac{1}{k^2} \sum_{k=1}^{2^n} k^2 (k-1)^2 \delta_k^2 \leq O(1) m^{-4}$$

$$\| B_m \|^2 \leq \frac{1}{2^{2n}} \| \sum_{k=1}^{2^n} \eta_k \|^2$$

$$\leq \frac{1}{2^{2n}} \sum_{k=1}^{2^n} \frac{1}{k^2} \sum_{k=1}^{2^n} k^2 \delta_k^2 \leq O(1) m^{-2}$$

and, finally,

$$\|C_m\|^2 \le \frac{1}{m^4} \Big\| \sum_{k=2^n+1}^{m} (m-k+1)\eta_k \Big\|^2$$

$$\le \frac{1}{m^3} \sum_{k=2^n+1}^{m} (m-k+1)^2 \, \delta_k^2$$

$$\le \frac{1}{m^3} \sum_{k=2^n+1}^{2^{n+1}} (2^{n+1}-k+1)^2 \, \delta_k^2 \le O(1)m^{-3}.$$

Thus, (6) holds.

(v) Let us arrange the sequences $\{\tau_{2^m}\}$ and $\{\tau_m - t_m \xi_o - \tau_{2^{n(m)}} + t_{2^{n(m)}} \xi_o\}$ into one sequence $\{\alpha_m\}$ with

(7) $$\sum_{m=1}^{\infty} \|\alpha_m\|^2 < \infty.$$

Applying Lemma 1, by (5) and (7), for every $\varepsilon > 0$, we can find a projection $p \in \mathrm{Proj}\, M$ with $\Phi(1-p) < \varepsilon$ such that

$$\|p D_n p\| \to 0$$

(8) $$\|\tau_{2^n}\|_p \to 0, \qquad n \to \infty$$

$$\|\tau_m - t_m \xi_o - \tau_{2^{n(m)}} + t_{2^{n(m)}} \xi_o\|_p \to 0, \quad m \to \infty.$$

For $2^n < m \le 2^{n+1}$ we have the following estimation

$$\|\tau_m\|_p \le \|\tau_{2^n}\|_p + \|(t_m - t_{2^n})\xi_o\|_p$$

$$+ \|\tau_m - t_m \xi_o - \tau_{2^n} + t_{2^n} \xi_o\|_p.$$

Moreover, by 4

$$\|(t_m - t_{2^n})\xi_o\|_p \leq \|(t_m - t_{2^n})\, p\|_\infty$$

$$= \|p(t_m - t_{2^n})^2\, p\|_\infty^{1/2} \leq \|p\, D_n\, p\|_\infty^{1/2}.$$

To finish the proof it is enought to apply (8).

5. Applying the Kronecker lemma to the Rademacher-Menshov theorem, we can easy obtain the following laws of large numbers.

THEOREM 2. Let (ξ_n) be an orthogonal sequence in $H = L_2(M,\Phi)$ such that

$$(9) \qquad \sum_{k=1}^{\infty} \|\xi_k\|^2 < \infty.$$

Then $u_n = \frac{1}{\log n} \sum_{k=1}^{n} \xi_k \to 0$ a.s.

P r o o f. By (9), Theorem 0 implies that the series

$$\sum_{k=1}^{\infty} \frac{1}{\log k}\, \xi_k$$

converges a.s. whence Lemma 5 yields $u_n \to 0$ a.s.

Similarly, we get

THEOREM 3. Let (ξ_n) be an orthogonal sequence in $H = L_2(M,\Phi)$ such that

$$(10) \qquad \sum_{k=1}^{\infty} \frac{1}{k^2} \| \xi_k \|^2 < \infty.$$

Then $w_n = \frac{1}{n \log n} \sum_{k=1}^{n} x_k \to 0$ a.s.

REFERENCES

[1] Hensz E.: On a Weyl theorem in von Neumann algebras, Bull. PAN, Math. Vol. 35, No 3-4, (1987), 193-201.

[2] Hensz E., Jajte R.: Pointwise convergence theorems in L_2 over a von Neumann algebra, Math. Z. 193, (1986), 413-429.

[3] Jajte R.: Strong limit theorems in non-commutative probability, Lect. Notes in Math., No 1110, Berlin-Heidelberg-New York-Tokyo; Springer 1985.

[4] Moricz F: On the Cesaro means of orthogonal sequences of random variables, Annals of Prob. 11 (1983), No. 3, 827-832.

[5] Moricz F., Tandori K.: Counterexamples in the theory of orthogonal series, Acta Math. Hung. 49 (1-2), 1987, 283-290.

[6] Révész P.: The laws of large numbers, Akademiai Kiado, Budapest, 1967.

[7] Takesaki M.: Theory of operator algebras I, Berlin-Heidelberg-New York; Springer 1979.

Institute of Mathematics
University of Łódź
ul. Banacha 22, 90-238 Łódź,
POLAND

Probability Theory on Vector Spaces IV
Lancut, June '87, Springer's LNM 1391

STRONG LAWS OF LARGE NUMBERS FOR SEVERAL CONTRACTIONS IN A VON NEUMANN ALGEBRA

R. Jajte

Institute of Mathematics
Łódź University
90-238 Łódź, Poland

1. Introduction. Recently, many fundamental pointwise convergence theorems in probability and ergodic theory have been extended to the operator algebra context. In particular, a remarkable progress has been made in the individual ergodic theory of semigroups of some positive contractions in von Neumann algebras [1], [7], [10], [11], [13], [14], [15], [16].

The study of such semigroups is motivated by the theory of open (irreversible) quantum-dynamical systems. From the physical point of view the most important are semigroups of completely positive maps on C*- or W*-algebras [8] but in the context of this paper it seems to be more natural to consider a larger class of positive contractions. We shall discuss the asymptotic behaviour of Schwarz maps in von Neumann algebras. More exactly we are going to prove some strong laws of large numbers (individual ergodic theorems) for several contractions in L_2 over a von Neumann algebra M (with respect to a faithful normal state). These contractions in L_2 are generated by some Schwarz maps in M. The results presented here are related to [3], [4], [1], [9], [17].

The main result is in the spirit of Dunford-Schwartz-Zygmund type ergodic theorem for several kernels [17].

Let us begin with some notation and definitions. Throughout the paper, M will denote a σ-finite von Neumann algebra with a faithful normal state ϕ. We asume that M acts, in a standard way, on the Hilbert space $H = L_2(M,\phi)$ the completion of M under the norm $x \to \phi(x^*x)^{1/2}$ with a cyclic and separating vector Ω such that $\phi(x) = (x\Omega,\Omega)$, for $x \in M$. We shall identify M with the subset $M\Omega = \{x\Omega : x \in M\}$ of H. We put, for $x \in M$, $|x|^2 = x^*x$. We call $\alpha \in L(M)$ a Schwarz map if α satisfies the inequality $|\alpha(x)|^2 \leq \alpha(|x|^2)$, for $x \in M$. Note that α is necessarily a contraction. A map $\alpha \in L(M)$ is said to be ϕ-contractive if $\phi(\alpha x) \leq \phi(x)$, for all $x \in M_+$, where $M_+ = \{|z|^2 : z \in M\}$ is the positive cone in M. The norm in H will be denoted by $\|.\|$ and the norm in M by $\|.\|_\infty$.

For a $\xi \in H$ and a projection $p \in M$, we set

$$S_{\xi,p} = \{(x_k) \subset M : \sum_{k=1}^{\infty} x_k\Omega = \xi \text{ in } H \text{ and } \sum_{k=1}^{\infty} x_k p \text{ converges in norm in } M\}$$

and

$$\|\xi\|_p = \inf\{\|\sum_{k=1}^{\infty} x_k p\|_\infty : (x_k) \in S_{\xi,p}\}.$$

Obviously, for all $\xi,\eta \in H$, we have

$$\|\xi + \eta\|_p \leq \|\xi\|_p + \|\eta\|_p,$$

and, for $x \in M$

$$\|x\Omega\|_p \leq \|xp\|_\infty.$$

We adopt the following

1.1. Definition. A sequence (ξ_n) in $H = L_2(M,\phi)$ is said to be almost surely (a.s) convergent to $\xi \in H$ if for every $\varepsilon > 0$, there exists a projection $p \in M$ such that $\phi(1 - p) < \varepsilon$ and $\| \xi_n - \xi \|_p \to 0$ as $n \to \infty$.

In other words, $\xi_n \to 0$ a.s. if for every strong neighbourhood U of the unity in M, there is a projection $p \in U$ and a matrix $(x_{n,k})$ with entries in M such that $\sum_{k=1}^{\infty} x_{nk}\Omega = \xi$ in H and $\| \sum_{k=1}^{\infty} x_{nk}p \|_\infty \to 0$ as $n \to \infty$.

It is easily seen that in the classical commutative case of $M = L_\infty$ (over a probability space) the convergence just defined coincides with the usual almost everywhere convergence (via Egorov´s theorem). For the elements of the algebra M the following kind of convergence (introduced by E.C. Lance) is mostly used. A sequence $(x_n) \subset M$ is said to be almost uniformly convergent to $x \in M$ if, for every $\varepsilon > 0$, there exists a projection $p \in M$ with $\phi(1 - p) < \varepsilon$ such that $\|(x_n - x)p\|_\infty \to 0$ as $n \to \infty$. Obviously, the almost uniform convergence of (x_n) implies the almost sure convergence of $(x_n\Omega)$. Let us recall that M.S. Goldstein [2] uses the following notion of convergence in H. namely, for $\xi_n, \xi \in H$, $\xi_n \to \xi$ almost everywhere if, for every $\varepsilon > 0$ there exists a projection $p \in M$ and $(x_n) \subset M$ such that $\phi(1 - p) < \varepsilon$, $p(\xi_n - \xi) = x_n\Omega$ for n large enough, and $\|x_n\|_\infty \to 0$ as $n \to \infty$.

2. Let $\alpha \in L(M)$ be a ϕ-contractive Schwarz map: Put $\beta(x\Omega) = \alpha(x)\Omega$, for $x \in M$. Then, after a standard unique extension to H, we obtain a contraction β in H. We say that the contraction β is generated by the ϕ-contractive Schwarz map α in M.

Let us recall the following result of M.S. Goldstein [2] which will be frequently used.

2.1. Theorem [2]. Let α be a normal positive map of M such that

$$\phi(\alpha x) \leq \phi(x), \quad \text{for all } x \in M_+$$

and

$$\alpha 1 \leq 1.$$

Put

$$s_n(x) = n^{-1} \sum_{k=0}^{n-1} \alpha^k x.$$

Assume that a sequence (x_n) of positive elements from M and a sequence (ϵ_n) of positive numbers are given. Then there exists a projection $p \in M$ with

$$\phi(1 - p) \leq \sum_{n=1}^{\infty} \epsilon_n^{-1} \phi(x_n)$$

and such that

$$\| p s_m(x_n) p \|_\infty \leq 2\epsilon_n, \quad \text{for } n,m = 1,2,\dots .$$

In the sequel we shall consider several Schwarz maps $\alpha_1, \alpha_2, \dots, \alpha_k$ in M, preserving the state ϕ. From this assumption it follows in particular that all α_j's are positive normal maps.

2.2. Theorem. Let us put $s_{n_i} = n_i^{-1} \sum_{\nu=0}^{n_i - 1} \alpha_i^{\nu}$ $(i = 1,2,\dots,k)$.

Let $(x_m) \subset M_+$, $\epsilon_m > 0$, for $m = 1,2,\dots$. Then there exists a projection $p \in M$ such that we have

$$\phi(1 - p) \leq 2^{k+1} \sum_{m=1}^{\infty} \epsilon_m^{-k} \phi(x_m) \tag{1}$$

and

$$\sup_{1 \leq \nu \leq k} \| p s_{n_k}^{(k)} \dots s_{n_\nu}^{(\nu)}(x_m) p \|_\infty \leq (A_k + B_k \| x_m \|) \epsilon_m, \tag{2}$$

for some constants A_k, B_k depending only on k, and all m and all $n_1, n_2, \dots n_k$.

Proof. For $k = 1$ the result reduces to Theorem 2.1. Assume that (2) holds for $k - 1$. Putting $x_s = 0$ for $s \neq m$, we obtain

$$(3) \qquad \sup_{1 \le \nu \le k-1} \| p_m s_{n_{k-1}}^{(k-1)} \dots s_{n_\nu}^{(\nu)}(x_m) p_m \|_\infty \le (A_{k-1} + B_{k-1} \|x_m\|)\varepsilon_m$$

and

$$(4) \qquad \phi(1 - p_m) \le 2^k \varepsilon_m^{-k+1} \phi(x_m).$$

Applying Theorem 2.1 with $\alpha = \alpha_k$, $x_m = 1 - p_m$ and ε_m, we find a projection $p \in M$ such that

$$(5) \qquad \| p s_{n_k}^{(k)} (1 - p_m) p \|_\infty < 2\varepsilon_m$$

and

$$\phi(1 - p) \le 2 \sum_{m=1}^{\infty} \varepsilon_m^{-1} \phi(1 - p_m) \le 2^{k+1} \sum_{m=1}^{\infty} \varepsilon_m^{-k} \phi(x_m).$$

Using the inequality

$$z \le 2p_m z p_m + 2(1 - p_m) z (1 - p_m)$$

for $z = s_{n_{k-1}}^{(k-1)} \dots s_{n_\nu}^{(\nu)}(x_m) \in M_+$ (with $1 \le \nu \le k-1$), we obtain

$$p s_{n_k}^{(k)} \dots s_{n_\nu}^{(\nu)}(x_m) p \le$$

$$\le p s_{n_k}^{(k)} (2 p_m s_{n_{k-1}}^{(k-1)} \dots s_{n_\nu}^{(\nu)}(x_m) p_m +$$

$$+ 2(1 - p_m) s_{n_{k-1}}^{(k-1)} \dots s_{n_\nu}^{(\nu)}(x_m)(1 - p_m)) p \le$$

$$\le 2(A_{k-1} + B_{k-1}\|x_m\|)\varepsilon_m I + 2 p s_{n_k}^{(k)} (1 - p) p \|x_m\|.$$

Now, taking into account (5) it is enough to put

$$A_k = 2A_{k-1} \quad \text{and} \quad B_k = 2B_k + 4,$$

to obtain (2), which completes the proof.

Let us denote $\tilde{\alpha}_j$ the extension of α_j to a contraction in H via the formula

$$\tilde{\alpha}_j(x\Omega) = \alpha_j(x)\Omega, \quad \text{for} \quad x \in M \quad (j = 1,2,\ldots).$$

Let us put

$$\tilde{s}_{n_i}^{(i)} = n_i^{-1} \sum_{k=0}^{n_i-1} \tilde{\alpha}_i^k \qquad (i = 1,2,\ldots,k).$$

2.3. Theorem. Let $s_{n_i}^{(i)}$ and $\tilde{s}_{n_i}^{(i)}$ $(i = 1,2,\ldots,k)$ denote the averages as above. Then, for every $(\xi_n) \subset H$, every sequence (δ_n) of positive numbers and $(x_n) \subset M^+$, there exists a projection $p \in M$ and some positive constants A_k and B_k such that

$$\sup_{1 \le \nu \le k} \| p s_{n_k}^{(k)} \ldots s_{n_\nu}^{(\nu)}(x_m) p \| \le (A_k + B_k \| x_m \|)\delta_m$$

and

$$\sup_{1 \le \nu \le k} \| \tilde{s}_{n_k}^{(k)} \ldots \tilde{s}_{n_\nu}^{(\nu)}(\xi_m) \|_p < (A_k + B_k \| \xi_m \|)\delta_m^{1/2},$$

for all $m, n_1, n_2, \ldots, n_k$.

Morover, we have

$$\phi(1 - p) \le 2^{k+3} \sum_{m=1}^{\infty} \delta_m^{-1}(\phi(x_m) + \|\xi_m\|^2).$$

Proof. Let $(x_{m,r})$ be a matrix with entries in M and such that

$$\xi_m = \sum_{r=1}^{\infty} x_{m,r}\Omega, \quad \text{for} \quad m = 1,2,\ldots,$$

and

$$\| x_{mr}\Omega \| < 2^{-rk+1} \| \xi_m \|, \quad \text{for} \quad m,r = 1,2,\ldots .$$

Put $\varepsilon_{mr} = \delta_m 2^{-r}$ for $r \geq 1$ and $\varepsilon_{mo} = \delta_m$ and $y_{mr} = |x_{mr}|^2$ for $r \leq 1$ and $y_{mo} = x_m$. Then we have

$$\sum_{r=0}^{\infty} \sum_{m=1}^{\infty} \varepsilon_{mr}^{-k} \phi(y_{mr}) \leq \sum_{m=1}^{\infty} \delta_m^{-k} \phi(x_m) +$$

$$+ \sum_{r=1}^{\infty} \sum_{m=1}^{\infty} \delta_m^{-k} 2^{rk} 2^{-rk+2} \|\xi_m\|^2 \leq$$

$$\leq 4 \sum_{m=1}^{\infty} \delta_m^{-k} (\phi(x_m) + \|\xi_m\|^2).$$

By Theorem 2.2 there exists a projection $p \in M$ such that

$$\sup_{1 \leq \nu \leq k} \| p s_{n_k}^{(k)} \cdots s_{n_\nu}^{(\nu)} (x_m) p \|_\infty \leq (A_k + B_k \|x_m\|) \delta_m$$

and

$$\sup_{1 \leq \nu \leq k} \| p s_{n_k}^{(k)} \cdots s_{n_\nu}^{(\nu)} (|x_{mr}|^2) p \| \leq (A_k + B_k \| |x_{mr}|^2 \|) 2^{-r} \delta_m .$$

It is easy to verify that $\alpha_j^{(i)}$ are all Schwarz maps for $j = 1,2,\ldots,k$ and $i = 1,2,\ldots$. Moreover, for two Schwarz maps α and β, we have

$$|(n^{-1} \sum_{k=0}^{n-1} \alpha^k)(\frac{1}{m} \sum_{i=0}^{m-1} \beta^i(x))|^2 = \frac{1}{n^2 m^2} | \sum_{k=0}^{n-1} \alpha^k (\sum_{i=0}^{m-1} \beta^i(x))|^2 \leq$$

$$\leq \frac{1}{nm^2} \sum_{k=0}^{n-1} |\alpha^k (\sum_{i=0}^{m-1} \beta^i(x))|^2 \leq \frac{1}{nm^2} \sum_{k=0}^{n-1} \alpha^k (| \sum_{i=0}^{m-1} \beta^i(x)|^2) \leq$$

$$\leq \frac{1}{nm} \sum_{k=0}^{n-1} \alpha^k \sum_{i=0}^{m-1} \beta^i(|x|^2) = (\frac{1}{n} \sum_{k=0}^{n-1} \alpha^k)(\frac{1}{m} \sum_{i=0}^{m-1} \beta^i(|x|^2)).$$

Thus we have, in particular,

$$| s_{n_2}^{(2)} s_{n_1}^{(1)} (x)|^2 \leq s_{n_2}^{(2)} s_{n_1}^{(1)} (|x|^2),$$

and also more general formula (for several contractions). This implies the following estimation for $1 \leq \nu \leq k$

$$\| s_{n_k}^{(k)} \cdots s_{n_\nu}^{(\nu)}(x_{mr})p\|_\infty^2 = \| p|s_{n_k}^{(k)} \cdots s_{n_\nu}^{(\nu)}(x_{mr})|^2 p\|_\infty \le$$

$$\le \| ps_{n_k}^{(k)} \cdots s_{n_\nu}^{(\nu)}(|x_{mr}|^2)p\|_\infty \le (A_k + B_k\||x_{mr}|^2\|)2^{-r}\delta_m.$$

Since

$$\tilde{s}_{n_k}^{(k)} \cdots \tilde{s}_{n_\nu}^{(\nu)}(\xi_m) = \sum_{r=1}^{\infty} \tilde{s}_{n_k}^{(k)} \cdots \tilde{s}_{n_\nu}^{(\nu)}(x_{mr}\Omega) =$$

$$= \sum_{r=1}^{\infty} s_{n_k}^{(k)} \cdots s_{n_\nu}^{(\nu)}(x_{mr})\Omega,$$

we have, for $1 \le \nu \le k$,

$$\| \tilde{s}_{n_k}^{(k)} \cdots \tilde{s}_{n_\nu}^{(\nu)}(\xi_m)\|_p \le \sum_{r=1}^{\infty} \| s_{n_k}^{(k)} \cdots s_{n_\nu}^{(\nu)}(x_{mr})p\|_\infty \le$$

$$\le \delta_m^{1/2}(A_k + B_k\|\xi_m\|) \sum_{r=1}^{\infty} 2^{-(r/2)} < 5\delta_m^{1/2}(A_k + B_k\|\xi_m\|),$$

for all $m = 1,2,\ldots$ and $n_s = 1,2,\ldots;\ 1 \le s \le k$. The proof is completed.

For $\xi \in H$ and $i = 1,2,\ldots,k$, let us put

$$\psi_i(\xi) = \lim_{n\to\infty} n^{-1} \sum_{k=0}^{n-1} \tilde{\alpha}_i^k \xi, \tag{6}$$

where the limit in (6) is taken in the norm topology in H (mean ergodic theorem for the contraction $\tilde{\alpha}_i$).

We shall prove the following main result.

2.4. Theorem. let $\phi, \alpha_i, \tilde{\alpha}_i, s_{n_i}^{(i)}, \tilde{s}_{n_i}^{(i)}$ and ψ_i be as above. Then, for every $\xi \in H$, we have that

$$\tilde{s}_{n_k}^{(k)} \cdots \tilde{s}_{n_1}^{(1)}(\xi) \to \psi_k\psi_{k-1}\cdots\psi_1(\xi) \quad \text{a.s.} \tag{7}$$

as $n_1 \to \infty, \ldots, n_k \to \infty$ independently.

Proof. We shall confine ourselves to the case $k = 2$. The proof for an arbitrary k is similar.

Let us write

$$\tilde{s}_{n_2}^{(2)}\tilde{s}_{n_1}^{(1)}(\xi) - \psi_2\psi_1(\xi) =$$

$$= \tilde{s}_{n_2}^{(2)}(\tilde{s}_{n_1}^{(1)}(\xi) - \psi_1(\xi)) + (\tilde{s}_{n_2}^{(2)} - \psi_2)\psi_1(\xi) =$$

$$= \tilde{s}_{n_2}^{(2)}\tilde{s}_{n_1}^{(1)}(\xi - \psi_1(\xi)) + s_{n_2}^{(2)}(\psi_1(\xi) - \psi_2(\psi_1(\xi))$$

(since $\psi_i(\xi)$ is $\tilde{\alpha}_i$-invariant).

We have $\tilde{s}_{n_i}^{(i)}\eta - \psi_i(\eta) \to 0$ in H. Let us define four closed linear subspaces $H_i^{(j)}$ of H, putting, for $j = 1,2$,

$$H_1^{(j)} = [\xi \in H : \tilde{\alpha}_j\xi = \xi]$$

and

$$H_2^{(j)} = [(y - \alpha_j y)\Omega : y \in M]^-.$$

Then, we get

$$H = H_1^{(j)} \bullet H_2^{(j)} \qquad (j = 1,2,).$$

For $\eta \in H$, $\eta - \psi_j(\eta) \in H_2^{(j)}$. Let us fix some $\varepsilon > 0$ and take $\delta_k > 0$ with $\sum_{k=1}^{\infty} \delta_k < \varepsilon/64$. Then there exist $\eta_m^{(i)} \in H$ with $\|\eta_m^{(i)}\| < \delta_m$ and $y_m \in M$ such that

$$\xi - \psi_1(\xi) = \eta_m^{(1)} + (y_m^{(1)} - \alpha_1 y_m^{(1)})\Omega \quad \text{for all} \quad m = 1,2,...$$

and

$$\psi_1(\xi) - \psi_2\psi_1(\xi) = \eta_m^{(2)} + (y_m^{(2)} - \alpha_2 y_m^{(2)})\Omega \quad \text{for} \quad m = 1,2,\ldots .$$

Then, by Theorem 2.3, there is a projection $p \in M$ such that

$$(8)\qquad \| \tilde{s}^{(2)}_{n_2}(\eta^{(2)}_m)\|_p \leq (A_2 + B_2\|\eta^{(2)}_m\|)\delta^{1/2}_m \quad \text{for} \quad m = 1,2,\ldots$$

and

$$(9)\qquad \| \tilde{s}^{(2)}_{n_2}\tilde{s}^{(1)}_{n_1}(\eta^{(1)}_m)\|_p < (A_2 + B_2\|\eta^{(1)}_m\|)\delta^{1/2}_m \quad \text{for} \quad m = 1,2,\ldots$$

and

$$(10)\qquad \phi(1-p) \leq 2^5 \sum_{m=1}^{\infty} \delta^{-1}_m(\|\eta^{(1)}_m\|^2 + \|\eta^{(2)}_m\|^2) \leq$$

$$\leq 2^6 \sum_{m=1}^{\infty} \delta_m < \varepsilon .$$

Let us fix m in formulae (8) and (9) in such a way that

$$\| \tilde{s}^{(2)}_{n_2}(\eta^{(2)}_m)\|_p < \varepsilon/3$$

and

$$\| \tilde{s}^{(2)}_{n_2}\tilde{s}^{(1)}_{n_1}(\eta^{(1)}_m\|_p < \varepsilon/3$$

(with (10), i.e. $\phi(1-p) < \varepsilon$). Then we have

$$\| \tilde{s}^{(2)}_{n_2}\tilde{s}^{(1)}_{n_1}(\xi) - \psi_2\,\psi_1(\xi)\|_p \leq$$

$$\leq 2\varepsilon/3 + \|s^{(2)}_{n_2}s^{(1)}_{n_1}(y^{(1)}_m - \alpha_1 y^{(1)}_m)p\|_\infty +$$

$$+ \|s^{(2)}_{n_2}(y^{(2)}_m - \alpha_2 y^{(2)}_m)p\|_\infty < \varepsilon$$

for n_1 and n_2 large enough. This shows that

$$\tilde{s}^{(2)}_{n_2}\tilde{s}^{(1)}_{n_1} \to \psi_2\psi_1(\xi) \qquad \text{almost surely}$$

as $n_1 \to \infty$, $n_2 \to \infty$ independently, which ends the proof.

3. In this section we prove a continuous version of Theorem 2.4. Let us consider several weak*-continuous semigroups $\alpha^{(i)} = (\alpha^i_t)_{t\geq 0}$ $(i = 1,2,..,k)$ of linear maps of M with $\alpha^{(i)}_o = I$ (identity maps), $\alpha_t(1) = 1$, and such that $\alpha^{(i)}_t$ are Schwarz maps. We assume that the

state Φ is $\alpha_t^{(i)}$-invariant for all t and i. In particular, it follows that $\alpha_t^{(i)}$ are positive normal maps of M.

With the dynamical systems $(M, \alpha^{(i)}, \phi)$ we associate the semigroups $\beta^{(i)} = (\beta_t^{(i)})_{t\geq 0}$ $(i = 1,2,\ldots,k)$ of contractions in H. Namely, we set $\beta_t^{(i)}(x\Omega) = \alpha_t^{(i)}\Omega$, for $x \in M$, $t \geq 0$ and $i = 1,2,\ldots,k$.

Let us put , for $i = 1,2,\ldots,k$,

$$
(11) \qquad s_{T_i}^{(i)}(x) = T_i^{-1} \int_0^{T_i} \alpha_t^{(i)}(x)dt, \qquad x \in M,
$$

and

$$
(12) \qquad \tilde{s}_{T_i}^{(i)}(\xi) = T_i^{-1} \int_0^{T_i} \beta_t^{(i)}(\xi)dt, \qquad \xi \in H.
$$

Obviously, the above integrals (11) and (12) exist in the weak and weak* sense, respectively.

For $\xi \in H$ and $i = 1,2,..,k$, we put

$$
(13) \qquad \omega_i(\xi) = \lim_{T_i\to\infty} s_{T_i}^{(i)}(\xi),
$$

where the limit in (13) is taken in the norm topology in H (mean ergodic theorem for the contraction semigroup $(\beta_t^{(i)})_{t\geq 0}$).

The main result in this section is the following individual ergodic theorem for several contraction semigroups.

3.1. Theorem. For every $\xi \in H$, we have that

$$
(14) \qquad \tilde{s}^{(k)}_{T_k}\ldots\tilde{s}^{(1)}_{T_1}(\xi) \to \omega_k\ldots\omega_1(\xi) \quad \text{almost surely,}
$$

as $T_1 \to \infty, \ldots, T_k \to \infty$ independently.

The proof of the above theorem is similar to the proof of Theorem 2.4 and is based on a few lemmas which are natural extension of the previous results discussed in section 2. Let us formulate them with some indications concerning the proofs.

3.2. Lemma. Let, for $i = 1,2,\dots,k$, $(\alpha_t^{(i)})_{t\geq 0}$ be semigroups as defined in the beginning of this section. Let $(x_m) \subset M_+$, $\varepsilon_n > 0$ $(n = 1,2,\dots)$. Then there exists a projection $p \in M$ such that

$$\phi(1 - p) < 2 \sum_{n=1}^{\infty} \varepsilon_n^{-1} \phi(x_n)$$

and

$$\| p \int_0^{T_i} \alpha_t^{(i)}(x_n)dt\, p \|_\infty < 4\, \varepsilon_n T_i,$$

for $T_i \geq 1$, $n = 1,2,\dots;$ $i = 1,2,\dots,k$.

This lemma can be reduced to Theorem 2.1 by considering the averages

$$A_n^{(i)} = \int_0^1 \alpha_t^{(i)}(x_n)dt$$

and noting that, for $N \leq T_i < N + 1$, we have

$$\int_0^{T_i} \alpha_t^{(i)}(x_n)dt = A_n^{(i)} + \alpha_1^{(i)}(A_n^{(i)}) + \dots + (\alpha_1^{(i)})^{N-1}(A_n^{(i)}) + + \int_N^{T_i} \alpha_t^{(i)}(x_n)dt.$$

3.3. Lemma. Let $(x_m) \subset M_+$, $\varepsilon_m > 0$ $(m = 1,2,\dots)$. Then there exists a projection $p \in M$ with

$$\phi(1 - p) < 2^{k+1} \sum_{m=1}^{\infty} \varepsilon_m^{-k} \phi(x_m)$$

and such that

$$\sup_{1\leq \nu\leq k} \| p s_{T_k}^{(k)} \dots s_{T_\nu}^{(\nu)}(x_m) p \|_\infty \leq (A_k + B_k \| x_m \|)\varepsilon_m$$

for some constants A_k and B_k (depending only on k) and all $m = 1,2,\dots$ and all $T_1, T_2, \dots T_k \geq 1$.

To prove the above lemma it is enough to combine the general idea of the proof of Theorem 2.2 with Lemma 3.2.

3.4. Lemma. Let $s_{T_i}^{(i)}$ and $\tilde{s}_{T_i}^{(i)}$ $(i = 1,2,\ldots,k)$ denote the averages defined above. Then, for every $(\xi_m) \subset H$, every sequence (δ_m) of positive numbers and $(x_m) \subset M_+$, there exist a projection $p \in M$ and some positive constants A_k and B_k such that

$$\sup_{1\le\nu\le k} \| p s_{T_k}^{(k)} \ldots s_{T_1}^{(1)}(x_m) p \|_\infty \le (A_k + B_K \|x_m\|)\delta_m$$

and

$$\sup_{1\le\nu\le k} \| \tilde{s}_{T_k}^{(k)} \ldots \tilde{s}_{T_\nu}^{(\nu)}(\xi_m) \|_p \le (A_k + B_k \|\xi_m\|)\delta_m^{1/2},$$

for all $m = 1,2,\ldots,$ $T_1,\ldots,T_k \ge 1$.

Moreover, we have

$$\phi(1-p) \le 2^{k+3} \sum_{m=1}^{\infty} \delta_m^{-1}(\phi(x_m) + \|\xi_m\|^2).$$

In the proof of this lemma we have to use the following inequalities. Namely, for $(\xi_k) \subset H$ such that $\xi_k = \sum_{m=1}^{\infty} x_{km}\Omega$ with $(x_{km}) \subset M$, we have

$$\| \int_0^{T_i} \beta_t^{(i)}(\xi_k)dt \|_p \le \sum_{s=1}^{\infty} \| \int_0^{T_i} \alpha_t^{(i)}(x_{ks})dt\, p \|_\infty =$$

$$= \sum_{s=1}^{\infty} \| p \,| \int_0^{T_i} \alpha_t^{(i)}(x_{ks})dt |^2 p \|_\infty^{1/2} \le$$

$$\le T_i^{1/2} \sum_{s=1}^{\infty} \| p \int_0^{T_i} \alpha_t^{(i)}(|x_{ks}|^2)dt\, p \|_\infty^{1/2}.$$

To conclude this section let us formulate an individual local ergodic theorem for several contraction semigroups in H generated by Schwarz maps in M. Let $(\alpha_t^{(i)})_{t\ge0}$, $(\beta_t^{(i)})_{t\ge0}$, $s_{T_i}^{(i)}$ and $\tilde{s}_{T_i}^{(i)}$ $(i = 1,2,\ldots,k)$ denote the contraction semigroups and their averages as considered before. Then, we have the following result.

3.5. Theorem. For every $\xi \in H$,

$$\tilde{s}_{T_k}^{(k)} \ldots \tilde{s}_{T_1}^{(1)}(\xi) \to \xi \quad \text{almost surely,}$$

as $T_1 \to 0, \ldots, T_k \to 0$ independently.

We omit the proof which can be obtained by a careful inspection of the methods developed in this section and combining them with the ideas presented in [6], proof of Theorem 2.

References

[1] I.P. Conze and N. Dang-Ngoc, Ergodic theorems for non-commutative dynamical systems, Invent. Math. 46 (1978), 1-15.

[2] M.S. Goldstein, Theorems in almost everywhere convergence in von Neumann algebras (Russian), J. Oper. Theory 6 (1981), 233-311.

[3] E. Hensz and R. Jajte, Pointwise convergence theorems in L_2 over a von Neumann algebra, Math. Z. 193 (1986), 413-429.

[4] R. Jajte, Ergodic theorems in von Neumann algebras, Semesterbericht Funktionalanalysis, Tübingen, Sommersemester 1986, 135--144.

[5] R. Jajte, Strong limit theorems in non-commutative probability, Lecture Notes in Math., vol. 1110, Springer-Verlag, Berlin-Heidelberg-New York-Tokyo 1985.

[6] R. Jajte, Contraction semigroups in L^2 over a von Neumann algebra, Proc. Oberwolfach Conference on Quantum Probability, Lecture Notes in Math., to appear.

[7] E.C. Lance, Ergodic theorem for convex sets and operator algebras, Invent. Math. 37 (1976), 201-211.

[8] G. Lindblad, On the generators of quantum dynamical semi-groups, Comm. Math. Phys. 48 (1970), 119-130.

[9] D. Petz, Ergodic theorems in von Neumann algebras, Acta. Sci. Math. 46 (1983), 329-343.

[10] D. Petz, Quasi-uniform ergodic theorems in von Neumann algebras, Bull. London math. Soc. 16 (1984), 151-156.

[11] Y.G. Sinai, V.V. Anshelevich, Some problems of non-commutative ergodic theory, Russian Math. Surveys 31 (1976), 157-174.

[12] M. Takesaki, Theory of operator algebras I, Springer Verlag, Berlin-Heidelberg-New York 1979.

[13] S. Watanabe, Ergodic theorems for dynamical semigroups on operator algebras, Hokkaido Math. Journ. 8 (1979), 176-190.

[14] S. Watanabe, Asymptotic behaviour and eigenvalues of dynamical semi-groups on operator algebras, Journ. of Math. Anal. and Applic. 86 (1982), 411-424.

[15] F.I. Yeadon, Ergodic theorems for semifinite von Neumann algebras I, J. London Math. Soc. 16 (1977), 326-332.

[16] F.I. Yeadon, Ergodic theorems for semifinite von Neumann algebras II, Math. Proc. Cambridge Phil. Soc. 88 (1980), 135-147.

[17] A. Zygmund, An individual ergodic theorem for noncommutative transformations, Acta Sci. Math. 14 (1951), 105-110.

Probability Theory on Vector Spaces IV
Lancut, June'87, Springer's LNM 1391

LINEAR SUPPORTS AND ABSOLUTE CONTINUITY OF CERTAIN RANDOM INTEGRALS

by

Zbigniew J. Jurek (Wroclaw University).

Let Q be a linear bounded operator on a real separable Banach space E such that $t^Q := \exp(Q \ln t) \to 0$, in the operator topology, as $t \to 0$. By $U(Q;E)$, or simply $U(Q)$, we denote the class of all probability distributions of the following random integrals

$$\int_{(0,1]} t^Q dY(t), \tag{1}$$

where Y is an arbitrary $D_E[0,1]$-valued random variable with stationary independent increments and $Y(0) = 0$ a.s. Hence the integrals (1) have infinitely divisible distributions. In Jurek (1985 b) the class $U(Q)$ is identified as a class of limit distributions for some particular type of summation and is characterized in many different ways; cf. Theorem 2.3. The class $U(I;E)$, where I is the identity operator, coincides with limit distributions of sequences of independent random variables deformed by certain non-linear transformations; cf. Jurek (1985 a), Remark 2.1. In this particular case ($Q = I$) distributions from $U(I;E)$ are called s-selfdecomposable.

Let $L(Q;E)$, or simply $L(Q)$, be a class of all probability distributions of the following random integrals

$$\int_{(0,\infty)} e^{-tQ} dZ(t), \tag{2}$$

where Z is $D_E[0,\infty)$-valued rv with stationary independent increments such that

$\mathbb{E}\log(1+\|Z(1)\|) < \infty$ and $Z(0) = 0$ a.s. By Corollary 2.4. in Jurek (1985 b) we know that $L(Q) \subseteq U(Q)$ and consequently we get appropriate corollaries for distributions of (2).

The paper is organized as follows: In Section 1 we briefly discuss linear supports of arbitrary measures and then, in Section 2, we deal with supports of Gaussian and Poissonian parts of (1). Finally, Section 3 contains partial results concerning the absolute continuity of distributions of (1).

1. LINEAR SUPPORTS. For a Borel measure μ on E, let $\hat{\mu}$ be its characteristic functional and supp μ its support, i.e. the smallest closed subset of E whose complement has μ-measure zero. Further, let lin(supp μ) denotes the smallest closed linear subspace of E containing supp μ. By $<\cdot,\cdot>$ we mean the bilinear dual pair between E^* (the topological dual to E) and E. For a subset A of E let $A^{\perp} := \{y \in E^* : <y,a> = 0 \text{ for all } a \in A\}$. If V is a closed linear subspace of E then $(V^{\perp})^{\perp} = V$ and $V^{\perp}$ coincides with V°, the polar of V; cf. Schaefer (1966), Chapter IV, Section 1.

LEMMA 1.1. *For a probability measure* μ *on* E *we have*

$$\text{lin(supp } \mu) = \{y \in E^* : \hat{\mu}(ty) = 1 \text{ for all } t \in \mathbb{R}\}^{\perp}.$$

Proof. For $y \in (\text{lin(supp } \mu))^{\perp}$ we get $<ty,x> = 0$ for all $x \in \text{supp } \mu$ and all $t \in \mathbb{R}$. Therefore

$$\hat{\mu}(ty) = \int_{\text{supp } \mu} \exp i <ty,x> \mu(dx) = 1$$

for $t \in \mathbb{R}$.

Conversely, let $\hat{\mu}(ty) = 1$ for all $t \in \mathbb{R}$. Then

$$\int_{\text{supp } \mu} (1-\cos t<y,x>)\mu(dx) = 0 \qquad \text{for all } t \in \mathbb{R}.$$

Hence $t<y,x> = 0 \pmod{2\Pi}$ for all $t \in \mathbb{R}$ and $x \in \text{supp } \mu$. Consequently, $<y,x> = 0$ for $x \in \text{supp } \mu$ and hence for all $x \in \text{lin(supp } \mu)$, i.e. $y \in (\text{lin(supp } \mu))^{\perp}$. Therefore we have shoved $(\mathrm{lin}(\mathrm{supp}\ \mu))^{\perp} = \{y \in E^{*} : \hat{\mu}(ty) = 1 \text{ for } t \in \mathbf{R}\}$, which completes the proof.

In the sequel we will write $\nu = [a,R,M]$ if ν is infinitely divisible, i.e. $\nu \in ID(E)$ and its characteristic functional $\hat{\nu}$ is of the following form

$$\hat{\nu}(y) = \exp\{i\langle y,a\rangle - 1/2\langle y,Ry\rangle + \int_{E \setminus \{0\}} [\exp i\langle y,x\rangle - 1 - i\langle y,x\rangle 1_{B}(x)]M(dx)\},$$

where $a \in E$, R is a Gaussian covariance operator and M is a Lévy measure corresponding to ν; cf. Araujo-Giné (1980). We refer to $[0,R,0]$ as a Gaussian part and to $[0,0,M]$ as a Poissonian part of ν. Of course,

$$\nu = \delta(a) * [0,R,0] * [0,0,M].$$

LEMMA 1.2. *For an infinitely divisible measure* $\nu = [0,0,M]$ *we have*

$$\mathrm{lin}(\mathrm{supp}\ \nu) = \mathrm{lin}(\mathrm{supp}\ M).$$

Proof. Let $y \in (\mathrm{lin}(\mathrm{supp}\ \nu))^{\perp}$. Then, by Lemma 1.1, we have $\nu(ty) = 1$ for all $t \in \mathbf{R}$. Hence

$$\int_{\mathrm{supp}\ M} (1 - \cos t\langle y,x\rangle) M(dx) = 0$$

and similary as in the proof of Lemma 1.1 we conclude that $y \in (\mathrm{lin}(\mathrm{supp}\ M))^{\perp}$. In other words we obtained the inclusion $\mathrm{lin}(\mathrm{supp}\ M) \subseteq \mathrm{lin}(\mathrm{supp}\ \nu)$. Suppose that $\mathrm{lin}(\mathrm{supp}\ M)$ is a proper subspace of $\mathrm{lin}(\mathrm{supp}\ \nu)$. By Hahn-Banach Theorem there exists $0 \neq y_0 \in (\mathrm{lin}(\mathrm{supp}\ \nu))^{*}$ such that $\langle y_0, x\rangle = 0$ for $x \in \mathrm{lin}(\mathrm{supp}\ M)$. Hence $\langle ty_0, x\rangle = 0$ for $t \in \mathbf{R}$ and $x \in \mathrm{supp}\ M$ and therefore

$$\hat{\nu}(ty_0) = \exp \int_{\mathrm{supp}\ M} [\exp i\langle ty_0,x\rangle - 1 - i\langle ty_0,x\rangle 1_{B}(x)]M(dx) = 1$$

for all $t \in \mathbf{R}$. By Lemma 1.1, $y_0 \in (\mathrm{lin}(\mathrm{supp}\ \nu))^{\perp}$ and consequently $y_0 = 0$, which contradicts our assumption. Thus the Lemma is proved completely.

2. SUPPORTS OF MEASURES FROM $U(Q;E)$. In this Section we will examine support of Gaussian and Poissonian part of distributions of random integrals

(1), i.e. elements from the class $U(Q;E)$. Let $\nu = [a,R,M] := L(Y(1))$ (= probability distribution of E-valued *rv* $Y(1)$), $Y(0) = 0$ a.s. and Y be a $D_E[0,1]$-valued *rv* with stationary independent increments. Then let us define $J_Q : ID(E) \to U(Q;E)$ as follows

$$J_Q(\nu) = L\Big(\int_{(0,1]} t^Q dY(t) \Big) . \tag{3}$$

In Jurek (1985 b) has been proved that J_Q is a topological isomorphism between the topological semigraphs, $ID(E)$, of all infinitely divisible measures and the semigraphs $U(Q;E)$. Furthermore, if $[a',R',M'] := J_Q[a,R,M]$ then we have the following relations

$$a' = (Q+I)^{-1}a + \int_{E\setminus\{0\}} \int_0^1 t^Q x[1_B(t^Q x) - 1_B(x)]dtM(dx) \ , \tag{4}$$

$$R' = \int_0^1 t^Q R t^{Q^*} dt \ , \tag{5}$$

$$M'(F) = \int_0^1 (t^Q M)(F)dt, \text{ for all Borel subsets } F \text{ of } E\setminus\{0\}. \tag{6}$$

As before lin(A) denotes the smallest closed linear subspace of E containing set A and $\mathrm{lin}_Q(A)$ denotes the smallest closed Q-invariant subspace containing A. Of course, lin(A) is a subspace of $\mathrm{lin}_Q(A)$.

LEMMA2.1. *For measures* M' *and* M *related by* (6) *we have the following*

(i) $\mathrm{supp}\, M' = \overline{\Big(\bigcup_{0<t\le 1} t^Q \mathrm{supp}\, M \Big)}$,

(ii) $\mathrm{lin}(\mathrm{supp}\, M') = \mathrm{lin}_Q(\mathrm{supp}\, M)$,

(iii) *M' is atomless, i.e., $M'(\{a\}) = 0$ for all $a \in E\setminus\{0\}$.*

Proof. (i) Suppose that $x \in \mathrm{supp}\, M'$ and $x \notin \overline{\Big(\bigcup_{0<t\le 1} t^Q \mathrm{supp}\, M \Big)}$. Then there is an open set $G \ni x$ such that G is disjoint with $t^Q \mathrm{supp}\, M$ for all $0 < t \le 1$. Hence $t^{-Q}G$ are open and disjoint with supp M and consequently

$$M'(G) = \int_0^1 M(t^{-Q}G)dt = 0,$$

which contradicts that $x \in \text{supp } M'$. Conversely, let $x = t^Q u$ for some $0 < t \le 1$ and $u \in \text{supp } M$. For each open set $G \ni x$ there exist an open interval $(t-\varepsilon, t+\varepsilon)$ in positive half-line and an open set $V \ni u$ such that $s^Q v \in G$ for all $s \in (t-\varepsilon, t+\varepsilon)$ and all $v \in V$. Since $s^{-Q}G$ are open and contain u for $s \in (t-\delta, t)$ we obtain

$$M'(G) \ge \int_{t-\delta}^{t} M(s^{-Q}G)ds > 0,$$

i.e. $x \in \text{supp } M'$. Consequently, $\overline{\left(\bigcup_{0<t\le 1} t^Q \text{supp } M \right)} \subseteq \text{supp } M'$ which proves (i).

(ii) Since $Q^k(\text{supp } M) \subseteq \text{lin}_Q(\text{supp } M)$ for $k = 0,1,2, \ldots$ we infer that $t^Q(\text{supp } M) \subseteq \text{lin}_Q(\text{supp } M)$ for all $t > 0$. Finally by (i) we obtain $\text{lin}(\text{supp } M') \subseteq \text{lin}_Q(\text{supp } M)$.

Conversely, if we put $A := \bigcup_{0<t\le 1} t^Q \text{supp } M$ then $s^Q A \subseteq A$ for any $0 < s \le 1$. Since $Qx = \lim_{t\to 1}(t^Q x - x)/(t-1)$ therefore $Qx \in \text{lin } A$ for all $x \in A$. Thus $\text{lin } A$ is Q-invariant subspace containing supp M. Hence

$$\text{lin}_Q(\text{supp } M) \subseteq \text{lin } A \subseteq \text{lin}(\text{supp } M'),$$

which completes the proof of (ii).

(iii) Any $a \in E \setminus \{0\}$ has a unique polar coordinate representation as $a = s^Q u$ for some $s > 0$ and u from a unit sphere in E with respect to the new (equivalent) norm $|| \cdot ||_Q$; cf. Jurek (1984), Proposition 2. Thus we get

$$M'(\{s^Q u\}) = \int_{E\setminus\{0\}} \int_0^1 1_{\{s^Q u\}}(t^Q x) dt M(dx) =$$

$$= \int_{E\setminus\{0\}} l_1\{0 < t \le 1 : x = (s/t)^Q u\} M(dx) = 0$$

where l_1 is the Lebesque measure on $\mathbb{R}$. This completes the proof of Lemma 2.1.

LEMMA 2.2. *For Gaussian covariance operators R' and R related by* (5) *we have the following*

(i) $\ker R' = \bigcap_{0<t\le 1} t^{-Q^*} \ker R$ *and* $\ker R'$ *is* Q^**-invariant subspace.*

(ii) $\overline{R'(E^*)} = \mathrm{lin}_Q(\overline{R(E^*)})$.

Proof. (i) Let T be a linear operator from E^* into E with properties: $\langle y_1, Ty_2\rangle = \langle y_2, Ty_1\rangle$ for $y_1, y_2 \in E^*$ (symmetry) and $\langle y, Ty\rangle \geq 0$ (nonnegativity). Then for all $y_1, y_2 \in E^*$

$$|\langle y_1, Ty_2\rangle|^2 \leq \langle y_1, Ty_1\rangle\langle y_2, Ty_2\rangle \quad \text{(Schwarz inequality)},$$

and consequently $\ker T = \{y \in E^* : \langle y, Ty\rangle = 0\}$. Hence and (5) we get $y \in \ker R'$ iff $\langle y, t^Q R t^{Q^*} y\rangle = 0$ for all $0 < t \leq 1$ iff $y \in \ker(t^Q R t^{Q^*}) = \ker(R t^{Q^*}) = t^{-Q^*} \ker R$ for all $0 < t \leq 1$, which gives (i). From the established equality we get $s^{Q^*} \ker R' \subseteq \ker R'$ for all $0 < s \leq 1$. Hence for each $y \in \ker R'$, $Q^* y = \lim_{s\to 1}(s^{Q^*}y - y)/(s-1) \in \ker R'$, which shows the second part of (i).

(iii) For a bilinear form $\langle\cdot,\cdot\rangle$ between E^* and E and the operation $\perp$ we have the following properties:

(1) for a closed linear subspace V (of E or E^*) we have $(V^\perp)^\perp = V$, cf. Schaefer (1966) Chapter 1, Theorem 1.5;

(2) for a symmetric operator T from E^* into E, $\ker T = (\overline{T(E^*)})^\perp$;

(3) for an invertible operator A on E and a subset W of E^* we have $(A^* W)^\perp = A^{-1}(W^\perp)$. Hence and the equality (i) we obtain

$$\overline{R'(E^*)} = \Big(\bigcap_{0<t\leq 1} t^{-Q^*}\ker R\Big)^\perp \supseteq \bigcup_{0<t\leq 1} (t^{-Q^*}\ker R)^\perp = \bigcup_{0<t\leq 1} t^Q(\overline{R(E^*)}).$$

So, the subspace $\overline{R'(E^*)}$ contains $\overline{R(E^*)}$ and is Q-invariant. Consequently, $\overline{R'(E^*)} \supseteq \mathrm{lin}_Q(\overline{R(E^*)})$. Suppose the inclusion is proper. Then by Hahn-Banach Theorem there exists $0 \neq y \in (\overline{R'(E^*)})^*$ such that y vanishes on $\mathrm{lin}_Q(\overline{R(E^*)})$. On the other hand, for any $u \in E^*$, $t^Q R t^{Q^*} u \in \mathrm{lin}_Q(\overline{R(E^*)})$ for all $t > 0$ and hence

$$R'u = \int_0^1 t^Q R t^{Q^*} u\,dt \in \mathrm{lin}_Q(\overline{R(E^*)}) \quad \text{for } u \in E^*.$$

Therefore $\langle y, R'u\rangle = 0$ for all $u \in E^*$, which implies $y = 0$ and contradicts the assumption $y \neq 0$. Thus the Lemma is completely proved.

3. ABSOLUTE CONTINUITY. In this section we will prove that some of distributions of (1) are absolutely continuous with respect to Lebesque measure. From (6) we see that M' is finite measure whenever so is M. Furthermore for finite M the compound Poisson measure $e(M)$ has an atom at zero of the magnitude greater or equal $\exp(-M(E))$. Thus we will deal with infinite measures M and the proof is based on the following fact.

LEMMA 3.1. (Sato (1982)). *Let* μ *be an infinitely divisible on* $\mathbb{R}^d$ *with infinite Lévy measure* M. *If* $(\tilde{M})^{*k}$, *for some* $k \in \mathbb{N}$, *is absolutely continuous with respect to d-dimensional Lebesque measure* ℓ_d *then so is* μ. [*Here* $\tilde{M}(dx) = \|x\|^2/(1+\|x\|^2)M(dx)$].

Our partial result concerning the absolute continuity of s-selfdecomposable measures is the following

LEMMA 3.2. *All s-selfdecomposable measures* μ *on* $\mathbb{R}$ (*i.e.,* $\mu \in U(I;\mathbb{R})$) *with non-zero Gaussian part or with infinite Lévy measure are absolutely continuous.*

Proof. Of course, we need to discuss only the case of $\mu \in U(I;\mathbb{R})$ with infinite Lévy measure and zero Gaussian part. Thus by (1) and (6) we get $\mu = [a',0,M'] = J[0,0,M]$, M is infinite Lévy measure and

$$M'(F) = \int_{\mathbb{R}\setminus\{0\}} \int^1 1_F(tx)dxM(dx) = \int_{\mathbb{R}\setminus\{0\}} l_1\{s\in(0,1]:s\in x^{-1}F\}M(dx).$$

Hence if $l_1(F) = 0$ then $l_1(x^{-1}F) = 0$ for each $x \neq 0$ and consequently $M'(F) = 0$, i.e., M' is absolutely continuous on $\mathbb{R}$. This with Lemma 3.1 completes the proof of the lemma.

REFERENCES

[1] Araujo, A. and Giné, E. (1980). *The central limit theorem for real and Banach valued random variables*; John Wiley, New York.

[2] Jurek, Z.J. (1984). Polar coordinates in Banach spaces; *Bull. Pol. Ac : Math.* 32, pp. 61-65.

[3] Jurek, Z.J. (1985 a). Relations between the *s*-selfdecomposable and self-decomposable measures; *Ann.Probab.* 13, pp. 592-608.

[4] Jurek, Z.J.(1985 b). Random integral reprezentation for another class of limit laws; *Lecture Notes in Math.* 1153, pp. 297-309.

[5] Sato, K.(1982). Absolute continuity of multivariate distributions of class *L*; *J. Multivar. Anal.* 12, pp. 89-94.

[6] Schaefer, H. (1966). *Topological vector spaces*; The Macmillan Co., New York.

Author's address: Institute of Mathematics
Wroclaw University
Pl. Grunwaldzki 2/4
50-384 WROCLAW, Poland.

Probability Theory on Vector Spaces IV
Lancut, June '87, Springer's LNM 1391

CONDITIONAL MOMENTS FOR COORDINATES OF STABLE VECTORS

Raoul LePage

Michigan State University

ABSTRACT.

An arbitrary symmetric α–stable (SαS) random vector has exactly the law of a normal random vector whose covariance is itself random and possesses the appropriate $(\alpha/2)$–stable law (LePage, 1980). This fact is exploited in connection with the problem of prediction for SαS random vectors. The $(\alpha/2)$–stable measure on covariances can be treated as an a–priori measure on nuisance parameters. It is found that the conditional expectation of one stable r.v. given another can (unexpectedly) be a.s. finite even for $\alpha \leq 1$. This leads to predictors which take the form of a weighted average of predictors that would be used for the normal case. Such weighted averages are taken over the space of the covariances, according to an a–posteriori measure obtained by conditioning on the observations.

0. INTRODUCTION.

The work presented here is directed toward extending classical theories of prediction, detection, and smoothing to encompass models in which normal noise is replaced by symmetric α–stable (SαS) noise, where $0 < \alpha \leq 2$ is the index of stability. This is consistent with the fuller implications of the theory of errors (Levy, 1925) and includes normal noise as the case $\alpha = 2$. We make use of the fact, proved in [4], that every SαS law is in general a probability mixture of symmetric normal laws having different covariance functions.

Bochner's Theorem (Doob, 1953) implies that, for normal noise, frequency domain models are identical with time domain models. The same is not true for $\alpha < 2$. The class FT–SαS of processes

$$X_t = \int e^{it\lambda}\, Z(d\lambda)\, ,\ t \in \mathbb{R},$$

which are Fourier transforms of SαS independent increments noise Z for some $\alpha < 2$, will not be the same as the class of processes that are the outputs of linear systems (such as ARMA models)

AMS 1980 Subject Classifications. Primary 60E07; Secondary 60G10.

Research partially supported by the Office of Naval Research under grant USN N000014–85–K–0150 and Air Force Office of Scientific Research F49620 85 C 0144.

additively driven by SαS noise (Cambanis and Soltani, 1984). This has a direct bearing on the problem of prediction since the dependency between coordinates of a SαS random vector will be shown to differ greatly in the two cases as regards the existence of conditional expectations.

For FT–SαS processes X having a vector index $t \in T$ we prove, in Proposition 2.1 below, that the conditonal expectations

$$E(|X_t| \mid X_{t-\delta}) \; ; \; E(X_t^2 \mid X_{t-\delta}, X_{t-2\delta})$$

are almost surely finite for every $\alpha \leq 2$ and $\delta \neq 0$. Since the came is obviously not true for ARMA processes driven by independent SαS noise, we obtain another proof that such processes cannot be FT–SαS. Existence of the conditional expectations is somewhat unexpected in view of the fact that for $\alpha \leq 1$ the expectations $E\,|X_t|$ are infinite, while for all $\alpha < 2$ the second moments are infinite. As a consequence, in the case of FT–SαS processes, the conditional expectation predictor of X_t, regardless of the number of predicting variables, is well–defined and optimal for conditional mean squared error, provided the predicting variables include two time points of the form $t-\delta$, $t-2\delta$ for some $\delta \neq 0$. The identity

$$E(X_t \mid X_s) = X_s \, E \cos(t-s, \Lambda) ,$$

where Λ is a random sample from the normalized spectral measure, was known for $1 < \alpha \leq 2$ and $T = \mathbb{R}$ (Kanter and Steiger, 1974). Proposition 3.1 below proves this result for $0 < \alpha \leq 2$ by a new direct calculation not requiring existence of the unconditional expectation and bypassing differentiation of the characteristic function altogether.

1. STABLE VECTORS.

From [4], every SαS random vector X has the law of an a.s. norm convergent series of the form

$$V = \Sigma_{j=1}^{\infty} \Gamma_j^{-1/\alpha} U_j Z_j , \tag{1.1}$$

where the random sequences Γ, U, Z are mutually independent and

$\Gamma_j = \Sigma_{i=1}^{j} \xi_i$, with $\{\xi_i\}$ i.i.d. unit exponential r.v.,

$\{Z_j\}$ are i.i.d. centered normal r.v.,

$\{U_j\}$ are i.i.d. from a distribution on the sphere of norm one.

Furthermore, every a.s. convergent series of the above type is SαS.

Because finiteness of conditional expectation is an hereditary property for increasing σ–algebras, our results need involve only two or three coordinates of a random vector. We therefore require the series construction (1.1) only for SαS laws on $\mathbb{R}^2$ or $\mathbb{R}^3$.

Conditional on the sequences Γ, U, the random vector V is conditionally distributed according to a centered normal law with conditional covariance function given by the series

$$\theta(x, y) = E^{\Gamma, U} x(V)\, y(V) = \Sigma_{j=1}^{\infty} \Gamma_j^{-2/\alpha} x(U_j)\, y(U_j) , \tag{1.2}$$

which is absolutely convergent for continuous linear functions x, y. For details, see [6].

2. CONDITIONAL MOMENTS FOR FT–SαS.

For V satisfying (1.1), and continuous linear functions w, x, y, set W = w(V) (etc). It is seen from (1.2) that $P(\theta(x,x) = 0) = 0$ unless X is degenerate at zero. Suppose none of W, X, Y is identically zero. The next proposition suffices to establish conditional expectations for FT–SαS processes with a vector time paramter, even for $\alpha \leq 1$.

Proposition 2.1. For V, Γ, U, θ, as in Sec. 1.,

a. If a.s. $\theta(x,x) = \theta(y,y)$ (i.e. θ is a stationary covariance matrix for arguments x, y) then $E^X |Y|$ is a.s. finite.

b. There exist cases in which $E^X Y^2$ is a.s. infinite.

c. If θ is a.s. a stationary covariance matrix for arguments w, x, y then a.s. $E^{W,X} Y^2$ is finite.

Proof. a) If a.s. $\theta(x,x) = \theta(y,y)$ then for each fixed θ the conditional distribution of Y given X is normal with mean X $\rho(x,y)$ and variance $\theta(x,x)\ (1-\rho(x,y))^2$ where $\rho(x,y) = \theta(x,y)/\theta(x,x) = \theta(x,y)/\theta^{1/2}(x,x)\theta^{1/2}(y,y)$ is the correlation. Then,

$$E^X|Y| = E^X E^{X,\theta}|Y| \leq E^X\ (\ |X| + \theta^{1/2}(x,x)\ )\ . \tag{2.1}$$

Now apply Bayes theorem using the fact that the conditional density of X given $\theta(x,x)$ is

$$(2\pi)^{-1/2}\ \theta^{-1/2}(x,x)\ \exp(-X^2/2\theta(x,x))\ , \tag{2.2}$$

from which it is seen that the θ–term of (2.1) is a.s. finitely integrable E^X.

b) Define $H(X) = E^X\ \rho(x,y)$, which exists by the above. That is, $E^X\ Y = H(X)\ X$. Then, letting ρ denote $\rho(x,y)$,

$$E^X\ (Y-E^X\ Y)^2 = E^X\ \theta(x,x)\ (1-\rho^2) + E^X\ X^2\ (H(X)-\rho)^2 \tag{2.3}$$

The second integral on the right of (2.3) is a.s. finite since the quadratic term is after all bounded. We now show that there are cases in which the first integral on the right of (2.3) is divergent. An example may be constructed with $\theta(x,y) = 0$ a.s. (see below), leaving onlt $E^X\ \theta(x,x)$ to be dealt with in (2.3). From (1.2) it is seen that $\theta(x,x)$ possesses a stable distribution of index $\alpha/2$. Combining this with the conditional distribution (2.2) in another application of Bayes theorem gives an integral which is asymptotically of order $r^{-(\alpha+1)/2}$ at infinity. This fails to be integrable if $\alpha \leq 1$.

c) Now suppose θ is a.s. stationary in w, x, y. In Bayes theorem, replace (2.2) by the conditional density of (W,X) given θ, which is bounded above by

$$(2\pi)^{-1}\ \theta^{-1}(x,x)\ (1-\rho^2)^{-1}. \tag{2.4}$$

From (i) we may write

$$E^{W,X}\ Y^2 = E^{W,X}\ [\ E^{W,X,\theta}\ (Y - E^{W,X,\theta}\ Y)^2 + (E^{W,X,\theta}\ Y)^2\ ] \tag{2.5}$$

The first term in the brackets of (2.5) is less than or equal to $\theta(x,x)\ (1-\rho^2)$ (the θ–conditional mean squared error for prediction of Y from X, θ) and is therefore integrable according to the marginal distribution for θ when multiplied by (2.4). The second term in the brackets of (2.5) may, on ($\rho^2 \neq 1$), be written

$$[\rho X + \frac{\theta(2) - \rho\theta(1)}{\theta(0)\ (1-\rho^2)} (W - \rho X)]^2 , \tag{2.6}$$

where we have used $\theta(2)$ for $\theta(w,y)$; $\theta(1)$ for $\theta(x,y)$, etc. The first component in (2.6) is ρX whose square is integrable $E^{W,X}$. It is therefore enough to bound

$$D(\theta) = \left[\frac{\theta(2) - \rho\theta(1)}{\theta(0)\ (1-\rho^2)}\right] . \tag{2.7}$$

Let $||\quad||$ denote L_2 norm with respect to the zero-mean stationary Gaussian process with covariance function θ. Then

$$\theta(0) = ||Y||^2 \geq ||\rho X||^2 + ||D(\theta)\ (W-\rho X)||^2$$
$$= \rho^2\theta(0) + D^2(\theta)\ \theta(0)\ (1-\rho^2). \tag{2.8}$$

Therefore $D^2(\theta) \leq 1$ on $\rho^2 \neq 1$. But, on $\rho^2 = 1$, (2.6) is replaced by X. Thus (2.5) is a.s. finite. Note that the Cauchy–Schwartz bound splitting the numerator from the denominator of (2.7) is inadequate. □

3. STRUCTURE OF FT–SαS.

To verify the existence of an example with a.s. $\theta(x,y) = 0$, as used in the proof of (ii) above, we require more detailed information concerning the representation of FT–SαS processes. These take the form (from [4])

$$X_t = \Sigma_{j=1}^{\infty} \Gamma_j^{-1/\alpha} \cos((\Lambda_j, t) + \psi_j)\ \beta_j\ , t \in T\ , \tag{3.1}$$

where $\{\Lambda_j\}$ are i.i.d. random linear functions on a linear space T, $\{\psi_j\}$ are i.i.d. uniform r.v. on $[0, 2\pi]$, $\{\Gamma_j\}$ are as above, $\{\beta_j\}$ are i.i.d r.v. with finite α–th absolute moment, and Γ, Λ, ψ, β are mutually independent. The series (2.9) converges a.s. for each t and

$$\log E \exp i\ \Sigma_{k=1}^{n} r_k X_{t_k}$$
$$= 2^{-\alpha}B(\alpha)C(\alpha)\ E|\beta_1|^{\alpha}\ E|\Sigma_{k=1}^{n}\ r_k \exp i(\Lambda_1, t_k)|^{\alpha}\ , \tag{3.2}$$

where $B(\alpha) = \alpha \int_0^{\infty} (\cos(r)-1) \frac{dr}{r^{1+\alpha}}$, and $C(\alpha) = \int_{-\pi}^{+\pi} |1+e^{i\eta}|^{\alpha} d\eta$.

From (1.3) it may be seen that the law of X depends on the law of β_1 only through the α–th absolute moment. Taking $\{\beta_j\}$ to be Rayleigh distributed, equivalent to letting $\{\beta_j\}$ be complex Gaussian and taking the real part of $\{X_t, t\epsilon T\}$, yields a process X which is conditionally Gaussian and stationary given the sequences Λ, Γ. From (3.1) it is seen that the conditional covariance function of X given Λ, Γ, is given by

$$\theta(s, t) = \Sigma_{j=1}^{\infty} \Gamma_j^{-2/\alpha} \cos(\Lambda_j, s-t)/2, \quad t \in T\ . \tag{3.3}$$

<u>Proposition 3.1.</u> (Cambanis–LePage). For a process of the form (3.1) with $0 < \alpha \leq 2$,

$$E^{X_s} X_t = X_s\ E^{X_s}\ \theta(s, t)/\theta(s, s) = X_s\ E \cos(\Lambda_1, t-s)\ . \tag{3.4}$$

Proof. The case $\alpha = 2$ is obvious. Applying Proposition 2.1 the left side of (3.4) is well defined. By stationarity we may take s = 0. For $\alpha < 2$, write $\theta(1)$ for $\theta(s, t)$, $\theta(0)$ for $\theta(s, s)$, then

$$X_0 = \Sigma_{j=1}^{\infty} \cos(\psi_j)\, \Gamma_j^{-1/2}\, \beta_j$$

$$\theta(1)/\theta(0) = \Sigma_{j=1}^{\infty} \Gamma_j^{-2/\alpha} \cos(\Lambda_j, t)/\Sigma_{j=1}^{\infty} \Gamma^{-2/\alpha}.$$

Since Λ, ψ, Γ, β are independent and X_0 is measurable $\sigma\{\psi,\Gamma,\beta\}$,

$$E^{X_0}(\theta(1)/\theta(0)) = E^{X_0} E^{\psi,\Gamma,\beta}\, \theta(1)/\theta(0)$$

$$= \Sigma_{j=1}^{\infty} \Gamma_j^{-2/\alpha} E \cos(\Lambda_j, t)/\Sigma_{j=1}^{\infty} \Gamma_j^{-2/\alpha} = E \cos(\Lambda_1, t). \quad \square$$

This result extends (Kanter and Steiger, 1974), proven for the case $1 < \alpha \leq 2$, to the case $0 < \alpha \leq 2$ by a new method. Certain other conditional and unconditional integrals have been directly calculated by the same method, including the integer moments of the characteristic function of the processs conditioned on the invariant sigma algebra (Cambanis and LePage, 1987 unpublished).

4. CONCLUDING REMARKS.

Consideration of even the simplest ARMA model, X consisting of i.i.d. SαS, shows that we cannot in general hope to have finite conditional expectations. However, by conditioning on a random covariance θ which makes X Gaussian

$$E^{\theta,X_s} | X_t | < \infty ; \quad E^{X_s} | E^{\theta,X_s} X_t | < \infty \tag{4.1}$$

are true under conditions general enough to include ARMA models.

These conditioning methods also have application to the problem of estimating location.

REFERENCES.

1. Cambanis, S. and Soltani, A.R. (1984). Prediction of stable processes: Spectral and moving average representations. Z. W. verw. Geb. 66, 593–612.

2. Doob, J. L. (1953) Stochastic Processes. John Wiley & Sons Inc., New York.

3. Kanter, M. and Steiger, W.L., (1974). Regression and autoregression with infinite variance, Adv. Appl. Prob., 6, 768–783.

4. LePage, R. (1980). Multidimensional infinitely divisible variables and processes, Part I: Stable case. Tech. Rept. No. 292, Statistics Department, Stanford University.

5. LePage, R. (1981). Multidimensional infinitely divisible variables and processes, Part II. Lecture Notes in Mathematics, 860, 279–284.

6. LePage, R., Woodroofe, M., and Zinn, J. (1981). Convergence to a stable distribution via order statistics. Ann. Prob. 9, 624–632.

7. Levy, P. (1925). Calcul des Probabilites, Gauthier–Villars, Paris.

appendix

MULTIDIMENSIONAL INFINITELY DIVISIBLE VARIABLES AND PROCESSES. PART I: STABLE CASE

(Technical Report #292, Statistics Department, Stanford, 1980)

Raoul LePage

Michigan State University

ABSTRACT.

Elementary series constructions, involving a Poisson process, are obtained for multi–dimensional stable variables and random functions. Symmetric stable laws are shown to be mixtures of Gaussian laws.

1. INTRODUCTION.

Series decompositions, involving the arrival times of a Poisson process, have been given by Ferguson and Klass [1] for the non–Gaussian component of an arbitrary (real–valued) independent increments random function on the unit interval. LePage and Woodroofe and Zinn [4] have rediscovered a variant of this decomposition in connection with their study, via order statistics, of the limit distribution for self–normalized sums (e.g. Student's–t), when sampling from a distribution in the domain of attraction of an arbitrary stable law of index $\alpha < 2$.

The present paper obtains a characterization of stable laws on spaces of dimension greater than one. This characterization is formally like that of Ferguson–Klass for dimension one, but with i.i.d. vector multipliers on the Poisson terms. The law of these coefficients may be chosen proportional to the Lévy measure, although this is not necessary. These results take a particularly elegant form in the case of symmetric stable laws, where something of a calculus is developed showing: (i) which Lévy measure associates with vector coefficients other than the aforementioned ones, (ii) what happens when independent stables are linearly combined as in a weighted sum, (iii) how to construct an arbitrary multidimensional independent–increments symmetric stable set function, and (iv) how to construct an arbitrary harmonizable stationary symmetric stable random function having multidimensional domain and/or range. Symmetric stable laws are shown to be mixtures of Gaussian laws.

Partly because of the self–contained character of Kuelbs' paper [3], in which the characterization of the log–characteristic function of a stable law is extended to real separable Hilbert space, the Hilbert space level of generality has been chosen for this paper. Later extensions

---

AMS 1970 Subject Classifications. Primary 60E07; Secondary 62E10

Partially supported by National Science Foundation MCS 78–26143, and the Office of Naval Research N00014–76–C–0475.

of Kuelbs' result to Banach and more general spaces support a corresponding generalization of these results. In addition to Kuelbs' result we need a method employed by Ferguson and Klass to transform certain dependent series into eventually identical independent ones. We also require standard results giving conditions under which an independent series in Hilbert space converges almost surely (e.g. [2], Theorem 5.3). The rest of the paper is basically self–contained and affords a surprisingly accessible and clear view of $\alpha < 2$ stable laws and random functions, based on elementary series constructions.

Part II of this paper will generalize these results to the infinitely divisible case.

2. NOTATION.

The following symbols and conventions will be in force throughout this paper.

$$\begin{array}{ll} \approx & \text{"is asymptotic with"} \\ \triangleq & \text{"equals by definition"} \\ \overset{D}{=} & \text{"has the same distribution as"} \\ \Rightarrow & \text{"converges in distribution to"} \\ \alpha & 0 < \alpha < 2, \text{ an index of stability} \\ \{\Gamma_j, j \geq 1\} & \text{arrival times of a Poisson process with unit rate} \\ H & \text{a real separable Hilbert space} \end{array} \tag{2.1}$$

The material of the next section is drawn from [4].

3. MOTIVATION.

Limit theorems are not the subject of this paper. However, we should not proceed without benefit of the following example, which exposes some connections between $\alpha < 2$ stable r.v. and the Poisson process.

Let $\{\epsilon_j, j \geq 1\}$ be independent of the sequence Γ and i.i.d. with $P(\epsilon_1 = 1) = P(\epsilon_1 = -1) = 1/2$, and define $G(x) \triangleq x^{-\alpha}$, $\forall x \geq 1$. Think of $G(x) = P(|X| \geq x)$ $\forall x \geq 1$, where X is a r.v. symmetrically distributed about zero. We will construct particular r.v. $X_1,\dots,X_n$, i.i.d. as X, whose normalized sums converge in distribution to the symmetric stable law of index α. To do this, we use the arrival times of a Poisson process to generate uniform order statistics, apply G^{-1} to these, multiply by the signs ϵ, and randomly permute. As constructed, the normalized sums will actually converge almost surely to $\Sigma_1^\infty \epsilon_j \Gamma_j^{-1/\alpha}$ (see (3.1) below), a series possessing the symmetric stable law of index α. In fact, a direct proof of the stability of $\Sigma_1^\infty \epsilon_j \Gamma_j^{-1/\alpha}$ follows easily from the observation that the arrival times of several (say $K > 1$) independent unit rate Poisson processes (run simultaneously) constitute K^{-1} times the arrival times of a unit rate Poisson process. This argument works just as well for ϵ replaced by any vector sequence (provided the series converges) and suggests the multivariate extensions of sections 4 and 5.

For each $n > 1$ let $U_{(1)} \leq U_{(2)} \leq \cdots \leq U_{(n)}$ denote the order statistics of i.i.d. random variables $U_1,\ldots,U_n$ which are uniformly distributed on [0,1]. Then for each fixed $n > 1$, letting $X_j \triangleq \epsilon_j G^{-1}(U_j)$, we have that,

$$n^{-1/\alpha}\Sigma_1^n \epsilon_j X_j \stackrel{D}{=} n^{-1/\alpha}\Sigma_1^n \epsilon_j G^{-1}(U_{(j)}) = n^{-1/\alpha}\Sigma_1^n \epsilon_j U_{(j)}^{-1/\alpha} \tag{3.1}$$

$$\stackrel{D}{=} n^{-1/\alpha}\Sigma_1^n \epsilon_j(\Gamma_j/\Gamma_{n+1})^{-1/\alpha} = \Sigma_1^n \epsilon_j \Gamma_j^{-1/\alpha}(\Gamma_{n+1}/n)^{1/\alpha}$$

$$\downarrow \text{a.s. (3 series)}, \quad \downarrow \text{a.s. (SLLN)}$$

$$\Sigma_1^\infty \epsilon_j \Gamma_j^{-1/\alpha} \qquad 1$$

It is convenient to refer to $\{\epsilon_j\Gamma_j, j \geq 1\}$ as the residual order statistics, keeping in mind that the ordering is on decreasing absolute values.

The same example suggests an invariance principle (proved in [4]) for self–normed sums such as $\Sigma_1^n X_j/\sqrt{\Sigma_1^n X_n^2}$ which, regardless of α, converge in distribution to a limit law depending only on the stable attracting X_1. For the r.v. constructed above,

$$\frac{\Sigma_1^n X_j}{\sqrt{\Sigma_1^n X_j^2}} \stackrel{D}{=} \frac{\Sigma_1^n \epsilon_j\Gamma_j^{-1/\alpha}}{\sqrt{\Sigma_1^n \Gamma_j^{-2/\alpha}}} \stackrel{\text{a.s.}}{\rightarrow} \frac{\Sigma_1^\infty \epsilon_j\Gamma_j^{-1/\alpha}}{\sqrt{\Sigma_1^\infty \Gamma_j^{-2/\alpha}}} \tag{3.2}$$

That is, the limit law of the t–statistic (the square of this t–statistic is simply related to the square of Student's–t and both have the same limit law) is that of the t–statistic calculated on the residual order statistics (see also [6]).

Even the construction of stable independent–increments processes can be motivated by means of the same example. We restrict our attention to the homogeneous increments case. Let $\{T_j, j \geq 1\} \stackrel{D}{=} \{U_j, j \geq 1\}$, and suppose the sequences T, ϵ, Γ are mutually independent. The partial sum processes $\Sigma_1^{[nt]} X_j$, $0 \leq t \leq 1$, $n \geq 1$, can be effected by independent selections of $X_1,\ldots,X_n$ into subsets of sizes $[nt]$ using multiplication by indicators:

$$I_1^{(n)}(t) \triangleq I(T_1 \leq \tfrac{[nt]}{n}) \tag{3.3}$$

$$I_j^{(n)}(t) \triangleq I\left(T_j \leq \frac{[nt] - \Sigma_{i<j} I_i^{(n)}(t)}{n+1-j}\right), \forall\, 1 \leq j \leq n\,.$$

Then for each $n \geq 1$,

$$\{n^{-1/\alpha}\Sigma_1^{[nt]}X_j,\ t \in [0,1]\} \stackrel{D}{=} \{\Sigma_1^n I_j^{(n)}(t)\,\epsilon_j\Gamma_j^{-1/\alpha}(\Gamma_{n+1}/n)^{1/\alpha},\ t \in [0,1]\}$$

$$\downarrow \text{a.s. in } D[0,1]$$

$$\{\Sigma_1^\infty I(T_j \leq t)\epsilon_j\Gamma_j^{-1/\alpha},\ t \in [0,1]\}\,. \tag{3.4}$$

Details of this argument are unpublished.

4. STABLE LAWS ON H.

Suppose $\{X_j, j \geq 1\}$ are i.i.d. random vectors in a real separable Hilbert space H and that the sequences X, Γ are independent. For each $n \geq 1$ denote by K_n the number of arrival times Γ in the interval $[0, a_n]$, $a_n \triangleq \Sigma_1^n j^{-1}$. This choice of a_n is from [1]. Its advantages will be apparent in what follows.

Remark. Sums of the kind $\Sigma_1^{K_n}$ are for each $n \geq 1$ defined to zero on the event $K_n = 0$. Use (,), || ||, to denote H inner product and norm.

For each $n \geq 1$, $x \in H$, $c > 0$, (see also [1], pg. 1639),

$$E\, e^{i\,(x, c\Sigma_1^{K_n} X_j \Gamma_j^{-1/\alpha})} = E\, E^{K_n}\, e^{i\, c\Sigma_1^{K_n}(x,X_j)\Gamma_j^{-1/\alpha}} \tag{4.1}$$

$$= E\big(a_n^{-1} \int_0^{a_n} E e^{ic(x,X_1)t^{-1/\alpha}}\, dt\big)^{K_n}$$

$$= e^{E \int_0^{a_n} (e^{ic(x,X_1)t^{-1/\alpha}} - 1)\, dt}$$

$$= e^{E\, c^\alpha \alpha \int_{ca_n^{-1/\alpha}}^{\infty} (e^{i(x,X_1)r} - 1)\, r^{-(1+\alpha)}\, dr}$$

Kuelbs ([3], lemma 2.2) has proved that the log–characteristic function of a (non–Gaussian) stable probability measure on H is necessarily of the following form, for a unique $\beta \in H$ and finite Borel measure σ on $S \triangleq \{x \in H \,\|x\| = 1\}$,

$$i(x, \beta) + \int_S \int_0^\infty (e^{i(x,s)r} - 1 - \frac{i(x,s)r}{1+r^2}) \frac{dr}{r^{1+\alpha}}\, \sigma(ds),\ \forall x \in H \tag{4.2}$$

It is convenient to refer to expression (4.2) with $\beta = 0$ as (4.3). Define $\delta \triangleq \sigma(s)$, $\mu \triangleq \sigma/\delta$, and

$$\beta_n \triangleq \int_S \int_{(\alpha a_n/\delta)^{-1/\alpha}}^\infty \frac{sr}{1+r^2} \frac{dr}{r^{1+\alpha}}\, \sigma(ds),\ \forall n \geq 1 .$$

Lemma 4.4 If (4.3) is the characteristic function of a stable law on H and if the sequence $\{X_j, j \geq 1\}$ is i.i.d. μ and independent of the sequence Γ, define

$$X^{(n)} \triangleq \alpha^{-1/\alpha} \delta^{1/\alpha} \Sigma_1^{K_n} X_j \Gamma_j^{-1/\alpha} - \beta_n,\ \forall n \geq 1 .$$

Then $\forall x \in H$, $E \exp i(x, X_n)$ converges to (4.3).

Remark. This result is not altogether satisfactory since convergence of the series $\{X^{(n)}, n \geq 1\}$ is through stochastic times K and is not yet a.s. in H. These defects are remedied in Theorem 4.8 below.

Proof. For each $x \in H$, $n \geq 1$, by (4.1),

$$\log_e E e^{i(x,X^{(n)})} = -i(x,\beta_n) + E\,\delta \int_{\alpha^{-1/\alpha}\delta^{1/\alpha}a_n^{-1/\alpha}}^{\infty} (e^{i(x,X_1)r} - 1)\frac{dr}{r^{1+\alpha}}$$

$$= \int_S \int_{\alpha^{-1/\alpha}\delta^{1/\alpha}a_n^{-1/\alpha}}^{\infty} (e^{i(x,s)r} - 1 - \frac{i(x,s)r}{1+r^2})\frac{dr}{r^{1+\alpha}}\,\sigma(ds) \tag{4.5}$$

$\to$ (4.3) . $\square$

From ([1], lemma 2), we conclude that

$$\Sigma_{K_n}^{K_{n+1}} X_j \Gamma_j^{-1/\alpha},\ n \geq 1,$$

are independent. Furthermore, using an argument drawn from [4],

$$E\left\| \Sigma_{K_n}^{K_{n+1}} X_j \Gamma_j^{-1/\alpha} \right\|^2 \leq a_n^{-2/\alpha} E(K_{n+1} - K_n)^2 \tag{4.6}$$

$$\approx < \frac{2}{n(\log_e n)^{2/\alpha}},$$

which is summable in n. Therefore,

$$\Sigma_1^{K_n} X_j \Gamma_j^{-1/\alpha} - E\,\Sigma_1^{K_n} X_j (\Gamma_j^{-1/\alpha} \wedge 1)$$

converges in probability in H. A short calculation gives

$$A_n(\alpha,\delta) \triangleq E\alpha^{-1/\alpha}\delta^{1/\alpha}\Sigma_1^{K_n}(\Gamma_j^{-1/\alpha}\wedge 1) - \delta\int_{\alpha^{-1/\alpha}\delta^{1/\alpha}a_n^{-1/\alpha}}^{\infty} \frac{r}{1+r^2}\frac{dr}{r^{1+\alpha}}$$

$$\overset{(3.1)}{=} \alpha^{-1/\alpha}\delta^{1/\alpha}\int_0^{a_n}((t^{-1/\alpha}\wedge 1) - \frac{t^{-1/\alpha}}{1+\alpha^{-2/\alpha}\delta^{2/\alpha}t^{-2/\alpha}})\,dt \tag{4.7}$$

$\longrightarrow$ finite limit $\triangleq A(\alpha,\delta)$ as $n \to \infty$.

Therefore $X^{(n)}$ converges in probability. $\square$

Theorem 4.8. If (4.3) is the log characteristic function of a stable law on H then the series

$$\alpha^{-1/\alpha}\delta^{1/\alpha}\,\Sigma_1^n \{ X_j\Gamma_j^{-1/\alpha} - (EX_1)\int_{j-1}^{j} \frac{t^{-1/\alpha}}{1+\alpha^{-2/\alpha}\delta^{2/\alpha}t^{-2/\alpha}}\,dt \} \tag{4.8}$$

converges a.s. in H to a random vector with log characteristic function (4.3).

Remark. Centering is not needed for the case $\alpha < 1$, nor is it needed for the symmetric case which will appear in [4].

Proof. Since $X^{(n)}$ converges in probability and $X^{(n)}$ is an independent series, we conclude by ([2], Theorem 5.3(6)) that $X^{(n)}$ converges a.s. in H. Recall that with probability one $\exists$ finite M such that $(n \geq M) \Rightarrow (\exists$ smallest $N(n)$ with $n = K_{N(n)})$. Then $\forall n \geq M$,

$$X^{(N(n))} = \alpha^{-1/\alpha}\,\delta^{1/\alpha}\{\Sigma_1^n X_j\Gamma_j^{-1/\alpha} - (EX_1)\int_0^{a_{N(n)}}(t^{-1/\alpha}\wedge 1)dt\} + (EX_1)A_n(\alpha,\delta) + o(1)\,. \tag{4.9}$$

Since for $n \geq M$, $(n = K_{N(n)}) \Rightarrow (a_{N(n)-1} < \Gamma_n \leq a_{N(n)})$,

$$\left|\int_{\Gamma_n}^{a_{N(n)}}(t^{-1/\alpha}\wedge 1)dt\right| \leq (N(n))^{-1} \xrightarrow{a.s.} 0\,. \tag{4.10}$$

By the law of the iterated logarithm, a.s. eventually as $n \to \infty$

$$\left|\int_n^{\Gamma_n}(t^{-1/\alpha}\wedge 1)dt\right| = \int_n^{\Gamma_n} t^{-1/\alpha}\,dt \tag{4.11}$$

$$\leq |\Gamma_n - n|\,(n\wedge\Gamma_n)^{-1/\alpha} \leq 2\sqrt{n\log\log n}\,(n+o(n))^{-1/\alpha} \to 0\,.$$

Therefore,

$$X^{(N(n))} = \alpha^{-1/\alpha}\,\delta^{1/\alpha}\{\Sigma_1^n X_j\Gamma_j^{-1/\alpha} - (EX_1)\int_0^n (t^{-1/\alpha}\wedge 1)dt\} - A(\alpha,\delta)(EX_1) + o(1)\,. \tag{4.12}$$

which converges a.s. in H to a random vector with log characteristic function (4.3). □

Remark. The centerings used above also have an interpretation involving (EX_1) $E(\Gamma_j^{-1/\alpha}\wedge 1)$, which will not be given here.

5. SYMMETRIC CASE, MULTIPLE REPRESENTATIONS.

In this section we do not assume that X_1 is distributed according to the measure μ, or even restrict its distribution to S. Suppose X, ϵ, Γ are mutually independent sequences, with ϵ, Γ as in section 3, and $\{X_j, j\geq 1\}$ i.i.d on H with $E\|X_1\|^\alpha < \infty$.

Remark. Series of form $\Sigma_1^n \epsilon_j X_j \Gamma_j^{-1/\alpha}$ will be termed symmetric.

Lemma 5.1. The symmetric series $\Sigma_1^n \epsilon_j X_j \Gamma_j^{-1/\alpha}$ converges a.s. in H and the log characteristic function of its limit is $E|(x, X_1)|^\alpha B(\alpha)$, $\forall x \in H$, where $B(\alpha) \triangleq \int_0^\infty \alpha(\cos(r)-1)\frac{dr}{r^{1+\alpha}}$.

Proof. The arguments needed are similar to those of theorem 4.8, but easier in the symmetric case. For each $x\in H$,

$$\log_e Ee^{i(x,\Sigma_1^{K_n}\epsilon_j X_j\Gamma_j^{-1/\alpha})} \overset{(4.1)}{=} \alpha E\int_{a_n^{-1/\alpha}}^\infty (e^{i\epsilon_1|(x,X_1)|r}-1)\frac{dr}{r^{1+\alpha}}$$

$$= \alpha E\int_0^\infty (e^{i\epsilon_1 r}-1)\, I(|(x,X_1)| \leq r a_n^{1/\alpha})\,|(x,X_1)|^\alpha \frac{dr}{r^{1+\alpha}}$$

$$= \alpha \, E \int_0^\infty (\cos(r)-1) \, I(|(x,X_1)| \le r a_n^{1/\alpha}) |(x, X_1)|^\alpha \frac{dr}{r^{1+\alpha}}$$

$$= \alpha \int_0^\infty (\cos(r)-1) \frac{dr}{r^{1+\alpha}} E|(x, X_1)|^\alpha . \tag{5.2}$$

From ([3], corollary 2.1), the limit (5.2) is the characteristic function of a (symmetric) stable law on H. Since the sums $\Sigma_{K_n}^{K_{n+1}} \epsilon_j X_j \Gamma_j^{-1/\alpha}$ are independent and symmetric, we have from ([2], theorem 5.3 (1)) that $\Sigma_1^{K_n} \epsilon_j X_j \Gamma_j^{-1/\alpha}$ converges a.s. in H. Since eventually a.s. $K_{n+1} \le K_n+1$ as $n \to \infty$, we conclude $\Sigma_1^n \epsilon_j X_j \Gamma_j^{-1/\alpha}$ converges a.s. in H. □

Several series may represent the same stable law.

<u>Theorem 5.2</u>. If $E\|X_1\|^\alpha < \infty$ then for every $x \in H$, $E|(x, X_1)|^\alpha = E\|X_1\|^\alpha E|(x, X_1^*)|^\alpha$ where X_1^* is distributed on S according to the measure:

$$P(X_1^* \in A) = EI\left(\frac{X_1}{\|X_1\|} \in A\right) \frac{\|X_1\|^\alpha}{E\|X_1\|^\alpha}$$

<u>Proof</u>. For every $x \in H$,

$$E|(x, X_1)|^\alpha = E|(x, \frac{X_1}{\|X_1\|})|^\alpha \|X_1\|^\alpha$$

$$= E\|X_1\|^\alpha E|(x, X_1^*)|^\alpha. \quad \square$$

As an example of the above, every symmetric stable law on H has a construction of the form $\Sigma_1^\infty Z_j X_j \Gamma_j^{-1/\alpha}/E(|Z_1|^\alpha)^{1/\alpha}$ in terms of an independent sequence Z of i.i.d. standard normal r.v.. Conditional on the sequences Γ and X the symmetric stable is Gaussian. That is, symmetric stable laws are particular mixtures of Gaussian laws with zero means and differing covariance kernels. The latter will not in general differ only by scale, though this is necessarily true for $H = \mathbb{R}$.

6. SYMMETRIC STABLE RANDOM SET FUNCTIONS.

Let $\{\tau_j, j \ge 1\}$ be i.i.d taking values in a measurable space with measurable sets generically denoted A. Suppose $\tau, \epsilon, X, \Gamma$ are mutually independent, where the latter three sequences are as in section 5. Define

$$X(A) \triangleq \Sigma_1^\infty I(\tau_j \in A) X_j \epsilon_j \Gamma_j^{-1/\alpha}, \ \forall A . \tag{6.1}$$

<u>Theorem 6.2</u>. The series (6.1) is a.s. convergent for each A, is jointly symmetric stable for finitely many A at a time, and $X(A_1),\dots,X(A_n)$ are for each $n \ge 1$ mutually independent if $A_1,\dots,A_n$ are mutually disjoint.

<u>Proof</u>. For each $n \ge 1$, $x \in H$, real numbers $r_1,\dots,r_n$ and measurable sets $A_1,\dots,A_n$

$$\log_e E e^{i(x,\, \Sigma_1^n r_k X(A_k))} = \log_e E\, e^{i \sum_{j=1}^{\infty} (\Sigma_{k=1}^n r_k I(\tau_j \in A_k) \} (x, X_j) \epsilon_j \Gamma_j^{-1/\alpha}}$$

$$\overset{(5.1)}{=} B(\alpha)\, E\, |(x, X_1)|^{\alpha}\, E\, |\Sigma_{k=1}^n r_k I(\tau_1 \in A_k)|^{\alpha}. \tag{6.2}$$

If $A_1,\ldots,A_n$ are mutually disjoint then

$$E|\Sigma_{k=1}^n r_k I(\tau_1 \in A_k)|^{\alpha} = \Sigma_{k=1}^n |r_k|^{\alpha} P(\tau_1 \in A_k). \quad \square \tag{6.3}$$

Remark. The simplicity of this construction is interesting, as is the way in which α–dependence, dimensional structure, and functional dependence are identified with mutually independent coefficient sequences $\Gamma^{-1/\alpha}$, X, I.

Remark. Schilder [8] and Kuelbs [3] have explored a representation of multidimensional symmetric stable r.v. by means of a stochastic integral with respect to a one–dimensional stable independent increments process. Theorem (4.8) and lemma (5.1) sharpen and extend such representations by connecting them with the Ferguson–Klass representation, making explicit the choice of coefficients required to obtain each stable law[4.], and establishing H convergence of the indicated series.

Remark. Suppose $K > 1$ and $Y_k = \Sigma_1^{\infty} \xi_{kj} \Gamma_{kj}^{-1/\alpha}$, $1 \le k \le K$, are independently constructed (as per (5.1)) symmetric stable r.v. taking values in H. Then for an arbitrary choice of real number $r_1,\ldots,r_K$ the sum $\Sigma_1^K r_k Y_k$ is representable $\Sigma_1^{\infty} \xi_j \Gamma_j^{-1/\alpha}$ where $\{\xi_j, j \ge 1\}$ are i.i.d. and $\xi_1 \overset{D}{=}$ an equi–probable random selection from $K^{1/\alpha} r_1 \xi_{11},\ldots,K^{1/\alpha} r_n \xi_{n1}$. This uses the property (discussed in section 3) of K Poisson processes run simultaneously.

7. HARMONIZABLE STATIONARY SYMMETRIC STABLE RANDOM FUNCTIONS.

Basically, we seek to construct the stable analogies of Gaussian stationary random functionharmonic decomposition. The characteristic function of such a Gaussian random function involves

$$E\left|\Sigma_{k=1}^n r_k e^{i(\Lambda_1, t_k)_o}\right|^2,$$

where t is generic for a point of the domain, and $(\Lambda_1, \;)_0$ is a random linear function on the domain. The stable analogues of these Gaussian random functions have characteristic functions that employ an $\alpha^{\underline{th}}$ –power in this integral instead of the 2, but are otherwise the same. Define $\forall t$,

$$X(t) \overset{\Delta}{=} \Sigma_1^{\infty} \cos\left((\Lambda_j, t)_o + \Theta_j\right) X_j \epsilon_j \Gamma_j^{-1/\alpha} \tag{7.1}$$

where Λ are i.i.d., Θ are i.i.d. uniforms on $[-\pi, \pi]$, $\{\epsilon, X, \Gamma\}$ are as in section 6, and Λ, Θ, ϵ, X, Γ are mutually independent sequences. The series (7.1) is a.s. convergent in H for each t by lemma 5.1. The random function X is clearly stationary because Θ are uniform on $[-\Pi, \Pi]$, but this will also be a simple consequence of the form of the characteristic function which we now compute.

$$\log_e E\, e^{i(x,\Sigma_{k=1}^n r_k X(t_k))} = \log_e E\, e^{i\Sigma_{j=1}^\infty\{\Sigma_{k=1}^n r_k \cos((\Lambda_j,t_k)_o+\Theta_j)\}\epsilon_j(x,X_j)\Gamma_j^{-1/\alpha}}$$

$$\overset{(5.1)}{=} B(\alpha)\, E|(x, X_1)|^\alpha\, E|\Sigma_{k=1}^n r_k \cos((\Lambda_1, t_k)_o + \Theta_1)|^\alpha . \tag{7.2}$$

The final term in the right side of (7.2) reduces as follows, letting $z \triangleq \Sigma_1^n r_k e^{i(\Lambda_1,t_k)_o}$,

$$E|\Sigma_1^n r_k \cos((\Lambda_1, t_k)_o + \Theta_1)|^\alpha = E\, 2^{-\alpha}|ze^{i\Theta_1} + \bar{z}e^{-i\Theta_1}|^\alpha \tag{7.3}$$

$$= E\,|z|^\alpha 2^{-\alpha} E^z\, |1 + z^{-1}\bar{z}e^{-2i\Theta_1}|^\alpha$$

$$= E\,|z|^\alpha 2^{-\alpha} C(\alpha)$$

where

$$C(\alpha) = \int_{-\pi}^{\pi} |1 + e^{i\eta}|^\alpha d\eta = \int_{-\pi}^{\pi} |1 + e^{i\psi-2i\eta}|^\alpha d\eta$$

for all real ψ. We have therefore proved,

Theorem 7.4. The random function defined by (7.1) converges a.s. in H for each t, and has log characteristic function

$$2^{-\alpha}B(\alpha)C(\alpha)\, E\, |\Sigma_{k=1}^n r_k e^{i(\Lambda_1,t_k)_o}|^\alpha$$

for all $n \geq 1$, $r_1,\dots,r_n$, $t_1,\dots,t_n$.

Corollary 7.5. The random function (7.1) is non–ergodic for each $\alpha < 2$.

Proof. By using (7.2) the construction (7.1) remains valid if ϵ_j are replaced by $Z_j/(E|Z_1|^\alpha)^{1/\alpha}$, $j \geq 1$, where Z is an independent i.i.d. standard normal sequence. Conditional on the sequences Λ, X, Γ, the process (7.1) is a.s. stationary Gaussian with discrete spectrum, therefore conditionally non–ergodic a.s.. .

8. OPERATOR STABLE AND INFINITELY DIVISIBLE MULTIDIMENSIONAL LAWS.

Infinitely divisible laws, of which the operator stable laws are a special case with particularly interesting structure, are treated in Part II. In brief, this is what happens: A construction of infinitely divisible random vectors is given by $\Sigma\{X_j H(\Gamma_j, X_j) - \gamma_j\}$, in which the real function H is monotone decreasing and positive for each value X_j, and is determined from the Lévy measure. A construction of full operator stable random vectors in a finite dimensional real vector space R^d is $\Sigma\{A(\Gamma_j^{-1})X_j - \gamma_j\}$, in which the vectors γ are non–stochastic centerings, $\{A(t) = \exp(B\log t), t > 0\}$ is the group of linear transformations figuring in the definition of operator stability (e.g. Sharpe [9]), and the vectors X are i.i.d. from a probability measure (a factor of the Lévy measure) on a set of generators of the subgroups induced by A. The methods of sections 4 and 5 carry over, as will now be indicated. If X is any i.i.d. sequence in R^d, and X is independent of Γ, then $\forall x \in R^d$, $n \geq 1$,

$$\log_e E\, e^{i(x,\, \Sigma_1^{K_n} A(\Gamma_j^{-1})X_j)} = \int_{a_n^{-1}}^{\infty} E\,(e^{i(x,A(t)X_1)}-1)\,\frac{dt}{t^2}\,. \tag{8.1}$$

As usual, the symmetric case is simplest. If we examine Sharpe's Theorem 5, we discover that the limit of (8.1) is precisely the form taken by the operator stable in this case, provided we choose for the distribution of X_1 the probability measure figuring in Sharpe's representation of the Lévy measure as a mixture, this measure being placed on (Sharpe's notation) generators θ characterized by $s\ M_\theta\{t^B\theta\colon t > s\} = s^{-1}$, $\forall\, s > 0$. Arguing as in section 5, we conclude $\Sigma_1^\infty A(\Gamma_j^{-1})X_j$ converges a.s. in R^d and has the log–characteristic function which is the limit of (8.1).

9. PRIORITY OF P. LEVY.

P. Lévy has anticipated the series constructions of one dimensional stable r.v. with $\alpha < 2$. For the case of a positive stable with $\alpha < 1$, up to scale and location, this construction is $\sum_{j=1}^{\infty} \Gamma_j^{-1/\alpha}$, with $\{\Gamma_j,\ j > 1\}$ being the arrival times of a Poisson process (on R^+) having unit intensity function. Lévy writes the series in the form $\sum_x U_x$, where

$$\{U_x,\ x > 0\} \text{ are independent r.v.} \tag{9.1}$$

and

$$P(U_x = x) = \frac{\alpha dx}{x^{1+\alpha}} = 1 - P(U_x = 0)\,.$$

Here is my abstract of the key parts of Lévy's (1935) arguments for the above case:

$$\left[\int_{x_0}^{\infty} \frac{\alpha dx}{x^{1+\alpha}} = x_0^{-\alpha} < \infty\right] \Rightarrow \left[\ \{U_x\colon U_x \neq 0,\ x > x_0\} \text{ is finite}\ \right]$$

for $x_0 > 0$. Also, as $x_0 \downarrow 0$,

$$\left[\int_0^{x_0} \frac{\alpha dx}{x^{\alpha}} = \frac{\alpha}{1-\alpha}\, x_0^{(1-\alpha)} \downarrow 0\right] \Rightarrow \left[E \sum_{x<x_0} U_x \downarrow 0\right].$$

Therefore, for arbitrary $c_1, c_2 > 0$ (defining $c_3^\alpha = c_1^\alpha + c_2^\alpha$ and taking independent copies),

$$c_1 \sum_x U_x^{(1)} + c_2 \sum_x U_x^{(2)} \stackrel{D}{=} \sum_x Y_x^{(1)} + \sum_x Y_x^{(2)} \stackrel{D}{=} \sum_x Y_x^{(x)} \stackrel{D}{=} c_3 \sum_x U_x \quad (\Rightarrow \text{stable})\,,$$

where $\{Y_x^{(k)},\ x > 0\}$ have respective intensities $c_k^\alpha\, \alpha\, dx/x^{1+\alpha}$, $k = 1, 2, 3$ and are independent for $k = 1, 2$.

The above arguments do yield a proof of the representation if we apply them to the independent sub–sums $\Sigma\,\{U_x\colon\ x \in [b_{n+1}, b_n]\}$, $n \geq 1$, where $b_n^{-\alpha} = \log n$. This is essentially the argument of Ferguson–Klass (1972). The particular choice of b_n, $n \geq 1$ is one which ensures that eventually as $n \to \infty$ each sub–sum contains at most one summand, so it really is (almost) as though one could add independent U_x one at a time toward $x \to 0$. A quite different justification is to interpret $\sum_x U_x$ as a generalized process driven by "white noise" $\{U_x, x > 0\}$.

Lévy's observations are easily overlooked. Ferguson and Klass, Vervaat (1979), LePage and Woodroofe and Zinn (1979) (in manuscript form), rediscover the Lévy construction as byproducts of the following independent pursuits respectively: (F–K)–representing the positive non–Gaussian part of an independent increments random function as the sum of its ordered jumps. (V)–examining a shot–noise associated with the asymptotic behavior of the solution of a stochastic difference equation as time is increased, (L–W–Z)–studying the limit behavior of the normalized order statistics from a distribution attracted to a stable. Resnick (1976) reconciles the Ferguson–Klass construction with the Ito representation, meaning by the latter Ito's generalization of Lévy's stochastic integral construction by a Poisson random measure.

Acknowledgement. The estimate (4.6) was suggested by J. Zinn. Helpful discussions were also held with V. Mandrekar, M. Woodroofe, and M. Steele.

REFERENCES.

[1] Ferguson, T. and Klass, M. (1972). A representation of independent increments processes without Gaussian components. Ann. Math. Statist., 43, 5, 1634–1643.

[2] Hoffman–Jϕrgensen, J. (1977). Probability in B–spaces. Lecture note series, No. 45, Aarhus Universitet.

[3] Kuelbs, J. (1973). A representation theorem for symmetric stable processes and stable measures on H. Z. W. Verw. Geb. 26, 259–271.

[4] LePage, R., Woodroofe, M., and Zinn, J. (1980). Convergence to a stable limit distribution via order statistics. Ann. Prob. 9, 624–632.

[5] Lévy, P. (1935). Properties asymptotiques des sommes de variables aleatoires independantes ou enchainees. Journ. de Math., tome XIV.–Fasc. IV.

[6] Logan, B. F., Mallows, C. F., Rice, S. O., and Shepp, L. A. (1973). Limit distributions of self–normalized sums. Ann. Prob., 1, 5, 788–809.

[8] Resnick, S. (1976). Anextermal decomposition of a process with stationary, independent increments. Tech. Rep. No. 79, Department of Statistics, Stanford University, Stanford, California.

[9] Sharpe, M. (1969). Operator–stable probability distributions on vector groups. Transactions Amer. Math. Soc., 135, 51–65.

[10] Vervaat, W. (1979). On a stochastic difference equation and a representation of non–negative infinitely divisible random variables. Adv. Appl. Prob., 11, 750–783.

Probability Theory on Vector Spaces IV
Lancut, June '87, Springer's LNM 1391

REAL TRANSLATES OF COMPLEX MEASURES

by

M. Lewandowski and W. Linde

In 1976 W.Rudin (cf. [Ru]) proved the following remarkable uniqueness theorem for measures on the complex plane (cf. [Pl 1] for a slighty weaker result): Let μ and ν be two finite measures on the complex plane $\mathbb{C}$ satisfying

$$(*) \qquad \int_{\mathbb{C}} |z+w|^p d\mu(z) = \int_{\mathbb{C}} |z+w|^p d\nu(z) < \infty$$

for all $w \in \mathbb{C}$ and some $p > 0$, $p \neq 2k$. Then necessarily $\mu = \nu$. This theorem has been extended into several different directions. For instance, the function $|\cdot|^p$ has been replaced by more general functions (cf. [Go/Ko], [St]) or the same question was investigated for Banach spaces with the norm instead of the absolute value (cf. [Ko 1], [Ko 2], [Li 2]). Sometimes only special measures a Gaussian ones have been treated (cf. [Le], [Li 3]). The aim of this paper is to extend Rudin's theorem into another direction. We investigate measures μ and ν on $\mathbb{C}$ which satisfy $(*)$ only for all real numbers w. Such questions arose during our investigations in [Li 3]. Easy examples show that we cannot expect to obtain $\mu = \nu$ assuming $(*)$ only for real numbers w and some $p > 0$, $p \neq 2k$. But we characterize pairs of measures satisfying this weaker property for some $p > 0$. Futhermore, we prove $\mu = \nu$ under some additional assumptions about μ and ν.

We state several interesting open questions related to our problem.

1. NOTATION

For each $p > 0$ we denote by $M_p(\mathbb{C})$ or $M_p(\mathbb{R})$ the set of finite measures μ on $\mathbb{C}$ or $\mathbb{R}$ with

$$\int_{\mathbb{C}} |z|^p d\mu(z) < \infty \quad \text{or} \quad \int_{\mathbb{R}} |u|^p d\mu(u) < \infty ,$$

respectively. Given $\mu, \nu \in M_p(\mathbb{C})$ we write $\mu \sim_p \nu$ whenever

$$\int_{\mathbb{C}} |z+t|^p d\mu(z) = \int_{\mathbb{C}} |z+t|^p d\nu(z)$$

for all real numbers t .

If μ is a finite measure on $\mathbb{C}$, its characteristic function $\hat{\mu}: \mathbb{C} \to \mathbb{C}$ is defined by

$$\hat{\mu}(u + iv) = \int_{\mathbb{C}} e^{i(ux+vy)} d\mu(x+iy) .$$

Recall that μ is uniquely determined by $\hat{\mu}$. During our subsequent investigations a special measure σ_p on $\mathbb{R}$ will play an important role. It is defined by

$$d\sigma_p(u) := \frac{C_p du}{(1+u^2)^{p/2+1}}$$

where

$$C_p := \left[\int_{-\infty}^{\infty} \frac{du}{(1+u^2)^{p/2+1}}\right]^{-1} = \left[\int_{-\pi/2}^{\pi/2} \cos^p\theta d\theta\right]^{-1} = \frac{\Gamma(\frac{p}{2}+1)}{\sqrt{\pi}\ \Gamma(\frac{p+1}{2})} .$$

Its characteristic function $\hat{\sigma}_p$ is denoted by k_p and

$$k_p(s) := \int_{-\infty}^{\infty} \cos(su) d\sigma_p(u) = C_p \int_{-\infty}^{\infty} \frac{\cos(su)}{(1+u^2)^{p/2+1}} du$$

$$= 2s^{p+1} C_p \int_0^{\infty} \frac{\cos u}{(s^2+u^2)^{p/2+1}} du$$

$$= \frac{2C_p s^{p+1}\sqrt{\pi}}{\Gamma(\frac{p}{2}+1)(2s)^{p/2+1/2}} K_{p/2+1/2}(s)$$

$$= \frac{s^{p/2+1/2}}{2^{p/2-1/2}\Gamma(\frac{p+1}{2})} K_{p/2+1/2}(s) , \quad s > 0 ,$$

where $K_{p/2+1/2}$ denotes the modified Bessel function of the third kind (cf. [Ba/Er]). Clearly, $k_p(0) = 1$ and $k_p(-s) = k_p(s)$. We shall use the following properties of k_p :

(1) Since $K_{p/2+1/2}(s) > 0$, $s > 0$, it follows

$$k_p(s) > 0 , \quad s \in \mathbb{R} \quad \text{(cf. [Wa])} .$$

(2) Because of

$$K_{p/2+1/2}(s) = \sqrt{\frac{\pi}{2s}}\, e^{-s}(1 + O(\tfrac{1}{s})) , \quad s \to \infty \quad \text{(cf. [Ba/Er])}$$

we have

$$k_p(s) = \frac{\sqrt{\pi}}{\Gamma(\frac{p+1}{2})\, 2^{p/2}} s^{p/2} e^{-s}(1 + O(\tfrac{1}{s})) , \quad s \to \infty .$$

2. FORMULATION OF THE PROBLEM

In view of Rudin's theorem one could hope that $\mu \sim_p \nu$ for $p \neq 2k$, $k \in \mathbb{N}$, implies $\mu = \nu$. We want to show now that this is not so by trivial reasons. Let us first mention that $\mu \sim_p \nu$ is equivalent to

$$\int_{\mathbb{C}} [(x+t)^2 + y^2]^{p/2} d(\mu - \nu)(x+iy) = 0$$

for all $t \in \mathbb{R}$. Thus, if $f : \mathbb{R} \to \mathbb{R}$ is any measurable function with $|f(y)| = |y|$, $y \in \mathbb{R}$, then $\mu \sim_p \mu_f$ where

$$\mu_f(B) := \mu\{x+iy \in \mathbb{C}\ ,\ x + if(y) \in B\} .$$

Observe that there are a lot of such functions f as, for instance, $f(y) := |y|\chi_{\mathbb{C}}(y) - y\chi_{\mathbb{R}\setminus\mathbb{C}}(y)$ for some $\mathbb{C} \in \mathbb{R}$, measurable. Consequently, one should correctly ask whether or not $\mu \sim_p \nu$, $p \neq 2k$, implies $\mu_+ = \nu_+$ where $\mu_+(B) := \mu\{x+iy \in \mathbb{C};\ x+i|y| \in B\}$ and ν_+ is defined in the same way. Since $\mu \sim_p \mu_+$, this question is equivalent to the following. Does $\mu \sim_p \nu$, $p \neq 2k$, imply $\mu = \nu$ provided that μ and ν are concentrated on $\overline{\mathbb{C}^+} := \{z \in \mathbb{C};\ \mathrm{Im}(z) \geq 0\}$?

We shall show now that even this is not so.

PROPOSITION 1. Let $\mu \in M_p(\mathbb{C})$ be arbitrary. Then there exists a measure $\chi_\mu \in M_p(\mathbb{R})$ with $\mu \sim_p \chi_\mu$. Moreover, if $p \neq 2k$, then χ_μ is uniquely determined by these properties.

PROOF. We start with the equality

$$(a^2 + b^2)^{p/2} = C_p \int_{-\pi/2}^{\pi/2} |a\cos\theta + b\sin\theta|^p d\theta$$

valid for $a,b \in \mathbb{R}$. Here $C_p > 0$ is defined as above. Using this we obtain

$$\begin{aligned}
\int_{\mathbb{C}} |z+t|^p d\mu(z) &= \int_{\mathbb{C}} [(x+t)^2 + y^2]^{p/2} d\mu(x+iy) \\
&= C_p \int_{\mathbb{C}} \int_{-\pi/2}^{\pi/2} |(x+t)\cos\theta + y\sin\theta|^p d\theta d\mu(x+iy) \\
&= C_p \int_{\mathbb{C}} \int_{-\pi/2}^{\pi/2} |x+y\cdot\tan\theta + t|^p \cos^p\theta d\theta d\mu(x+iy) \\
&= \int_{\mathbb{C}} \int_{-\infty}^{\infty} |x + uy + t|^p d\sigma_p(u) d\mu(x+iy) .
\end{aligned}$$

Let $\Phi : \mathbb{C} \times \mathbb{R} \to \mathbb{R}$ be defined by $\Phi(x+iy, u) := x + uy$. Then, if $\chi_\mu := \Phi(\mu \otimes \sigma_p)$, it follows

$$\int_{\mathbb{C}} |z+t|^p d\mu(z) = \int_{-\infty}^{\infty} |v+t|^p d\chi_\mu(v)\ ,\quad t \in \mathbb{R}\ ,$$

i.e. we have $\mu \sim_p \chi_\mu$ as asserted. To prove the uniqueness of χ_μ assume $\mu \sim_p \lambda$ for some $\lambda \in M_p(\mathbb{R})$. Then $\chi_\mu \sim_p \lambda$ as well, and because of the real version of Rudin's theorem (cf. [Ha] or [Li 1]) we conclude $\chi_\mu = \lambda$ provided that $p \neq 2k$. This completes the proof.

COROLLARY. We have $\mu \sim_p \nu$ iff $\chi_\mu \sim_p \chi_\nu$. Consequently, if $p \neq 2k$, then $\mu \sim_p \nu$ is equivalent to $\chi_\mu = \chi_\nu$.

3. A CHARACTERIZATION OF P-EQUIVALENT MEASURES.

As we have seen the property $\mu \sim_p \nu$ is equivalent to $\chi_\mu \sim_p \chi_\nu$ or $\chi_\mu = \chi_\nu$, $p \neq 2k$, respectively. Thus we have to investigate those measures χ_μ .

PROPOSITION 2.

(i) If $\alpha_1, \alpha_2 \geq 0$ and $\mu_1, \mu_2 \in M_p(\mathbb{C})$, then

$$\chi_{\alpha_1\mu_1+\alpha_2\mu_2} = \alpha_1 \chi_{\mu_1} + \alpha_2 \chi_{\mu_2} .$$

(ii) We have $\chi_\mu = \mu$ for each μ concentrated on $\mathbb{R}$.

(iii) If $\mu \in M_p(\mathbb{C})$ with $\mu(\mathbb{R}) = 0$, then χ_μ is absolutely continuous with respect to the Lebesque measure on $\mathbb{R}$ and

$$d\chi_\mu(v) = C_p \int_{\mathbb{C}} \frac{|y|^{p+1}}{(y^2+(x-v)^2)^{p/2+1}} d\mu(x+iy) .$$

(iv) $$\hat{\chi}_\mu(s) = \int_{\mathbb{C}} e^{ixs} k_p(ys)\, d\mu(x+iy) = \int_{-\infty}^{\infty} \hat{\mu}(s+isu)\, d\sigma_p(u) .$$

PROOF. (i) and (ii) are obvious. To prove (iii) we can and do assume that μ is concentrated on $\mathbb{C}^+ = \{z \in \mathbb{C},\ \mathrm{Im}(z) > 0\}$. Otherwise replace μ by μ_+ . If $f : \mathbb{R} \to \mathbb{R}$ is measurable and bounded, then

$$\int_{-\infty}^{\infty} f(v)\, d\chi_\mu(v) = \int_{\mathbb{C}^+} \int_{-\infty}^{\infty} f(x+uy)\, d\sigma_p(u)\, d\mu(x+iy)$$

$$= \int_{-\infty}^{\infty} f(v) \int_{\mathbb{C}^+} \frac{C_p}{y(1+(\frac{x-v}{y})^2)^{p/2+1}} d\mu(x+iy)\, dv$$

proving

$$d\chi_\mu(v) = \int_{\mathbb{C}^+} \frac{C_p y^{p+1}}{(y^2+(x-v)^2)^{p/2+1}} d\mu(x+iy) .$$

Let h_μ be this density. Then

$$\int_{-\infty}^{\infty} h_\mu(v)\, dv < \infty$$

which yields $h_\mu(v) < \infty$ for almost all $v \in \mathbb{R}$. But observe that there are easy examples with $h_\mu(v) = \infty$ for some $v \in \mathbb{R}$. To verify (iv) we use

$$\hat{\chi}_\mu(s) = \int_{-\infty}^{\infty} e^{ivs} d\chi_\mu(v) = \int_{\mathbb{C}^+} \int_{-\infty}^{\infty} e^{i(x+uy)s} d\sigma_p(u) d\mu(x+iy)$$

$$= \int_{\mathbb{C}} e^{ixs} \hat{\sigma}_p(ys) d\mu(x+iy) = \int_{\mathbb{C}} e^{ixs} k_p(ys) d\mu(x+iy)$$

with $k_p = \hat{\sigma}_p$. On the other hand,

$$\widehat{\chi_\mu}(s) = \int_{-\infty}^{\infty} \int_{\mathbb{C}} e^{i(x+uy)s} d\mu(x+iy) d\sigma_p(u) = \int_{-\infty}^{\infty} \hat{\mu}(s+ius) d\sigma_p(u)$$

completing the proof.

Combining all previous results we obtain the following theorem:

THEOREM 3. Let μ, ν be measures in $\mathcal{M}_p(\mathbb{C})$ and suppose that $p \neq 2k$. Then the following are equivalent:

(i) $\mu \sim_p \nu$

(ii) $\chi_\mu = \chi_\nu$, i.e. we have

$$\int_{\mathbb{C}} \int_{-\infty}^{\infty} f(x+uy) d\sigma_p(u) d(\mu - \nu)(x+iy) = 0$$

for each measurable bounded function f on $\mathbb{R}$.

(iii) For each $s \in \mathbb{R}$,

$$\int_{\mathbb{C}} e^{ixs} k_p(sy) d(\mu - \nu)(x+iy) = 0$$

(iv) If $s \in \mathbb{R}$, then

$$\int_{-\infty}^{\infty} \hat{\mu}(s+isu) d\sigma_p(u) = \int_{-\infty}^{\infty} \hat{\nu}(s+isu) d\sigma_p(u) .$$

Moreover, if $\mu(\mathbb{R}) = \nu(\mathbb{R}) = 0$, then this is also equivalent to

(v) $$\int_{\mathbb{C}} \frac{|y|^{p+1}}{(y^2+(x-u)^2)^{p/2+1}} d\mu(x+iy) = \int_{\mathbb{C}} \frac{|y|^{p+1}}{(y^2+(x-u)^2)^{p/2+1}} d\nu(x+iy)$$

for almost all $u \in \mathbb{R}$.

REMARKS.

(1) In the general case (not necessarely $\mu(\mathbb{R}) = \nu(\mathbb{R}) = 0$) we can write $\mu = \mu_1 + \mu_2$ and $\nu = \nu_1 + \nu_2$ with $\mu_1, \nu_1 \in \mathcal{M}_p(\mathbb{R})$ and $\mu_2(\mathbb{R}) = \nu_2(\mathbb{R}) = 0$. Then we have $\mu \sim_p \nu$ iff $\mu_1 + \chi_{\mu_2} = \nu_1 + \chi_{\nu_2}$ and this is equivalent to the following:

The signed measure $\mu_1 - \nu_1$ is absolutely continuous with respect to the Lebesque measure and its density is

$$d(\mu_1 - \nu_1)(v) = (h_{\nu_2} - h_{\mu_2})(v)$$

where h_{ν_2} and h_{μ_2} are the densities of χ_{ν_2} and χ_{μ_2}, respectively.

(2) Define two measures $\tilde{\mu}$ and $\tilde{\nu}$ on $\mathbb{C}$ by

$$d\tilde{\mu}(z) = |\mathrm{Im}(z)|^{p+1} d\mu(z) \quad \text{and} \quad d\tilde{\nu}(z) = |\mathrm{Im}(z)|^{p+1} d\nu(z) .$$

Then (v) is equivalent to

$$\int_{\mathbb{C}} \frac{d\tilde{\mu}(z)}{|z-u|^{p+2}} = \int_{\mathbb{C}} \frac{d\tilde{\nu}(z)}{|z-u|^{p+2}}$$

for almost all $u \in \mathbb{R}$. This is related to the problem of uniqueness of Riesz potentials (cf. [La], p.99). But observe, that there $u \in \mathbb{C}$ (not only $u \in \mathbb{R}$).

We want to investigate χ_μ assuming $\mu = \mu_1 \otimes \mu_2$, i.e.

$$\int_C f(z)d\mu(z) = \int_{-\infty}^{\infty} \int_{-\infty}^{\infty} f(x+iy)d\mu_1(x)d\mu_2(y) .$$

To do so let $\lambda \in M_p(\mathbb{R})$ and define $(\lambda)_p$ by

$$\int_{-\infty} f(v)d(\lambda)_p(v) := \int_{\mathbb{R}} \int_{\mathbb{R}} f(uv)d\sigma_p(u)d\lambda(v) .$$

That is, $(\lambda)_p$ is the σ_p-mixture of dilations of λ.

PROPOSITION 4. If $\mu = \mu_1 \otimes \mu_2 \in M_p(\mathbb{C})$, then $\chi_\mu = \mu_1 * (\mu_2)_p$.

PROOF. This easily follows from $\chi_\mu = \Phi(\mu \otimes \sigma_p)$ with $\Phi(x+iy,u) = x + uy$.

COROLLARY. If $p \neq 2k$, then $\nu_1 \otimes \nu_2 \sim_p \mu_1 \otimes \mu_2$ iff $\mu_1 * (\mu_2)_p = \nu_1 * (\nu_2)_p$. Let us state some properties of the mapping $\lambda \to (\lambda)_p$. Observe that $(\lambda)_p$ is nothing else as $\chi_{\delta_0 \otimes \lambda}$.

PROPOSITION 5. Suppose $\lambda \in M_p(\mathbb{R})$.

(i) $\widehat{(\lambda)}_p(s) = \int_{-\infty}^{\infty} k_p(sy)d\lambda(y) = \int_{-\infty}^{\infty} \hat{\lambda}(su)d\sigma_p(u)$

(ii) If $\lambda\{0\} = 0$, then $(\lambda)_p$ has density

$$d(\lambda)_p(v) = \int_{-\infty}^{\infty} \frac{|y|^{p+1}}{(y^2+v^2)^{p/2+1}} d\lambda(y)$$

(iii) $(\lambda)_p$ is a symmetric measure on $\mathbb{R}$ and

$$(\lambda)_p = (\lambda_+)_p \quad \text{with} \quad \lambda_+(B) := \lambda\{y \in \mathbb{R},\ |y| \in B\} .$$

(iv) If $p \neq 2k$ and $(\lambda_1)_p = (\lambda_2)_p$, then $(\lambda_1)_+ = (\lambda_2)_+$.

PROOF. We only have to prove (iv). Moreover in view of (iii) we may assume that λ_1 and λ_2 are concentrated on $[0,\infty)$. Then $(\lambda_1)_p = (\lambda_2)_p$ is equivalent to $\delta_0 \otimes \lambda_1 \sim_p \delta_0 \otimes \lambda_2$,

i.e. $\int_0^\infty (t^2+y^2)^{p/2} d\lambda_1(y) = \int_0^\infty (t^2+y^2)^{p/2} d\lambda_2(y)$

for all $t \in \mathbb{R}$. Since $p/2 \notin \mathbb{N}$ we obtain $\lambda_1\{y^2 \in B\} = \lambda_2\{y^2 \in B\}$ for all measurable sets $B \subseteq [0,\infty)$ (cf. [Pl 2] or [Li 2]), i.e. $\lambda_1 = \lambda_2$ as asserted.

REMARKS.

(1) (iv) could also be proved via properties of Riesz potentials.

(2) A reformulation of (iv) is as follows:

$$\text{If} \int_0^\infty k_p(sy)d\lambda_1(y) = \int_0^\infty k_p(sy)d\lambda_2(y)$$

for all $s \in \mathbb{R}$, then $\lambda_1 = \lambda_2$ provided that $p \neq 2k$. It would be interesting to know whether or not this remains true assuming this equatity only for s in some open interval.

4. APPLICATIONS AND EXAMPLES.

As we saw above the relation $\mu \sim_p \nu$ does in general not imply $\mu = \nu$. But in the cases $\mu \sim_p \nu$ and $\mu \neq \nu$ at least one of these measures was not concentrated on $\mathbb{C}^+$ $(p \neq 2k)$ as for instance μ_f or χ_μ. This it is natural to ask whether $\mu \sim_p \nu$, $p \neq 2k$, combined with $\mu(\mathbb{C} \setminus \mathbb{C}^+) = \nu(\mathbb{C} \setminus \mathbb{C}^+) = 0$ yields $\mu = \nu$. That this is not so follows froma general construction. Let $\lambda \in M_p(\mathbb{C})$ be given. Then we define two measures $\mu = \mu(\lambda,p)$ and $\nu = \nu(\lambda,p)$ in the following way:

$$\int_{\mathbb{C}} f(z)d\mu(z) := \int_{\mathbb{C}} \int_{-\infty}^{\infty} f(vx+iy)d\sigma_p(v)d\lambda(x+iy) \quad \text{and}$$

$$\int_{\mathbb{C}} f(z)d\nu(z) := \int_{\mathbb{C}} \int_{-\infty}^{\infty} f(vy+ix)d\sigma_p(v)d\lambda(x+iy) .$$

Since

$$\int_{\mathbb{C}} |z|^p d\mu(z) = \int_{\mathbb{C}} \int_{-\infty}^{\infty} (v^2x^2+y^2)^{p/2} d\sigma_p(v)d\lambda(x+iy)$$

$$\leq \int_{-\infty}^{\infty} \max\{1,|v|^p\}d\sigma_p(v) \cdot \int_{\mathbb{C}} |z|^p d\lambda(z) < \infty$$

we conclude $\mu \in M_p(\mathbb{C})$ and, analogously, $\nu \in M_p(\mathbb{C})$. Let us summarize some properties of $\mu = \mu(\lambda,p)$ and $\nu = \nu(\lambda,p)$.

PROPOSITION 6.

(i) We have

$$\hat{\mu}(s+it) = \int_{\mathbb{C}} e^{ity}k_p(sx)d\lambda(x+iy) = \int_{-\infty}^{\infty} \hat{\lambda}(vs+it)d\sigma_p(v)$$

$$\hat{\nu}(s+it) = \int_{\mathbb{C}} e^{itx}k_p(sy)d\lambda(x+iy) = \int_{-\infty}^{\infty} \hat{\lambda}(t+isv)d\sigma_p(v)$$

(ii) If $\lambda = \lambda_1 \otimes \lambda_2$, then $\mu = (\lambda_1)_p \otimes \lambda_2$ and $\nu = (\lambda_2)_p \otimes \lambda_1$

(iii) If λ is concentrated on $\{z \in \mathbb{C}\ ,\ \mathrm{Re}(z)\ ,\ \mathrm{Im}(z) > 0\}$, then μ and ν are concentrated on $\mathbb{C}^+$.

(iv) We always have $\mu(\lambda,p) \sim_p \nu(\lambda,p)$.

PROOF. We only prove (iv). All other properties are obvious. To verify (iv) we have to show $\chi_\mu = \chi_\nu$. But this follows from

$$\hat{\chi}_\mu(s) = \int_{\mathbb{C}} e^{ixs} k_p(sy)\,d\mu(x+iy) = \int_{\mathbb{C}} \int_{-\infty}^{\infty} e^{ivxs} k_p(sy)\,d\sigma_p(v)\,d\lambda(x+iy)$$

$$= \int_{\mathbb{C}} k_p(sx) k_p d\lambda(x+iy)$$

and

$$\hat{\chi}_\mu(s) = \int_{\mathbb{C}} e^{ixs} k_p(sy)\,d\nu(x+iy) = \int_{\mathbb{C}} \int_{-\infty}^{\infty} e^{ivys} k_p(sx)\,d\sigma_p(v)\,d\lambda(x+iy)$$

$$= \int_{\mathbb{C}} k_p(sy) k_p(sx)\,d\lambda(x+iy)\ ,$$

i.e. we have $\chi_\mu = \chi_\nu$ as asserted.

REMARK. If $\lambda = \lambda_1 \otimes \lambda_2$, then

$$\chi_\mu = (\lambda_1)_p * (\lambda_2)_p = (\lambda_2)_p * (\lambda_1)_p = \chi_\nu\ .$$

EXAMPLES.

(1) If $\lambda = \delta_{z_o}$, $z_o = x_o + iy_o \in \mathbb{C}$, then we have

$$\mu = (\delta_{x_o})_p \otimes \delta_{y_o} \quad \text{and} \quad \nu = (\delta_{y_o})_p \otimes \delta_{x_o}\ .$$

Thus, both measures are concentrated on the lines $\{z;\ \mathrm{Im}(z) = y_o\}$ and $\{z;\ \mathrm{Im}(z) = x_o\}$, respectively. Observe that $\widehat{(\delta_{x_o})}_p = k_p(x_o s)$ and $\widehat{(\delta_{y_o})}_p = k_p(y_o s)$. Moreover, if $x_o \neq \pm y_o$, then $\mu \neq \nu$ as well as $\mu_+ \neq \nu_+$, and μ and ν are concentrated on $\mathbb{C}^+$ with $\mu \sim_p \nu$ and $\mu \neq \nu$ provided that $x_o\ ,\ y_o > 0$ and $x_o \neq y_o$.

(2) Let γ be the standard Gaussian measure on $\mathbb{R}$ and set $\lambda = \gamma \otimes \delta_{y_o}$ for some $y_o \in \mathbb{R}$. Then $\mu = (\gamma)_p \otimes \delta_{y_o}$ and $\nu = (\delta_{y_o})_p \otimes \gamma$. It's interesting that μ is singular with respect to the Lebesque measure on $\mathbb{C}$ while ν is equivalent to it.

(3) If $\gamma_1 \neq \gamma_2$ are two symmetric Gaussian measures on $\mathbb{R}$, then for $\lambda = \gamma_1 \otimes \gamma_2$ we obtain $\mu = (\gamma_1)_p \otimes \gamma_2$ and $\nu = (\gamma_2)_p \otimes \gamma_1$, i.e. both measures possess a Gaussian component.

PROBLEM. Assume $p \neq 2k$ and let $\mu,\nu \in M_p(\mathbb{C})$ be concentrated on $\mathbb{C}^+$. Does $\mu \sim_p \nu$ imply the existence of a measure $\lambda \in M_p(\mathbb{C})$

with $\mu = \mu(\lambda,p)$ and $\nu = \nu(\lambda,p)$?

REMARK. After completing this paper, we found two measures which are p-related but not λ-related for any λ from $\mathcal{M}_p(\mathbb{C})$. Thus the answer to the problem is negative.

Next we want to show that we can conclude $\mu = \nu$ whenever $\mu \sim_p \nu$ provided that μ and ν are molecular. A measure μ is said to be molecular if

$$\mu = \sum_{j=1}^{n} \lambda_j \delta_{z_j}$$

for some $\lambda_j \geq 0$ and $z_j \in \mathbb{C}$.

PROPOSITION 7. Let μ and ν be molecular measures on $\mathbb{C}$ with $\mu \sim_p \nu$. If $p \neq 2k$, then $\mu_+ = \nu_+$.

PROOF. We have to show the following. Let $\alpha_1,\dots,\alpha_N \in \mathbb{R}$ and $z_1,\dots,z_N \in \mathbb{C}$ with $\mathrm{Im}(z_j) \geq 0$ and $z_i \neq z_j$, $i \neq j$. If for all $t \in \mathbb{R}$

$$(*) \qquad \sum_{j=1}^{N} \alpha_j |z_j + t|^p = 0 ,$$

then $\alpha_1 = \dots = \alpha_N = 0$. In view of Th. 3, (*) is equivalent to

$$\sum_{j=1}^{N} \alpha_j e^{isx_j} k_p(sy_j) = 0 , \quad s \in \mathbb{R} ,$$

where $z_j = x_j + iy_j$, $y_j \geq 0$. Changing the numeration we may write this as

$$\sum_{j=1}^{N'} \sum_{k=1}^{n_j} \alpha_{jk} e^{isx_{jk}} k_p(sy_j) = 0$$

with $0 \leq y_1 < y_2 < \dots < y_{N'}$ and $x_{jk} \neq x_{jl}$ whenever $k \neq l$. We obtain

$$\sum_{k=1}^{n_1} \alpha_{1k} e^{isx_{1k}} = - \sum_{j=2}^{N'} \alpha_{jk} e^{isx_{jk}} \frac{k_p(sy_j)}{k_p(sy_1)} .$$

The right hand side tends to zero if $s \to \infty$. This follows from the asymptotically behaviour of k_p . But the left hand side is an almost periodic function in the sense of [Bo]. Thus,

$$\sum_{k=1}^{n_1} \alpha_{1k} e^{isx_{1k}} = 0$$

for all $s \in \mathbb{R}$, and because of $x_{1k} \neq x_{1l}$, $k \neq l$, we conclude that $\alpha_{11} = \dots = \alpha_{1n_1} = 0$. Repeating this procedure N'-1-times we arrive at $\alpha_{jk} = 0$, $1 \leq j \leq N'$, $1 \leq k \leq n_j$, which completes the proof.

PROBLEMS.

(1) Does Prop. 7 remain true for discrete measures? In different words, does

$$\sum_{j=1}^{\infty} \alpha_j e^{isx_j} k_p(sy_j) = 0 , \quad s \in \mathbb{R},$$

with $x_k + iy_k \neq x_l + iy_l$, $k \neq l$, and $y_l \geq 0$ imply $\alpha_1 = \alpha_2 = \ldots = 0$? The answer is affirmative provided that $\{\mathrm{Im}(z_j)$, $j = 1,2,\ldots\}$ consists only of isolated points.

Then one can apply the same arguments as in the preceding proof.

(2) Let μ be molecular (or discrete) on $\mathbb{C}^+$ with $\mu \sim_p \nu$ for some ν concentrated on $\mathbb{C}^+$ as well. Does this imply $\mu = \nu$, $p \neq 2k$? Observe that this does not hold without the assumption that ν is concentrated on $\mathbb{C}^+$. Indeed, if $\mu = \delta_o \otimes \delta_{y_o}$ for some $y_o > 0$, then $\mu \sim_p (\delta_{y_o})_p$ with $(\delta_{y_o})_p = \chi_\mu$ concentrated on $\mathbb{R}$.

Next we prove a weak general uniqueness theorem.

PROPOSITION 8. Let $\mu,\nu \in M_p(\mathbb{R})$ be concentrated on $[0,\infty)$ and assume that for some $\rho \in M_p(\mathbb{R})$ with $\hat{\rho}(s) \neq 0$, $s \in \mathbb{R}$, $\rho \otimes \mu \sim_p \rho \otimes \nu$. If $p \neq 2k$, then $\mu = \nu$.

PROOF. Since $\chi_{\rho\otimes\mu} = \rho * (\mu)_p$ and $\chi_{\rho\otimes\nu} = \rho * (\nu)_p$ the assumption $\rho \otimes \mu \sim_p \rho \otimes \nu$ implies $\rho * (\nu)_p = \rho * (\mu)_p$. Since $\hat{\rho}(s) \neq 0$ we conclude that $(\nu)_p = (\mu)_p$. Then (iv) of Prop. 5 proves $\nu = \mu$ as asserted.

PROBLEM. Does Prop. 8 remain true without the assumption $\hat{\rho}(s) \neq 0$? To prove this one has to show that

$$\int_0^\infty k_p(sy)d\mu(y) = \int_0^\infty k_p(sy)d\nu(y)$$

for $s \in (-s_o,s_o)$, $s_o > 0$, implies $\mu = \nu$.

5. GAUSSIAN AND STABLE MEASURES.

We want to investigate α-stable, $0 < \alpha \leq 2$, symmetric measures μ,ν on $\mathbb{C}$ satisfying $\mu \sim_p \nu$ for some $p > 0$. Recall that we have to assume $0 < p < \alpha$ provided that $\alpha < 2$. Here μ is α-stable symmetric (Gaussian for $\alpha = 2$) provided that $\hat{\mu}$ maps $\mathbb{C}$ into $\mathbb{R}$ and satisfies $\hat{\mu}(n^{1/\alpha}z) = \hat{\mu}(z)^n$, $n \in \mathbb{N}$.

PROPOSITION 9. If $0 < p < \alpha < 2$ or $\alpha = 2$ and $p \neq 2k$, then two α-stable symmetric measures μ and ν on $\mathbb{C}$ satisfy $\mu \sim_p \nu$ iff

$$\sigma_p\{u \in \mathbb{R}; \hat{\mu}(1+iu) \in B\} = \sigma_p\{u \in \mathbb{R}; \hat{\nu}(1+iu) \in B\}$$

for $B \subseteq \mathbb{R}$ measurable.

PROOF. Let us first assume $\mu \sim_p \nu$ for some pair of α-stable symmetric measures. Then for $s \in \mathbb{R}$

$$\int_{-\infty}^{\infty} \hat{\mu}(s+isu)d\sigma_p(u) = \int_{-\infty}^{\infty}\hat{\nu}(s+isu)d\sigma_p(u)$$

because of Th. 3. Choosing $s = n^{1/\alpha}$, $n \in \mathbb{N}$, we arrive at

$$\int_{-\infty}^{\infty} \hat{\mu}(1+iu)^n d\sigma_p(u) = \int_{-\infty}^{\infty} \hat{\nu}(1+iu)^n d\sigma_p(u) ,$$

$n \in \mathbb{N}$. Observe that $0 < \hat{\mu}(1+iu)$, $\hat{\nu}(1+iu) \le 1$ and we can apply Theorem 1 ch. VII, § 3 in [Fe] to see that $\hat{\mu}(1+iu)$ and $\hat{\nu}(1+iu)$ possess the same distribution with respect to σ_p. Conversely, if $\hat{\mu}(1+iu)$ and $\hat{\nu}(1+iu)$ are equidistributed with respect to σ_p, then for each $t \ge 0$

$$\int_{-\infty}^{\infty} \hat{\mu}(1+iu)^t d\sigma_p(u) = \int_{-\infty}^{\infty} \hat{\nu}(1+iu)^t d\sigma_p(u)$$

yielding

$$\int_{-\infty}^{\infty} \hat{\mu}(s+isu)d\sigma_p(u) = \int_{-\infty}^{\infty} \hat{\nu}(s+isu)d\sigma_p(u)$$

whenever $s \ge 0$. Since μ and ν are symmetric, this is also valid for $s < 0$ proving $\mu \sim_p \nu$ via Th. 3.

PROPOSITION 10. Let μ and ν be Gaussian on $\mathbb{C}$ with $\mu \sim_p \nu$, $p \neq 2k$. Then either $\mu = \nu$ or $\mu = \bar{\nu}$ with $\bar{\nu}(B) := \nu\{z \in B , \bar{z} \in B\}$.

PROOF. Define

$$g(u) := -2 \log\hat{\mu}(1+iu) \quad \text{and} \quad h(u) := -2 \log\hat{\nu}(1+iu) , \quad u \in \mathbb{R}.$$

By assumption,

$$\sigma_p\{u \in \mathbb{R};\ g(u) < t\} = \sigma_p\{u \in \mathbb{R};\ h(u) < t\} , \quad t \in \mathbb{R}.$$

Set

$$\alpha_{11} := \int_{\mathbb{C}} x^2 d\mu(x+iy) , \quad \alpha_{12} := \int_{\mathbb{C}} xy d\mu(x+iy) \quad \text{and} \quad \alpha_{22} := \int_{C} y^2 d\mu(x+iy)$$

and define β_{11}, β_{12} and β_{22} in the same way with μ replaced by ν. Then

$$g(u) = \alpha_{11} + 2\alpha_{12}u + \alpha_{22}u^2 \quad \text{and} \quad h(u) = \beta_{11} + 2\beta_{12}u + \beta_{22}u^2 , \ u \in \mathbb{R},$$

implying

$$\alpha_{11} - \frac{\alpha_{12}^2}{\alpha_{22}} = \inf_u g(u) = \inf_u h(u) = \beta_{11} - \frac{\beta_{12}^2}{\beta_{22}} .$$

Consequently,

$$\sigma_p\{u \in \mathbb{R}, \frac{1}{\alpha_{22}} (\alpha_{12} + \alpha_{22}u)^2 < t^2\}$$
$$= \sigma_p\{u \in \mathbb{R}, \frac{1}{\beta_{22}} (\beta_{12} + \beta_{22}u)^2 < t^2\} , \quad t \in \mathbb{R} ,$$

i.e.
$$\sigma_p(- \frac{t}{\sqrt{\alpha_{22}}} - \frac{\alpha_{12}}{\alpha_{22}} , \frac{t}{\sqrt{\alpha_{22}}} - \frac{\alpha_{12}}{\alpha_{22}})$$
$$= \sigma_p(- \frac{t}{\sqrt{\beta_{22}}} - \frac{\beta_{12}}{\beta_{22}} , \frac{t}{\sqrt{\beta_{22}}} - \frac{\beta_{12}}{\beta_{22}})$$

for all $t \geq 0$. Differentiating this w.r.t. t leads to

$$\frac{1}{\sqrt{\alpha_{22}}} \left\{\frac{1}{\left[1+\left(\frac{t}{\sqrt{\alpha_{22}}} + \frac{\alpha_{12}}{\alpha_{22}}\right)^2\right]^{p/2+1}} + \frac{1}{\left[1+\left(\frac{t}{\sqrt{\alpha_{22}}} - \frac{\alpha_{12}}{\alpha_{22}}\right)^2\right]^{p/2+1}}\right\} =$$
$$= \frac{1}{\sqrt{\beta_{22}}} \left\{\frac{1}{\left[1+\left(\frac{t}{\sqrt{\beta_{22}}} + \frac{\beta_{12}}{\beta_{22}}\right)^2\right]^{p/2+1}} + \frac{1}{\left[1+\left(\frac{t}{\sqrt{\beta_{22}}} - \frac{\beta_{12}}{\beta_{22}}\right)^2\right]^{p/2+1}}\right\}$$

Multiplying both terms by t^{p+2} and taking the limit $t \to \infty$ yields $\alpha_{22} = \beta_{22}$. Consequently, choosing $t = 0$ we obtain $|\alpha_{12}| = |\beta_{12}|$ which proves $\alpha_{11} = \beta_{11}$ because of

$$\alpha_{11} - \frac{\alpha_{12}^2}{\alpha_{22}} = \beta_{11} - \frac{\beta_{12}^2}{\beta_{22}} .$$

This completes the proof.

REMARKS.

(1) We used $\alpha_{22} \neq 0$ as well as $\beta_{22} \neq 0$. If, for instance, $\alpha_{22} = 0$, then g is constant and h to be constant as well. Clearly, $g = h$ which also proves the theorem in this case.

(2) Prop. 10 fails for $p = 2 , 4$ (cf. [Li 3]). We do not know what happens in the remaining cases $p = 6,8,\ldots$.

(3) The case of arbitrary Gaussian measures (non-necessarily symmetryc) seems to be open.

Next we investigate the α-stable case, $\alpha < 2$. Here several questions remain open.

PROPOSITION 11. Let μ and ν be two α-stable symmetric measures on C and assume $p < \alpha$ and $1 \leq \alpha < 2$. If $\mu = \bar{\mu}$ as well as $\nu = \bar{\nu}$, then $\mu \sim_p \nu$ yields $\mu = \nu$.

PROOF. As above set

$$g(u) := -\log \hat{\mu}(1+iu) \ , \quad h(u) := -\log \hat{\nu}(1+iu) \ , \quad u \in \mathbb{R} \ .$$

Then $\mu \sim_p \nu$ implies

$$(*) \qquad \sigma_p\{u, g(u) < t\} = \sigma_p\{u \ , h(u) < t\} \ , \quad t \in \mathbb{R}.$$

By assumption g and h are symmetric functions. Moreover, because of $1 \le \alpha < 2$ and Levy's spectral representation theorem they are also convex. Consequently, there are some $u_o \ge 0$ and some $v_o \ge 0$ such that $g(u) = c_o$, $|u| \le u_o$, $h(u) = d_o$, $|u| \le v_o$ and g and h are strictly increasing on (u_o, ∞) and (v_o, ∞), respectively. It is easy to see that $(*)$ implies $c_o = d_o$ as well as $u_o = v_o$ and, consequently, $h = g$. This completes the proof.

<u>REMARK</u>. If $\alpha > 1$, then necessarily $u_o = v_o = 0$.

<u>PROBLEM</u>. Does $\mu \sim_p \nu$, μ, ν α-stable (symmetric) with $0 < p < \alpha < 2$ always imply $\mu = \nu$ or $\mu = \overline{\nu}$?

6. CONCLUDING REMARKS.

The problem investigated in this paper is part of a much more general question. To make this more precise we need the following definition.

A set $A \subseteq \mathbb{C}$ is said to be determining whenever

$$\int_{\mathbb{C}} |z+w|^p d\mu(z) = \int_{\mathbb{C}} |z+w|^p d\nu(z) < \infty$$

for all $w \in A$ and some $p \neq 2k$, $p > 0$, implies $\mu = \nu$. As we have seen above, the real line is not determining while $A = \mathbb{C}$ is so (Rudin's theorem). Let us state some obvious properties of determining sets:

(1) $A \subseteq \mathbb{C}$ is determining iff $A+z_o$ is so for some (each) $z_o \in \mathbb{C}$.
(2) $A \subseteq \mathbb{C}$ is determining iff $z_o \cdot A$ is so for some (each) $z_o \in \mathbb{C}\setminus\{0\}$.
(3) A is determining iff its closure $\overline{A}$ possesses this property.
(4) If $A \subseteq B$ and A is determining, then B is determining as well.

The most interesting question seems to be whether or not $U = \{z \in \mathbb{C} \ , |z| \le 1\}$ is determining. If the answer is yes, then every set A with $\mathring{\overline{A}} \neq \emptyset$ (the interior of $\overline{A}$) is determining. If U is not determining, then there do not exist determining bounded sets in $\mathbb{C}$. Thus our conjecture is:

<u>CONJECTURE</u>. $A \subseteq \mathbb{C}$ is determining if $\mathring{\overline{A}} \neq \emptyset$.

REFERENCES

[Ba/Er] H.Bateman/A.Erdelyi: Higher transcendental function, Vol.2. New York - Toronto - London 1953.

[Bo] H.Bohr: Fastperiodische Funktionem. Berlin 1932.

[Fe] W.Feller: An introduction to probability theory and its applications, Vol.2. New York - London - Sydney - Toronto 1971.

[Go/Ko] E.A.Gorin/A.L.Koldobskii: About potentials of measures in Banach spaces (Russ.), Sibirsk Mat. Sh.28 (1987), 65-80.

[Ha] C.D.Hardin: Isometries on subspaces of L^p, Indiana Univ. Math. J. 30 (1981), 449-465.

[Ko 1] A.L.Koldobskii: On isometric operators in some vector-valued L^p-spaces (Russ.), Zap. Nauchn., Sem. Leningrad Otdel Mat. Inst. Steklov (LOMI) 107 (1982), 198-203.

[Ko 2] A.L.Koldobskii: Uniqueness theorem for measures in $C(K)$ and its application in the theory of random processes, J. Soviet Math. 27 (1984), 3095-3102.

[La] N.S.Landkof: Foundations of modern potential theory (Russ.). Moscow 1966.

[Le] M.Lewandowski: On the shifted moment problem for Gaussian measures in L_p , $p \geq 2$.

[Li 1] W.Linde: Moments of Measures in Banach spaces, Math. Ann. 258 (1982), 277-287.

[Li 2] W.Linde: Uniqueness theorems for measures in L_r and $C_o(\Omega)$. Math. Ann. 274 (1986), 617-626.

[Li 3] W.Linde: Uniqueness theorems for Gaussian measures in l_q , $1 \leq q < \infty$, to appear Math. Z.

[Pl 1] A.I.Plotkin: Continuation of L^p-isometries, J. Sov. Math. 2 (1974), 143-165.

[Pl 2] A.I.Plotkin: A algebra generated by translation operators and L^p-norms. (Russ.), Functional analysis No 6: Theory of operators in linear spaces, 112-121. Uljanovsk Gos. Ped. Inst. 1976.

[Ru] W.Rudin: L^p-isometries and equimeasurability, Indiana Univ. Math. J. 25 (1976), 215-228.

[St] K.Stephenson: Certain integral equalities which imply equimeasurability of functions, Can. J. Math. 29 (1977), 827-844.

[Wa] G.N.Watson: A treatise on the theory of Bessel functions. Cambridge 1922.

Maciej Lewandowski
Institute of Mathematcs
Technical University
Wrocław, Poland

Werner Linde
Sektion Mathematik
FSU Jena, DDR

Probability Theory on Vector Spaces IV
Lancut, June'87, Springer's LNM 1391

BANACH SPACE-VALUED GLEASON MEASURES ON VON NEUMANN ALGEBRAS

Andrzej Łuczak (Łódź)

Introduction. This paper aims at giving a short survey of some results in the theory of vector-valued Gleason measures. Many of these results have been obtained earlier under special assumptions and our attempt here is to present them in full generality, usually with even simpler proofs. In this presentation we make use of the close connection between Gleason measures and normal fields; in fact, under mild assumptions, each Gleason measure has such a field as its unique extension. This, in turn, allows for considering the fields instead of measures and in such a context the description of the so-called "density function" of a Gleason measure becomes simple and natural. The importance of the fields in our considerations (even though the primary goals are Gleason measures) requires paying them some attention; in particular, we give a solution to the problem when an orthogonally scattered field originates from a Jordan morphism.

It should be noted here that there are many things of interest in vector-valued Gleason measures or fields that have not been dealt with in this article. In particular, neither dilation problems nor stationary properties are discussed.

1. Preliminaries. Let M be a σ-finite von Neumann algebra, $\mathbb{P}\mathrm{roj}\, M$ - the lattice of projections of M and X - a Banach space. A mapping $\lambda : \mathbb{P}\mathrm{roj}\, M \to X$ is called a Gleason measure if

(i) $\sup_{p \in \mathbb{P}\mathrm{roj}\, M} \|\lambda(p)\| = C < \infty$;

(ii) $\lambda(\sum_{n=1}^{\infty} p_n) = \sum_{n=1}^{\infty} \lambda(p_n)$ for any sequence $\{p_n\}$ of pairwise orthogonal projections in M, where the series on the right hand side is convergent in the weak topology on X.

This definition generalizes the classical notion of the Gleason measure in which $X = \mathbb{R}$, $M = \mathbb{B}(H)$ - the algebra of all bounded linear operators on a separable Hilbert space H and λ is positive valued. The celebrated Gleason theorem states that in this case, if $\dim H \geq 3$, then λ has a unique extension to a mapping $\Lambda : M \to \mathbb{C}$ being, in fact, a normal positive linear functional on $\mathbb{B}(H)$. Recently, the Gleason theorem has been extended to the case of an arbitrary von Neumann algebra having no type I_2 direct summand (cf. [3], [8], [11]). For short, we shall refer to such an algebra as having the Gleason property. In other words, M has the Gleason property if any positive valued Gleason measure on $\mathbb{P}\mathrm{roj}\, M$ has a unique extension to a normal positive linear functional on M.

Generalizing a little the terminology of [5], we shall call a linear mapping from M into X a field (over M). A field which is continuous in the σ-weak topology on M and the weak topology on X will be called normal. The set of all normal fields over M with values in X will be denoted by $F(M,X)$ and $L(M,X)$ will stand for bounded fields.

The following two simple results give necessary and sufficient conditions for a bounded field to be normal. They are proved in a general setting without the assumption that M is σ-finite.

LEMMA 1. Let $\Phi \in L(M,X)$. Then $\Phi \in F(M,X)$ iff for each $\{p_i\} \subset \subset \mathbb{P}\mathrm{roj}\, M$, $p_i p_j = 0$, $i \neq j$,

$$\Phi(\Sigma p_i) = \Sigma\, \Phi(p_i),$$

where the series on the right hand side is convergent in the weak topology on X.

P r o o f. We need only to prove the "if" part, the other being simply an immediate consequence of the continuity.

Consider the dual mapping $\Phi^* : X^* \to M^*$. For each $x^* \in X^*$ and each family $\{p_i\}$ of pairwise orthogonal projections in M, we have

$$< \Sigma p_i, \Phi^*(x^*)> = < \Phi(\Sigma p_i), x^*> =$$

$$= \Sigma <\Phi(p_i), x^*> = \Sigma <p_i, \Phi^*(x^*)>$$

which means that $\Phi^*(x^*)$ is normal (cf. [10; Cor. 3.11, p. 136]). Consequently, $\Phi^*(X^*) \subset M_*$ which shows that Φ is continuous in the σ-weak topology on M and the weak topology on X. ∎

As an easy consequence of the above lemma we have

COROLLARY 2. Let $\Phi \in L(M,X)$. Then $\Phi \in F(M,X)$ iff for each net $\{a_i\}$ in M, $\|a_i\| \leq 1$, $a_i \to a$ σ-weakly implies $\Phi(a_i) \to \Phi(a)$ weakly in X. ∎

2. Gleason measures and density functions. We begin with the following counterpart of the Gleason theorem for vector-valued measures due to Ylinen [12; Th. 2.2].

THEOREM 3. Let the algebra M have the Gleason property and let λ be a Gleason measure on $\mathbb{P}\mathrm{roj}\, M$ with values in a Banach space X.

Then λ has a unique extension to a normal field over M.

For the sake of completeness we give a proof of this theorem. The proof is different from that of Theorem 2.2 in [12] and seems to be simpler.

P r o o f. We preserve the notation of Section 1. Let x^* be an arbitrary fixed element of X^*. Consider a mapping $\tilde{\lambda}_{x^*} : \mathbb{P}\mathrm{roj}\, M \to \mathbb{C}$ defined as

$$\tilde{\lambda}_{x^*}(p) = \langle \lambda(p), x^* \rangle .$$

Since

$$\sup_{p \in \mathbb{P}\mathrm{roj}\, M} |\tilde{\lambda}_{x^*}(p)| \le \sup_{p \in \mathbb{P}\mathrm{roj}\, M} \|\lambda(p)\| \, \|x^*\| = C\|x^*\| < \infty$$

and $\tilde{\lambda}_{x^*}$ is clearly σ-additive on orthogonal projections, thus, according to [8; Th. 6.1], $\tilde{\lambda}_{x^*}$ has a unique extension to a normal linear functional $\tilde{\Lambda}_{x^*}$ on M. Now, we may define a mapping $\tilde{\Lambda} : X^* \to \tilde{\Lambda}_{x^*}$. It is easily seen that $\tilde{\Lambda}$ is linear; moreover, for $\phi \in M_*$ we have $\|\phi\| \le 4 \sup_{p \in \mathbb{P}\mathrm{roj}\, M} |\phi(p)|$ and therefore

$$\|\tilde{\Lambda}(x^*)\| = \sup_{\|a\| \le 1} |\langle \tilde{\Lambda}(x^*), a \rangle| = \sup_{\|a\| \le 1} |\tilde{\Lambda}_{x^*}(a)| \le \|\tilde{\Lambda}_{x^*}\| \le$$

$$\le 4 \sup_{p \in \mathbb{P}\mathrm{roj}\, M} |\tilde{\Lambda}_{x^*}(p)| = 4 \sup_{p \in \mathbb{P}\mathrm{roj}\, M} |\tilde{\lambda}_{x^*}(p)| =$$

$$= 4 \sup_{p \in \mathbb{P}\mathrm{roj}\, M} |\langle \lambda(p), x^* \rangle| \le 4C\|x^*\|$$

which means that $\tilde{\Lambda}$ is bounded. For the dual mapping $\tilde{\Lambda}^* : M \to X^{**}$ we have

$$\langle \tilde{\Lambda}^*(p), x^* \rangle = \langle p, \tilde{\Lambda}(x^*) \rangle = \tilde{\Lambda}_{x^*}(p) = \tilde{\lambda}_{x^*}(p) = \langle \lambda(p), x^* \rangle$$

for each $x^* \in X^*$, $p \in \mathbb{P}\text{roj } M$, so $\tilde{\Lambda}^*(p) = \lambda(p) \in X$. From the linear density of $\mathbb{P}\text{roj } M$ in M, we obtain $\tilde{\Lambda}^*(M) \subset X$. Finally, for pairwise orthogonal projections $\{p_n\}$ in M, we have

$$\tilde{\Lambda}^*(\sum_n p_n) = \lambda(\sum_n p_n) = \sum_n \lambda(p_n) = \sum_n \tilde{\Lambda}^*(p_n)$$

yielding, on account of Lemma 1, the continuity of $\tilde{\Lambda}^*$ in the σ-weak topology on M and the weak topology on X. Thus $\tilde{\Lambda}^*$ is an extension of λ to a normal field. The uniqueness of such an extension follows from the continuity properties and the fact that any two extensions must agree on the set $\mathbb{P}\text{roj } M$. ◻

In the rest of this section we assume that the algebra M has the Gleason property. For a Gleason measure λ on $\mathbb{P}\text{roj } M$ we shall denote by Λ its extension to the normal field. The notion of the density function for λ as the function $\Phi : X^* \to M_*$ satisfying

$$\langle \Phi(x^*), p\rangle = \langle x^*, \lambda(p)\rangle, \qquad p \in \mathbb{P}\text{roj } M, \qquad x^* \in X^*,$$

was discussed in [1], [6], [7]. Let us consider the dual pairs (M,M_*) and (X,X^*) with their duality topologies. By virtue of Theorem 3, Λ is a continuous mapping of M into X, thus Λ^* maps X^* into M_* and is continuous in the topologies $\sigma(X^*,X)$ on X^* and $\sigma(M_*,M)$ on M_*. Moreover, for $p \in \mathbb{P}\text{roj } M$, $x^* \in X^*$, we have

$$\langle \Lambda^*(x^*), p\rangle = \langle x^*, \Lambda(p)\rangle = \langle x^*, \lambda(p)\rangle,$$

so the density function of λ is nothing else than the dual map Λ^*. The following simple result characterizes the density funciton as a mapping from X^* to M_* (cf. [1] for the case $M = \mathbb{B}(H)$).

THEOREM 4. Let Φ be a linear mapping from X^* into M_*. Φ is the density function of a Gleason measure iff Φ is continuous in the topologies $\sigma(X^*,X)$ on X^* and $\sigma(M_*,M)$ on M_*.

P r o o f. Clearly, we have only to prove the "if" part. In this case, considering again the dual pairs (M,M_*) and (X,X^*), we infer that the dual mapping Φ^* maps M into X and is continuous in the topology $\sigma(M,M_*)$ on M and $\sigma(X,X^*)$ on X, thus it is a normal field whose restriction to $\mathbb{P}\mathrm{roj}\, M$ is, by virtue of Lemma 1 and the inequality $\|\Phi^*(p)\| \le \|\Phi\|$, a Gleason measure. The density function of this measure is equal, by our preceding considerations, to $\Phi^{**} = \Phi$ which proves the claim. ◻

COROLLARY 5. If X is reflexive, then the set of density functions coincides with the set of all linear bounded mappings from X^* into M_* (cf. [6], [7; Th. 2.7] for the case $M = \mathbb{B}(H)$).

P r o o f. By the Banach-Steinhaus theorem each linear $\sigma(X^*,X)$ - $\sigma(M_*,M)$ continuous map of X^* into M_* is bounded. Now, if Φ : : $X^* \to M_*$ is bounded, then the dual mapping Φ^* maps M into X^{**}. As X is reflexive, this means that $\Phi^*(M) \subset X$, showing the continuity of Φ in the $\sigma(X^*,X)$ topology on X^* and the $\sigma(M_*,M)$ topology on M_* which, by Theorem 4, implies that Φ is the density function. ◻

<u>3. The fields</u>. As seen in the previous section, for an algebra having the Gleason property, there is a one-one correspondence between Gleason measures and normal fields. This inclines one to investigate the fields rather than the measures and in this section we adopt such an attitude. In doing this, the assumption about the Gleason property of the algebra may be dropped, so we shall deal with the fields over an arbitrary von Neumann algebra M taking values in a Banach space X.

Let us observe that by the Banach-Steinhaus theorem we have

$$F(M,X) \subset L(M,X).$$

LEMMA 6. $F(M,X)$ is a closed subspace of $L(M,X)$.

P r o o f. Let $\Phi_n \in F(M,X)$ and $\Phi_n \to \Phi$ in norm. Take an arbitrary net $\{a_i\}$ in M, $\|a_i\| \le 1$, $a_i \to a$ σ-weakly. For any fixed $x^* \in X^*$ we have

$$\begin{aligned} |\langle x^*, \Phi(a_i) - \Phi(a_i)\rangle| &\le |\langle x^*, \Phi(a_i) - \Phi_n(a_i)\rangle| + \\ &+ |\langle x^*, \Phi_n(a_i) - \Phi_n(a)\rangle| + |\langle x^*, \Phi_n(a) - \Phi(a)\rangle| \le \\ &\le \|x^*\| \, \|\Phi_n - \Phi\| + |\langle x^*, \Phi_n(a_i) - \Phi_n(a)\rangle| + \\ &+ \|x^*\| \, \|\Phi_n - \Phi\| \, \|a\|. \end{aligned}$$

The first and third terms can be made arbitrarily small by taking sufficiently large n, and the second term, having chosen and fixed this n, will be small for each sufficiently large i. On account of Corollary 2, we infer that Φ is in $F(M,X)$. □

THEOREM 7. $F(M,X) = L(M,X)$ iff M is finite dimensional.

P r o o f. If M is finite dimensional, then the σ-weak and norm topologies on M coincide and therefore $F(M,X) = L(M,X)$.

Now, assume that $F(M,X) = L(M,X)$ and choose a non-zero element $x_0 \in X$. For an arbitrary $\phi \in M^*$ put

$$\Phi_\phi(a) = \phi(a)x_0, \qquad a \in M.$$

Φ_ϕ is bounded thus, by assumption, it is continuous in the σ-weak topology on M and the weak topology on X. Let $\{a_i\}$ be a net in M converging σ-weakly to a. For each $x^* \in X^*$ $\langle x^*, \Phi_\phi(a_i)\rangle \to$ $\to \langle x^*, \Phi_\phi(a)\rangle$ which means that

$$\langle x^*,(\phi(a_i) - \phi(a))x_o\rangle = [\phi(a_i) - \phi(a)]\langle x^*,x_o\rangle \to 0.$$

But this implies that $\phi(a_i) \to \phi(a)$, so ϕ is normal. We have thus obtained that $M^* = M_*$, hence M is finite dimensional (cf. [10; Ex. 5, p. 130]). ∎

The last part of our considerations is devoted to orthogonally scattered fields. Let X be a Hilbert space with an inner product $(\bullet,\bullet)$ and Φ - a field. Φ is called orthogonally scattered if, for each pair p,q of orthogonal projections in M,

$$(\Phi(p),\Phi(q)) = 0.$$

There is an interesting problem when an orthogonally scattered field Φ originates from some Jordan morphism, i.e. when there exist a Jordan morphism $\pi : M \to \mathbb{B}(X)$ and $\xi \in X$ such that

$$\Phi(a) = \pi(a)\xi, \qquad a \in M.$$

This problem was considered in [7] in the case $M = \mathbb{B}(X)$ and π being a *-representation of M.

THEOREM 8. Let Φ be an orthogonally scattered field over M. Φ originates from a Jordan morphism π iff there are a projection e in X and positive linear functionals ϕ,ψ on M such that the following conditions hold:

$$(*)\quad \begin{aligned} &\text{(i)} \quad \|e\Phi(a)\|^2 = \phi(a^*a);\\ &\text{(ii)} \quad \|e^{\perp}\Phi(a)\|^2 = \psi(aa^*); \end{aligned}$$

for all a in M. Moreover, Φ is normal iff ϕ and ψ are normal as well (cf. [7; Thms 5.11 and 8.9]).

P r o o f. Sufficiency. Let us assume that (*) holds. By using the polarization identity we obtain equivalent conditions:

$$(**)\quad \begin{aligned} &\text{(i)} \quad (e\Phi(a),\Phi(b)) = \phi(b^*a);\\ &\text{(ii)} \quad (e^{\perp}\Phi(a),\Phi(b)) = \psi(ab^*);\end{aligned}$$

for all a,b in M. Restricting our attention to the subspace $\overline{\Phi(M)}$, we may assume the later to be equal to X. For each $a \in M$ we define an operator $\pi(a)$ on a dense subspace of X by

$$(***) \qquad \pi(a)(\Phi(b)) = e\Phi(ab) + e^{\perp}\Phi(ba), \qquad b \in M.$$

We have

$$\begin{aligned}\|\pi(a)\Phi(b)\|^2 &= \|e\Phi(ab)\|^2 + \|e^{\perp}\Phi(ba)\|^2 =\\ &= \phi(b^*a^*ab) + \psi(baa^*b) \leq\\ &\leq \|a^*a\|\,\phi(b^*b) + \|aa^*\|\,\psi(bb^*) =\\ &= \|a\|^2[\phi(b^*b) + \psi(bb^*)] = \|a\|^2\,\|\Phi(b)\|^2,\end{aligned}$$

which shows that $\pi(a)$ is well-defined and bounded. Clearly, $\pi(a)$ is linear, thus π maps M into $\mathbb{B}(X)$ and again it is easily seen to be linear and bounded. Furthermore, for all a,b,c in M, using (**) we get

$$\begin{aligned}(\pi(a^*)(\Phi(b)),\Phi(c)) &= (e\Phi(a^*b) + e^{\perp}\Phi(ba^*),\Phi(c)) =\\ &= \phi(c^*a^*b) + \psi(ba^*c^*) =\\ &= (\Phi(b),e\Phi(ac)) + (\Phi(b),e^{\perp}\Phi(ca)) =\\ &= (\Phi(b),\pi(a)(\Phi(c)))\end{aligned}$$

showing that $\pi(a^*) = \pi(a)^*$. Now,

$$\begin{aligned}(\pi(a)(e\Phi(b)),\Phi(c)) &= (e\Phi(b),\pi(a^*)(\Phi(c))) =\\ &= (e\Phi(b),e\Phi(a^*c)) = \phi(c^*ab) = (e\Phi(ab),\Phi(c))\end{aligned}$$

thus

$$\pi(a)(e\Phi(b)) = e\Phi(ab)$$

and, analogously,

$$\pi(a)(e^{\perp}\Phi(b)) = e^{\perp}\Phi(ba).$$

So, we finally have

$$\pi(ab)(\Phi(c)) = e\Phi(abc) + e^{\perp}\Phi(cab),$$

$$\pi(ba)(\Phi(c)) = e\Phi(bac) + e^{\perp}\Phi(cba),$$

and

$$\pi(a)\pi(b)(\Phi(c)) = \pi(a)(e\Phi(bc) + e^{\perp}\Phi(cb)) = \\ = e\Phi(abc) + e^{\perp}\Phi(cba),$$

$$\pi(b)\pi(a)(\Phi(c)) = \pi(b)(e\Phi(ac) + e^{\perp}\Phi(ca)) = \\ = e\Phi(bac) + e^{\perp}\Phi(cab),$$

which shows that $\pi(ab + ba) = \pi(a)\pi(b) + \pi(b)\pi(a)$, and thus π is a Jordan morphism. Putting $\xi = \Phi(\mathbb{1})$, we obtain

$$\Phi(a) = \pi(a)\xi$$

ending the proof of sufficiency.

Necessity. Let us assume that Φ originates from a Jordan morphism π, i.e. $\Phi(a) = \pi(a)\xi$. Then there is a projection e in X such that putting

$$\pi_1(a) = e\pi(a), \qquad \pi_2(a) = e^{\perp}\pi(a), \qquad a \in M,$$

we obtain that π_1 is a *-representation of M in the Hilbert space eX and π_2 is a *-antirepresentation of M in the Hilbert space $e^{\perp}X$ (cf. [2; Prop. 3.2.2, p. 207]). Define ϕ and ψ by

$$\phi(a) = (\pi_1(a)\xi,\xi), \qquad \psi(a) = (\pi_2(a)\xi,\xi), \qquad a \in M.$$

Then ϕ and ψ are positive linear functionals on M; moreover,

$$\|e\Phi(a)\|^2 = \|e\pi(a)\xi\|^2 = (\pi_1(a)\xi, \pi_1(a)\xi) =$$

$$= (\pi_1(a^*a)\xi, \xi) = \phi(a^*a)$$

and

$$\|e^{\perp}\Phi(a)\|^2 = \|e^{\perp}\pi(a)\xi\|^2 = (\pi_2(a)\xi, \pi_2(a)\xi) =$$

$$= (\pi_2(aa^*)\xi, \xi) = \psi(aa^*)$$

which shows that conditions (*) hold and completes the proof of necessity.

Now, if Φ is normal, then on account of (**) we have

$$\phi(a) = (e\Phi(a), \Phi(\mathbb{1})), \qquad \psi(a) = (e^{\perp}\Phi(a), \Phi(\mathbb{1})),$$

so ϕ and ψ are normal. Conversely, if ϕ and ψ are normal, then for each fixed $b \in M$ and a net $\{a_i\}$ in M converging σ-weakly to a, we get

$$(\Phi(a_i), \Phi(b)) = (e\Phi(a_i), \Phi(b)) + (e^{\perp}\Phi(a_i), \Phi(b)) =$$

$$= \phi(b^*a_i) + \psi(a_ib^*) \to \phi(b^*a) + \psi(ab^*) =$$

$$= (\Phi(a), \Phi(b)),$$

which means that Φ is normal. Let the Jordan morphism π be defined as in (***). According to [9; Ex. 5.17, p. 132], π is normal iff for any net $\{p_i\}_{i \in I}$ of pairwise orthogonal projections in M,

$$\pi\Big(\sum_{i \in I} p_i\Big) = \sum_{i \in I} \pi(p_i),$$

where the series on the right hand side is convergent in the σ-weak topology on $\mathbb{B}(X)$. Since Jordan morphisms map orthogonal projections into orthogonal projections, we need only to show the convergence in the weak operator topology, and the norm-boundedness of the net $\{\sum_{i\in J}\pi(p_i),\ J\subset I,\ J$ - finite$\}$ allows for proving this on the dense subset $\Phi(M)$ of X. By virtue of the normality of Φ and Lemma 1, we have

$$\begin{aligned}(\sum_{i\in I}\pi(p_i)(\Phi(a)),\Phi(b)) &= \sum_{i\in I}(\Phi(p_i a),e\Phi(b)) + \\ &+ \sum_{i\in I}(\Phi(ap_i),e^{\perp}\Phi(b)) = (\Phi((\sum_{i\in I}p_i)a),e\Phi(b)) + \\ &+ (\Phi(a\sum_{i\in I}p_i),e^{\perp}\Phi(b)) = (e\Phi((\sum_{i\in I}p_i)a) + \\ &+ e^{\perp}\Phi(a\sum_{i\in I}p_i),\Phi(b)) = (\pi(\sum_{i\in I}p_i)(\Phi(a)),\Phi(b))\end{aligned}$$

and the normality of π follows. For the sake of completeness, let us observe that if π is normal, then Φ is immediately seen to be normal too. □

REMARK 8. We have the following situation, regarding the projection e in (*):

1. If $e = \mathbb{1}$, then Φ originates from a *-representation of M.
2. If $e = 0$, then Φ originates from a *-antirepresentation of M.
3. If $0 < e < \mathbb{1}$, then Φ originates from a proper Jordan morphism.

The following corollary can be easily deduced from Theorem 7; however, its "only if" part was found earlier (cf. [4; Prop. 8.10, p.78]).

COROLLARY 9. Let Φ be an orthogonally scattered field over M. Φ originates from a Jordan morphism iff there exist orthogonally scattered fields Φ_1,Φ_2 and positive linear functionals ϕ,ψ such that

$\Phi = \Phi_1 + \Phi_2$, the subspaces $\overline{\Phi_1(M)}$ and $\overline{\Phi_2(M)}$ are orthogonal, and

$$(\Phi_1(a),\Phi_1(b)) = \phi(b^*a), \qquad (\Phi_2(a),\Phi_2(b)) = \psi(ab^*)$$

for all a,b in M. ∎

REFERENCES

1. A. Bartoszewicz, S. Goldstein, On density function of vector--valued Gleason measure, Bull. Pol. Acad. Sci. Math. 28 (1980), 331-336.

2. O. Bratelli, D.W. Robinson, Operator Algebras and Quantum Statistical Mechanics I, Springer-Verlag, New York - Heidelberg - - Berlin, 1979.

3. E. Christensen. Measures on projections and physical states, Comm. Math. Phys. 86 (1982), 529-538.

4. S. Goldstein, Orthogonal Forms on von Neumann Algebras (in Polish), Acta Univ. Lodziensis, Łódź, 1987.

5. S. Goldstein, R. Jajte, Second-order fields over W* -algebras, Bull. Pol. Acad. Sci. Math. 30 (1982), 255-260.

6. E. Hensz, Remarks on vector Gleason measures, Preprint, 1977.

7. R. Jajte, Gleason measures, Probabilistic Analysis and Related Topics, vol. 2; ed. A.T. Bharucha-Reid, Academic Press, New York--London 1979, 69-104.

8. A. Paszkiewicz, Measures on projections in W*-factors, J. Funct. Anal. 62 (1985), 87-117.

9. S. Strătilă, L. Zsidó, Lectures on von Neumann Algebras, Abacus Press, Tunbridge Wells, Kent 1979.

10. M. Takesaki, Theory of Operator Algebras I, Springer-Verlag, New York - Heidelberg - Berlin, 1979.

11. F.J. Yeadon, Measures on projections in W*-algebras of type II_1, Bull. Lond. Math. Soc. 15 (1983), 139-145.

12. K. Ylinen, Vector measures on the projections of a W* -algebra, Ann. Univ. Turkuensis, Ser. AI 186 (1984), 129-135.

Institute of Mathematics
Łódź University
ul. Banacha 22
90-238 Łódź, POLAND

Probability Theory on Vector Spaces IV
Lancut, June '87, Springer's LNM 1391

ON THE p-LÉVY-BAXTER PROPERTY AND ITS APPLICATIONS.

by

ANDRZEJ MĄDRECKI

Abstract. We consider the p-stable motion of index p ($0 < p < 1$) for which we establish the property analogous to the Lēvy-Baxter one of the Brownian motion. We call this property the p-Lēvy-Baxter property and using it we find a sufficient condition for perpendicularity of measures induced by the processes, which are Riemann-Stieltjes stochastic integrals with respect to the standard p-stable motions .

1. Introduction. Let $S^p = \{ S_t^p : 0 \le t \le 1 \}$ be a standard p-stable motion with $0 < p < 1$ (cf. Sect. 2). Then (cf. Lemma 2.1) we can define a new stochastic process $X = \{ X_t : 0 \le t \le 1 \}$ by the formula

$$(1.1) \qquad X_t = X_0 + \int_0^t f(u)dS_u^p ,$$

where X_0 is some random variable (r.v.) , $f = \{ f_t : 0 \le t \le 1 \}$, is some stochastic process, and finally, the integral on the right-hand side of (1.1) is a Riemann-Stieltjes stochastic integral with respect to the S^p.

Note that $\int_0^t f(u)dS_u^p$ is a r.v. defined as follows :

1980 Mathematics Subject Classification, 60G17 ,60G30 .

Key words and phrases : p-stable motion ($0 < p < 1$), modular functions of p-stable processes, perpendicularity of measures , Levy-Baxter type property .

$$\left(\int_0^t f(u)\,dS_u^p\right)(\omega) := \int_0^t f(u,\omega)\,dS_u^p\ , \quad \omega \in \Omega \ .$$

If (1.1) holds, then we write $X = J_p(X_0, \{f_t\})$.

This paper deals with the following problem : is it possible to give "functional conditions" for singularity of measures induced by the processes of the form (1.1) , similar to those given by Slepian [8] and Wong [11] for processes being Ito stochastic integrals with respect to the Wiener (and generally Gaussian) processes .

Before we will answer the above question , in Sect.2 we establish the so-called p-Lévy-Baxter property of S^p ($0 < p < 1$), which is basic tool for proving the mentioned above functional conditions for singularity. The p-Lévy-Baxter property of S^p is an immediate consequence of the Lévy-Baxter property of the Wiener process $S^2 = W$, and the following fact : $(S_t^p - S_s^p)/(t - s)$ has the same tailbehaviour (near zero) as $((W_t - W_s)/(t - s))^{-2(1-p)/p}$ where $0 < p < 1$ and $0 \leq s < t \leq 1$ (cf.[7]) .

Let $K_p = \{ x : x$ is a non-decreasing function on $[0, 1], x \geq 0$ and $\rho_p(x) = \int_0^1 \dot{x}(t)^{\lambda_p}\,dt < +\infty \}$

,where $\lambda_p = p/(p-1)$ (notice that λ_p is conjugate to p in the sense that $1/p + 1/\lambda_p = 1$) .

Such sets were considered in [6] and [10] . It is shown , that K_p are the sets of all limit points in the functional law of the iterated logarithm.

The main result of this paper states that the condition $\rho_p(f_X) \neq \rho_p(f_Y)$ implies that the distribution L(X) of $X = J_p(X_0, f_X)$ is singular to the (measure) distribution L(Y) of $Y = J_p(Y_0, f_Y)$ (cf. (1.1)). (Here and in the sequel L(X) denotes the measure induced by a process X) .

Finally , we remark, that p-stable distributions and measures with $0 < p < 1$ are important and have applications in physic (cf.e.g.[9]), (in fact only p-stable distributions with $0 < p < 1$ and $p = 2$).

2. <u>The modular cone</u> $L_+{}^{\lambda}p[0\ ,\ 1]$ <u>with</u> $0 < p < 1$.

Let (Ω, Σ, P) be a probability space. A stochastic process $S^p = \{ S_t^p : 0 \le t \le 1 \}$ on (Ω, Σ, P) is said to be a p-<u>stable motion</u> (cf.e.g. [3] , [4]) where $0 < p \le 2$, if all finite-dimensional distributions of S^p are stable (of index p) and

(SM1) S^p has stationary and independent increments , and

(SM2) $Ee^{iuS_t^p} = e^{-t|u|^p}$ for $1 \le p \le 2$, and

$Ee^{-uS_t^p} = e^{-t|u|^p}$ for $0 < p < 1$,

(Here $EX = \int_\Omega X dP$) .

For the existence of such a process , cf. [3] , [4] .

In the sequel we assume that S^p is a right-continuous version of S^p. It is well-known (cf.[2],[4]), that for all p with $0< p <2$, the paths of S^p belong to the <u>Skorohod space</u> $D[0\ ,\ 1]$ endowed with the <u>Skorohod topology</u> (cf.[2]) . Moreover , for p with $0 < p < 1$, we have the following , easy verificable fact

<u>Lemma 2.1</u>. For a.e.$\omega \in \Omega$ a trajectory $[0\ ,\ 1]\ t \longmapsto S_t^p(\omega)$ is a positive real valued strictly increasing function on $[0\ ,\ 1]$.

By $I[0\ ,\ 1]$ we denote the set of all real valued strictly increasing functions f on $[0\ ,\ 1]$ with $f(0) = 0$. We also assume that the zero function 0 belongs to $I[0\ ,\ 1]$. It is obvious that $I[0\ ,\ 1]$ is a cone , that is $0 \in I[0\ ,\ 1]$ and for all $x_1\ ,\ x_2 \in I[0\ ,\ 1]$ and $\alpha_1\ ,\ \alpha_2 \ge 0\ ,\ \alpha_1 x_1 + \alpha_2 x_2 \in I[0\ ,\ 1]$.

Following Lemma 2.1 , for each p with $0 <p <1$, the stochastic process S^p induces a probability measure s^p on $(I[0\ ,\ 1],\ \Sigma_I)$, where Σ_I is the σ-field in $I[0\ ,\ 1]$ being the trace of the Borel σ-field in $D[0\ ,\ 1]$.

The measure s^p is completely determined by its values on the cylinder sets C of the form

(2.1) $\{ x \in I[0\ ,\ 1] : (x(t_1)\ ,\ \ldots\ ,\ x(t_n)) \in E \}$,

where $0 \le t_1 < \ldots < t_n = 1$ and E is a Borel set in

The value $s^p(C)$ on a cylinder C of the form (2.1) is equal to $P\{(S^p(t_1), \dots, S^p(t_n)) \in E\}$. Let us mention that a stable process $\{S^p(t)\}$ has stationary independent increments and further that $S^p(t)$ has the same distribution as $t^{1/p} S_1^p$.

Let x be a real function on $[0, 1]$ such that $\{t : x(t) < +\infty\} = [0, 1]$. The <u>derivative</u> of x is the almost surely unique function $\dot{x}$ on $[0, 1]$ such that

$$x(t) = \int_0^t \dot{x}(s)ds \quad , \text{ for all } t \text{ in } [0, 1] .$$

Now, let p be a number from $[0, 1]$. Put $\lambda_p = p/(p - 1)$, and note that $\lambda_p < 0$. We define

$$\dot{L}_+^{\lambda_p}[0, 1] = \{x \text{ is an absolutely continuous function and } \dot{\rho}_p(x) = \int_0^1 (\dot{x}(t))^{\lambda_p} dt < +\infty , x \in I[0,1]\}$$ ([10, Sect. 2, (2.23)]).

Observe that the functions $g_p(u) = u^{\lambda_p}$ $(0 < p < 1)$, $u \geq 0$, are convex, because $\ddot{g}_p(u) = \lambda_p(\lambda_p - 1)u^{(\lambda_p - 2)} > 0$ for $u > 0$ $(\lambda_p < 0)$.

Thus we have $\dot{\rho}_p(0) = 0$ and $\dot{\rho}_p(\alpha_1 x_1 + \alpha_2 x_2) \leq \alpha_1 \dot{\rho}_p(x_1) + \alpha_2 \dot{\rho}_p(x_2)$ for all $x_1, x_2 \in \dot{L}_+^{\lambda_p}[0, 1]$ and $\alpha_1, \alpha_2 \geq 0$ with $\alpha_1 + \alpha_2 = 1$.

Therefore $\dot{L}_+^{\lambda_p}[0, 1]$ is the subcone of $I[0, 1]$ and $\dot{\rho}_p$ is a <u>convex modular</u> on it. In this paper we will need the following version of $\dot{L}_+^{\lambda_p}[0, 1]$: by $L_+^{\lambda_p}[0, 1]$ we denote the set of all $\mathbb{R}$-<u>valued positive and continuous</u> functions f on $[0, 1]$ with the property:

$$\rho_p(f) = \int_0^1 (f(t))^{\lambda_p} dt < +\infty .$$

Obviously $\rho_p : L_+^{\lambda_p}[0, 1] \longmapsto [0, +\infty)$ is the <u>convex modular</u> on $L_+^{\lambda_p}[0, 1]$.

3. <u>The</u> p-<u>Lévy-Baxter property with</u> $0 < p < 1$.

Let S be a standard $\frac{1}{2}$-stable r.v. with the density

$$f_S(t) = (1/\sqrt{2\pi t^3})\exp\{-\tfrac{1}{2}t\} \quad , \quad t > 0 \quad (\text{cf.}[4]) .$$

Let $X = 1/S$ .Then it is easy to check that the density $f_X(t)$ of X has the form

$$f_X(t) = f_S(1/t)(1/t^2) = (1/\sqrt{2\pi t})\exp\{-t/2\} \ , \ t> 0 .$$

Thus , the r.v. $X = 1/S$ has the standard χ^2-distribution. Let $\{W_t : 0 \leq t \leq 1\}$, be the standard Brownian motion . By the above, we state the following result : $(S_t^{\frac{1}{2}} - S_s^{\frac{1}{2}})/(t - s)$ has the same distribution as $((W_t - W_s)/(t - s))^{-2}$, or equivalently (puting t=1 and s=0) we obtain

$$P(S_1^{\frac{1}{2}} \leq x) = P(1/U^2 \leq x) = 2P(U \geq x^{-\frac{1}{2}}) \qquad ; \qquad x \geq 0 ,$$

where U denotes the r.v. with standard normal distribution N(0 , 1).

Since $\lambda_{\frac{1}{2}} = -1$, then the above equality has the final form

(3.1) $$P(S_1^{\frac{1}{2}} \leq x) = 2P(U \geq x^{(\lambda_{\frac{1}{2}}/2)}) \quad , \quad \text{for} \quad x \geq 0 .$$

The relation (3.1) has the following generalization for p with $0< p < 1$:there exist such positive constants c_1 , c_2 that

(3.2) $$\lim_{x \to 0} P(S_1^p \leq x)/c_1 P(U \geq c_2 x^{\lambda_p/2}) = 1 .$$

The above expression follows from the expansions given in [7] .The relation (3.2) implies the following

<u>Lemma 3.1</u>. For each p with $0 < p < 1$, $d(p) = E(S_1^p)^{2\lambda_p} < +\infty$.

<u>Theorem 3.1</u>. (p-<u>Lévy-Baxter property</u> , $0 < p < 1$) . Let

$$T_n = \{ a = t_0(n) < t_1(n) < \dots < t_{N(n)}(n) = b \} , n = 1,2, \dots ,$$

be a sequence of subdivisions of [a , b] , such that

$$\Delta_n := \underset{1 \leq i \leq N(n)}{\text{maximum}} (t_i(n) - t_{i-1}(n)) \underset{n}{\longrightarrow} 0 .$$

Let $\{ S_t^p : a \leq t \leq b \}$ be the p-stable motion with $0 <p <1$. Then

(3.3) $$\underset{n}{\text{l.i.m.}} \sum_{i=1}^{N(n)} (S^p_{t_i(n)} - S^p_{t_{i-1}(n)})^{\lambda_p} / (t_i(n) - t_{i-1}(n))^{\alpha_p}$$

$$= c(p) (b - a) ,$$

where $\alpha_p = (\lambda_p/p - 1)$ and $c(p) = E(S_1^p)^{\lambda_p}$ is some positive constant dependent on p only .

If $\sum_{n=1}^{+\infty} \Delta_n < +\infty$, then the above convergence is with probability 1 .

Proof. The proof utilizes similar arguments to these from the proof of the L.B.property for the Wiener process (cf.[1],[5],[11]) . To show(3.3) we write

$$S_n(p) = \sum_{i=1}^{N(n)} (S^p_{t_i(n)} - S^p_{t_{i-1}(n)})^{\lambda_p}/(t_i(n) - t_{i-1}(n))^{\alpha_p} - c(p) .$$

$$(b - a) = \sum_{i=1}^{N(n)} \{(S^p_{t_i(n)} - S^p_{t_{i-1}(n)})^{\lambda_p}/(t_i(n) - t_{i-1}(n))^{\alpha_p} - c(p)(t_i n - t_{i-1} n$$

Since $S_n(p)$ is the sum of the independent r.v.'s ,each of which has the expected value equal to zero , then the following equalities hold:

$$ES_n(p) = 0 \quad \text{and} \quad ES_n^2(p) = \sum_{i=1}^{N(n)} E\{(S^p_{t_i(n)} - S^p_{t_{i-1}(n)})^{\lambda_p}/(t_i(n)-t_{i-1}n)^{\alpha_p}$$

$$- c(p)(t_i(n) - t_{i-1}(n))\}^2 = (d(p) - c^2(p)) \sum_{i=1}^{N(n)} (t_i(n) - t_{i-1}(n))^2$$

$\leq (d(p) - c^2(p))\,\Delta_n$, where $d(p) = E(S_1^p)^{2\lambda_p}$ $(E(S_1^p)^{\lambda_p})^2 = c(p)^2$

(in fact $d(p) > c(p)^2$ since $ES_n^2(p) > 0$) .

Since $\Delta_n \to 0$ if $n \to +\infty$, then $\underset{n}{\text{l.i.m.}}\, S_n(p) = 0$.

For the proof of the second assertion of Th.3.1 cf.e.g. [11] .

4. The application of Theorem 3.1.

Now , we prove a sufficient condition for singularity of some class of processes , which indicates an interesting connection between the p-Lévy-Baxter property ($0 < p < 1$) and the cone $L_+{}^{\lambda_p}[0 , 1]$.

Lemma 4.1. (1) Let $\{r_n(t)\}$ be a broken line (a stochastic process for a given n) with the vertices in the points

$(k/2^n , \sum_{i=1}^{N(n)} (S^p((i+1/2^n) - S^p(i/2^n))^{\lambda_p}\, 2^{n\alpha_p})$; $k,n \geq 0$ and $0 \leq k \leq 2^n$. Then , the sequence $\{r_n(t)\}$ converges uniformly to the function $r(t) = t$, for $t \in [0 , 1]$, on the every finite segment $[0 , T]$, with probability 1 .

(2) Let $\{\beta_n(t)\}$ be a sequence of monotonic (deterministic) functions, which uniformly converges to the function $\beta(t) = t$, for $t \in [a , b]$. Let $\{G_n(t)\}$ be a sequence of functions which converges uniformly to a continuous function $G(t)$ on $[a , b]$. Then

$$\lim_n \int_a^b G_n(t)d\beta_n(t) = \int_a^b G(t)dt \quad .$$

Proof. (1) All paths of the processes $\{r_n\}$, $n \geq 1$, are nondecreasing. It is known that from the convergence of the sequence of nondecreasing functions to a continuous nondecreasing function on a dense subset of a segment (which contains the ends) , it follows the uniform convergence. Hence , it suffices to show , that for all $t = k/2^m$, we have $P\{r_n(t) \to_n t\} = 1$ .For $n \geq m$, we have $E\,(r_n(t) - t)^2 = 2t2^{-n}c(p)$,

$$E(\sum_n (r_n(t) - t)^2) = c(p)(2t \sum_n 2^{-n}) < +\infty \quad .$$

Since the series $\sum_n (r_n(t) - t)^2$ has a finite expected value, then it converges with probability 1 , i.e. $(r_n(t) - t) \xrightarrow[n]{} 0$ a.e.

(2) It is the well-known fact from the elementary analysis.

Lemma 4.2. Let $\{S_t^p : 0 \leq t \leq 1\}$ be the p-stable motion with $0 <p< 1$. Let $f = \{ f_t : 0 \leq t \leq 1 \}$ be a separable and measurable stochastic process , such that the paths of f belong to the cone $L_+{}^\lambda p[0 , 1]$. Let $X = \{ X_t : 0 \leq t \leq 1 \}$ be a stochastic process defined as follows:

$$X_t = X_0 + \int_0^t f(u)dS_u^p \quad ,$$

where X_0 is an arbitrary r.v. and the integral on the right-hand side is the Riemann-Stieltjes stochastic integral with respect to S^p .Then

$$\text{(GLB)} \lim_n \sum_{k=0}^{(2^n-1)} 2^{n\alpha_p}(X((k+1)/2^n) - X(k/2^n))^{\lambda_p} = \int_0^1 f(u)^\lambda p du =$$

$$= \rho_p(f) \quad , \quad \text{a.e.}$$

Proof. Under our assumptions we have

$$(4.1) \qquad A_n = \sum_{k=0}^{(2^n-1)} 2^{n\alpha_p}(X((k+1)/2^n - X(k/2^n))\lambda p =$$

$$\sum_{k=0}^{(2^n-1)} 2^{n\alpha_p} \left(\int_{k/2^n}^{(k+1)/2^n} f(u)dS_u^p \right) \lambda p \quad .$$

Since f has the continuous paths, then by the mean value theorem for the Riemann-Stieltjes integrals , for each k,n , we have

$$(4.2) \quad \int_{k/2^n}^{(k+1)/2^n} f(u)\,dS^p_u = f(\tilde{u}_k)\left(S^p((k+1)/2^n) - S^p(k/2^n) \right) ;$$

$$k/2^n \le \tilde{u}_k < (k+1)/2^n .$$

Combining (4.1) and (4.2) yields

$$(4.3) \quad A_n = \sum_{k=0}^{(2^n-1)} \left(2^{n\alpha_p} f(\tilde{u}_k)^{\lambda_p} (S^p((k+1)/2^n) - S^p(k/2^n))^{\lambda_p} \right) .$$

The sum in the formula (4.3) is nothing other than the integral $\int_0^1 G_n(t,\omega)\,dr_n(t,\omega)$, where $r_n(t,\cdot)$ are defined in the Lemma 4.1 (1) and $G_n(t,\omega) = f^{\lambda_p}(\tilde{u}_k)(\omega)$ for $k/2^n \le \tilde{u}_k < (k+1)/2^n$.

From the our assumption , Lemma 4.1 and (4.3) we conclude that

$$\lim_n A_n = \int_0^1 f(t)^{\lambda_p}\,dt = \rho_p(f) ,$$

which completes the proof (obviously $G_n(t,\cdot) \underset{n}{\to} f^{\lambda_p}(t)$, $t \in [0,1]$, a.e.uniformly) .

Theorem 4.1. Let $f_1 = \{f_1(t) : 0 \le t \le 1\}$ and $f_2 = \{f_2(t) : 0 \le t \le 1\}$ be two stochastic processes , such that for a.e. $\omega \in \Omega$ the paths of f_1 and f_2 belong to the cone $L_+^{\lambda_p}[0,1]$. Let

$$X_i^p(t) = X_0^i + \int_0^t f_i(u)\,dS^p_u , \quad (i=1,2 \text{ and } 0<p<1, \text{ and } 0 \le t \le 1)$$

, where X_0^i are the r.v. s and the integrals on the right-hand side are the Riemann-Stieltjes stochastic integrals. Finally, let μ_i be the probability measure induced by the process X_i^p (i = 1 , 2) .

If there exists $t \in [0,1]$ such that

$$(SC) \quad \rho_p(f_1,t) := \int_0^t f_1(u)^{\lambda_p}\,du \ne \int_0^t f_2(u)^{\lambda_p}\,du =: \rho_p(f_2,t) \text{ a.e.}$$

then the measure μ_1 is perpendicular to the measure μ_2 on the measurable space (I [0 , 1] , Σ_I) i.e. $\mu_1 \perp \mu_2$ (for short) .

Proof. Note that without loss of generality we can assume that t=1.

Under our assumptions , from Lemma 4.2 we get

$$P(A_i) := P\{\omega : \lim_n \sum_{k=0}^{(2^n-1)} 2^{n\alpha}p(X_i^p((k+1)/2^n) - X_i^p(k/2^n))^\lambda p=\int_0^1 f_i(u)^\lambda pdu\}$$

$= 1$, $(i=1,2)$.Let $B_i = \{ f\in I[0 , 1]: f(t) = X_i^p(t ,\omega)$, $t\in[0 , 1]$ and $\omega \in A_i \}$. Obviously $\mu_1(B_1) = P(A_1) = 1$ and $\mu_2(B_2) = P(A_2) = 1$. Therefore , if (SC) holds , then $B_1\cap B_2 = \emptyset$, and $\mu_1(B_1) = \mu_2(B_2) = 1$, which shows the perpendicularity of measures μ_1 and μ_2 .

References

[1] Baxter G., A strong limit theorem for Gaussian processes , Proc. Amer.Math.Soc.7(1956) 522-527 .

[2] Billingsley P., Convergence of probability measures, John Wiley and Sons , Inc., New York-London-Sydney-Toronto, 1968 .

[3] Breiman L., Probability , Addison-Wesley Reading, Mass, 1968 .

[4] Feller W., An introduction to probability theory and its applications, vol.II, John Wiley and Sons, Inc.,New York, 1966 .

[5] Lévy P., Le mouvement brownien plan, Amer.J.Math.62(1940),487-550.

[6] Mijnheer J.L., Sample paths properties of stable processes, Math. Centre Tract.59(1975).

[7] Skorohod A.V., Asymptotic formulas for stable distribution laws , Selected Translations in Mathematical Statistics and Probability1, (1961), 157-161 .

[8] Slepian D., Some comments on the detection of Gaussian signals in Gaussian noise , IRE Trans.Inf.Th.4(1958) , 65-68 .

[9] Weron K., Relaxation in glassy materials from Levy stable distribution , Acta Physica Polonica 5 A70(1986) .

[10] Wichura M.J., Functional laws of the iterated logarithm for partial sums of i.i.d. random variables in the domain of attraction of a completely asymetric stable law , Ann.Probability 2(1974) , 1108-1138 .

[11] Wong E., Stochastic processes in information and dynamical systems , McGraw - Hill ,Inc., 1971 .

Andrzej Mądrecki ,
Institute of Mathematics,
Technical University of Wroclaw ,
50-370 Wroclaw ,
Wybrzeze Wyspianskiego 27 ,
POLAND .

Probability Theory on Vector Spaces IV
Lancut, June '87, Springer's LNM 1391

NOTES ON INFINITE DIMENSIONAL STATIONARY SEQUENCES

A. Makagon
Wroclaw Technical University

and

H. Salehi
Michigan State University

CONTENT

1. INTRODUCTION

These notes are the outgrowth of a seminar: Dilation theory and stationary processes, which was conducted in the Department of Statistics and Probability at Michigan State University during the winter and spring terms of 1985. The intention here is to collect in one place some recent advances which are made in interpolation, extrapolation and the angle between the past and future in the theory of infinite dimensional stationary sequences. No attempt has been made to include all topics related to stationary sequences. We have merely touched upon the topics which we have been working on recently. Proofs are compact, short or even sometimes omitted in order to keep the paper in moderate length. However the essential steps are included in proofs so that a reader could continue reading the paper without much of the need to consult the references.

The notes are mostly expository. However it contains a few results which to our knowledge have not been published before. Special efforts have been made to point out the difficulties involved in passing to the infinite dimensional case in extending the finite dimensional results.

The content of these notes may be summarized as follows. Chapter 2 contains the bulk of basic theoretical materials on stationary sequences. It consists of the analysis of correlation operator and spectral measure; Wold and Lebesgue decomposition; the formulation of the prediction problem; multiplicity and rank; models for stationary sequences with or without densities including the discussion of Hellinger integrals. Section 3 deals with interpolation of stationary sequences. A review of the known results on this topic is given. In comparison with the subject of extrapolation and angle treated in sections 4 and 5, the problem of interpolation reviewed here is nearly complete. Section 4 treats some selected aspects of extrapolation theory in the infinite dimensional case. Section 5 is a discussion of angle between the past and future of a stationary sequence. The relationship between the angle, the existence of "basis" and relevant Fourier series expansion is examined.

Many topics such as nonstationary sequences (cf. [1], [4], [38]), continuous parameter process (cf. [47], [56]), random fields (cf. [30], [20]), non–Hilbertian space valued processes (cf. [37], [3], [33], [54]) etc. are not treated in this paper. The reference list is not exhaustive either. We have merely compiled a list which will make the reading of the paper easier.

The following notations will be used throughout the paper.

(1.1) $\mathbb{C}$, $\mathbb{R}$, Z and N will denote the sets of complex numbers, real numbers, integers and positive integers, respectively. $\mathbb{C}$ also will be regarded as a complex Hilbert space of dimension one.

(1.2) H, K, $\mathcal{H}$, $\mathcal{K}$ will be complex Hilbert spaces with an inner product (,) and a norm $|\ |$.

(1.3) For any subset $S \subset H$, spS will stand for the set of all linear combinations of elements from S, $\bar{S}$ will denote the closure of S. In particular $\overline{\text{sp}}\ S$ will denote the smallest closed subspace of H containing the set S.

(1.4) For any closed subspace $M \subset H$ the symbol $H \ominus M$ or simply $M^\perp$ will stand for the orthogonal complement of M in H. $M_1 \oplus M_2$ will denote the orthogonal sum of M_1 and M_2.

(1.5) $L(H,K)$ will stand for the space of all bounded linear operators from H into K equipped with the Banach norm $\|T\| = \sup\{|Tx|: |x| \leq 1\}$.

(1.6) For any $T \in L(H,K)$, T^* will denote the adjoint operator and Null(T) will be the null space of T defined by $\text{Null}(T) = \{x \in H: Tx = 0\}$. Recall that $\text{Null}(T^*) = (TH)^{\perp}$, $\overline{TH} = [\text{Null}(T^*)]^{\perp}$.

(1.7) If $H = K$, then $L(H,K)$ will be abbreviated by $L(H)$. I will denote the identity operator and P_M will denote the orthogonal projection onto a closed subspace M.

(1.8) By $L^+(H)$ we will denote the subset of $L(H)$ consisting of all positive operators, i.e., all bounded linear operators T in $L(H)$ such that $(Tx,x) \geq 0$ for all $x \in H$. Every positive operator is self adjoint. If $T \in L^+(H)$ then $T^{1/2}$ will denote its square–root, i.e., the unique positive operator whose square is equal to T. Recall that $T^{1/2}H \supset TH$, $\overline{T^{1/2}H} = \overline{TH}$, provided $T \in L^+(H)$.

(1.9) For any $T \in L(H,K)$ the symbol $T^{\#}$ will stand for the generalized inverse of T, that is a linear operator from TH onto $\text{Null}(T)^{\perp}$ defined by the formula: $T^{\#}y = x$ if and only if $x \in \text{Null}(T)^{\perp}$ and $Tx = y$. If TH is closed then $T^{\#}$ restricted to the subspace TH is bounded and its norm will be denoted by $\|T^{\#}\|$. In fact in this case $T^{\#}$ can be extended to a bounded linear operator from K into H by setting $T^{\#} = T^{\#}P_{TH}$. If H and K are finite dimensional Hilbert spaces then, in order to be consistent with the commonly used notation, $T^{\#}$ will always mean the extension of $T^{\#}$ to K as described above.

(1.10) $\mathcal{B}$ will be the Borel σ–algebra of subset of the interval $(-\pi,\pi]$. If μ is a nonnegative Borel measure on $(-\pi,\pi]$, H is a Hilbert space and $p \geq 1$ then by $L^p(\mu,H)$ we will denote the Banach space of all H–valued μ–measurable (in the Bochner sense) functions f such that

$$\int |f(t)|^p \mu(dt) < \infty,$$

where here and in the sequel the integral $\int$ will denote the Lebesgue integral over the interval $(-\pi,\pi]$. If $H = \mathbb{C}$, we will write $L^p(\mu)$ instead of $L^p(\mu,\mathbb{C})$.

(1.11) The normalized Lebegue measure $\frac{dt}{2\pi}$ on the interval $(-\pi,\pi]$ will be denoted by σ. If $\mu = \sigma$, then we will write $L^p(H)$ instead of $L^p(\sigma,H)$.

(1.12) For any two nonegative measures μ and ν we will write $\mu \sim \nu$, $\mu << \nu$ or $\mu \perp \nu$ depending on whether μ is equivalent to ν, μ is absolutely continuous (a.c.) with respect to (w.r.t.) ν or μ is singular w.r.t. ν, respectively.

(1.13) μ–ess sup and μ–ess inf as usual will stand for an essential supremum and an essential infinum of function w.r.t. μ, respectively.

(1.14) For a set Δ, Δ^c will denote its complement and 1_{Δ} will stand for the indicator of Δ.

(1.15) Σ without limits will indicate that except finitely many terms the rest are zero.

2. ELEMENTS OF THEORY OF STATIONARY SEQUENCES

In this section we introduce the notion of a stationary sequence and we state some basic facts concerning such sequences.

2.1. Stationary sequences, correlation and spectral measure.

2.1.1. Definition. Let H and K be any two Hilbert spaces. A function $X: Z \to L(H,K)$ is said to be an L(H,K)–valued stationary sequence ("L(H,K)–valued" is occasionally omitted if no confusion may arise) if for any $x, y \in H$ the function $(X(n)x,X(m)y) = K(n,m,x,y)$ depends only on $n-m$, $n, m \in Z$.

Let X be a stationary sequence. The L(H)–valued function $\Gamma(n) = X(0)^*X(n)$ is called the correlation of X. Note that from stationarity it follows that $X(m)^*X(n) = \Gamma(n-m)$, $n, m \in Z$. The smallest closed subspace of K containing all elements of the form $X(n)x$, $x \in H$, $n \in Z$ is denoted by M(X) and is called the time domain of X. For A, any subset of Z, let

(2.1.2) $$M_A(X) = \overline{sp}\{X(n)x: \ x \in H, \ n \in A\}$$

(2.1.3) $$N_A(X) = M(X) \ominus M_A(X).$$

The unitary operator acting in M(X) defined on a dense subset of M(X) by the formula

(2.1.4) $$U(\Sigma X(n)x_n) = \Sigma X(n+1)x_n$$

is called the shift operator of X. From spectral representation theorem (e.g. [7] X.2) it follows that there exists an L(M(X))–valued function $E(\Delta)$, $\Delta \in \mathcal{B}$ (called the spectral measure of U) with the following properties:

(2.1.5)
- (i) $E(\Delta)$ is an orthogonal projection operator in M(X), $\Delta \in \mathcal{B}$,
- (ii) $E(\Delta_1 \cap \Delta_2) = 0$, provided $\Delta_1 \cap \Delta_2 = \emptyset$, $\Delta_1, \Delta_2 \in \mathcal{B}$,
- (iii) $E((-\pi,\pi]) = I$,
- (iv) for any $x \in H$ the vector valued function $\mathcal{B} \ni \Delta \to E(\Delta)x$ is a countably additive H–valued measure,

(2.1.6) for every $x \in H$ and $n \in Z$,

$$Ux = \int e^{itn}E(dt)x,$$

where the integral above is defined as in [7], IV. 10.

Since $X(n) = U^n X(0)$, from (2.1.5) we derive the following representation for X

(2.1.7) $$X(n)x = \int e^{int} E(dt)X(0)x, \ x \in H, \ n \in Z.$$

Consequently the correlation function Γ of X has the form

(2.1.8) $$(\Gamma(n)x,y) = \int e^{int} (X(0)^*E(dt)X(0)x,y), \ x, y \in H, n \in Z.$$

If the need arises we will write Γ_X, U_X and E_X instead of Γ, U and E to avoid any confusion.

2.1.9. Definition. Let X be an L(H,K)–valued stationary sequence and let $E(\cdot)$ be the spectral measure of the shift operator of X. The $L^+(H)$–valued function defined on $\mathcal{B}$ by the formula

$$F(\Delta) = X(0)^* E(\Delta) X(0)$$

is said to be the spectral measure of the sequence X.

Note, that by [7], IV.10.1, $F(\cdot)x$ is a countably additive H–valued set function. Thus for every bounded Borel measurable function f the integral $\int f(t)F(dt)x$ is well defined. In particular (2.1.8) implies that

$$\Gamma(n)x = \int e^{int} F(dt)x$$

for every $x \in H$ and $n \in Z$. The following theorem which is due to Naimark gives the description of spectral measure of a stationary sequence.

2.1.10. Theorem. An L(H)–valued set function $F(\Delta)$, $\Delta \in \mathcal{B}$, is the spectral measure of a stationary sequence if and only if for every $x \in H$, the set function $B \ni \Delta \to (F(\Delta)x,x)$ is a nonnegative (scalar valued) measure.

There are several proofs of this theorem. The proof sketched below was given by Sz. Nagy ([52]) in a more general setting for positive definite functions defined on semigroups. For more details, other references and generalizations we refer the reader to [36], [9], [26].

Proof. ($\Leftarrow$). Let K_0 be the set of all H–valued set functions on $\mathcal{B}$ of the form $\mu(\cdot) = \sum_{k=1}^{n} F(\cdot \cap \Delta_k)x_k$, where $\Delta_1,\dots,\Delta_n \in \mathcal{B}$, $x_1,\dots,x_n \in H$, $n \geq 1$, equipped with the inner product

$$\Big(\sum_{k=1}^{n} F(\cdot\cap\Delta_k)x_k, \sum_{j=1}^{m} F(\cdot\cap\Delta_j')x_j'\Big)_0 = \sum_{k=1}^{n}\sum_{j=1}^{m} (F(\Delta_k \cap \Delta_j')x_k, x_j').$$

It is easy to see that $(\mu,\nu)_0$ is a sesquilinear form on $K_0 \times K_0$. One can also show (see [36] p. 30–31) that $(\mu,\mu)_0 \geq 0$ for every $\mu \in K_0$. This implies that $|(\mu,\nu)_0| \leq |\mu|_0 |\nu|_0$, $\mu, \nu \in K_0$, where by definition $|\mu|_0 = \sqrt{(\mu,\mu)_0}$. In particular, setting $\nu(\cdot) = F(\cdot \cap \Delta)x$ in the inequality above we obtain

$$|(\mu(\Delta),x)| = |(\mu,\nu)_0| \leq \sqrt{(F(\Delta)x,x)}\,|\mu|_0. \tag{2.1.11}$$

This yields that $|\mu|_0 = 0$ if and only if $\mu(\Delta) = 0$ for every $\Delta \in \mathcal{B}$. Thus $(\mu,\nu)_0$ is a legitimate inner product in K_0. Let K be the completion of K_0 with respect to $|\ |_0$–norm. Define the operators R: $H \to K_0$ and $E(\Delta)$: $K_0 \to K_0$ by the formulas

$$(Rx)(\cdot) = F(\cdot)x,\ x \in H \ \text{ and } \ (E(\Delta)\mu)(\cdot) = \mu(\cdot \cap \Delta),\ \Delta \in \mathcal{B},\ \mu \in K_0 .$$

Then $|Rx|_0 \leq \sqrt{(F((-\pi,\pi])x,x)}$, $x \in H$; $|E(\Delta)\mu|_0 \leq |\mu|_0$ $\mu \in K_0$, (see [36], p. 31) and $(F(\Delta)x,y) = (E(\Delta)Rx,Ry)_0$.

Now, keeping in mind that weak countably additivity implies countably additivity for Hilbert space valued additive set function it is easy to check that the set function $F(\cdot)$ satisfies properties (i) – (iv) of (2.1.5) and that the sequence defined by the formula

$$X(n)x = \int e^{int} E(dt)Rx, \quad x \in H, \; n \in Z \tag{2.1.12}$$

is an $L(H,K)$–valued stationary sequence with the spectral measure $F(\cdot)$ and the time domain $M(X) = K$. □

From Theorem 2.1.9 it follows that an $L(H)$–valued function Γ is the correlation of a stationary sequence if and only if it admits the representation

$$(\Gamma(n)x,y) = \int e^{int}(F(dt)x,y), \quad x, y \in H, \; n \in Z, \tag{2.1.13}$$

where $F(\cdot)$ is an $L(H)$–valued set function such that for every $x \in H$, $(F(\cdot)x,x)$ is a nonnegative measure (such an F is called <u>a semi–spectral measure</u>). If Γ is a <u>weakly positive definite</u> function, i.e., if Γ satisfies the condition

$$\sum_{k,i} (\Gamma(n_k-n_i)x,x)\lambda_k\bar{\lambda}_i \geq 0, \text{ for every } x \in H, \; \lambda_1,\dots,\lambda_m \in C, \; n_1,\dots,n_m \in Z, \; m \geq 1, \tag{2.1.14}$$

then using the Bochner theorem one can easily find that Γ is the Fourier transform of a semispectral measure. On the other hand if Γ admits the representation (2.1.12) then Γ is <u>positive definite</u>, i.e.,

$$\sum_{k,i} (\Gamma(n_k-n_i)x_k,x_i) \geq 0, \quad \text{for every } x_1,\dots,x_m \in H, \; n_1,\dots,n_m \in Z, \tag{2.1.15}$$

In fact, from Theorem 2.1.10 it follows that $F(\Delta) = R^* E(\Delta)R$, where E satisfies (2.1.4), and thus

$$\Sigma(\Gamma(n_k-n_i)x_k,x_i) = |\sum_k \int e^{in_kt} E(dt)Rx_k|^2 \geq 0.$$

Hence we have the following theorem.

2.1.16. <u>Theorem</u>. Let Γ be an $L(H)$–valued function on Z. The following conditions are equivalent:

(i) Γ is weakly positive definite,

(ii) Γ is positive definite,

(iii) there exists a semispectral measure $F(\Delta)$, $\Delta \in B$, such that (2.1.12) holds,

(iv) Γ is the correlation of a stationary sequence.

For a different proof of the theorem 2.1.15 in more general setting we refer the reader to [3]. The relationship between positive and weak positive definiteness for general semigroup case is discussed in [36].

2.1.17. <u>Definition</u>. Let H, K_1, K_2 be three Hilbert spaces and let X_i be $L(H,K_i)$–valued stationary sequences $i = 1, 2$. The sequences X_1 and X_2 are said to be equivalent if their correlations are equal.

It is easy to see that if sequences X_1, X_2 are equivalent, then the operator V defined by the formula

$$V(\Sigma X_1(n)x_n) = \Sigma X_2(n)x_n$$

extends to a unitary operator from $M(X_1)$ onto $M(X_2)$ such that for every $B \subset Z$

$$V(M_B(X_1)) = M_B(X_2).$$

2.2. Lebesgue decomposition. Let $X = \{X(n): n \in Z\} \subset L(H,K)$ be a stationary sequence and let F be its spectral measure. If μ is a nonnegative σ–finite measure on $(-\pi,\pi]$ then we will say that

(2.2.1) F is absolutely continuous (a.c.) w.r.t. μ, if for every $x \in H$, $(F(\cdot)x,x) << \mu$ (write $F << \mu$))

(2.2.2) F is singular w.r.t. μ if for every $x \in H$ $(F(\cdot)x,x) \perp \mu$ (write $F \perp \mu$).

If H is separable, then the nonnegative measure

$$\nu_F(\Delta) = \sum_{k=1}^{\infty} (F(\Delta)e_k,e_k)2^{-k}$$

has the same null sets as $F(\Delta)$ (here $\{e_k: k = 1,2,...\}$ is any orthonormal basis in H). Thus from the Lebesgue decomposition theorem it follows that there exists a set $\Delta_0 \in \mathcal{B}$ such that

$$\nu_F{}^a(\cdot) \stackrel{\mathrm{df}}{=} \nu_F(\cdot \cap \Delta_0) << \mu \text{ and } \nu_F{}^s(\cdot) \stackrel{\mathrm{df}}{=} \nu_F(\cdot \cap \Delta_0{}^c) \perp \mu.$$

This decomposition yields the decomposition of X into the sum $X = X^a + X^s$, called the Lebesgue decomposition of the sequence X, where

$$X^a(n)x = \int e^{int} E(dt \cap \Delta_0)X(0)x,\ x \in H\ X^s(n)x = \int e^{int} E(dt \cap \Delta_0^c)X(0)x,\ x \in H.$$

The Lebesgue decomposition also exists in the nonseparable case and is stated below.

2.2.3. Theorem. Let $X = \{X(n): n \in Z\} \subset L(H,K)$ be a stationary sequence with the spectral measure F and let μ be a nonnegative (scalar) σ–finite measure on $\mathcal{B}$. Then there exist two unique stationary sequences X^a and X^s with the properties:

(i) $X^a(n) + X^s(n) = X(n),\ n \in Z$

(ii) $M(X^a) \oplus M(X^s) = M(X)$

(iii) $U X^a(n) = X^a(n+1)$, $U X^s(n) = X^s(n+1)$, where U is the shift of X, $n \in Z$,

(iv) the spectral measure F^a of X^a is absolutely continuous w.r.t. μ

(v) the spectral measure F^s of X^s is singular w.r.t. μ.

Proof: We sketch the proof of the existence and leave the proof of the uniqueness to the reader. Let E be the spectral measure of U, $M^a(X)=\{x \in M(X): (E(\cdot)x,x) << \mu\}$, $M^s(X)=\{x \in M(X): (E(\cdot)x,x) \perp \mu\}$. Let P^a and P^s be orthogonal projections onto $M^a(X)$ and $M^s(X)$, respectively. Both $M^a(X)$ and $M^s(X)$ are invariant under the family $E(\Delta)$, $\Delta \in \mathcal{B}$. Note that $M(X) \ominus M^s(X) = M^a(X)$. In fact if $x \in M(X) \ominus M^s(X)$ and $\sigma(\Delta) = 0$ then for all $y \in M(X)$, $(E(\Delta)x,y) = (x, E(\Delta)y) = 0$, since $E(\Delta)y \in M^s(X)$. Thus $(E(\Delta)x,x) = |E(\Delta)x|^2 = 0$. Conversly, if $y \in M^a(X)$ and $x \in M^s(X)$ then there exists a set Δ_x such that $\sigma(\Delta_x) = 0$ and $E(\Delta_x)x = x$. Thus for every $x \in M^s(X)$, $(x,y) = (x,E(\Delta_x)y) = 0$ so $y \in M(X) \ominus M^s(X)$. Let us define

$$X^a(n) = P^a X(n), \ \ X^s(n) = P^s X(n).$$

Since P^a and P^s commute with U, conditions (i) – (v) are satisfied. □

2.3. Prediction problem and the Wold decomposition. Suppose that X is an $L(H,K)$–valued stationary sequence and let A be a nonempty subset of Z. The general prediction problem can be formulated as follows:

(2.3.1) for a given n not in A and $x \in H$ find $y(x,n)$ in $M_A(X)$ such that

$$|y(x,n) - X(n)x| = \inf\{|y-X(n)x|: \ y \in M_A(X)\}.$$

As we know, $y(x,n) = P_{M_A(X)}X(n)x$. In the sequel the operator $P_{M_A(X)}X(n)$ is denoted by $\tilde{X}(A,n)$ and is called the predictor of $X(n)$ based on A.

Two kinds of observation sets play important role in prediction theory, namely

$$A = \{n \leq -1\} \ \text{ and } \ A = \{n \neq 0\}.$$

In the first case we deal with extrapolation problem, in the second with interpolation.

2.3.2. Definition. The operator

$$\Sigma(A,n) = (X(n) - \tilde{X}(A,n))^*(X(n) - \tilde{X}(A,n))$$

is called the prediction error operator (of $X(n)$ based upon A).

The following lemma due to Yaglom ([58], p. 175) gives a certain characterization of the prediction error operator. As we see later it works very nicely in the interpolation case. The proof of the lemma is based on the fact that if M is a closed subspace of a Hilbert space K then

$$|P_M x|^2 = \sup\{|y|^2: \ y \in M, \ (x-y,y) = 0\}.$$

For the proof the reader may consult [18] Lemma 3.2, where further details are also available.

2.3.3. Lemma. $\Sigma(A,n) = \max\{T^*T: \ \overline{TH} \subset N_A(X), \ T^*T = X^*(n)T\}$, where by definition $S = \max \mathcal{A}$, $\mathcal{A} \subset L^+(H)$, if $S \in \mathcal{A}$ and for every $R \in \mathcal{A}$, $S - R \in L^+(H)$.

Let $A \subset Z$ be given. A stationary sequence X is said to be

(2.3.4) A–deterministic, if $M_A(X) = M(X)$,

(2.3.5) A–completely nondeterministic (or A–regular), if $\bigcap_{n \in Z} M_{A+n}(X) = \{0\}$.

A stationary sequence X is A–deterministic if and only if each $X(n)$ admits a perfect prediction. Every stationary sequence can be decomposed into mutually orthogonal A–deterministic and A–completely nondeterministic parts. This is stated below. For the case $A = \{n \leq -1\}$ this is precisely the theorem on the structure of an isometry in a Hilbert space which in prediction theory is known as the Wold decomposition. The proof for a general A involves no additional difficulties (see e.g. [51]).

2.3.6. Wold decomposition theorem. Let X be an $L(H,K)$– valued stationary sequence and let A be a nonempty proper subset of Z. Then there exists a unique decomposition of the sequence

X in the form

$$X(n) = X^d(n) + X^r(n)$$

satisfying the following properties

(i) $U X^d(n) = X^d(n+1)$, $U X^r(n) = X^r(n+1)$, where U is the shift of X,

(ii) $M(X) = M(X^d) \oplus M(X^r)$ and for every $n \in Z$ $M_{A+n}(X^d) \oplus M_{A+n}(X^r) = M_{A+n}(X)$,

(iii) X^d is A–deterministic,

(iv) X^r is A–regular.

2.4. Three term decomposition. As we will see in Section 2.6 if the set A contains a half–line $\{n \geq n_0\}$ or $\{n \leq n_0\}$ then A–regularity of a stationary sequence X implies that its spectral measure is equivalent to the Lebesgue measure (i.e., for each $x \in H$, $(F(\cdot)x,x) \sim \sigma$ or $(F(\cdot)x,x) \equiv 0$). By the Lebesgue decomposition theorem (Theorem 2.2.3) X can be written as

$$X(n) = X^a(n) + X^s(n), \quad n \in Z,$$

where the spectral measures of X^a and X^s are a.c. and singular w.r.t. the Lebesgue measure, respectively, and $M(X^a) \oplus M(X^s) = M(X)$. Since $M(X^a) = \{x \in M(X): (E(\cdot)x,x) << \sigma\}$, $M(X^r) \subset M(X^a)$, where X^r is the A–regular component in the Wold decomposition of X. From the uniqueness in the Wold decomposition theorem it follows that

$$X^a = X^r + X^{ad}$$

where X^r is A–regular and X^{ad} is the A–deterministic part of X^a. This yields the following theorem.

2.4.1. Theorem. Let X be an $L(H,K)$–valued stationary sequence. If A contains a half–line $\{n \leq n_0\}$ or $\{n \geq n_0\}$, then X admits a unique decomposition in the form

$$X(n) = X^a(n) + X^{ad}(n) + X^s(n)$$

where:

(i) $M(X^a) \oplus M(X^{ad}) \oplus M(X^s) = M(X)$ and for every $n \in Z$
$M_{A+n}(X^a) \oplus M_{A+n}(X^{ad}) \oplus M_{A+n}(X^s) = M_{A+n}(X)$,

(ii) $U X^a(n) = X^a(n+1)$, $U X^{ad}(n) = X^{ad}(n+1)$, $U X^s(n) = X^s(n+1)$, $n \in Z$, where U is the shift operator of X,

(iii) X^a is A–regular,

(iv) X^{ad} is A–deterministic and its spectral measure is a.c. w.r.t. the Lebesgue measure,

(v) the spectral measure of X^s is singular w.r.t. the Lebesgue measure and X^s is A–deterministic.

2.5. Dimension, multiplicity and rank. In this section we assume that H is separable.

2.5.1. Definition. Let X be an $L(H,K)$–valued stationary sequence. Then

(i) the dimension of H is called the dimension of the sequence X,

(ii) the smallest number of elements $x_1, x_2, \ldots$ in $X(0)H$ such that $\overline{sp}\{U^n x_k, k \geq 1, n \in Z\} = M(X)$ is called the multiplicity of the sequence X and is by denoted $m(X)$, Note that $m(X) \leq$ dimension of X.

The main focus of this paper is on the analysis of infinite dimensional sequences. For excellent treatments of finite dimensional case we refer the reader to [24], [27], [28] and [45].

If a stationary sequence X has dimension $q < \infty$ (i.e., if $H = \mathbb{C}^q$) then there exist a probability space $(\Omega, \mathcal{F}, P)$ and a stationary sequence $x_n(\omega)$ of H–valued random variables such that the $L(H, L^2(\Omega, \mathcal{F}, P))$–valued operator sequence defined by the formula

$$(Y_n x)(\omega) = (x, x_n(\omega)) \tag{2.5.2}$$

is equivalent to X (Defintion 2.1.15). If $\dim H = \infty$, then this is not true in general. However if $X(0)$ is a Hilbert–Schmidt class operator then X is equivalent to the sequence Y generated via formula (2.5.2) by a sequence of H–valued random variables with $\int |x_n(\omega)|^2 P(d\omega) < \infty$. More detailed discussion of the relationship between stochastic processes and their operator models can be found in [8], [40], [14] and [3].

The notion of multiplicity of a stationary sequence has it roots in the theory of spectral representation of operators in a Hilbert space. It is easy to see that $m(X)$ equals the smallest number $n \in N \cup \{+\infty\}$ such that there exists a set $\{x_k : k < n+1\} \subset M(X)$ with the property that

$$M(X) = \overline{sp}\{E(\Delta)x_k : \Delta \in \mathcal{B}, k < n+1\},$$

where E is the spectral measure of the shift U. Thus the whole theory of multiplicity and Hellinger types can be applied in this case. In particular the following theorem follows from [7], p. 914–918.

2.5.3. Theorem. Suppose that X is an $L(H,K)$–valued stationary sequence such that $X(0)H$ is separable. Then there exists a sequence $x_k \in M(X)$, $x_k \neq 0$, $k = 1, 2, \ldots, m(X)$, $k \in N$ such that

(i) $$M(X) = \bigoplus_{k=1}^{m(X)} \overline{sp}\{U^n x_k : n \in Z\}$$

(ii) $$|E(\cdot)x_1|^2 >> |E(\cdot)x_2|^2 >> \ldots$$

Moreover if y_k, $k < n+1$, $n \in N\cup\{+\infty\}$ is any other sequence in $M(X)$ such that $M(X) = \bigoplus_{k=1}^{n} \overline{sp}\{U^n y_k : n \in Z\}$ and $|E(\cdot)y_1|^2 >> |E(\cdot)y_2|^2 >> \ldots$, then $n = m(X)$ and for every k, $|E(\cdot)y_k|^2 \sim |E(\cdot)x_k|^2$.

The measures $\mu_k = |E(\cdot)x_k|^2$ appearing in (ii) are called the spectral types of X.

The theorem below proved in [19] gives some relationship among the multiplicities of components in the three term decomposition of a stationary sequence as it relates to the multiplicity of X. Note that in the Lebesgue decomposition always $m(X) = \max(m(X^a), m(X^s))$.

2.5.4. Theorem. Let X be an $L(H,K)$–valued stationary sequence such that $X(0)H$ is separable. Let A be a proper non–empty subset of Z which contains a half–line $\{n_0 \leq n\}$ or $\{n \leq n_0\}$ and let $X = X^r + X^{ad} + X^s$ be the three term decomposition of X as defined in Theorem 2.4.1. Then

$$m(X) = \max(m(X^s),\ m(X^r) + m(X^{ad}))$$

Proof. First note that since X^r is necessary $(-\infty,-1]-$ regular all spectral types of X^r are equivalent to the Lebesgue measure (see Section 4.2). On the other hand all spectral types of X^{ad} are absolutely continuous w.r.t. σ. Since $M(X^a) = M(X^r) \oplus M(X^{ad})$ and both components are invariant under U, we have that if $m(X^r) < \infty$ then

(i) $\quad M(X^a) = \bigoplus_{k=1}^{m(X^r)} \overline{sp}\,\{U^n x_k\colon n \in Z\} \oplus \bigoplus_{k=1}^{m(X^{ad})} \overline{sp}\,\{U^n y_k\colon n \in Z\}$ and

(ii) $\quad |E(\cdot)x_1|^2 \sim \dots \sim |E(\cdot)x_m|^2 >> |E(\cdot)y_1|^2 >> \dots$

Thus $m(X^a) = m(X^r) + m(X^{ad})$, which is obvious also if $m(X^r) = \infty$.

Now suppose that x_k^a, $k < m(X^a) + 1$ and x_k^s: $k < m(X^s) + 1$ are such that conditions (i) and (ii) in Theorem 2.5.3 hold for X^a and X^s, respectively. Suppose that $m(X^a) < \infty$ and $m(X^a) \le m(X^s)$. Let

$$z_k = \begin{cases} x_k^a + x_k^s, & 1 \le k \le m(X^a) \\ x_k^s, & m(X^a) < k < m(X^s) + 1. \end{cases}$$

Then it is easy to check that

$$M(X) = \bigoplus_{k=1}^{m(X^s)} \overline{sp}\,\{U^n z_k\colon n \in Z\} \text{ and } |E(\cdot)z_1|^2 >> |E(\cdot)z_2|^2 >> \dots$$

Thus $m(X) = m(X^s)$. By symmetry, if $m(X^s) \le m(X^a)$ we obtain that $m(X) = m(X^a)$ which together prove that $m(X) = \max(m(X^s), m(X^a)) = \max(m(X^s), m(X^r) + m(X^{ad}))$. □

Theorem 2.5.4 pertains to the problem of the concordance of th Wold and Lebesgue decompositions. Note that if $m(X) = 1$ then either X^r or X^{ad} must vanish. As a consequence we have (cf. [15] Theorem 23, [45] Theorem 8.3)

2.5.5. Corollary. Suppose that X has multiplicity 1 (which always holds if X is a nonzero one dimensional stationary sequence) and that A contains a half–line. Then

(i) the Wold and Lebesgue decompositions coincide provided X is not A–deterministic,

(ii) X is either A–deterministic or A–regular provided the spectral measure of X is absolutely continuous w.r.t. σ.

Finally we introduce the notion of the rank of a stationary sequence. This notion has played an important role in the study of extrapolation and innovation problems.

2.5.6. Definition. The rank $r(X)$ of a stationary sequence X is the dimension of the subspace $M_{(-\infty,0]}(X) \ominus M_{(-\infty,-1]}(X)$.

Note that $r(X) = m(X^r)$, where X^r is the $(-\infty,-1]$–regular component of X (see Section 4.2). We remark that Rozanov in his book [45] uses different terminology, e.g., he calls a sequence X to be of rank n if $\dim(\frac{dF}{dt}(t)H) = n$ σ a.e., which corresponds in our notation to multiplicity n.

More about construction of the Wold and other orthogonal decompositions of a stationary processes and about sequences of rank 1 can be found in [42], [39], [19]. We will return to this topic in the sequel.

2.6. General model for a stationary sequence. Suppose that X is a given $L(H,K)$–valued stationary sequence. Since all properties related to the prediction of X depend only on the correlation function of X, usually it is convenient to consider an equivalent sequence Y with well described time domain $M(Y)$, which may serve as a model for X.

Consider first the case of one dimensional stationary sequences. Note that if $T \in L(H,K)$ and $\dim H = 1$ (i.e., $H = \mathbb{C}$), then T is determined by the value of T at 1, so a one dimensional stationary sequence $X(n)$ can be identified with a sequence of vectors $x(n) \in K$ (namely $x(n) = X(n)1$), and then $F(\Delta) = (E(\Delta)x(0),x(0))$ is a nonnegative scalar measure. For this case the following models are of common use.

2.6.1. Model 1. Suppose that $x(n) \in K$ is a one dimensional stationary sequence and F is its spectral measure. Then a sequence of functions in $L^2(F)$ defined by $y_n(t) = e^{int}$, $n \in Z$, is a stationary sequence equivalent to $x(n)$ (to see this, it is enough to evaluate the correlation of $\{y_n\}$ and use the definition of the spectral measure). Thus $\{y_n\}$ serves as a model for $\{x(n)\}$. It is easy to see that $M(y) = L^2(F)$, $(Uf)(t) = e^{it}f(t)$, and $(E(\Delta)f)(t) = 1_\Delta(t)f(t)$, $f \in L^2(F)$. This model was introduced by Kolmogorov [15].

2.6.2. Model 2. Let $x(n)$ be as above, F be its spectral measure and let $H_{2,F}$ be the set of all finite complex measures ν on $(-\pi,\pi]$ such that $\nu << F$ and

$$\sup_{\mathcal{P}} \sum_{\Delta \in \mathcal{P}} \left|\frac{\nu(\Delta)}{F(\Delta)}\right|^2 F(\Delta) \overset{\mathrm{df}}{=} \|\nu\|_F^2 < \infty, \tag{2.6.3}$$

where the supremum is taken over the set of all finite partitions $\mathcal{P}$ of $(-\pi,\pi]$. One can prove that $\nu \in H_{2,F}$ if and only if $\frac{d\nu}{dF} \in L^2(F)$ and that $\|\nu\|_F^2 = \int |\frac{d\nu}{dF}(t)|^2 F(dt)$. Therefore, in view of Model 2.6.1, the stationary sequence $\mu_n \in H_{2,F}$, defined by the formula

$$\mu_n(\Delta) = \int_\Delta e^{int}\, F(dt)$$

serves as a model for $x(n)$. It is easy to see that $M(\mu) = H_{2,F}$, $(E_\mu(\Delta)\nu)(\cdot) = \nu(\cdot \cap \Delta)$ and $(\nu_1,\nu_2)_F = \int (\frac{d\nu_1}{dF})(\overline{\frac{d\nu_2}{dF}})dF$.

Both models can be extended to the case of finite dimensional sequences ([48], [43]). Unfortunately Model 2.6.1 does not lend itself to extension in the infinite dimensional case as it stands. Roughly speaking the reason is that the set of all H–valued measurable functions φ for which $\int (\frac{dF}{d\mu}(t)\varphi(t),\varphi(t))\mu(dt) \overset{\mathrm{df}}{=} \|\varphi\|^2 < \infty$ need not be complete in this given norm. It is complete if $\frac{dF}{d\mu}(t)H$ is a closed subspace a.e.μ. This condition seems to be also necessary (for more detailed discussion see [25], [9] and also [23], [23a]).

Nevertheless, under an appropriate definition of $H_{2,F}$–space, Model 2.6.2 can serve as a model for any infinite dimensional stationary sequence. The detailed proof of the theorem below can be found in [17] or [9] for a more general case of a locally convex space.

2.6.4. Theorem. Let X be an L(H,K)–valued stationary sequence and F be its spectral measure. Let $H_{2,F}$ be the set of all H–valued measures m such that

(i) $m(\Delta) \in F(\Delta)^{1/2}H$ for all $\Delta \in \mathcal{B}$,

(ii) $\sup_{\mathcal{P}} \sum_{\Delta \in \mathcal{P}} |F(\Delta)^{1/2\#} m(\Delta)|^2 = \|m\|_F^2 < \infty$,

where the supremum is taken over all finite Borel partitions $\mathcal{P}$ of $(-\pi,\pi]$. Then

(a) $(H_{2,F}, \| \ \|_F)$ is a Hilbert space,

(b) the $L(H, H_{2,F})$–valued sequence $Y(n)$ defined by the formula $(Y(n)x)(\Delta) = \int_\Delta e^{int}F(dt)x$, $x \in H$, $\Delta \in \mathcal{B}$ is a stationary sequence equivalent to X,

(c) $M(Y) = H_{2,F}$, $(E_Y(\Delta)m)(\cdot) = m(\cdot \cap \Delta)$, $(U_Y m)(\Delta) = \int_\Delta e^{it}\, m(dt)$,

(d) for any subset $A \subset Z$ we have

$$M(Y) \ominus M_A(Y) = \{m \in H_{2,F}: \hat{m}(n) = \int e^{-int} m(dt) = 0 \text{ for all } n \in A\}.$$

Proof. From the definition of the spectral measure F it follows that for every Δ, $F(\Delta)$ has the factorization shown in Figure 1, where $E(\Delta)$ is the spectral measure of the shift of X. For any fixed x in M(X) let us define a measure Vx with values in H by the formula $(Vx)(\Delta) = (E(\Delta)X(0))^* x = X(0)^* E(\Delta)x$.

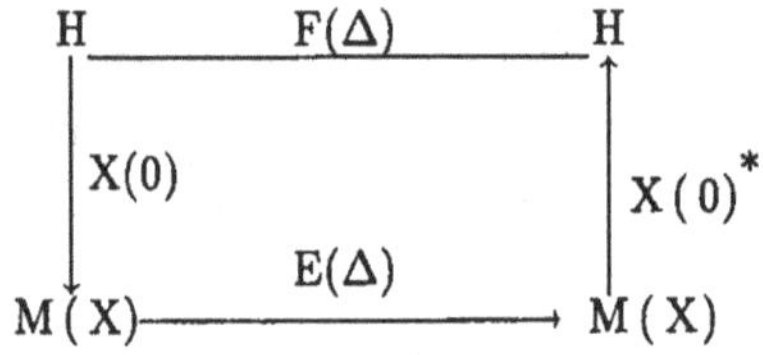

Figure 1

Using the facts that $(E(\Delta)X(0))^* M(X) = F(\Delta)^{1/2}H$, $|(E(\Delta)X(0))^* y|^2 = |F(\Delta)^{1/2\#}y|^2$ one can prove that V maps M(X) onto $H_{2,F}$ and $\|Vx\|_F = |x|$ for any $x \in M(X)$. This proves (a) and that V is unitary from M(X) onto $H_{2,F}$. Note that for every Δ

$$V(X(n)x)(\Delta) = V(\int e^{int}E(dt)X(0)x)(\Delta) = \int e^{int}(E(\Delta)X(0))^* E(dt)X(0)x = (Y(n)x)(\Delta),$$

which proves (b) since V is unitary. Part (c) of the theorem is obvious. To see (d) we observe that $m \in M(Y) \ominus M_A(Y)$ if and only if there exists $x \in N_A(X)$ such that $m = Vx$. Since $x \in N_A(X)$ if and only if $(X(n)y,x) = 0$ for all $y \in H$ and $n \in A$, we obtain that $m \in N_A(Y)$ iff $0 = (x,X(n)y) = \int e^{-int}(x,E(dt)X(0)y) = \int e^{-int}(m(dt),y)$ for all y and $n \in A$. □

2.6.5. Remark. It should be noticed that the space K_0 defined in the proof of Theorem 2.1.9 is a dense subset of $H_{2,F}$ and $\|m\|_0 = |m|_F$ for $m \in K_0$.

Using the model described above it is obvious that for any $A \subset Z$, a stationary sequence with a spectral measure F is:

(2.6.6) not A–deterministic if and only if there exists a nonzero measure $m \in H_{2,F}$ such that $\hat{m}(n) = 0$ for every $n \in A$; ($\hat{m}$ denote the Fourier transform of m).

(2.6.7) A–regular if and only if there exists a family $\{m_\alpha: \alpha \in I\} \subset H_{2,F}$ such that $\overline{sp}\{m_\alpha(\cdot \cap \Delta): \alpha \in I, \Delta \subset \mathscr{B}\} = H_{2,F}$, (where the closure is taken in $\| \ \|_F$–norm) and $\hat{m}_\alpha(n) = 0$ for all $\alpha \in I$ and $n \in A$.

Also it is easy to justify the fact, which we have already used in Section 2.4, that if A contains a half–line then the spectral measure of any A–regular stationary process X is equivalent to the Lebesgue measure. Note that if A contains a half–line, then $\hat{m}(n) = 0$ for all $n \in A$ implies that $(m(\cdot), x)$ is either equivalent to the Lebesgue measure or is zero. Thus if X is A–regular, then by (2.6.7) there exists a nonzero m such that $m(\Delta) \neq 0$ iff $\sigma(\Delta) \neq 0$ and $m(\Delta) \in F(\Delta)^{1/2}H$. It implies that $\sigma << F$ (note that we used only the fact that X is not A–deterministic). On the other hand from (2.1.11) and (2.6.5) it follows that for every $m \in H_{2,F}$, $|m(\Delta)| \le \sqrt{\|F(\Delta)\|}\,\|m\|_F$, $\Delta \in \mathscr{B}$. Therefore if X is A–regular then by (2.6.7) any $m \in H_{2,F}$ is a.c. w.r.t. the Lebesgue measure. In particular $(F(\cdot)x,x) << \sigma$ for all $x \in H$, since $F(\cdot)x \in H_{2,F}$.

In fact one can prove a more general theorem ([17]), which we state here.

2.6.8. Theorem. Suppose that X is an L(H,K)–valued stationary sequence with the spectral measure F. If X is A–regular then for every $\Delta \in \mathscr{B}$

$$F(\Delta)^{1/2}H = \overline{sp}^{\Delta}\{m(\Delta'): \Delta' \subset \Delta, \Delta' \in \mathscr{B}, m \in N_A(F)\},$$

where $N_A(F)$ is the set of all measures in $H_{2,F}$ such that $\hat{m}(n) = \int e^{-int} m(dt) = 0$ for all $n \in A$ and the closure is taken in the $| \ |_\Delta$–norm defined on $F(\Delta)^{1/2}H$ by $|y|_\Delta = |F(\Delta)^{1/2\#}y|$.

Note that if $y = F(\Delta)x$, then $|F(\Delta)x|_\Delta^2 = (F(\Delta)x,x)$.

2.6.9. Corollary. If an L(H,K)–valued stationary sequence X is $\{0\}^c$–regular then $\overline{F(\Delta)X} = \text{const}$ for all $\Delta \in B$ such that $\sigma(\Delta) \neq 0$ and $\overline{F(\Delta)X} = \{0\}$ otherwise.

Proof. Note that in this case with $A = \{0\}^c$, $N_A(F)$ is equal to $\{m = x\,\sigma: x \in F(\Delta)^{1/2}H$ for all Δ with $\sigma(\Delta) \neq 0$ and $\sup_{\mathscr{P}} \sum_{\Delta} |F(\Delta)^{1/2\#} x\sigma(\Delta)|^2 < \infty\}$. Thus from Theorem 2.6.8 and the preceding discussion it follows that $F(\Delta)^{1/2}H = \{0\}$ if $\sigma(\Delta) = 0$ and $F(\Delta)^{1/2}H = \overline{sp}^{\Delta}\{x: x\,\sigma \in N_A(F)\}$ if $\sigma(\Delta) \neq 0$. Since $|y|_\Delta = |F(\Delta)^{1/2\#}y| \ge |F(\Delta)^{1/2}F(\Delta)^{1/2\#}y| = |y|$, $y \in F(\Delta)^{1/2}H$, $\overline{F(\Delta)H} = \overline{sp}\{x: x\sigma \in N_A(F)\}$ and the right hand side does not depend on Δ if $\sigma(\Delta) \neq 0$ □

2.7. <u>Stationary sequences with operator densities</u>. In this section we assume that H is a separable Hilbert space and F is the spectral measure of an L(H,K)–valued stationary sequence X.

2.7.1. <u>Definition</u>. Let μ be a σ–finite nonnegative Borel measure on $(-\pi,\pi]$. A function $\frac{dF}{d\mu}$: $(-\pi,\pi] \to L^{+}(H)$ is said to be an <u>operator spectral density</u> of the sequence X if $(\frac{dF}{d\mu}(\cdot)x,y)$ is μ–measurable, and for every $\Delta \in \mathcal{B}$, x, y ∈ H

$$(F(\Delta)x,y) = \int_{\Delta} (\frac{dF}{d\mu}(t)x,y)\mu(dt)$$

An operator spectral density may not exist. If it does it is μ–almost everywhere unique. Note that if X(0) is a Hilbert– Schmidt operator then $\frac{dF}{d\mu}$ exists with μ = trace F ([40]). In particular any finite dimensional sequence admits operator densty.

Sequences with operator densities admit model simpler than the model introduced in Theorem 2.6.4. This model was first discussed by Rozanov [46]. With a slight variation we state Rozanov's result as follows:

2.7.2. <u>Theorem</u>. Let X be an L(H,K)–valued stationary sequence and let $\frac{dF}{d\mu}$ be an operator spectral density of X with respect to a nonnegative σ–finite scalar measure μ. Let L_F be the set of all weakly integrable functions $\varphi:(-\pi,\pi] \to H$ such that $\varphi(t) \in [\frac{dF}{d\mu}(t)]^{1/2}H$ μ a.e., and

$$|\varphi|_L^2 \overset{df}{=} \int |\frac{dF}{d\mu}(t)^{1/2\#}\varphi(t)|^2\mu(dt) < \infty$$

(two functions φ_1, φ_2 in L_F are considered to be equal if $\varphi_1(t) = \varphi_2(t)$ μ a.e.). Then

(a) $(L_F, |\ |_L)$ is a Hilbert space,

(b) an $L(H, L_F)$–valued stationary sequence $Y = \{Y(n), n\in Z\}$ defined by the formula

$$(Y(n)x)(\cdot) = e^{in\cdot}\frac{dF}{d\mu}(\cdot)x, \quad x \in H, \quad n \in Z$$

is equivalent to X,

(c) for any subset $A \subset Z$,

$$N_A(Y) = \{\varphi \in L_F\colon \int e^{-int}(\varphi(t),x)\mu(dt) = 0 \quad n \in A, \quad x \in H\}.$$

Proof. Since for any $\varphi, \psi \in L_F$

$$(\varphi,\psi)_L = \int(\frac{dF}{d\mu}(t)^{1/2\#}\varphi(t), \frac{dF}{d\mu}(t)^{1/2\#}\psi(t))\mu(dt) \tag{2.7.3}$$

and the function $\frac{dF}{d\mu}(t)^{1/2\#}\varphi(t)$ is Bochner μ–measurable for $\varphi \in L_F$, it is clear that $(L_F, |\ |_L)$ is a Hilbert space and that the mapping

$$L^2(\mu, H) \ni f(\cdot) \to \left[\frac{dF}{d\mu}(\cdot)\right]^{1/2} f(\cdot)$$

is an isometry from $\{f \in L^2(\mu, H)\colon\ f(t) \in \overline{\frac{dF}{d\mu}(t)H}\ \mu \text{ a.e.}\}$ onto L_F. Thus (a). Part (b) follows immediately from (2.7.3). To prove (c) let us first note that $M(Y) = L_F$. In fact it is easy to check that functions of the form $e^{in\cdot}\frac{dF}{d\mu}(\cdot)^{1/2}x$, $x \in H$, $n \in Z$ form a linearly dense subset of the space $\{f\in L^2(\mu, H)\colon\ f(t) \in \overline{\frac{dF}{d\mu}(t)H}\ \mu \text{ a.e.}\}$. Therefore the functions $e^{in\cdot}\frac{dF}{d\mu}(t)x$ are linearly dense in L_F. Let now $\varphi \in L_F$. The function φ belongs to $M(Y) \ominus M_A(Y)$ if and only if $(\varphi, Y(n)x)_L = 0$ for all $n \in A$ and $x \in H$, i.e., if $\int(\frac{dF}{d\mu}(t)^{1/2\#}\varphi(t), \frac{dF}{d\mu}(t)^{1/2\#}e^{int}\frac{dF}{d\mu}(t)x)\mu(dt) = \int(\varphi(t),x)e^{-int}\mu(dt)=0$, $n \in A$, $x \in H$. □

Similarily like the $|\ |_\Delta$–norm we may introduce the $|\ |_t$–norm in $\frac{dF}{d\mu}(t)H$ by the formula

$$|\frac{dF}{d\mu}(t)x|_t = (\frac{dF}{d\mu}(t)x,x)^{1/2}, \quad x \in H$$

(t is fixed). It is clear that the completion of $\frac{dF}{d\mu}(t)H$ with respect to this norm can be identified with $\frac{dF}{d\mu}(t)^{1/2}H$ and for any $y \in \frac{dF}{d\mu}(t)^{1/2}H$

$$|y|_t = |\frac{dF}{d\mu}(t)^{1/2\#}y|. \tag{2.7.4}$$

Now we are ready to express A–determinism and A–regularity of a stationary sequence in terms of L_F space.

2.7.5. <u>Theorem</u>. Suppose that X is an $L(H,K)$–valued stationary sequence and $\frac{dF}{d\mu}$ is its operator spectral density with respect to a nonnegative scalar measure μ (we assume that $\frac{dF}{d\mu}$ exists). Let $A \subset Z$. Then

(a) X is not A–deterministic if and only if there is a non– zero $\varphi \in L_F$ such that

$$\int e^{-int}(\varphi(t),x)\mu(dt) = 0$$

for all $x \in H$ and $n \in A$,

(b) X is A–regular if and only if there exists a sequence $\{\varphi_k\}$ in L_F such that

(i) $\int e^{-int}(\varphi_k(t),x)\mu(dt) = 0$ for all $n \in A$, $x \in H$, $k = 1, 2,\dots$,

(ii) $\frac{dF}{d\mu}(t)^{1/2}H = \overline{sp}^t\{\varphi_k(t)\colon\ k = 1, 2,\dots\}$ μ a.e.,

where the closure is taken in $|\ |_t$–norm (see (2.7.4)).

Proof. Part (a) is obvious. We prove (b). In view of Theorem 2.7.2 X is A–regular iff there exists a sequence $\{\varphi_k\}$ such that (i) holds and $\overline{sp}\{e^{in\cdot}\varphi_k,\ k \ge 1,\ n \in Z\} = L_F$. We now show that the last equality holds iff (ii). Let $\varphi_k \in L_F$ and let $f_k(t) = \frac{dF}{d\mu}(t)^{1/2\#}\varphi_k(t)$, $k = 1, 2,\dots$ Since the mapping

$$L_F \ni \varphi \to \frac{dF}{d\mu}(t)^{1/2\#}\varphi(\cdot)$$

is an isometry from L_F onto $M \overset{df}{=} \{f \in L^2(\mu, H)\colon\ f(t) \in \overline{\frac{dF}{d\mu}(t)H}\ \mu \text{ a.e.}\}$, then $sp\{e^{in\cdot}\varphi_k(\cdot)\colon k = 1, 2,\dots,\ n \in Z\} = L_F$ if and only if $\overline{sp}\{e^{in\cdot}f_k(\cdot)\colon\ n \in Z,\ k = 1, 2,\dots\} = M$. But it is easy to check that the last equality holds if and only if

$$\overline{\mathrm{sp}}\{f_k(t):\ k = 1, 2, ...\} = \overline{\frac{dF}{d\mu}(t)H}, \ \mu \text{ a.e.}$$

(cf [47], p 89). Applying the definition of the $|\ |_t$–norm completes the proof. □

There are other possibilities for modeling stationary sequences having operator spectral densities. A more detailed discussion of this subject can be found in [9]. One can also find related materials in [23], [55], [57].

We end this section with a theorem proved recently in [19], which gives a description of the multiplicity of a stationary sequence in terms of its operator spectral density (this problem remains open for the case where there is no spectral density).

2.7.6. Theorem. Suppose that an $L(H,K)$–valued stationary sequence X has the operator spectral density $\frac{dF}{d\mu}$ with respect to a nonnegative scalar measure μ. Then the multiplicity $m(X)$ of X (Definition 2.5.1 part (ii)) is given by

$$m(X) = \mu\text{–ess sup } [\dim(\overline{\frac{dF}{d\mu}(t)})].$$

3. INTERPOLATION PROBLEM

3.1. Introductory remarks. In this section we review the interpolation problem in the infinite dimensional case. As it was mentioned in Section 2.3 we will assume that $X(n)$, $n \neq 0$, are known. According to commonly used notation we will call an $L(H,K)$–valued stationary sequence X

(3.1.1) J_0–regular, if it is $\{0\}^c$–regular, and

(3.1.2) minimal, if it is not $\{0\}^c$–deterministic

From (2.6.6) it follows that a stationary sequence X is minimal if and only if there exists a nonzero $x \in H$ such that

(3.1.3)
$$\text{(i) } x \in F(\Delta)^{1/2}H \text{ for all } \Delta \in \mathscr{B} \text{ with } \sigma(\Delta) \neq 0,$$
$$\text{(ii) } \sup_{\mathscr{P}} \sum_{\Delta \in \mathscr{P}} |F(\Delta)^{1/2\#}\, x\, \sigma(\Delta)|^2 < \infty,$$

where F is the spectral measure of X. Let $\mathscr{M}$ be the set of all x's statifying (3.1.3). By (2.6.7) the sequence X is J_0–regular if and only if $\mathscr{M}$ is "large" enough in the sense that measures of the form $m(\cdot) = x\,\sigma(\cdot \cap \Delta)$, $\Delta \in \mathscr{B}$, $x \in \mathscr{M}$, generate $H_{2,F}$. Note that from Theorem 2.6.8 it follows that if X is J_0–regular then for every $\Delta \in \mathscr{B}$ either $F(\Delta)H = 0$ or $F(\Delta)^{1/2}H = \overline{\mathscr{H}}^{\Delta}$, where the closure is taken in the $|\,|_\Delta$–norm defined in Theorem 2.6.8, depending on whether $\sigma(\Delta)$ is zero or not (in particular it implies that $\overline{F(\Delta)H} = \overline{\mathscr{H}}$ for all Δ whenever $\sigma(\Delta) \neq 0$, see Corollary 2.6.9). The open question is whether the property $F(\Delta)^{1/2}H = \overline{\mathscr{H}}^{\Delta}$ for all Δ with $\sigma(\Delta) \neq 0$ and $F(\Delta)H = 0$ otherwise implies J_0–regularity.

We have also shown that X admits the three term decom– position Theorem 2.4.2, and that the singular part X^s of X is always $\{0\}^c$–deterministic. Thus without any loss of generality we may assume that the spectral measure F of a stationary sequence X is a.c. w.r.t. the Lebesgue measure.

This is almost all that can be said in the general case. More satisfactory results can be obtained in the case of a sequence admitting an operator density. This will be discussed in the next section.

3.2. J_0–regularity of stationary sequences with spectral densities.

In this section we will assume that H is a separable Hilbert space and that an $L(H,K)$–valued stationary sequence X has the operator spectral density $\frac{dF}{dt}$ w.r.t. the Lebesgue measure.

The following theorem which is a slight reformulation of a result in [46] is an immediate consequence of Theorem 2.7.5 applied to the case $A = \{0\}^c$.

3.2.1. Theorem. Suppose that $\frac{dF}{dt}$ is the spectral density of an $L(H,K)$–valued stationary sequence X with respect to the Lebesgue measure. The sequence X is J_0–regular if and only if there exists a sequence $x_n \in H$, $n = 1, 2, \ldots$, such that

(i) $\quad x_n \in \left[\frac{dF}{dt}(t)\right]^{1/2} H \qquad \sigma$ a.e., $n = 1, 2, \ldots$,

(ii) $\quad \int \left|\frac{dF}{dt}(t)^{1/2\#} x_n\right|^2 dt < \infty \qquad n = 1, 2, \ldots$,

(iii) $\quad \overline{\mathrm{sp}}^t\{x_n \colon n = 1, 2, \ldots\} = \frac{dF}{dt}(t)^{1/2} H \qquad \sigma$ a.e.,

where the closure is taken in the $|\ |_t$–norm (see (2.7.4)).

Note that, by Theorem 2.7.5 (a), the sequence X is minimal if there is at least one nonzero $x \in H$ satisfying (i) and (ii).

Below we list some consequences of Theorem 3.2.1. Easy proofs are not included.

3.2.2. Corollary. Let X be an $L(H,K)$–valued stationary sequence with the density $\frac{dF}{dt}$ and let $\mathcal{M}$ be the set of all $x \in H$ such that

(i) $\quad x \in \frac{dF}{dt}(t)^{1/2} H \qquad \sigma$ a.e.

(ii) $\quad \int \left|\frac{dF}{dt}(t)^{1/2\#} x\right|^2 dt < \infty.$

If X is J_0–regular then $\frac{dF}{dt}(t)^{1/2} H = \overline{\mathcal{M}}^t$, σ a.e.. In particular, if X is J_0–regular then $\overline{\frac{dF}{dt}(t)H}$ = const σ a.e.

3.2.3. Corollary. Suppose that the operator spectral density $\frac{dF}{dt}$ of a stationary sequence X satisfies

$$\frac{dF}{dt}(t)H = \overline{\frac{dF}{dt}(t)H} \qquad \sigma \text{ a.e.}$$

Then the sequence X is J_0–regular if and only if $\frac{dF}{dt}(t)H = \text{const}$ σ a.e. and there exists a sequence $x_n \in \frac{dF}{dt}(t)H$ σ a.e., $n = 1, 2, \ldots$, such that

(i) $\overline{sp}\{x_n : n = 1, 2, \ldots\} = \frac{dF}{dt}(t)H$ σ a.e.,

(ii) $\int(\frac{dF}{dt}(t)^{\#}x_n, x_n)dt < \infty$, $n = 1, 2, \ldots$

3.2.4. Corollary. Suppose that the spectral density $\frac{dF}{dt}$ of a stationary sequence X has a bounded inverse $\frac{dF}{dt}(t)^{-1}$ for almost all $t \in (-\pi, \pi]$. Then the sequence X is J_0–regular if and only if there exists a sequence $x_n \in H$ $n = 1, 2, \ldots$, linearly dense in H, and such that for all $n \geq 1$,

$$\int(\frac{dF}{dt}(t)^{-1}x_n, x_n)dt < \infty.$$

In particular, if $\int\|\frac{dF}{dt}(t)^{-1}\|dt < \infty$ then X is J_0–regular.

3.2.5. Corollary. ([24], see also [51], [56] and [22]). Suppose that $H = \mathbb{C}^q$, $q < \infty$. An $L(\mathbb{C}^q, K)$–valued (i.e. a q–variate) stationary sequence X is J_0–regular if and only if

(1) $F << \sigma$,

(2) $\frac{dF}{dt}(t)\mathbb{C}^q = \text{const.}$ σ a.e.

(3) $\int\frac{dF}{dt}(t)^{\#}dt$ exists.

Note that if $q = 1$ this Corollary yields the well–known Kolmogorov's condition.

3.2.6. Example. From the reading of Corollary 3.2.4. one might suspect that in the presence of the existence of $\frac{dF}{dt}(t)^{-1}$ σ a.e., the J_0–regularity would imply the finiteness of the intergral $\int(\frac{dF}{dt}(t)^{-1}x, x)dt$ for all $x \in H$. The following example shows the contrary. Let $H = \ell^2$ be the Hilbert space of all square summable sequences $(d_k)_{k=1}^{\infty}$, and let the spectral density of an $L(H,K)$–valued stationary sequence X be given by the infinite dimensional matrix with zero off diagonal entries and whose diagonal elements are given by

$$f_k(t) = \begin{cases} \frac{1}{k^3}, & |t| \leq \frac{\pi}{k}, \\ 1, & \text{elsewhere,} \end{cases} \quad , \quad k \geq 1.$$

Obviously $\frac{dF}{dt}(t)$ is a bounded operator in H and it has a bounded inverse. Since $\int\frac{1}{f_k(t)}dt < \infty$ for every k, by Corollary 3.2.4 the sequence X is J_0–regular. Nevertheless for $x = (\frac{1}{k})_{k=1}^{\infty} \in \ell^2$ we have

$$\int(\frac{dF}{dt}(t)^{-1}x, x)dt = \infty.$$

In fact if $\frac{\pi}{n+1} < |t| \leq \frac{\pi}{n}$, then $(\frac{dF}{dt}(t)^{-1}x, x) = \sum_{k=1}^{n}\frac{k^3}{k^2} + \sum_{n=n+1}^{\infty}\frac{1}{k^2} \geq \frac{n(n+1)}{2}$, and so

$\int(\frac{dF}{dt}(t)^{-1}x,x)dt \geq \sum_{n=1}^{\infty} \frac{n(n+1)}{2}(\frac{\Pi}{n} - \frac{\Pi}{n+1}) = \infty$. Of course, this is strictly the infinite dimensional effect.

3.3. Characterization of the prediction error. The following theorem proved in [18] gives some description of the interpolation error in the case of stationary sequences with operator densities. For simplicity we will write $\Sigma_0 = \Sigma(\{0\}^c,0)$ (see Definition 2.3.2)

3.3.1. Theorem. Let X be a stationary sequence such that its spectral measure has an operator density $\frac{dF}{dt}$ w.r.t. the Lebesgue measure. Let $\mathcal{D}_0$ be the set of all $S \in L^+(H)$ such that

(i) $Sx \in \frac{dF}{dt}(t)^{1/2}H$ σ a.e.,

(ii) $(Sx,x) = \int|\frac{dF}{dt}(t)^{1/2\#}Sx|^2dt$ for all $x \in H$.

Then $\Sigma_0 = \max \mathcal{D}_0$, i.e., $\Sigma_0 \in \mathcal{D}_0$ and for every $S \in \mathcal{D}_0$, $\Sigma_0 - S \in L^+(H)$.

Proof. Let Y be the stationary sequence constructed in Theorem 2.7.2 and let $\mathcal{D} = \{T^*T\colon TH \subset N_A(Y),\ T^*T = Y(0)^*T\}$, where $A = \{0\}^c$. According to Lemma 2.3.2 it suffices to prove that $\mathcal{D} = \mathcal{D}_0$. From Theorem 2.7.2(c) it follows that $N_A(Y)$ is equal to

$$\{\varphi = \text{const.} = y\colon y \in \frac{dF}{dt}(t)^{1/2}H\ \sigma \text{ a.e.}, \int|\frac{dF}{dt}(t)^{1/2\#}y|^2dt < \infty\}.$$

Let $S = T^*T \in \mathcal{D}$. Then $Tx = \text{const.}$ for every $x \in H$ and $(Sx,y) = (Tx,Ty)_L = (Tx,Y(0)y)_L = \int(\frac{dF}{dt}(t)^{1/2\#}Tx, \frac{dF}{dt}(t)^{1/2}y)dt = (Tx(\cdot),y)$, where L_F is as in Theorem 2.7.2. Thus $Sx = Tx(t) \in \frac{dF}{dt}(t)^{1/2}H$ σ a.e. and $(Sx,x) = (Tx,Tx)_L = \int|\frac{dF}{dt}(t)^{1/2\#}Sx|^2dt$. Hence $S \in \mathcal{D}_0$. Conversly, suppose that $S \in \mathcal{D}_0$. Let T be a mapping from H into $N_A(Y)$ assigning to each $x \in H$ a constant function equal to Sx σ a.e. Then properties (i) and (ii) show that T is a bounded operator and that

$$(T^*Tx,x) = \int|\frac{dF}{dt}(t)^{1/2\#}Sx|^2dt = (Sx,x) = (Tx,Y(0)x)_L.$$

Thus $S \in \mathcal{D}$. □

Note that if F is not a.c. w.r.t. the Lebesgue measure then by Theorem 2.4.1, the conclusion of Theorem 3.3.1 holds true with $\frac{dF}{dt}$ being replaced by $\frac{dF^a}{dt}$ (F^a stands for the a.c. part of F).

The problem of finding the maximal nonnegative solution of equation (ii) in Theorem 3.3.1 under the constraint (i) is still open. It is not even known whether the range of Σ_0 is equal to $\mathcal{M} = \{x \in H\colon x \in \frac{dF}{dt}(t)^{1/2}H\ \sigma \text{ a.e.}, \int|\frac{dF}{dt}(t)^{1/2\#}x|^2dt < \infty\}$, as it could be suspected. The only case in which we can obtain an explicit expression for Σ_0 is where $\mathcal{M}$ is a closed subspace of H (which is necessary satisfied in the finite dimensional case). Let us remark that Example 3.2.6 shows that the closedness of the range of $\frac{dF}{dt}$ σ a.e. is not sufficient for $\mathcal{M}$ to be closed.

3.3.2. Corollary. If $\mathcal{M} = \{x \in H\colon x \in \frac{dF}{dt}(t)^{1/2}H\ \sigma \text{ a.e.}, \int|\frac{dF}{dt}(t)^{1/2\#}x|^2dt < \infty\}$ is closed, then $\Sigma_0 = [\int P_{\mathcal{M}}\frac{dF}{dt}(t)^{\#}P_{\mathcal{M}}dt]^{\#}$ (for definition of the right side see the proof given below).

Proof. If $\mathcal{M} = \{0\}$, then the sequence $\{X(n), n \in Z\}$ is deterministic and $\Sigma_0 = 0$. Suppose that $\mathcal{M} \neq \{0\}$ and consider the mapping ϕ from $\mathcal{M}$ into $L^2(H)$ defined by the formula

$$(\phi x)(\cdot) = \frac{dF}{dt}(\cdot)^{1/2\#}x, \quad x \in \mathcal{M}.$$

Since for every $t \in (-\pi,\pi]$, $\frac{dF}{dt}(t)^{1/2\#}$ is a closed operator, from the closed graph theorem it follows that ϕ is bounded. Thus the operator $R: H \to H$ defined by

$$(3.3.3) \qquad (Rx,y)=(\phi^*\phi P_{\mathcal{M}}x,P_{\mathcal{M}}y)=\int(\frac{dF}{dt}(t)^{1/2\#}P_{\mathcal{M}}x, \frac{dF}{dt}(t)^{1/2\#}P_{\mathcal{M}}y)dt$$

is bounded and nonnegative. The linear operator R serves as the definition of the integral operator $\int P_{\mathcal{M}}\frac{dF}{dt}(t)^{\#}P_{\mathcal{M}}dt$. Since $|x|^2 = |\frac{dF}{dt}(t)^{1/2}\frac{dF}{dt}(t)^{1/2\#}x|^2 \leq |\frac{dF}{dt}(t)||\frac{dF}{dt}(t)^{1/2\#}x|^2$, $x \in \mathcal{M}$, then $(Rx,x) \geq (\int|\frac{dF}{dt}(t)^{-1}|dt)|P_{\mathcal{M}}x|^2$ for all $x \in H$. Moreover $Rx = 0$ if and only if $x \in \mathcal{M}^\perp$. Thus $\mathrm{Null}(R) = \mathcal{M}^\perp$ and $RH \subset \mathcal{M}$. The operator R restricted to $\mathcal{M}$ is then one to one, nonnegative and $(Rx,x) \geq c|x|^2$, $x \in \mathcal{M}$. Thus $RH = \mathcal{M}$ and $R^{\#}|_{\mathcal{M}}$ is bounded. Hence $R^{\#}$ can be extended to H as indicated in (1.9).

Let $S = R^{\#} = [\int P_{\mathcal{M}}\frac{dF}{dt}(t)^{\#}P_{\mathcal{M}}dt]^{\#}$. Then

(i) $\quad Sx \in \frac{dF}{dt}(t)^{1/2}H$, and

(ii) $\quad \int|\frac{dF}{dt}(t)^{1/2\#}Sx|^2dt = (RSx,Sx) = (Sx,x)$

This proves that $S \in \mathcal{D}_0$ (for the definition of $\mathcal{D}_0$ see Theorem 3.3.1). Now, let T be any other element of $\mathcal{D}_0$ and let $Q = R^{1/2}T$. The operator Q is bounded and from (3.3.3) it follows that

$$(RTx,Tx) = \int|\frac{dF}{dt}(t)^{1/2\#}Tx|^2dt = (Tx,x), \text{ i.e., } TRT = T.$$

Moreover we have

$$(S^{1/2} - Q)^*Q = S^{1/2}R^{1/2}T - TRT = 0.$$

Thus

$$\begin{aligned} S = S^{1/2}S^{1/2} &= [(S^{1/2} - Q) + Q]^*[(S^{1/2} - Q) + Q] \\ &= (S^{1/2} - Q)^*(S^{1/2} - Q) + Q^*Q \geq Q^*Q = T. \end{aligned}$$

Therefore S is maximal in $\mathcal{D}_0$ which completes the proof. □

3.3.4. Corollary. ([22]). Suppose that X is a $q-$ dimensional stationary sequence and let $\mathcal{M} = \{x \in \mathbb{C}^q: x \in \frac{dF^a}{dt}(t)\mathbb{C}^q$ σ a.e. and $\int(\frac{dF^a}{dt}(t)^{\#}x,x)dt < \infty\}$. Then, the interpolation error Σ_0 is given by

$$\Sigma_0 = [\int P_{\mathcal{M}}\frac{dF^a}{dt}(t)^{\#}P_{\mathcal{M}}dt]^{\#}.$$

3.3.5 Corollary. Let X be an $L(H,K)$–valued stationary sequence having an operator spectral density $\frac{dF}{dt}$ with respect to the Lebesgue measures. Let V be the isometry between $M(X)$ and $M(Y)$ defined by $V(X(n)x) = Y(n)x$, $x \in H$, $n \in Z$, where Y is the sequence constructed in

Theorem 2.7.2. Then the predictor $\tilde{X}(0) = \tilde{X}(\{\{0\}^c,0)$ is given by

$$\tilde{X}(0)x = V^{-1}(\frac{dF}{dt}(\cdot)x - \Sigma_0 x)$$

Proof. The proof follows immediately from the observation that if $\Sigma_0 = T^*T$, where $TH \subset N_A(X)$ with $A = Z - \{0\}$ and $T^*T = X(0)^*T$, then $T = X(0) - \tilde{X}(0)$. Details can be found in [22].

4. EXTRAPOLATION PROBLEM

Despite the vast literature devoted to the extrapolation problem the solution to extrapolation problem is far from complete. In this section we briefly discuss some aspects to the problem which relate to the models introduced earlier.

We have already observed that if an L(H,K)–valued stationary sequence is $(-\infty,-1]$–regular, then its spectral measure is a.c. w.r.t. the Lebesgue measure. Hence without loss of generality we may assume that the spectral measure is a.c. w.r.t. the Lebesgue measure. We will also assume that the operator density $\frac{dF}{dt}$ exists, since almost nothing is known otherwise, and since sequences of strongly square integrable H–valued random variables always admit an operator spectral density.

4.1. Regularity conditions. In this section we will always assume that H is a separable Hilbert space.

4.1.1. Definition. An L(H,K)–valued stationary sequence is called

(i) regular or completely nondeterminstic if it is $(-\infty,-1]$–regular,

(ii) deterministic if it is $(-\infty,-1]$–deterministic (see (2.3.4) and (2.3.5)).

The following theorem was proved first by Rozanov in [46] in a slightly different setting (see also [32]).

4.1.2. Theorem. Let X be an L(H,K)–valued stationary sequence and let $\frac{dF}{dt}$ be its spectral operator density w.r.t. the Lebesgue measure. The sequence X is

(a) nondeterministic if and only if there exists a weakly integrable function $\varphi\colon (-\pi,\pi] \to H$, $\varphi \neq 0$, such that

(i) $\int e^{-int}(\varphi(t),x)dt = 0$ for all $n \le -1$ and $x \in H$,

(ii) $\varphi(t) \in \frac{dF}{dt}(t)^{1/2}H$ σ a.e.,

(iii) $\int |\frac{dF}{dt}(t)^{1/2\#}\varphi(t)|^2 dt < \infty$,

(b) regular if and only if there exists a sequence of weakly integrable functions $\varphi_k\colon (-\pi,\pi] \to H$ such that for every $k = 1, 2,\ldots$

(i) $\int e^{-int}(\varphi_k(t),x)dt = 0$, $n \le -1$, $x \in H$,

(ii) $\varphi_k(t) \in \frac{dF}{dt}(t)^{1/2}H$ σ a.e.,

(iii) $\int |\frac{dF}{dt}(t)^{1/2\#}\varphi_k(t)|^2 dt < \infty$,

(iv) $\overline{sp}^t\{\varphi_k(t): k = 1,2,...\} = \frac{dF}{dt}(t)^{1/2}H$ σ a.e.,

where the closure is taken in the $|\ |_t$–norm introduced by (2.7.4).

Proof. Obvious by Theorem 2.7.5. □

Below we list a series of corollaries to Theorem 4.1.2, demonstrating some consequences of this theorem. Let us first introduce some auxiliary notations.

4.1.3. Definition. (i) By $L^{1,w}(H)$ we will denote the class of all H–valued weakly integrable function φ on $(-\pi,\pi]$ w.r.t. σ,

(ii) $L^p_+(H)$ ($L^{1,w}_+(H)$, resp.) will denote the subset of $L^p(H)$ ($L^{1,w}(H)$, resp.) consisting of all functions φ for which

$$\int e^{-int}(\varphi(t),x)dt = 0$$

for all $n \leq -1$ and $x \in H$,

(iii) $L^p_-(H)(L^{1.w}_-(H)$, resp.) will denote the subset of $L^p(H)(L^{1,w}(H)$, resp.) consisting of all functions φ such that for every $n \geq 1$ and $x \in H$

$$\int e^{-int}(\varphi(t),x)dt = 0.$$

We will also use the well known fact that a nonnegative function $f \in L^1(\mathbb{C})$ can be represented in the form $f(t) = |h(t)|^2$ with $h \in L^2_+(\mathbb{C})$ if and only if $\int \ln f(t)dt > -\infty$ (see [11]).

First by an application of Theorem 4.1.2 we deduce the classical Kolmogorov theorem [15], as was done in [47].

4.1.4. Corollary. A one dimensional stationary sequence $(x_n)_{n=-\infty}^{\infty} \subset K$ is regular if and only if its spectral measure F is a.c. w.r.t. the Lebesgue measure and $\int \ln \frac{dF}{dt}(t)\, dt > -\infty$.

Proof. If $(x_n)_{n=-\infty}^{\infty}$ is regular then $F \sim \sigma$ (see discussion preceding Theorem 2.6.8). From Theorem 4.1.2 there exists $\varphi \in L^1_+(\mathbb{C})$ such that $\int |\varphi(t)|^2 \frac{dF}{dt}(t)^{-1}dt < \infty$. Since $\int \ln|\varphi(t)|dt > -\infty$, we obtain that $-\int \ln \frac{dF}{dt}(t)dt + \int \ln|\varphi(t)|^2 dt < \infty$ which implies $\int \ln \frac{dF}{dt}(t)dt > -\infty$. Conversly, let $h(t)$ be such that $|h(t)|^2 = \frac{dF}{dt}(t)$ and $h(t) \in L^2_+(\mathbb{C})$. Then h satisfies (i), (ii), (iii) of Theorem 4.1.2(a). Thus, by Corollary 2.5.5 (ii), $(x_n)_{n=-\infty}^{\infty}$ is regular. □

4.1.5. Corollary. Suppose that the spectral density $\frac{dF}{dt}$ of a stationary sequence X satisfies

$$\overline{\frac{dF}{dt}(t)H} = \frac{dF}{dt}(t)H \quad \sigma \text{ a.e.}$$

Then X is regular if and only if there exists a sequence $\{\varphi_k\} \subset L^{1,w}_+(H)$ such that

(i) $\varphi_k(t) \in \frac{dF}{dt}(t)H$ σ a.e., $k = 1,2,...$

(ii) $\int |\frac{dF}{dt}(t)^{1/2\#}\varphi_k(t)|^2 dt < \infty$, $k = 1,2,...,$

(iii) $\overline{sp}\{\varphi_k(t): k = 1,2,...\} = \frac{dF}{dt}(t)H$ σ a.e.

Proof: In this case the $|\ |_t$–norm is equivalent to the original norm in H. □

4.1.6. Corollary. Suppose that the spectral density $\frac{dF}{dt}(t)$ has the bounded inverse $\frac{dF}{dt}(t)^{-1}$ σ a.e. If there exists a sequence $x_n \in H$ such that $\overline{sp}\{x_n: n = 1,2,...\} = H$ and such that for every n

$$-\infty < \int \ln[(\tfrac{dF}{dt}(t)^{-1}x_n,x_n)^{-1}]dt \le \int (\tfrac{dF}{dt}(t)^{-1}x_n,x_n)^{-1}dt < \infty,$$

then X is regular.

Proof. Let $f_k(t) = (\frac{dF}{dt}(t)^{-1}x_k,x_k)^{-1}$. Then by assumption $f_k \in L^1(\mathbb{C})$ and $\int \ln f_k(t)dt > -\infty$. Thus $f_k(t) = |h_k(t)|^2$ with $h_k \in L^2_+(\mathbb{C})$. Setting $\varphi_k(t) = h_k(t)x_k$ we easily see that φ_k satisfy (i), (ii) and (iii) of Corollary 4.1.5. □

4.1.7 Corollary. Suppose that the spectral density $\frac{dF}{dt}(t)$ satisfies $\overline{\frac{dF}{dt}(t)H} = \frac{dF}{dt}(t)H = \text{const} = M$ σ a.e. and that $\int \|\frac{dF}{dt}(t)\|dt < \infty$. If $\int \log\|\frac{dF}{dt}(t)^{\#}\|dt < \infty$, then X is regular.

Proof. Let $f(t) = \|\frac{dF}{dt}(t)^{\#}\|^{-1}$. Then $f(t) \le \|\frac{dF}{dt}(t)\|$ so $f \in L^1(\mathbb{C})$ and by assumption $\int \ln f(t)dt > -\infty$. Thus $f(t) = |h(t)|^2$ and it is easy to check that the sequence $\varphi_k(t) = h(t)e_k$, where $\{e_k: k = 1,2,...\}$ is a basis in M, satisfies (i), (ii), (iii) of Corollary 4.1.5. □

4.1.8. Corollary. ([5]). Suppose that $\frac{dF}{dt}(t)^{-1}$ exists and is bounded σ a.e. and that $\int \|\frac{dF}{dt}(t)\|dt < \infty$. If $\int \ln\|\frac{dF}{dt}(t)^{-1}\|dt < \infty$, then X is regular.

It should be noticed that if X is finite dimensional then $\int \|\frac{dF}{dt}(t)\|dt$ is always finite. Moreover in the finite dimensional case the condition $\int \ln\|\frac{dF}{dt}(t)^{-1}\|dt < \infty$ is also necessary for the regularity of a sequence X providing $\frac{dF}{dt}(t)$ is almost everywhere invertible (see [45]). In the infinite dimensional case the converse of Corollary 4.1.8 is not true as the following example shows.

4.1.9. Example. Let $g(t) = \exp\{-|\frac{\pi}{t}|\}$, $-\pi < t \le \pi$ and let

$$f_k(t) = \begin{cases} g(t) & , \quad \frac{\pi}{k} < |t| \le \pi \\ 1 & , \quad |t| \le \frac{\pi}{k}. \end{cases}, \qquad k = 1,2,...$$

Then (1) $\sup_k f_k(t) \in L^1(\mathbb{C})$, (2) $\int \ln f_k(t)dt > -\infty$ $k = 1,2,...$, and (3) $\int \ln[\inf_k f_k(t)]dt = -\infty$. Let X be an $L(\ell^2,K)$–valued stationary sequence whose spectral density is an infinite dimensional diagonal matrix with diagonal entries given by $f_1, f_2,...$. Then X is regular since it is the orthogonal sum of one dimensional regular stationary sequences (see Corollary 4.1.4) and $\int \|\frac{dF}{dt}(t)^{-1}\|dt < \infty$. Nevertheless

$$\int \ln\|\tfrac{dF}{dt}(t)^{-1}\|dt = \int \ln[\inf_k f_k(t)]dt = \infty.$$

4.1.10 Corollary. Let $\mathscr{R}(t)$ be a function on $(-\pi,\pi]$ whose values are closed subspaces of H such that for every $x \in H$, $P_{\mathscr{R}(t)}x$ is a measurable function of t, where $P_{\mathscr{R}(t)}$ denotes the orthogonal projection in H onto $\mathscr{R}(t)$ (any such a function will be called a measurable subspace–valued function). Let X be an L(H,K)–valued stationary sequence with the spectral density $\frac{dF}{dt}(t) = P_{\mathscr{R}(t)}$ σ a.e. Then X is regular if and only if there exists a sequence $\{\varphi_k\} \subset L^2_+(H)$ such that

$$\mathscr{R}(t) = \overline{sp}\{\varphi_k(t)\colon\ k = 1,2,...\}\ \ \sigma \text{ a.e.}$$

Proof. Proof follows immediately from Corollary 4.1.5 since $\int|\frac{dF}{dt}(t)^{1/2\#}\varphi_k(t)|^2dt = \int|\varphi_k(t)|^2dt$ in this case. □

Corollary 4.1.10 suggests the following definition which was introduced by Helson in [11].

4.1.11. Definition. A measurable subspace–valued function $\mathscr{R}(t)$ is analytic if there is a sequence of functions $\{\varphi_k\} \subset L^2_+(H)$ such that

$$\mathscr{R}(t) = \overline{sp}\{\varphi_k(t)\colon\ k = 1,2,...\}\ \ \sigma \text{ a.e.}$$

If $\mathscr{R}(t)$ is an analytic subspace–valued function then the functions φ_k in Definition 4.1.11 can be chosen in such a way that for almost all $t \in (-\pi,\pi]$, $\varphi_k(t)$, $k = 1,2,...$ forms an orthonormal basis in $\mathscr{R}(t)$ ([11]). In fact if $\mathscr{R}(t)$ is analytic then by Corollary 4.1.10 the sequence Y(n) constructed in Theorem 2.7.2 is regular. Let $\{\varphi_k\colon k = 1,2,...\}$ be an orthonormal basis in $M_{(-\infty,0]}(Y) \ominus M_{(-\infty,-1]}(Y)$. Then

$$(4.1.12)\qquad \begin{cases} (e^{in\cdot}\varphi_k, \varphi_l)_L = \delta_{n,0}\,\delta_{k,l} \text{ for all } n \in X,\ l, k \geq 1, \\ \text{where } \delta_{m,n} \text{ stands for the Kronecker symbol.} \end{cases}$$

Since $\frac{dF}{dt}(t)^{1/2\#} = P_{\mathscr{R}(t)}$ we have that for $\varphi, \psi \in L_F$

$$(\varphi,\psi)_L = \int(\varphi(t),\psi(t))dt.$$

Thus it easily follows from (4.1.12) that $(\varphi_k(t),\varphi_i(t)) = 0$ σ a.e. for all $k \neq i$, and that $|\varphi_k(t)|^2 = 1$ σ a.e. for all $k = 1,2,...$. In particular this proves that $\dim \mathscr{R}(t) = \text{const}$.

Using the fact mentioned above one can obtain a slight reinforcement of Corollary 4.1.7.

4.1.13. Corollary. Suppose that $\frac{dF}{dt}(t)H$ is closed σ a.e. and that $\int\|\frac{dF}{dt}(t)\|dt$. If

(i) $\frac{dF}{dt}(t)H$ is an analytic subspace–valued function and

(ii) $\int\ln\|\frac{dF}{dt}(t)^{\#}\|dt < \infty$,

then X is regular.

Proof. Let $f(t) = \|\frac{dF}{dt}(t)^{\#}\|^{-1}$ and let $h \in L^2_+(\mathbb{C})$ be such that $f(t) = |h(t)|^2$. Let η_k be a sequence of functions in $L^2_+(H)$ such that $\{\eta_k(t)\colon k = 1,2,...\}$ is an orthonormal basis in $\frac{dF}{dt}(t)H$ for almost all t (see the discussion above). Setting $\varphi_k(t) = h(t)\eta_k(t)$, $k = 1, 2,...$ it is easy to check that $\{\varphi_k\}$ satisfies the conditions (i), (ii), (iii) of Corollary 4.1.5. Therefore X is regular. □

If H is a finite dimensional space $\mathbb{C}^q$, then $\int\|\frac{dF}{dt}(t)\|dt < \infty$ for any $L(H,K)$–valued stationary sequence with the spectral measure a.c. w.r.t. σ. Moreover it is shown in [11] that the converse implication in Corollary 4.1.13 holds in the finite dimensional case. We feel that the converse of Corollary 4.1.13 is also true if the multiplicty of X, $m(X) < \infty$, provided that $\int\|\frac{dF}{dt}(t)\|dt < \infty$. In the case $m(X) = 1$ this problem was investigated in [19] where the following characterization was proved. Recall that in view of Theorem 2.7.6 a stationary sequence X having a spectral operator density $\frac{dF}{dt}$ has multiplicity 1 if and only if $\dim\left[\frac{dF}{dt}(t)H\right] \leq 1$ σ a.e. Note that in this case $\|\frac{dF}{dt}(t)^{\#}\| = \|\frac{dF}{dt}(t)\|^{-1}$ or 0. For the proof and more detailed discussion we refer the reader to [19].

4.1.14. <u>Theorem</u>. Let X be a nonzero $L(H,K)$–valued stationary sequence such that the operator spectral density $\frac{dF}{dt}(t)$ exists and $\dim\left[\frac{dF}{dt}(t)H\right] \leq 1$ σ a.e.

(a) If $\int\ln\|\frac{dF}{dt}(t)\|dt < \infty$ then X is regular if and only if

  (i) $\frac{dF}{dt}(t)H$ is an analytic subspace–valued function, and

  (ii) $\int\ln\|\frac{dF}{dt}(t)\|dt > -\infty$.

(b) The analyticity of $\frac{dF}{dt}(t)H$ is necessary for X to be regular if and only if $\int\ln\|\frac{dF}{dt}(t)\|dt < \infty$.

We should point out that from the regularity of X it does not follow that $\frac{dF}{dt}(t)H$ is an analytic subspace–valued function in the infinite dimensional case. An example demonstrating this point can be found in [19]. However it can be easily shown (see the note following Corollary 4.3.3) that if $\int\|\frac{dF}{dt}(t)\|dt < \infty$ and X is regular, then $\frac{dF}{dt}(t)H$ is an analytic subspace–valued function. In particular in the case $H = \mathbb{C}^q$, $q < \infty$, the range of the spectral density of a regular sequence is necessary analytic.

4.2. <u>Rank of a regular sequence</u>. If an $L(H,K)$–valued stationary sequence X is regular then all spectral types of X are equivalent to the Lebesgue measure and $r(X) = m(X)$ (see paragraph 2.5 for appropriate definitions). To prove this it suffices to choose an orthonormal basis $\{x_k\colon k=1,2,...\}$ in $M_{(-\infty,0]}(X) \ominus M_{(-\infty,-1]}(X)$ and observe that since $(U^k x_n, x_m) = \delta_{k,0}\,\delta_{n,m}$, $|E(\cdot)x_k|^2 = \sigma(\cdot)$ for each k. From the regularity of X if follows that $M(X) = \bigoplus_n \overline{sp}\{U^k x_n\colon k \in Z\}$. Thus from Theorem 2.5.3 it follows that $m(X) =$ cardinality of $\{x_n\colon n = 1,2,...\} = r(X)$, because the spectral measure of a regular sequence is equivalent to the Lebesgue measure. In general $r(X) \leq m(X)$. The following theorem shows the relations among rank, multiplicity and the dimension of the range of the density of X.

4.2.1. Theorem. Suppose that X is an L(H,K)–valued stationary sequence having a spectral density $\frac{dF}{dt}$ w.r.t. the Lebesgue measure. Then

$$r(X) \leq \sigma\text{–ess inf} \dim\left[\frac{dF}{dt}(t)H\right] \leq \dots$$
$$\leq \sigma\text{–ess sup} \dim\left[\frac{dF}{dt}(t)H\right] = m(X).$$

Proof. The equality σ–ess sup $\dim\left[\frac{dF}{dt}(t)H\right] = m(X)$ is the contention of Theofem 2.7.6. If $r(X) = 0$, then the conclusion is obvious. Suppose that $r(X) \neq 0$. Let $\{\varphi_k : 1 \leq k < r(X) + 1\}$ be an orthonormal basis in $M_{(-\infty,0]}(Y) \ominus M_{(-\infty,-1]}(Y) \subset L_F$, where Y is constructed as in Theorem 2.7.2. Then for almost all t, $\{f_k(t) = \frac{dF}{dt}(t)^{1/2\#}\varphi_k(t) : 1 \leq k < r(X) + 1\}$ forms an orthonormal system in $\frac{dF}{dt}(t)^{1/2}H$. Therefore $\dim\left[\frac{dF}{dt}(t)H\right] \geq r(X)$ σ a.e. □

As an immediate consequence of Theorem 4.2.1 we obtain the following corollaries.

4.2.2. Corollary. (cf. [11]). If X is regular then $\dim\left[\frac{dF}{dt}(t)H\right] = r(X) = m(X)$ σ a.e.

4.2.3. Corollary ([42], Theorem 2.5). Suppose that H is finite dimensional. Then the Wold and Lebesgue decompositions of X for $A = (-\infty,-1]$ (see Sections 2.2 and 2.3) coincide if and only if

$$\dim\left[\frac{dF^a}{dt}(t)H\right] = r(X) \quad \sigma \text{ a.e.}$$

4.3. Factorization of the density of a regular sequence and moving average representation. In this section we sketch briefly the approach to the extrapolation problem based on the notion of invariant subspaces of $L^2(H)$ as introduced by Helson in [11].

The following theorem was proved by several authors in different settings. For Banach space setting see [3], [54]. First we state some auxiliary definitions.

4.3.1. Definition. Let F be the spectral measure of a stationary L(H,K)–valued sequence X. Suppose that $F << \sigma$. Then

(i) the sesquilinear form $f(x,y): H \times H \to L^1(\sigma)$ defined by the formula

$$f(x,y) = \frac{d}{dt}(F(\cdot)x,y)$$

is called the density of F w.r.t. σ,

(ii) any bounded linear operator Q from H into $L^2(\mathcal{H})$, where $\mathcal{H}$ is a Hilbert space, such that for all $x, y \in H$

$$f(x,y)(t) = ((Qx)(t), (Qy)(t))_{\mathcal{H}} \quad \sigma \text{ a.e.}$$

is called a quasi square root of f.

One can prove that a quasi square root of f always exists (see [54] or [21]).

4.3.2. Theorem. Let H be a separable Hilbert space and let X be an L(H,K)–valued stationary sequence with the spectral measure F. Then the following are equivalent:

(1) X is regular,

(2) $F << \sigma$ and there exists a quasi square root Q of the density f such that for every $x \in H$, $Qx \in L^2_-(\mathscr{K})$,

(3) there exist a Hilbert space $\mathscr{K}$ and sequences of operators $W_k \in L(\mathscr{K},K)$, $k \in Z$, and $A_k \in L(H,\mathscr{K})$, $k = 0,1,\ldots$, such that

(i) $\sum_{k=0}^{\infty} |A_k x|^2 < \infty$ for all $x \in H$

(ii) $(W_k x, W_j y) = \delta_{k,j}(x,y)$, $k,j \in Z$, $x,y \in H$,

(iii) $X(n)x = \sum_{k=-\infty}^{n} W_k A_{n-k} x$, $n \in Z$, $x \in H$.

Proof. (1) ⇒ (2). From the discussion preceding Theorem 2.6.8 it follows that if X is regular then $F << \sigma$ so we may assume that $F << \sigma$. Let Q be any quasi square root of the density f of F w.r.t. the Lebesgue measure. Consider an $L(H,L^2(\mathscr{K}))$–valued sequence $W = \{W(n): n \in Z\}$ defined by the formula

$$W(n)x = e^{in\cdot}(Qx)(\cdot), \quad n \in Z, \quad x \in H.$$

The sequence $\{Z(n); n \in Z\}$ is regular if and only if the subspace M_Q of $L^2(H)$ defined by

$$M_Q = \overline{sp}\{e^{in\cdot}(Qx)(\cdot): n \le 0, x \in H\}$$

contains no subspace invariant under multiplication by e^{it}. But as shown in [11] this is true if and only if M_Q can be written as $UL^2_-(\mathscr{K}_0)$ for be some Hilbert space $\mathscr{K}_0$, where U is an isometry generated by some isometry–valued function $U_0(t)$, $t \in (-\pi,\pi]$, i.e., $(Uf)(t) = U_0(t)f(t)$ σ a.e., $f \in L^2(\mathscr{K}_0)$. Thus $(Q_0x)(t) = U_0(t)^{\#}(Q_x)(t)$, $x \in H$, is also a quasi square root of f and $Q_0x \in L^2_-(\mathscr{K}_0)$ for all $x \in H$. Note that implication (2) ⇒ (1) also has been proved in the meantime.

(2) ⇒ (3) From the discussion above it follows that we may assume that $M_Q = L^2_-(\mathscr{K})$. Setting $A_n x = \int e^{int}(Qx)(t)dt$, $x \in H$, $\mathscr{K} = \mathscr{K}$ and $W_n y = e^{in\cdot}y$, $y \in \mathscr{K}$ we obtain that sequence W has the properties (i), (ii), (iii) listed above. The implication (3) ⇒ (1) is obvious. □

If the spectral measure of a sequence X has an operator spectral density $\frac{dF}{dt}(t)$ w.r.t. the Lebesgue measure, then each quasi square root Q is generated by an $L(H,\mathscr{K})$–valued weakly measurable function $Q(t)$, namely for each $x \in H$, $(Qx)(t) = Q(t)x$ σ a.e. (see [34]). Thus Theorem 4.3.2 yields the following corollary.

4.3.3. <u>Corollary</u>. ([11]). Suppose that X has the operator spectral density $\frac{dF}{dt}$ w.r.t. the Lebesgue measure. Then X is regular if and only if there exist a Hilbert space $\mathscr{K}$ and an $L(H,\mathscr{K})$–valued function $Q(t)$, $t \in (-\pi,\pi]$ such that

(i) $Q(\cdot)x \in L^2_-(\mathscr{K})$ for all $x \in H$,

(ii) $\frac{dF}{dt}(t) = Q^*(t)Q(t)$ σ a.e.

Note that if $\int\|\frac{dF}{dt}(t)\|dt < \infty$ then $Q^*(\cdot)x \in L^2_+(H)$. Thus if $\|\frac{dF}{dt}(\cdot)\|$ is integrable, then the subspace–valued function $\overline{\frac{dF}{dt}(t)H}$ is analytic, provided X is regular.

4.4. Douglas comparison theorem. As we have mentioned earlier this paper is not intended to provide a complete account of material related to extrapolation problem. There are several topics not included in this report. We end this chapter with some result due to Douglas which has important consequences in prediction theory. A complete discussion of this can be found in [6] or [53].

4.4.1. Theorem ([6], Corollary 1). Let X^1 and X^2 be any two $L(H,K)$–valued stationary sequences having operator spectral densities $\frac{dF^1}{dt}$ and $\frac{dF^2}{dt}$, respectively. If $\psi(t)\frac{dF^1(t)}{dt} \leq \frac{dF^2}{dt}(t) \leq \frac{dF^1}{dt}(t)$ σ a.e. for some positive scalar function ψ and if X^2 is regular then X^1 is regular.

Proof. Let

$$W^1(n)x = e^{in\cdot}\frac{dF^1}{dt}(\cdot)^{1/2}x, \quad x \in H, \quad n \in Z \text{ and}$$
$$W^2(n)x = e^{in\cdot}\frac{dF^2}{dt}(\cdot)^{1/2}x, \quad x \in H, \quad n \in Z.$$

Then W^1 and W^2 are stationary $L(H,L^2(H))$–valued sequences equivalent to X^1 and X^2, respectively, and moreover

$$\begin{aligned}|\sum_{n\in I} W^2(n)x_n|^2 &= \int(\frac{dF^2}{dt}(t)p(t), p(t))dt \leq \int(\frac{dF^1}{dt}(t)p(t), p(t))dt \\ &= |\sum_{n\in I} W^1(n)x_n|^2,\end{aligned} \tag{4.4.2}$$

where $p(t) = \sum_{n\in I} e^{int}x_n$ and I is a finite subset of Z. Note that $M(W^1) = M(W^2) = \{\varphi \in L^2(H): \varphi(t) \in \overline{\frac{dF^1}{dt}(t)H}$ σ a.e.$\}$, since $\overline{\frac{dF^1}{dt}(t)H} = \overline{\frac{dF^2}{dt}(t)H}$ σ a.e.. Let $\varphi \in \bigcap_n M_{(-\infty,n]}(W^1)$. Then there exists a sequence of polynomials $p_n(t) = \sum_{k\leq -n} e^{ikt}x_k$, $n \geq 1$, such that

$$\frac{dF^1}{dt}(\cdot)^{1/2}p_n(\cdot) \to \varphi(\cdot)$$

in $L^2(H)$ and σ a.e.. Since W^2 is regular, from (4.4.2) it follows that

$$\frac{dF^2}{dt}(\cdot)^{1/2}p_n(\cdot) \to 0 \text{ in } L^2(H). \tag{4.4.3}$$

Let n_k be a subsequence such that (4.4.3) holds almost everywhere. Then $|\frac{dF^1}{dt}(t)^{1/2}p_{n_k}(t)|^2 = |\varphi_{n_k}(t)|^2 \leq \frac{1}{\psi(t)}|\frac{dF^2}{dt}(t)^{1/2}p_{n_k}(t)| \to 0$ σ a.e., which shows that $\varphi = 0$. □

4.4.4. Corollary. Let X be a stationary sequence with the spectral density $\frac{dF}{dt}$ and let X^1

be a stationary sequence with the spectral density $\frac{dF^1}{dt}(t) = \min(1,\|\frac{dF}{dt}(t)\|^{-1})\frac{dF}{dt}(t)$. If X^1 is regular then X is regular.

Proof. Note that

$$\psi(t)\frac{dF}{dt}(t) \le \frac{dF^1}{dt}(t) \le \frac{dF}{dt}(t)$$

where $\psi(t) = \min(1, \|\frac{dF}{dt}(t)\|^{-1})$ and $\psi(t) > 0$ σ a.e. since X^1 is regular. □

4.4.5. Corollary. ([6], Theorem 2). The conclusion of Corollary 4.1.13 holds without the assumption of integrability of $\|\frac{dF}{dt}(\cdot)\|$.

5. ANGLE BETWEEN PAST AND FUTURE

5.1. Introductory remarks. In this section we discuss briefly the third prediction problem, i.e., the problem of the angle between the past and the future for stationary sequences, as formulated by Helson and Szego in [12]. As we will see soon the problem is connected with the existence of a series representation of the linear predictor which exhibits its importance. We will assume that H is a separable Hilbert space.

5.1.1. Definiton. Let M and N be two closed subspace of a Hilbert space H. The cosine of the angle between M and N is defined by

$$\rho(M,N) = \sup\{|(x,y)| : x \in M, y \in N, |x| = |y| = 1\}$$

The subspaces M and N are said to be at positive angle if $\rho(M,N) < 1$.

It is easy to see (cf. [12]) that $\rho(M,N) < 1$ if an only if there exists a constant $C < \infty$ such that

$$|x| \le C|x+y| \quad \text{for all } x \in M_0 \; y \in N_0, \tag{5.1.2}$$

where M_0, N_0 are dense linear subspaces of M and N respectively.

5.1.3. Definition. Let X be an $L(H,K)$–valued stationary sequence. X is said to be of positive angle if $\rho(X) \overset{df}{=} \rho(M_{(-\infty,-1]}(X), M_{[0,+\infty)}(X)) < 1$.

From (5.1.2) it follows that X is of positive angle if and only if there exists a constant $C < \infty$ such that for every sum of the form $\Sigma X(n)x_n$, $x_n \in H$,

$$|\sum_{n\le -1} X(n)x_n| \le C|\Sigma X(n)x_n|, \tag{5.1.4}$$

where here and in the sequel the symbol Σ without limits stands for an infinite sum whose terms except possibly finitely many are zero. Since X is stationary, (5.1.4) yields that

$$|\sum_{n\le n_0} X(n)x_n| \le C|\Sigma X(n)x_n| \tag{5.1.5}$$

for any n_0 where C depends only on X. Thus

$$|\sum_{n=n_0}^{n_1} X(n)x_n| \leq 2C|\Sigma X(n)x_n|.$$

This bring us to the following lemma.

5.1.6. Lemma (cf. [12]). Let X be an $L(H,K)$–valued stationary sequence. Then $\rho(X) < 1$ if and only if there exists a constant $C < \infty$ such that for every sum of the form $\Sigma X(n)x_n$ and for every $-\infty \leq n_0 \leq n_1 \leq \infty$

$$|\sum_{n=n_0}^{n_1} X(n)x_n| \leq C|\Sigma X(n)x_n|.$$

5.2. Basis properties of a sequence with positive angle. Let P_m, $m \in Z$, denote the linear operator in $M(X)$ defined by $P_m(\Sigma X(n)x_n) = X(m)x_m$. Lemma 5.1.6 says that X is of positive angle if and only if all P_m's, $m \in Z$, are well–defined and the operators $P_k^m = \sum_{i=k}^{m} P_i$ are uniformly bounded, i.e.,

$$\sup\{\|\sum_{i=k}^{m} P_i\|: -\infty < k \leq m < +\infty\} < \infty.$$

This implies that for every $x \in M(X)$

$$x = \lim_{\substack{m\to\infty \\ k\to-\infty}} \sum_{i=k}^{m} P_i x \overset{df}{=} \sum_{i=-\infty}^{+\infty} P_i x. \tag{5.2.1}$$

The converse statement is also true. Namely, we have the following theorem.

5.2.2. Theorem. Let X be an $L(H,K)$–valued stationary sequence. Then $\rho(X) < 1$ if and only if each $x \in M(X)$ admits a unique representation of the form $x = \sum_{n=-\infty}^{+\infty} y_n$ with $y_n \in \overline{X(n)H}$.

Here and in the sequel $\sum_{n=-\infty}^{+\infty} y_n$ will stand for $\lim_{\substack{k\to-\infty \\ m\to+\infty}} \sum_{n=k}^{m} y_k$.

Proof. Suppose first that each $x \in M(X)$ admits a unique representation of the form $x = \sum_{-\infty}^{+\infty} y_n$, where $y_n \in \overline{X(n)H}$. Define a sequence of operators A_n: $H \to \overline{X(n)H}$ by the formula $A_n x = y_n$, where $x = \sum_{-\infty}^{+\infty} y_n$. Since the series representation of x is unique A_n is a well–defined operator from H onto $\overline{X(n)H}$, $n \in Z$. Using standard arguments (cf [16], p. 1) we show that A_n's, $n \geq 1$, are bounded and

$$\sup\{\|\sum_{n=k}^{m} A_n\|: -\infty < k \leq n < +\infty\} = C < \infty. \tag{5.2.3}$$

In fact, let us first note that for any fixed $x = \sum_{-\infty}^{+\infty} y_n \in M(X)$ there are constants $k_0 \leq 0 \leq n_0$ such that $|\sum_k^m y_n| \leq |x| + 1$ for all $k \leq k_0$ and $m \geq m_0$. Since $\sup\{|\sum_k^m y_n|: k_0 \leq k \leq m \leq n_0\} = B < \infty$, we obtain that

$$\|x\| \stackrel{df}{=} \sup\{|\sum_k^m y_n|: -\infty < k \leq m < +\infty\} \leq (|x| + 1 + B) < \infty.$$

It follows that $\| \ \|$ is a norm in H and $|x| \leq \|x\|$ for every $x \in H$. It is easy to check that H is complete if equipped with this norm. Thus, from the open mapping theorem, $\|x\| \leq C|x|$, $x \in H$ for some constant C which proves (5.2.3). Since $A_n X(n)x = X(n)x$, $n \in Z$, $x \in H$, the applications of Lemma 5.1.6 completes the proof of sufficiency.

We have already observed that if $\rho(X) < 1$, then any x in $M(X)$ admits a representation of the form (5.2.1). To prove the neccesity part it remains to prove that this representation is unique. Suppose that $x = \sum y_n$ with $y_n \in \overline{X(n)H}$, $n \in Z$. Since for every $m \neq n$ $P_n x = 0$ provided $x \in \overline{X(m)H}$, $y_n = P_n(\sum_{-\infty}^{+\infty} y_k) = P_n x$, $x \in H$, $n \in Z$ and the theorem is proved. □

5.2.4. <u>Corollary</u>. Suppose that X is an $L(H,K)$–valued stationary sequence such that $X(0)H$ is a closed subspace of K (note that this condition is always satisfied if X is a finite dimensional stationary sequence). Then the following two conditions are equivalent:

(1) $\rho(X) < 1$,

(2) each linear operator $Y \in L(H,M(X))$ admits a representation

$$Yx = \sum_{n=-\infty}^{+\infty} X(n)A_n x$$

where $A_n \in L(H)$ and the representation is unique in the sense that if $Yx = \sum_{n=-\infty}^{+\infty} X(n)B_n x$, $x \in H$ then $(P_{\mathrm{Null}(X(0))^{\perp}})A_n = (P_{\mathrm{Null}(X(0))^{\perp}})B_n$.

<u>Proof</u>. (1) ⇒ (2). By assumption $X(n)^{\#}$ is a bounded operator from $X(n)H$ onto $\mathrm{Null}(X(0))^{\perp}$. Thus if $\rho(X) < 1$ then from (5.2.1) it follows that $Yx = \sum_{-\infty}^{\infty} X(n)A_n x$, with $A_n = X(n)^{\#} P_n Y \in L(H)$. If $Yx = \sum_{-\infty}^{\infty} X(n)B_n x$ is another representation of Y, then $P_n Yx = X(n)B_n x = X(n)A_n x$ (see the proof of necessity in Theorem 5.2.2). Thus $(P_{\mathrm{Null}(X(0))^{\perp}})A_n = X(n)^{\#}X(n)A_n = (P_{\mathrm{Null}(X(0))^{\perp}})B_n$, $n \in Z$.

(2) ⇒ (1). In view of Theorem 5.2.2 this implication becomes obvious by considering one dimensional operators. □

In the case where X is a finite dimensional stationary sequence Corollary 5.2.4 was proved by Miamee and Pourahmadi in [31]. They referred to condition (2) as "X(n) being a generalized

Schauder basis in $M(X)$". In fact if $H = \mathbb{C}$ then A_n's are scalars and (2) means that the sequence $X_n = X(n)$ forms a Schaunder basis in $M(X)$ (see [16], p. 1).

Note that for the case $\rho(X) < 1$ if a subset A of Z is the union of finitely many intervals of the form (a,b), $-\infty \leq a < b \leq +\infty$ then the operator $P_A x = \sum_{i \in A} P_i x$ is bounded and $P_A P_i = 0$ if $i \notin A$. Thus from (5.2.1) it follows that if $x \in M_A(X)$, then

$$x = \sum_{i \in A} P_i x, \quad \text{provided } A \text{ is as above.} \tag{5.2.5}$$

We point out that condition $\rho(X) < 1$ does not imply that the right hand side of (5.2.5) converges for an arbitrary A (see [44]). From this discussion we arrive at the following corollary.

5.2.6. Corollary. Let X be an $L(H,K)$–valued stationary sequence such that $X(0)H$ is closed. Let A be the union of finitely many intervals of the form (a,b), $-\infty \leq a < b \leq +\infty$, and let $n_0 \in Z$. If $\rho(X) < 1$ then the predictor $\tilde{X}(n_0,A)$ of $X(n_0)$ based on A has the form

$$\tilde{X}(n_0,A)x = \sum_{n \in A} X(n)B_n x$$

for some sequence B_n, $n \in A$.

Another consequence of (5.2.5) is that a sequence with positive angle is necessary A–regular for any proper subset A of Z, a result which in the finite dimensional case was proved in [31].

5.2.7. Corollary. If $\rho(X) < 1$ then for every $A \subset Z$, $A \neq Z$, X is A–regular.

Proof. It suffices to prove that X is J_0–regular, since it implies A–regularity for all $A \neq Z$. Let $A = \{0\}^c$ and $x \in M_{A+n}(X)$ for all n. Then from (5.2.5) $x = \sum_{i \neq n} P_i x$, $n \in Z$. The uniqueness of this representation forces x to be zero. □

5.3. Spectral properties of a stationary sequence with positive angle. From Corollary 5.2.7 it follows that the spectral measure F of a stationary sequence X with positive angle is necessarily equivalent to the Lebesgue measure. If moreover $\frac{dF}{dt}$ exists then $\overline{\frac{dF}{dt}(t)H} = \text{const.}$ and $\frac{dF}{dt}(\cdot)^{1/2\#}x$ is square integrable for all x in some (not necessarily closed) subspace of $\frac{dF}{dt}(t)^{1/2}H$ (see Corollary 3.2.2). It is known that these conditions are not sufficient for X to be of positive angle, as J_0–regularity does not imply the positivity of the angle even in the one–dimensional case.

Below we present rather a vague characterization of sequences with positive angle by means of the behavior of the Fourier transforms of some class of functions connected with the spectral measure of the sequence.

5.3.1. Definition. Suppose that $\frac{dF}{dt}$ is an operator spectral density of an $L(H,K)$–valued stationary sequence. By $\mathcal{L}_F^2$ we will denote the set of all H–valued measurable functions f for which $f(t) \in \overline{\frac{dF}{dt}(t)H}$ σ a.e. and

$$\|f\|_F^2 = \int(\frac{dF}{dt}(t)f(t),f(t))dt < \infty.$$

It is clear that $|\ |_F$ is a Hilbertian norm in $\mathcal{L}_F^2$. However $\mathcal{L}_F^2$ need not be complete with respect to this norm (see [6]). Nevertheless if $\frac{dF}{dt}(t)H$ is closed σ a.e., then $\mathcal{L}_F^2$ becomes a Hilbert space (cf. [9]). In particular this is so if H is finite dimensional.

5.3.2. Lemma. Suppose that X is an L(H,K)–valued stationary sequence such that $\frac{dF}{dt}(t)$ exists and $\frac{dF}{dt}(t)H = \overline{\frac{dF}{dt}(t)H} = \text{const} \overset{df}{==} \mathcal{R}$ σ a.e.. If $\mathcal{L}_F^2 \subset L^1(\sigma,\mathcal{R})$ then X has a positive angle if and only if

(5.3.3) for each vector $f \in \mathcal{L}_F^2$ the Fourier series $\sum_{n=-\infty}^{+\infty} a_n(f)e^{int}$ of f converges to f in the $\|\ \|_F$–norm (where $a_n(f) = \int e^{-int}f(t)dt$ which is well defined by assumption).

Proof. Our proof uses ideas contained in [31]. First note that since $\mathcal{L}_F^2$ is complete in this case, it follows from the closed graph theorem that for some constant C

$$|\int f(t)dt| \le C\|f\|_F,\ f \in \mathcal{L}_F^2.$$

Moreover there exists a constant $D < \infty$ such that

$$\int(\frac{dF}{dt}(t)x,x)dt = (F(-\pi,\pi]x,x) \le D|x|^2,\ x \in H.$$

These two inequalities together imply that the mapping T_0 assigning a constant function $(T_0f)(\cdot) = \int f(t)dt$ to every $f \in \mathcal{L}_F^2$ is a bounded linear operator from $\mathcal{L}_F^2$ into $\mathcal{L}_F^2$.

Now consider an $L(H,\mathcal{L}_F^2)$–valued stationary sequence W defined by the formula

$$W(n)x = e^{in\cdot}P_{\mathcal{R}}x,\ x \in H,\ n \in Z.$$

Since $(W(n)x,W(m)y) = \int e^{(n-m)t}(\frac{dF}{dt}(t)x,y)dt$, W is equivalent to X. Clearly $M(W) = \mathcal{L}_F^2$. We will show that $\rho(W) < 1$ if and only if (5.3.3). Note that for every k and m with $k \le m$ and any linear combination $\Sigma W(j)x_j$

(5.3.4)
$$\sum_{n=k}^{m} W(n)x_n = (\sum_{n=k}^{m} T_n)(\Sigma W(j)x_j),$$

where $(T_nf)(t) = e^{int}(\int e^{-ins}f(s)ds) = a_n(f)e^{int}$. Thus if $\rho(W) < 1$ then $T_n = P_n$, where P_n is defined at the beginning of the Section 5.2, and by (5.2.1) for all $f \in \mathcal{L}_F^2$, $f = \sum_{-\infty}^{+\infty} T_nf = \sum_{-\infty}^{\infty} e^{in\cdot}a_n(f)$ in $\mathcal{L}_F^2$. Conversly, suppose that for every $f \in \mathcal{L}_F^2$, $f = \sum_{-\infty}^{+\infty} T_n(f)$. From the Banach–Steinhaus theorem it follows that $\sup\{\|\sum_{n=k}^{m} T_n\|: -\infty < k < m < \infty\} = C < \infty$ (we use the same argument as in the proof of Theorem 5.2.2). Thus

$$|\sum_{n=k}^{m} W(n)x_n| \leq C|\Sigma W(n)x_n|$$

for all $k < m$ and all finite linear combination $\Sigma W(n)x_n$. From Lemma 5.1.6 we conclude that $\rho(Z) < 1$. □

Note that under assumptions of Lemma 5.3.2 if $\int\|\frac{dF}{dt}(t)^{\#}\|dt$ exists, then by the use of Cauchy–Schwarz's inequality, $\mathcal{L}_F^2 \subset L^1(\sigma,\mathcal{R})$. Moreover we have already noticed that a stationary sequence X is J_0–regular, provided $\rho(X) < 1$ (Corollary 5.2.7). Thus from Corollary 3.2.5 it follows that if X is a finite dimensional stationary sequence with positive angle then the assumptions of Lemma 5.3.2 are satisfied and we obtain the following result due to Helson–Szego [12] for $q = 1$ and Miamee, Pourahmadi [31] for $q < \infty$.

5.3.5. Corollary. ([31]). Let X be a q–dimensional stationary sequence with the spectral measure F. Then $\rho(X) < 1$ if and only if

(i) $F << \sigma$

(ii) $\frac{dF}{dt}(t)\mathbb{C}^q = \text{const}$ σ a.e.

(iii) $\int\|\frac{dF}{dt}(t)^{\#}\|dt < \infty$

(iv) for all $f \in \mathcal{L}_F^2$, $f(t) = \sum_{-\infty}^{+\infty} a_n(f)e^{int}$ in $\mathcal{L}_F^2$, where $a_n(f) = \int e^{-int}f(t)dt$.

Let us note that (iii) and the Cauchy–Schwarz inequality guarantee that $a_n(f)$ are well–defined.

In the one dimensional case the celebrated works of Helson and Szego [12]; Hunt, Mackenhoupt and Wheeden [13] give a complete analytic characterization to the positivity of the angle. This result is stated below.

5.3.6. Theorem. Let $\{x_n: n \in Z\}$ be a one dimensional stationary sequence with the spectral measure F. Then the following are equivalent:

(1) $\rho(X) < 1$

(2) $F << \sigma$ and there exist bounded real functions φ and ψ such that $\sup\{|\psi(t)|: t \in (-\pi,\pi]\} < \frac{\pi}{2}$ and $\frac{dF}{dt}(t) = \exp\{\varphi(t) + \tilde{\psi}(t)\}$ where $\tilde{\psi}$ denotes the conjugate of ψ defined by the formula $\tilde{\psi}(t) = \lim_{\epsilon\to 0+} \int_{\epsilon<|s|\leq\pi} \frac{\psi(t-s)}{\tan s/2} ds$.

(3) $F << \sigma$ and there exist a constant C such that $(\int_I \frac{dF}{dt}(t)dt)(\int_I \frac{dF}{dt}(t)^{-1}dt) \leq C[\sigma(I)]^2$ for any $I \subset (-\pi,\pi]$, where I is an interval or the complement of an interval.

Attempts in [41] and [31] are made to extend the Helson–Szego result to the multivariate case. In our opinion the core of the problem still remains open. Some extension of the work of Helson– Szego and Hunt–Mackenhoupt–Wheeden for the one dimensional random fields is carried out in [30] and [20].

From Corollary 5.2.6 we conclude that in extrapolation theory the linear predictor admits the autoregressive representation of the form

$$\tilde{X}_0 = \sum_{-\infty}^{-1} X(n)\, B_n, \tag{5.3.7}$$

where $\tilde{X}_0 = \tilde{X}(0,(-\infty,-1]))$, provided that X is a finite dimensional sequence with positive angle. In prediction theory it is important to devise an algorithm for the determination of B_ns. Such an algorithm has been provided in [41]. The representation (5.3.7) also holds under a different set of conditions on the spectral density (see [24], [35]). In both cases the computation of B_ns involves the Fourier coefficients of the optimal factor of the spectral density and those of the inverse of the optimal factor. It seems to be worthwhile to explore the possibility of extension of the representation (5.3.7) to the infinite dimensional case (however see [29]).

REFERENCES

1. Abreu, J.L. (1970). A note on harmonizable and stationary sequences. Bol. Soc. Mat. Mexicana 15, 38–41.

2 Abreu, J.L., Salehi, H. (1984). Schauder basic measures in Banach and Hilbert spaces. Bol. Soc. Mat. Mexicana 15, 58–41

3 Chobanyan, S.A., Weron, A. (1975). Banach space valued stationary processes and their linear prediction. Dissertationes Math. 125, 1–45.

4. Cramer, H. (1962). On some class of non–stationary stochastic processes. Proc. IV Berkeley Sym. Math. Stat. Probab. 2, 57–78, Univ. Calif. Press.

5. Devinatz, A. (1961). The factorization of operator valued functions. Ann. of Math. 73, 458–495.

6. Douglas, R.G. (1966). On factoring positive operator functions. J. Math. and Mech. 16, 119–126.

7. Dunford, N., Schwartz, J.T. (1963). Linear operators. Part II: Spectral Theory. Interscience Publ. New York.

8. Gangolli, R. (1963). Wide–sense stationary sequences of distributions on Hilbert space and factorization of operator–valued functions. J. Math. Mech. 12, 893–910.

9. Gorniak, J., Makagon, A., Weron, A. (1983). An explicit form of dilation theorems for semispectral measures. Prediction Theory and Harmonic Analysis, The Pesi Masani Volume. Eds. V. Mandrekar and H. Salehi. North Holland Publ. Comp., 85–112.

10. Gorniak, J., Weron, A. (1976). An analogue of Sz.–Nagy's dilation theorem, Bull. Acad. Polon. Sci., Ser. Math. Phys. and Astronom. 24, 867–872.

11. Helson, H. (1964). Lectures on Invariant Subspaces. Academic Press. New York and London.

12. Helson, H., Szego, G. (1960). A problem in prediction theory. Ann. Mat. Pura Appl. 51, 107–138.

13. Hunt, R., Muckenhoupt, B., Wheedent, R.L. (1973). Weighted norm inequalities for the conjugate function and Hilbert transform. Trans. Amer. Math. Soc. 176, 227–251.

14. Kallianpur, G.P., Mandrekar, V. (1971). Spectral theory of stationary H–valued processes. J. Multivariate Anal. 1, 1–16.

15. Kolmogorov, A. (1941). Stationary sequences in Hilbert space. Bull. Math. Univ. Moscow. 2(6), 1–40.

16. Lindenstrauss, J., Tzafiri, L. (1977). Classical Banach spaces I. Springer–Verlag, New York.

17. Makagon, A. (1984). On the Hellinger square integral with respect to an operator–valued measure and stationary processes. J. Multivariate Anal. 14, 114–133.

18. Makagon, A. (1984). Interpolation error operator for Hilbert space valued stationary stochastic processes. Probab. Math. Stat. 4, 57–65.

19. Makagon, A., Salehi, H. (1987). Infinite dimensional stationary sequences with multiplicity one. Ann. Acad. Sci. Fernn. Ser. A. I. Math. 12, 135–150.

20. Makagon, A., Salehi, H. (1968). Stationary fields with positive angle. J. Multivariate Anal. 22, 106–125..

21. Makagon, A., Schmidt, F. (1980). A decomposition theorem for densities of positive operator–valued measures. Bull. Polon. Acad. Sci., Ser. Math. Phys. and Astronom. 28,41–43.

22. Makagon, A., Weron, A. (1976). q–variate minimal stationary processes. Stud. Math. 59, 41–52.

23. Mandrekar, V., Salehi, H. (1972). The square integrability of operator–valued functions with respect to a non–negative operator valued measure and Kolomogorov's isomorphic theorem. Indiana Univ. Math. J. 20, 545–563.

23a. Mandrekar, V., Salehi, H. (1973). On the structure of $L_{2,M}$, vector and operator–valued measures and applications. Academic Press, New York, 199–207.

24. Masani, P. (1960). The prediction theory of multivariate stochastic processes III. Acta. Math. 104, 142–162.

25. Masani, P. (1970). Quasi isometric measures. Bull. Amer. Math. Soc. 76, 427–528.

26. Masani, P. (1978). Dilations as propagators of Hilbertian varieties. SIAM J. Math. Anal. 9, 414–456.

27. Masani, P., Wiener, N. (1957). The prediction theory of multivariate stochastic processes I. Acta. Math. 98, 111–150.

28. Masani, P., Wiener, N. (1958). The prediction theory of multivariate stochastic processes II. Acta. Math. 99, 93–137.

29. Miamee, A.G. (1975). An algorithm for determining the generating function and the linear prediction of an infinite–dimensional stationary stochastic processes. Sankhya Ser. A, 175–188.

30. Miamee, A.G., Niemi, H. (1985). On the angle for stationary random fields. Tech. Report 92, Center for Stochastic Processes, Univ. of North Carolina.

31. Miamee, A.G., Pourahmadi, M. (1987). Degenerate multivariate stationary processes: Basicity, Past and Future and Autoregressive Representation; Sankhya Ser A., 316–334.

32. Miamee, A.G., Salehi, H. (1974). Necessary and sufficient conditions for factorability of non–negative operator valued functions on a Banach space, Proc. Amer. Math. Soc. 46(1), 43–50.

33. Miamee, A.G., Salehi, H. (1974). Factorization of positive operator–valued functions on a Banach space. Indiana Univ. Math. J. 25, 103–113.

34. Miamee, A.G., Salehi, H. (1977). On square root of a positive $B(\mathscr{X},\mathscr{X}^*)$–valued functions. J. Multivariate Anal. 7, 533–550.

35. Miamee, A.G., Salehi, H. (1983). On an explicit representation of the linear predictor of a weakly stationary stochastic sequence. Bol. Soc. Mat. Mexicana 28, 81–93.

36. Mlak, W. (1978). Dilations of Hilbert space operators (General theory). Dissertationes Math. CLIII, 1–61.

37. Nadkarni, M.G. (1970). Prediction theory of infinite variate weakly stationary stochastic processes. Sankhya, Ser. A 32, 145–172.

38. Niemi, H. (1975). On stationary dilations and the linear prediction of certain stochastic processes. Soc. Sci. Fenn. Comment. Phys.–Math. 45, 111–130.

39. Niemi, H. (1984). Subordination, rank and determinism of multivariate stationary sequences. J. Mult. Analysis 15,1, 99–123.

40. Payen, R. (1967). Functions aleatories du second ordre a valuers dans an espace de Hilbert. Ann. Inst. H. Poincare, Sect. B. 3, 323–396.

41. Pourahmadi, M. (1985). A matricial extension of the Helson–Szego theorem and its application in multivariate problem. J. Mult. Anal. 16, 265–275.

42. Robertson, J.B. (1968). Orthogonal decompositions of multivariate weakly stationary stochastic processes. Canad. J. Math. 20, 368–383.

43. Rosenberg, M. (1964). Square integrability of matrix–valued functions with respect to a nonnegative hermitian measures. Duke Math. J. 31, 291–298.

44. Rozanov, Yu. A. (1960). On stationary sequences forming a basis. Soviet Math. Dold. 1, 155–158.

45. Rozanov, Yu. A. (1963). Stationary random processes, Holden–Day.

46. Rozanov, Yu. A. (1972). Some approximation problems in the theory of stationary processes, J. Mult. Anal. 2, 135–144.

47. Rozanov, Yu. A. (1977). Innovation processes. V.H. Winston & Sons.

48. Salehi, H. (1967). Application of the Hellinger integrals to q–variate stationary stochastic processes. Ark. Mat. 7, 305–311.

49. Salehi, H. (1979). Algorithms for linear interpolation and interpolator error for minimal stationary stochastic processes. Ann Probability 7, 840–846.

50. Salehi, H. (1981). The continuation of Wiener's work on q–variate linear prediction and its extension to infinite–dimensional spaces. Norbert Wiener: Collected Works, vol. III, ed. P. Masani, MIT Press, 307–338.

51. Salehi, H., Scheidt, J.K. (1972). Interpolation of q–variate weakly stationary processes over locally compact Abelian group. J. Multivariate Anal. 2, 307–331.

52. Sz.–Nagy, B. (1965). Prolongements des transformations de l'espace de Hilbert qui sortent de cet espace. Appendix F. Riesz at B. Sz.–Nagy, Lecons d'analyse functionelle. Paris–Budapest.

53. Sz.–Nagy, B., Foias, C. (1970). Harmonic analysis of operators on Hilbert space. North–Holland. Amsterdam.

54. Schmidt, F. (1983). Banach space valued stationary processes on partially ordered groups. rediction Theory and Harmonic Analysis. The Pesi Masani Volume. Eds. V. Mandrekar and H. Salehi. North Holland. Amsterdam.

55. Truong–Van, B. (1985). Applications of autoreproducing kernel grammian moduli to $\mathcal{Y}$(U,H)–valued stationary random functions. J. Mult. Anal. 17(1), 56–75.

56. Weron, A. (1974). On characterization of interpolable and minimal stationary processes. Studia Math. 49, 165–183.

57. Weron, A. (1980). Second order stochastic processes and the dilation theory in Banach spaces. Ann. Inst. H. Poincare 16, 29–38.

58. Yaglom, A.M. (1949). On the problem of linear interpolation of stationary stochastic sequences and processes. Uspehi Mat. Nauk 4(4), 173–178, in Russian.

Probability Theory on Vector Spaces IV
Lancut, June '87, Springer's LNM

ADMISSIBLE AND SINGULAR TRANSLATES OF STABLE PROCESSES*

Mauro Marques and Stamatis Cambanis
Statistics Department, University of North Carolina
Chapel Hill, N.C. 27599-3260, U.S.A.

1. INTRODUCTION

This work investigates the equivalence and singularity of measures induced by non-Gaussian stable processes and their translates. For non-Gaussian measures in Hilbert space, these questions were studied in [15] for infinitely divisible measures and in [28] and [26] for stable measures. Sufficient conditions for an element to be an admissible translate of an infinitely divisible measure in a Hilbert space were obtained in [15]. However, as observed in [28], these conditions are difficult to verify and, as simplified for stable measures, they were found to be false. The structure of the set of admissible translates of symmetric stable measures in a Hilbert space was investigated in [28]. The admissible translates of symmetric stable measures with discrete spectral measures in a Banach space were characterized in [26].

For p^{th} order and for symmetric stable processes a function space is introduced in Section 2 which plays a role partly analogous to the reproducing kernel Hilbert space of a Gaussian or second order process. In particular this space provides an upper bound for the set of admissible translates, is a stochastic processes version of a space introduced in [28, p. 249], and extends the results of [28, Prop. 10] to general symmetric stable processes.

In Section 3, stable processes with an invertible spectral representation are considered. Their admissible translates are characterized, and a dichotomy is established: each translate is either admissible or singular. The result is applied to show that most continuous time moving averages, and all harmonizable processes with nonatomic spectral measure have no admissible translate. Thus these processes do not provide realistic models for additive noise, as every nonrandom signal can be perfectly detected in their presence. For general harmonizable processes and for invertible discrete time mixed autoregressive moving averages, the set of admissible translates is characterized.

*Research supported by the Air Force Office of Scientific Research Contract No. F49620-85-C-0144.

Background and Notation

The following setting is considered. $X = (X(t) = X(t,\omega);\ t \in \mathbf{T})$ is a stochastic process on a probability space $(\Omega, \mathcal{F}, P)$ with parameter set $\mathbf{T}$ and real or complex values, i.e., values in $\mathbf{X} = \mathbf{R}$ or $\mathbf{C}$. When $X(t) \in L_p(P)$ for all $t \in \mathbf{T}$ and some $p > 0$, X is called a pth order process. The linear space $\mathcal{L}(X)$ of a pth order process X is the $L_p(P)$ completion of the set of finite linear combinations of its random variables $l(X) \triangleq sp\{X(t);\ t \in \mathbf{T}\}$. $\overline{\mathbf{X}}^{\mathbf{T}}$ denotes the set of all extended $\mathbf{X}$-valued (i.e., real or complex valued) functions on $\mathbf{T}$, $\mathcal{C} = \mathcal{C}(\overline{\mathbf{X}}^{\mathbf{T}})$ the σ-field generated by the cylinder sets of $\overline{\mathbf{X}}^{\mathbf{T}}$, and μ_X the distribution of the process X, i.e. the probability induced on $\mathcal{C}$ by X: $\mu_X(C) = P(\{\omega;\ X(\cdot,\omega) \in C\})$.

For a nonrandom real or complex function s on $\mathbf{T}$, we are interested in the Lebesgue decomposition of the distribution μ_{s+X} of s+X with respect to the distribution μ_X of X, and in particular in conditions for μ_{s+X} and μ_X to be singular ($\mu_{s+X} \perp \mu_X$), and for μ_{s+X} to be absolutely continuous with respect μ_X ($\mu_{s+X} \ll \mu_X$). The function s is then called a singular or admissible translate of X respectively.

Here we focus on symmetric α-stable (SαS) processes. A real random variable X is SαS, $0 < \alpha \le 2$, with scale parameter $\|X\|_\alpha \in (0,\infty)$ if $E\{\exp(iuX)\} = \exp\{-\|X\|_\alpha^\alpha |u|^\alpha\}$. A real random vector $(X_1,\dots, X_n)$ is SαS if all linear combinations $\sum_{k=1}^n a_k X_k$ are SαS. Similarly a real stochastic process $X = (X(t);\ t \in \mathbf{T})$ is SαS if all linear combinations $\sum_{k=1}^n a_n X(t_k)$ are SαS random variables. When $\alpha = 2$ we have zero mean Gaussian random variables, vectors and processes respectively. When $0 < \alpha < 2$, the tails of the distributions are heavier and only moments of order $p \in (0,\alpha)$ are finite with $\{E(|X|^p)\}^{1/p} = C_{p,\alpha}\|X\|_\alpha$, where the constant $C_{p,\alpha}$ is independent of X. Thus a SαS process X is p^{th} order for all $0 < p < \alpha$, and its linear space $\mathcal{L}(X)$ does not depend on p and is the completion of $l(X)$ with respect to $\|\cdot\|_\alpha^{1\wedge\alpha}$, which in fact metrizes convergence in probability [23].

An important class of SαS processes consists of independently scattered SαS measures, which extend the concept of a stochastic process with independent increments to more general parameter spaces. Let $\mathbf{I}$ be an arbitrary set and $\mathcal{I}$ a δ-ring of subsets of $\mathbf{I}$ with the property that there exists an increasing sequence $(\mathbf{I}_n;\ n \in \mathbf{N})$ in $\mathcal{I}$ with $\cup_n I_n = \mathbf{I}$. A real process $Z = (Z(A);\ A \in \mathcal{I})$ is called an independently scattered SαS measure if for every sequence $(A_n;\ n \in \mathbf{N})$ of disjoint sets in $\mathcal{I}$, the random variables $\{Z(A_n);\ n \in \mathbf{N}\}$ are independent, and whenever $\cup_n A_n \in \mathcal{I}$ then $Z(\cup_n A_n) = \Sigma_n Z(A_n)$ a.s., and for every $A \in \mathcal{I}$, $Z(A)$ is a SαS random variable, i.e. $E\{\exp(iuZ(A))\} = \exp\{-m(A)|u|^\alpha\}$ where $m(A) = \|Z(A)\|_\alpha^\alpha$. Then m is a measure on $\mathcal{I}$ which extends uniquely to a σ-finite measure on $\sigma(\mathcal{I})$, and is called the control measure of Z. Conversely, the existence of a SαS independently scattered measure with a given control measure is a consequence of Komogorov's consistency theorem. When $\mathbf{I} = \mathbf{R}$ and the control measure m of Z is Lebesgue measure, then the process $X(t) = \text{sign}(t)$

$Z((0\wedge t, 0\vee t])$ has stationary independent increments and is called SαS motion.

For any function $f \in L_\alpha(\mathbf{I},\sigma(\mathfrak{I}),m) = L_\alpha(m)$ the stochastic integral $\int_{\mathbf{I}} f dZ$ can be defined in the usual way and is a SαS random variable with $\|\int_{\mathbf{I}} f dZ\|_\alpha = \|f\|_{L_\alpha(m)}$. The stochastic integral map $f \to \int_{\mathbf{I}} f dZ$ from $L_\alpha(m)$ into $\mathcal{L}(Z)$ is an isometry and

$$\mathcal{L}(Z) = \{ \int f dZ;\ f \in L_\alpha(m) \}. \tag{1.1}$$

It allows the construction of SαS processes with generally dependent values by means of the spectral representation

$$X(t) = \int_{\mathbf{I}} f(t,u) Z(du), \quad t \in \mathbf{T}, \tag{1.2}$$

where $\{f(t,\cdot);\ t \in \mathbf{T}\} \subset L_\alpha(m)$. In fact every S$\alpha$S process X has such a spectral representation in law (see, e.g., [20] and [17]), in the sense that for some family $\{f(t,\cdot),\ t \in \mathbf{T}\}$ in some $L_\alpha(m)$,

$$(X(t);\ t \in \mathbf{T}) \stackrel{\mathcal{L}}{=} (\int_{\mathbf{I}} f(t,u) Z(du);\ t \in \mathbf{T})\ . \tag{1.3}$$

The covariation $[X,Y]_\alpha$ of two jointly SαS random variables X and Y with $1 < \alpha \leq 2$ is defined by

$$\frac{[X,Y]_\alpha}{\|Y\|_\alpha^\alpha} = \frac{E(XY^{<p-1>})}{E(|Y|^p)}, \tag{1.4}$$

which holds for all $0 < p < \alpha$, where $y^{<q>} = |y|^{q-1}y$, $q > 0$ (see, e.g., [7]). It follows that $\|X\|_\alpha^\alpha = [X,X]_\alpha$. If X and Y have representations $\int_{\mathbf{I}} f dZ$ and $\int_{\mathbf{I}} g dZ$ respectively then $[X,Y]_\alpha = \int_{\mathbf{I}} f g^{<\alpha-1>} dm$.

In certain cases, such as when working with Fourier transforms, it is more natural and convenient to work with complex valued processes. A complex SαS random variable is defined as having jointly SαS real and imaginary parts. Except for the representation of the characteristic function, all concepts and results considered in this section for real SαS random variables and processes extend to the complex case (see, e.g., [5] and [7]).

2. AN UPPER BOUND FOR THE SET OF ADMISSIBLE TRANSLATES

A space of functions associated with a pth order, $0 < p \leq 2$, stochastic process is introduced and seen as a partial analog of the reproducing kernel Hilbert space (RKHS) of a second order process. We concentrate only on values of p in (0,2] because for $p \geq 2$ the second order theory is applicable.

The *function space* $\mathsf{F} = \mathsf{F}(X)$ of a pth order process $X = (X(t);\ t \in \mathbf{T})$ with $0 < p \leq 2$ is the set of all functions s on $\mathbf{T}$ such that

$$\|s\|_{\mathsf{F}} \triangleq \sup \frac{|\Sigma_{n=1}^{N} a_n s(t_n)|}{[E|\Sigma_{n=1}^{N} a_n X(t_n)|^p]^{1/p}} < \infty, \tag{2.1}$$

where the supremum is taken over all $N \in \mathbf{N}$, $a_1,\ldots,a_N \in \mathbf{X}$ and $t_1,\ldots,t_N \in \mathbf{T}$.

When $p = 2$, i.e., X is a second order process with correlation function $R(t,s) = E(X(t)\overline{X}(s))$, then F is precisely the RKHS of X (or of R), and consists of all functions s of the form $s(t) = E(X(t)\overline{Y})$, $t \in \mathbf{T}$, $Y \in \mathcal{L}(X)$. If $s_i(t) = E(X(t)\overline{Y}_i)$ then $\langle s_1,s_2\rangle_{\mathsf{F}} = E(Y_1\overline{Y}_2)$ defines an inner product and R is a reproducing kernel: $s(t) = \langle s,R(\cdot,t)\rangle_{\mathsf{F}}$, $t \in \mathbf{T}$, $s \in \mathsf{F}$ (See, e.g., [14]).

When $1 < p \leq 2$, a representation is known for the bounded linear functionals on the linear space of X, analogous to the Riesz representation for bounded linear functionals on a Hilbert space. This allows us to express the functions in F in terms of moments of the process X.

<u>Proposition 1</u>. Let $X = (X(t);\ t \in \mathbf{T})$ be a pth order process with $1 < p \leq 2$. Then the following are equivalent:

i) $s \in \mathsf{F}$.

ii) $s(t) = E(X(t)Y^{<p-1>})$ for all $t \in \mathbf{T}$ and some $Y \in \mathcal{L}(X)$.

iii) $s(t) = E(X(t)\overline{W})$ for all $t \in \mathbf{T}$ and some $W \in L_{p^*}(P)$ where $1/p+1/p^* = 1$.

Moreover the following properties hold.

a) If $s \in \mathsf{F}$, then Y in ii) is unique and satisfies $\|s\|_{\mathsf{F}} = \|Y\|_{L_p(P)}^{p-1}$, and there exists a unique $W \in L_{p^*}(P)$ (namely $\overline{W} = Y^{<p-1>}$) satisfying iii) and $\|s\|_{\mathsf{F}} = \|W\|_{L_{p^*}(P)}$.

b) $(\mathsf{F},\|\cdot\|_{\mathsf{F}})$ is a Banach space isometrically isomorphic to the quotient space $L_{p^*}(P)/l(X)^{\perp}$, where $l(X)^{\perp}$ denotes the annihilator of $l(X)$.

<u>Proof.</u> i) $\Rightarrow$ ii) follows by observing that if $\|s\|_{\mathsf{F}} < \infty$, then $\psi_s(\Sigma_{n=1}^{N} a_n X(t_n)) = \Sigma_{n=1}^{N} a_n s(t_n)$ defines a bounded linear functional on $\mathcal{L}(X)$ with norm $\|s\|_{\mathsf{F}}$. From [8, Prop. 2.1], there exists a unique $Y \in \mathcal{L}(X)$ such that $\psi_s(\cdot) = E(\,\cdot\, Y^{<p-1>})$ and $\|\psi_s\|_{\mathcal{L}(X)^*} = \|Y\|_{L_p(P)}^{p-1}$. Thus $s(t) = \psi_s(X(t)) = E(X(t)Y^{<p-1>})$.

ii) $\Rightarrow$ iii). If $Y \in \mathcal{L}(X)$ then $\overline{W} = Y^{<p-1>} \in L_{p^*}(P)$ and $\|W\|_{L_{p^*}(P)} = \|Y\|_{L_p(P)}^{p-1}$. Also $s(t) = E(X(t)Y^{<p-1>}) = E(X(t)\overline{W})$.

iii) $\Rightarrow$ i). If $s(t) = E(X(t)\overline{W})$ then it is clear from its definition that $\|s\|_F$ is finite.

a) Let $s \in F$. The part on Y follows as in the proof of i) $\Rightarrow$ ii). By iii) there exist $Z \in L_{p*}(P)$ such that $s(t) = E(X(t)\overline{Z})$. Let $l(X)^\perp$ be the closed linear space $\{Z' \in L_{p*}(\Omega);\ E(Z'\overline{Y}) = 0,\ Y \in l(X)\}$, and let Z_0 be the best approximation of X in $l(X)^\perp$, i.e., $\|Z-Z_0\|_{L_{p*}(P)} = \inf\{\ \|Z-Z'\|_{L_{p*}(P)};\ Z' \in l(X)^\perp\}$. Such a $Z_0 \in l(X)^\perp$ exists and is unique ([25, Cor. 3.5 and Th. 1.11). Set $W = Z-Z_0$. Then $E(Z\overline{Y}) = E(W\overline{Y})$ for all $Y \in l(X)$. If Z' is such that $E(Z'\overline{Y}) = E(Z\overline{Y})$ for all $Y \in l(X)$, then $Z-Z' \in l(X)^\perp$ and

$$\|W\|_{L_{p*}(P)} = \|Z-Z_0\|_{L_{p*}(P)} \leq \|Z-(Z-Z')\|_{L_{p*}(P)} = \|Z'\|_{L_{p*}(P)}.$$

Thus if $s(t) = E(X(t)\overline{W}')$, $W' \in L_{p*}(P)$, and $\|s\|_F = \|W'\|_{L_{p*}(P)}$ we must have $\|W\|_{L_{p*}(P)} \leq \|W'\|_{L_{p*}(P)}$. On the other hand

$$\|W'\|_{L_{p*}(P)} = \|s\|_F = \sup \frac{|E(\sum_{n=1}^N a_n X(t_n) W)|}{\|\sum_{n=1}^N a_n X(t_n)\|_{L_p(P)}} \leq \|W\|_{L_{p*}(P)}.$$

Therefore $\|W\|_{L_{p*}(P)} = \|W'\|_{L_{p*}(P)} = \|s\|_F$. Putting $V = Z-W'$ we have $V \in l(X)^\perp$ and

$$\|Z-Z_0\|_{L_{p*}(P)} = \|W\|_{L_{p*}(P)} = \|W'\|_{L_{p*}(P)} = \|Z-V\|_{L_{p*}(P)}.$$

Thus the unicity of Z_0 implies $W = W'$. Since $Y^{<p-1>} \in L_{p*}(P)$ and $\|s\|_F = \|Y\|^{<p-1>}_{L_{p*}(P)}$ for $s(t) = E(X(t)Y^{<p-1>})$ we must have $\overline{W} = Y^{<p-1>}$.

b) That $(F,\|\cdot\|_F)$ is a normed linear space is clear. To show that F is isometrically isomorphic to $L_{p*}(P)/l(X)^\perp$, let $s_i \in F$, $s_i(t) = E(X(t)\overline{W}_i)$, i=1,2, and $(s_1+s_2)(t) = E(X(t)\overline{W})$, where W_1, W_2 and W are the unique elements in $L_{p*}(P)$ such that $\|s_i\|_F = \|W_i\|_{L_{p*}(P)}$, and $\|s_1+s_2\|_F = \|W\|_{L_{p*}(P)}$. Since $E(X(t)W) = (s_1+s_2)(t) = E(X(t)(W_1+W_2))$, we have $W-(W_1+W_2) \in l(X)^\perp$, i.e., $[W] = [W_1+W_2] = [W_1]+[W_2]$, where $[\ \cdot\ ]$ denotes an equivalence class in $L_{p*}(P)/l(X)^\perp$. Similarly if $s(t) = E(X(t)\overline{W})$ and $(as)(t) = E(X(t)\tilde{W})$ we have $[\tilde{W}] = [aW] = a[W]$. Hence the map $s \to [W]$ is linear and since $\|[W]\|_{L_{p*}(P)/l(X)^\perp} = \|W\|_{L_{p*}(P)} = \|s\|_F$ it is an isometric isomorphism.

To finish the proof of c) we need to show that F is complete. Let $(s_k;\ k \in N)$ be a sequence in F such that $\sum_{k=1}^\infty \|s_k\|_F < \infty$ and let $W_k \in L_{p*}(P)$ be such that $\|W_k\|_{L_{p*}(P)} = \|s_k\|_F$. Hence $\sum_{k=1}^\infty \|W_k\|_{L_{p*}(P)} < \infty$ and $W = \sum_{k=1}^\infty W_k \in L_{p*}(P)$. Set $s(t) = E(X(t)\overline{W})$. Thus

$$|\Sigma_{n=1}^{N} a_n(\Sigma_{k=1}^{K} s_k - s)(t_n)| \le \|\Sigma_{k=1}^{K} W_k - W\|_{L_{p*}(P)} \|\Sigma_{n=1}^{N} a_n X(t_n)\|_{L_p(P)}$$

and

$$\|\Sigma_{k=1}^{K} s_k - s\|_{\mathsf{F}} \le \|\Sigma_{k=1}^{K} W_k - W\|_{L_{p*}(P)} \to 0 \quad \text{as } K \to \infty,$$

i.e., $\Sigma_{k=1}^{\infty} s_n \in \mathsf{F}$ proving that F is complete. □

If the process X is SαS with $1 < \alpha < 2$, then it is of pth order for each $p \in (1,\alpha)$ and its function space F does not depend on p but only on α. Furthermore the functions in F can be expressed in terms of the spectral representation of the process.

Corollary 1. Let $X = (X(t);\ t \in T)$ be a SαS process with $1 < \alpha < 2$ and spectral representation as in (1.2). Then the following are equivalent.

i) $s \in \mathsf{F}$.

ii) $s(t) = [X(t),Y]_\alpha$ for all $t \in T$ and some $Y \in \mathcal{L}(X)$.

iii) $s(t) = \int_I f(t,u)\bar{z}(u)m(du)$ for $t \in T$ and some $z \in L_{\alpha^*}(m)$ where $1/\alpha^* + 1/\alpha = 1$.

Moreover the following properties hold.

a) If $s \in \mathsf{F}$, then Y in ii) is unique and satisfies $\|s\|_{\mathsf{F}} = \|Y\|_\alpha^{\alpha-1}$, and there exists a unique $z \in L_{\alpha^*}(m)$ satisfying iii) and $\|s\|_{\mathsf{F}} = \|z\|_{L_{\alpha^*}(m)}$.

b) The map $s \to [z]$ from F into $L_{\alpha^*}(m)/l(f)^\perp$ is an isometric isomorphism, where $[\,\cdot\,]$ denotes equivalence class and $l(f) = sp\{f(t,\cdot);\ t \in T\}$.

Proof. i) ⇔ ii). It follows from 1.3 that for all $p \in (1,\alpha)$, $[\cdot, Y]_\alpha = E(\,\cdot\, Z^{<p-1>})$, where $Z = C_{p,\alpha}^{-p/(p-1)} \|Y\|_\alpha^{\alpha-p)/(p-1)} Y$ and $\|Z\|_{L_p(P)}^{p-1} = C_{p,\alpha}^{-1} \|Y\|_\alpha^{\alpha-1}$, so that $s \in \mathsf{F}$ if and only if $s(t) = [X(t),Y]_\alpha$ which does not depend on p.

ii) ⇒ iii). If $Y \in \mathcal{L}(X)$ then $Y = \int_I g dZ$ for some $g \in \mathcal{L}(f) = \overline{sp}\{f(t,\cdot);\ t \in T\}$, and

$$s(t) = [X(t),Y]_\alpha = [\int_I f(t)dZ, \int_I g dZ]_\alpha = \int_I f(t) g^{<\alpha-1>} dm = \int_I f(t) z dm,$$

where $z = g^{<\alpha-1>} \in L_{\alpha*}(m)$ and $\|z\|_{L_{\alpha*}(m)} = \|Y\|_\alpha^{\alpha-1}$.

The proofs of iii) ⇒ i), a) and b) are identical to those of Proposition 1. □

When X is second order, every translate $s \notin \mathsf{F}$ is singular [14, Th. 4.1]. Proposition 2 extends this to p^{th} order processes with $0 < p < 2$ (see also [12, Th. 1]).

Proposition 2. Let $X = (X(t);\ t \in T)$ be a pth order process with $0 < p \le 2$. If $s \notin \mathsf{F}$ then $\mu_{s+X} \perp \mu_X$. Consequently all admissible translates of X belong to F.

Proof. The proof is adapted from [22]. If $s \notin F$, then by (2.1), for each $n \in N$ we can choose N_n, $a_{n,k}$, $t_{n,k}$, $k = 1,\dots,N_n$, such that

$$\frac{|\Sigma_{k=1}^{N_n} a_{n,k} s(t_{n,k})|}{\|\Sigma_{k=1}^{N_n} a_{n,k} X(t_{n,k})\|_{L_p(P)}} \geq n^{1/p}.$$

Let $s_n = \Sigma_{k=1}^{N_n} a_{n,k} s(t_{n,k})$. Without loss of generality we can consider $s_n > 0$ for all n. Consider the random variables defined on $(X^T, \mathcal{C}, \mu_X)$ by $Y_n(x) = \Sigma_{k=1}^{N_n} a_{n,k} x(t_{n,k})$, $x \in X^T$. By the Markov inequality we have as $n \to \infty$,

$$\mu_X(Y_n \geq s_n/2) \leq \mu_X(|Y_n| \geq s_n/2) \leq 2^p \|Y_n\|_{L_p(P)}^p / s_n^p < 2^p/n \to 0,$$

$$\mu_{s+X}(Y_n \geq s_n/2) = \mu_X(Y_n \geq -s_n/2) \geq \mu_X(|Y_n| < s_n/2)$$

$$\geq 1 - 2^p \|Y_n\|_{L_p(P)}^p / s_n^p \geq 1 - 2^p/n \to 1.$$

Therefore $\mu_X \perp \mu_{s+X}$. □

In contrast with Gaussian processes whose set of admissible translates is always the entire RKHS F, the set of admissible translates of a SαS process with $0 < \alpha < 2$ may be as large as the entire function space F or as small as $\{0\}$, as is seen by the following examples.

Stable motion. If $X = (X(t);\ t \in [0,1])$ is a SαS motion, i.e., X has stationary independent SαS increments, it is known ([3, 15, 28]) that X has no nontrivial admissible translates for $0 < \alpha < 2$. On the other hand for $1 < \alpha < 2$, its function space F is the space of absolutely continuous functions with $s(0) = 0$ and derivative in $L_{\alpha^*}(\mathrm{Leb})$, with $\|s\|_F = \|s'\|_{L_{\alpha^*}(\mathrm{Leb})}$.

Sub-Gaussian processes. Let $X = (X(t);\ t \in T)$ be an α-sub-Gaussian process: $X(t) = A^{1/2}G(t)$, where A is a normalized positive $(\alpha/2)$-stable random variable independent of the Gaussian process $G = (G(t);\ t \in T)$ which has zero mean and covariance function R. It follows from [18] that the set of admissible translates of X coincides with the RKHS of G, once we observe that there the proof depends only on the representation of spherically invariant processes as scale mixtures of Gaussian processes and not on the existence of second moments. Moreover for any $Y \in \mathcal{L}(X)$, $[X(t),Y]_\alpha = 2^{\alpha/2}\{E(W^2)\}^{1-\alpha/2}E(G(t)W)$, where $W \in \mathcal{L}(G)$ is obtained from G by the same linear operation Y is obtained from X (see [8]). Therefore the function space F of X coincides with the RKHS of G and thus also with the set of admissible translates of X.

Stable processes as mixtures of Gaussian processes. It has been shown in [21] that every SαS process X is conditionally Gaussian with zero mean and random covariance function. Denoting by G_R a Gaussian process with mean zero and covariance function R, and by μ_{G_R}

its distribution, we have that for every SαS process X there exists a probability λ on the space $\Re$ of all covariance functions R such that $\mu_X(E) = \int_{\Re} \mu_{G_R}(E)\lambda(dR)$ for all $E \in \mathfrak{C}$, and then $\mu_{s+X}(E) = \int_{\Re} \mu_{s+G_R}(E)\lambda(dR)$.

It follows that if s is an admissible translate of almost all G_R's, then it is an admissible translate of X too. This gives a lower bound for the set of admissible translates of X, namely

$$\bigcup_{\lambda(\Lambda)=0} \bigcap_{R\in\Re\setminus\Lambda} \text{RKHS}(R).$$

Thus a SαS process will have admissible translates if it is a mixture of Gaussian processes whose RKHS's have a common part. The converse does not seem to be necessarily true, i.e. an admissible translate of X may not be an admissible translate of almost all the Gaussian processes whose mixture is X. It also follows that a singular translate of X is a singular translate of almost all the Gaussian processes whose mixture is X, and furthermore the same event separates them. This gives an upper bound for the set of singular translates of X, namely

$$\bigcup_{\lambda(\Lambda)=0} \bigcap_{R\in\Re\setminus\Lambda} \text{RKHS}(R)^c.$$

When a SαS process is a mixture of Gaussian processes having the same RKHS then a proof similar to that of Proposition 2 shows that the Gaussian dichotomy prevails and every translate is either admissible or singular.

<u>Proposition 3</u>. Let the SαS process $X = (X(t);\ t \in \mathbf{T})$ be the λ-mixture of Gaussian processes $G_R = (G_R(t);\ t \in \mathbf{T})$ such that $\text{RKHS}(R) = H$ a.e. (λ). Then a translate of X is admissible or singular according as $s \in H$ or $s \notin H$.

The assumptions of Proposition 3 are satisfied when X is sub-Gaussian, i.e., X is the mixture of the mutually singular Gaussian process $a^{1/2}G$, $a > 0$, which have identical RKHS; or in the more general case where X is the mixture of Gaussian processes with random covariance function of the form $\Sigma_{n=1}^N A_n R_n(t,s)$, where the R_n's are fixed (nonrandom) covariances such that $R_n - c_{nm}R_m$ are nonnegative definite for all $n \neq m$, and some $0 < c_{nm} < \infty$, and the positive random variables $A_1,\dots,A_n$, are jointly ($\alpha/2$)-stable. The usefulness of these general remarks is limited by the fact that the only SαS mixtures of Gaussian processes, which are currently known explicitly, are the sub-Gaussian processes, and the more general finite sums $\Sigma_{n=1}^N A_n^{1/2} G_n$, where $(A_1,\dots,A_N)$ is positive ($\alpha/2$)-stable and independent of the mutually independent Gaussian processes $G_1,\dots,G_N$.

Further examples where the set of admissible translates is trivial or a proper subset of the function space $\mathbf{F}$ are presented in the next section. It should finally be recalled that the set of admissible translates of a SαS process is always a linear space, even if it is not the entire function space F [28, Cor. 5.1]. Also, from the linear structure we have that $\mu_{s+X} \ll \mu_X \Rightarrow \mu_X \ll \mu_{s+X}$ (see, e.g., [26]) so that for every admissible translate s, μ_{s+X} and μ_X are equivalent.

3. TRANSLATES OF INVERTIBLE PROCESSES

In this section we present some general results on the admissible translates of certain SαS processes with invertible spectral representation.

Let $X = (X(t);\ t \in \mathbf{T})$ be a SαS stochastic process with spectral representation as in (1.3.2). It follows from the continuity of the stochastic integral map $f \to \int f dZ$ and (1.3.1) that the representing functions $\{(f(t,\cdot);\ t \in \mathbf{T})\}$ are linearly dense in $L_\alpha(m)$, i.e. that $\mathcal{L}(f) = L_\alpha(m)$, where $\mathcal{L}(f)$ is the completion of $l(f) = sp\{(f(t,\cdot);\ t \in \mathbf{T})\}$ in $L_\alpha(m)$, if and only if $\mathcal{L}(X) = \mathcal{L}(Z)$. Processes satisfying this condition will be said to have an *invertible spectral representation* or more simply to be *invertible.*

Every Gaussian process is invertible [4, Th. 2]. This is not generally true for non-Gaussian SαS processes as can be seen from the fact that the linear space of a sub-Gaussian process does not contain (nontrivial) independent random variables [9, Lem. 2.1]. Necessary and sufficient conditions for a general SαS process to have an invertible spectral representation are given in [5, Th. 5.1 and 5.5]. A stronger form of invertibility for a nonanticipating SαS moving average is considered in [9, Lem. 3.1]. SαS processes with invertible spectral representation in $L_2([0,1]$, Leb), i.e. $L_2\text{-}\overline{sp}\{f(t,\cdot);\ t \in [0,1]\} = L_2([0,1]$, Leb), are considered in [28]; clearly such a process has also invertible spectral representation in $L_\alpha([0,1]$, Leb). Examples of invertible SαS processes will be presented in the sequel.

For invertible processes the problem of finding their admissible translates can be reduced to finding the admissible translates of the independently scattered random measure Z, which we now consider first.

The next proposition is essentially based on [15, Th. 7.3] and on Shepp's characterization of the admissible translates of an i.i.d. SαS sequence [24]. It extends to independently scattered SαS random measures with non-atomic control measure the result in [3] and [28] on admissible translates of independent increments processes in [0,T] which are stochastically continuous and have no Gaussian component. It establishes a dichotomy for the translates of a general independently scattered SαS random measure and it characterizes its admissible translates as those of its atomic component.

The following notation will be used in Proposition 4. Recall that if a σ-finite measure space $(\mathbf{I},\sigma(\mathfrak{I}),m)$ is such that $\sigma(\mathfrak{I})$ contains all single points sets (e.g., $\mathbf{I}$ is a Polish space, $\sigma(\mathfrak{I})$

its Borel sets, and $\mathfrak{I}$ the δ-ring of Borel sets with finite m-measure) then $m = m_d + m_{at}$ where m_d is diffuse (non-atomic) and m_{at} is purely atomic with $N \leq \infty$ atoms, say $A = \{a_n; n = 1,2,...,N\}$ [19]. Thus if $Z = (Z(B); B \in \mathfrak{I})$ is an independently scattered SαS random measure with control measure m, it can be expressed as $Z = Z_d + Z_{at}$, where Z_d and Z_{at} are independent SαS independently scattered random measures defined for all $B \in \mathfrak{I}$ by $Z_d(B) = Z(A^c \cap B)$ and $Z_{at}(B) = Z(A \cap B)$, and have control measures m_d and m_{at} respectively. The atomic component has a series expansion: $Z_{at}(B) = \Sigma_{n=1}^{N} 1_B(a_n) Z(\{a_n\})$. This can be normalized by using the i.i.d. standard SαS random variables $Z_n \triangleq Z(\{a_n\}) m^{-1/\alpha}(\{a_n\})$ with $E\{\exp(iuZ_n)\} = \exp(-|u|^\alpha)$, as follows: $Z_{at}(B) = \Sigma_{n=1}^{N} 1_B(a_n) m^{1/\alpha}(\{a_n\}) Z_n$.

Proposition 4. Let $Z = (Z(B); B \in \mathfrak{I})$ be an independently scattered SαS random measure with $0 < \alpha < 2$ and control measure $m = m_d + m_{at}$, and let $S = (S(B); B \in \mathfrak{I})$ be a set function. Then the following are equivalent.

i) S is an admissible translate of Z.

ii) S is an admissible translate of Z_{at}.

iii) S is concentrated on A: $S(B) = \Sigma_{n=1}^{N} S(\{a_n\}) 1_B(a_n)$, and $\Sigma_{n=1}^{N} |S(\{a_n\})|^2 / m^{2/\alpha}(\{a_n\}) < \infty$.

Furthermore a translate which is not admissible is singular.

Proof. Let ζ_a and ζ_d be the stochastic processes with parameter set $\mathfrak{I}$ defined on the probability space $(\overline{X}^{\mathfrak{I}}, \mathcal{C}(\overline{X}^{\mathfrak{I}}), \mu_Z)$ by $\zeta_d(B,x) = x(A^c \cap B)$ and $\zeta_{at}(B,x) = x(A \cap B)$, $x \in X^{\mathfrak{I}}$, $B \in \mathfrak{I}$. Clearly

$$\zeta_d(B,Z(\cdot,\omega)) = Z_d(B,\omega) \quad \text{and} \quad \zeta_{at}(B,Z(\cdot,\omega)) = Z(A\cap B,\omega) = Z_{at}(B,\omega) \quad \text{a.s. (P)} \tag{3.1}$$

so that ζ_{at} and ζ_d are independently scattered SαS measures with control measures m_{at} and m_d. Let ζ_{at} and ζ_d also denote the corresponding linear maps $x \rightarrow \zeta_{at}(\cdot,x)$ and $x \rightarrow \zeta_d(\cdot,x)$ from $\overline{X}^{\mathfrak{I}}$ into $\overline{X}^{\mathfrak{I}}$.

i) ⇒ ii). Suppose $\mu_{S+Z} \ll \mu_Z$. Hence by Proposition 2, $S \in \mathbf{F}$ and by definition of $\mathbf{F}$ the map F: $\mathcal{L}(Z) \rightarrow \mathbf{X}$ defined by $F(\Sigma_{k=1}^{n} a_k Z(A_k)) = \Sigma_{k=1}^{n} a_k S(A_k)$ is a well defined linear functional so that S is a signed measures on $\mathfrak{I}$. Furthermore since $|S(B)| \leq C_{p,\alpha} \|S\|_F \|Z(B)\|_\alpha = C_{p,\alpha} \|S\|_F [m(B)]^{1/\alpha}$, S is absolutely continuous with respect to m, so that $S(B) = \int_B \bar{z} dm$ for some z locally in $L_1(m)$: $z 1_B \in L_1(m)$ for all $B \in \mathfrak{I}$.

It follows that $\mu_{S+Z} \zeta_d^{-1} \ll \mu_Z \zeta_d^{-1}$ or equivalently $\zeta_d(\cdot, S)$ is an admissible translate of the process ζ_d, since ζ_d is linear. Now

$$\zeta_d(B,S) = S(A^c \cap B) = \int_{A^c \cap B} \bar{z} dm = \int_B \bar{z} dm_d \triangleq S_d(B).$$

Since m_d is nonatomic it follows from a well known result [16, p. 174] that we can find measurable partitions $\{B_{j,k}(B)\colon k=1,2,\dots,K_j\}$, $j=1,2,\dots$, of each $B \in \mathcal{I}$ such that

$$\max_{1\le k\le K_j} m_d(B_{j,k}(B)) \to 0 \quad \text{as } j \to \infty. \tag{3.2}$$

For notational simplicity we will omit in the following the dependence on B. It follows that the triangular system of rowwise independent random variables $\{\zeta_d(B_{j,k});\ k=1,2,\dots,K_j,\ j=1,2,\dots\}$ is infinitesimal, i.e. for every $\epsilon > 0$,

$$\max_{1\le k\le K_j} \mu_Z(|\zeta_d(B_{j,k})| \ge \epsilon) \to 0 \quad \text{as } j \to \infty.$$

Hence, since for every j, $\zeta_d(B) = \Sigma_{k=1}^{K_j}\zeta_d(B_{j,k})$, we have, from the central limit theorem for triangular arrays and the fact that ζ_d has no Gaussian component, that

$$\liminf_{\epsilon\to 0}\ \liminf_{j\to\infty} \operatorname{Var}\{ \Sigma_{k=1}^{K_j}\zeta_d(B_{j,k})\, 1_{(-\epsilon,\epsilon)}(|\zeta_d(B_{j,k})|) \} = 0$$

(see, e.g., [1, Th. 4.7]). Thus by Chebyshev's inequality,

$$\Sigma_{k=1}^{K_j} \zeta_d(B_{j,k})\, 1_{(-\epsilon,\epsilon)}(|\zeta_d(B_{j,k})|)\} \to 0 \tag{3.3}$$

in μ_Z-probability (and in $L_p(\mu_Z)$, $p \in (0,\alpha)$) as $j\to\infty$ and $\epsilon\to 0$.

On the other hand, if $m_d(B_{j,k}) \to 0$ as $j \to \infty$ then $S_d(B_{j,k}) \to 0$, and hence for j large

$$S_d(B) = \Sigma_{k=1}^{K_j} S_d(B_{j,k}) = \Sigma_{k=1}^{K_j} S_d(B_{j,k})\, 1_{(-\epsilon,\epsilon)}(|S_d(B_{j,k})|).$$

Similarly

$$\Sigma_{k=1}^{K_j} [S_d(B_{j,k}) + \zeta_d(B_{j,k})]\, 1_{(-\epsilon,\epsilon)}(|S_d(B_{j,k}) + \zeta_d(B_{j,k})|) \to S_d(B). \tag{3.4}$$

in μ_Z-probability as $j\to\infty$ and $\epsilon\to 0$.

Define for $B \in \mathcal{I}$ the map $\phi(B,\cdot)\colon X^{\mathcal{I}} \to X^{\mathcal{I}}$ by

$$\phi(B,x) = \liminf_{\epsilon\to 0}\ \liminf_{j\to\infty} \Sigma_{k=1}^{K_j} x(B_{j,k})\, 1_{(-\epsilon,\epsilon)}(|x(B_{j,k})|). \tag{3.5}$$

Suppose S_d is not identically zero. Then there exists $B \in \mathcal{I}$ such that $S_d(B) \ne 0$. It follows from (3.3) and (3.4) that $\phi(B,\zeta_d(\cdot,x)) = 0$ and $\phi(B,S_d+\zeta_d(\cdot,x)) = S_d(B)$ a.e. (μ_Z). Thus $\mu_{S+Z}\phi^{-1}(B,\cdot) \perp \mu_Z\phi^{-1}(B,\cdot)$ and hence $\mu_{S+Z} \perp \mu_Z$ which is a contradiction. Therefore $S_d(B) = \int_B \bar{z}\, dm_d = 0$ for all $B \in \mathcal{I}$, i.e. $z = 0$ a.e. (m_d), so that

$$S(B) = \int_B \bar{z}\, dm_{at} = \Sigma_{n=1}^{N} \bar{z}(a_n) 1_B(a_n) m(\{a_n\}). \tag{3.6}$$

Reasoning as before we have $\mu_{S+Z}\zeta_{at}^{-1} \ll \mu_Z\zeta_{at}^{-1}$, i.e. $\zeta_{at}(\cdot, S)$ is an admissible translate of ζ_{at} (or Z_{at}), and by (3.6),

$$\zeta_{at}(B,S) = S(A\cap B) = \int_{A\cap B} \bar{z}dm = \int_B \bar{z}dm_{at} = S(B),$$

i.e., $S = \zeta_{at}(\cdot,S)$ is an admissible translate of Z_{at}.

ii) $\Rightarrow$ i). Suppose S is an admissible translate of Z_{at}. Since $Z = Z_d+Z_{at}$ and Z_{at} and Z_d are independent we have $\mu_Z = \mu_{Z_d} * \mu_{Z_{at}}$. Then $\mu_{S+Z_{at}} \ll \mu_{Z_{at}}$ implies $\mu_{S+Z} \ll \mu_Z$. Indeed $0 = \mu_Z(B) = \int \mu_{Z_{at}}(B-x)\mu_{Z_d}(dx)$ implies $\mu_{Z_{at}}(B-x) = 0$ a.e. (μ_{Z_d}), hence $0 = \mu_{S+Z_{at}}(B-x) = \mu_{Z_{at}}(B-S-x)$ a.e. (μ_{Z_d}) and thus $\mu_{S+Z}(B) = \mu_Z(B-S) = \int \mu_{Z_{at}}(B-S-x)\mu_{Z_d}(dx) = 0.$

ii) $\Rightarrow$ iii). Because $S \in \mathbf{F}$, S is absolutely continuous with respect to m_d, $S(B) = \Sigma_{n=1}^N S(\{a_n\})1_B(a_n)$. Let $\psi\colon \bar{X}^{\mathfrak{J}} \to \bar{X}^{\mathcal{N}}$, where $\mathcal{N} = \{1,\dots,N\}$ if $N < \infty$ and $\mathcal{N} = \mathbf{N}$ otherwise, be defined by

$$[\psi(x)](n) = \psi(n,x) = \zeta_{at}(\{a_n\},x)/m^{1/\alpha}(\{a_n\}).$$

Thus, $\psi(n,\cdot)$, $n \in \mathcal{N}$, are standard SαS i.i.d. random variables,

$$\psi(n,S) = S(\{a_n\})/m^{1/\alpha}(\{a_n\}), \qquad \psi(n,S+x) = \psi(n,S)+\psi(n,x).$$

Now $\mu_{S+Z_{at}} \ll \mu_{Z_{at}}$ implies $\mu_{S+Z_{at}}\psi^{-1} \ll \mu_{Z_{at}}\psi^{-1}$, i.e., $(S(\{a_n\})/m^{1/\alpha}(\{a_n\});$ $n \in \mathcal{N})$ is an admissible translate of $(\psi(n,\cdot);\ n \in \mathcal{N})$ defined on the probability space $(\bar{X}^{\mathfrak{J}}, \mathcal{C}(\bar{X}^{\mathfrak{J}}),\mu_Z)$. It follows from [24] if $N = \infty$ and trivially if $N < \infty$ that $\Sigma_{n=1}^N S^2(\{a_n\})/m^{2/\alpha}(\{a_n\}) < \infty$.

iii) $\Rightarrow$ ii). Conversely, if $\Sigma_{n=1}^N S^2(\{a_n\})/m^{2/\alpha}(\{a_n\}) < \infty$ it follows from [24] and the fact that stable densities have finite Fisher information [13] that $(S(\{a_n\})/m^{1/\alpha}(\{a_n\});\ n \in \mathcal{N})$ is an admissible translate of $(\psi(n,\cdot);\ n \in \mathcal{N})$ (the result is trivial if $N < \infty$). Therefore $\Sigma_{n=1}^N S(\{a_n\})1_B(\{a_n\}) = S(B)$ is an admissible translate of the process

$$\Sigma_{n=1}^N 1_B(\{a_n\})\, m^{1/\alpha}(\{a_n\})\, \psi(n,x) = \Sigma_{n=1}^N 1_B(\{a_n\})\, \zeta(\{a_n\},x) = \zeta_{at}(B,x)$$

and hence of Z_{at}.

To prove that a translate S which is not admissible is singular it suffices to take $S \in \mathbf{F}$, and from the proof of i) $\Rightarrow$ ii), of the form $S(B) = \int_B \bar{z}dm$. If $m_d(|z| > 0) > 0$ then $\mu_{S+Z} \perp \mu_Z$. Thus assume

$$S(B) = \int_B \bar{z}\,dm_{at} = \Sigma_{n=1}^{N} S(\{a_n\}) 1_B(a_n).$$

Since it is not admissible, by iii), $N = \infty$ and $\sum_{n=1}^{\infty} S^2(\{a_n\})/m^{2/\alpha}(\{a_n\}) = \infty$. Hence from [24], $(S(\{a_n\})/m^{1/\alpha}(\{a_n\});\ n \in \mathbb{N})$ is a singular translate of $(\psi(n,\cdot);\ n \in \mathbb{N})$, i.e. $\mu_{S+Z}\psi^{-1} \perp \mu_Z\psi^{-1}$, which implies $\mu_{S+Z} \perp \mu_Z$. □

It follows that the admissible translates of a SαS independently scattered random measure are quite different in the Gaussian and non-Gaussian cases. Indeed, for Z Gaussian ($\alpha = 2$) every element in its function space (i.e. its RKHS):

$$\begin{aligned} \mathsf{F}_2 &= \{\, S;\ S(B) = \int_B \bar{z}\,dm,\ \bar{z} \in L_2(m) \,\} \\ &= \{\, S;\ S \text{ signed measure on } \sigma(\mathfrak{I}),\ S \ll m,\ \tfrac{dS}{dm} \in L_2(m) \,\} \end{aligned}$$

(see, e.g., [10]) is an admissible translate. In contrast for Z non-Gaussian with $1 < \alpha < 2$ its only admissible translates are

$$\{\, S;\ S \text{ signed measure on } \sigma(\mathfrak{I}),\ S \ll m_{at},\ \frac{dS}{dm_{at}} \in L_2(m_{at}^{2/\alpha^*}) \,\},$$

and they form a proper subset of its function space:

$$\mathsf{F}_\alpha = \{\, S;\ S \text{ signed measure on } \sigma(\mathfrak{I}),\ S \ll m,\ \frac{dS}{dm} \in L_{\alpha^*}(m) \,\}.$$

In particular, while a diffuse Gaussian measure has a rich class of admissible translates, a diffuse non-Gaussian SαS measure has no admissible translate whatever. On the other hand, if m (or Z) is atomic ($m_d = 0$), Proposition 4 holds for $0 < \alpha \le 2$, since in the Gaussian case, $\alpha = 2$, the condition in iii) is equivalent to $\frac{dS}{dm} \in L_2(m)$.

The results of Proposition 4 can now be used to obtain a dichotomy for the translates of an invertible SαS process, and to characterize its admissible translates as those of its atomic component. In order to state the result for a SαS process $X(t) = \int_I f(t,u)Z(du)$ with control measure m, we introduce its independent SαS diffuse and atomic components:

$$X_d(t) = \int_I f(t,u)Z_d(t), \qquad X_{at}(t) = X(t) - X_d(t) = \int_I f(t,u)Z_{at}(du).$$

The atomic component X_{at} has a series expansion $X_{at}(t) = \Sigma_{n=1}^{N} f(t,a_n)Z(\{a_n\}) = \Sigma_{n=1}^{N} f_n(t)Z_n$, where $Z_n \triangleq Z(\{a_n\})/m^{1/\alpha}(\{a_n\})$ and $f_n(t) \triangleq f(t,a_n)m^{1/\alpha}(\{a_n\})$, so that the Z_n's are standard SαS i.i.d. random variables, and for all $t \in \mathbf{T}$, $\Sigma_{n=1}^{N} |f_n(t)|^\alpha < \infty$.

Proposition 5. Let $X = (X(t);\ t \in \mathbf{T})$ be a SαS process with $0 < \alpha < 2$, invertible spectral representation $X(t) = \int_I f(t,u)Z(du)$ and control measure m, and let $s = (s(t);\ t \in \mathbf{T})$ be a function on $\mathbf{T}$. Then the following are equivalent:

i) s is an admissible translate of X.

ii) s is an admissible translate of X_{at}.

iii) $s(t) = \Sigma_{n=1}^{N} s_n' f(t,a_n) = \Sigma_{n=1}^{N} s_n f_n(t)$ with $\Sigma_{n=1}^{N} |s_n'|^2/m^{2/\alpha}(\{a_n\}) < \infty$, $\Sigma_{n=1}^{N} |s_n|^2 < \infty$.

Furthermore a translate which is not admissible is singular.

Proof. Since $1_B \in L_\alpha(m) = \mathcal{L}(f)$, for any $B \in \mathcal{S}$, there exist $\phi_n(B,\cdot) \in \ell(f)$, $n = 1,2,\ldots$, i.e., $\phi_n(B,\cdot) = \Sigma_{n=1}^{N_n(B)} a_{n,k}(B)\, f(t_{n,k}(B),\cdot)$, such that $\phi_n(B,\cdot) \to 1_B(\cdot)$ in $L_\alpha(m)$ as $n \to \infty$. Define

$$\phi_n(B,x) = \sum_{k=1}^{N_n(B)} a_{n,k}(B)\, x(t_{n,k}(B)), \quad x \in X^T.$$

Then

$$\phi_n(B,X(\cdot,\omega)) = \sum_{n=1}^{N_n(B)} a_{n,k}(B)\, X(t_{n,k}(B),\omega) \tag{3.7}$$

$$= \int_I \phi_n(B,u) Z(du,\omega) \to \int_I 1_B(u) Z(du,\omega) = Z(B,\omega)$$

in L_p (hence in probability) as $n\to\infty$. Thus $(\phi_n(B,\cdot);\ n \in \mathbb{N})$ converges in μ_X-measure. Let $(\phi_{n_k}(B,\cdot);\ k \in \mathbb{N})$ be a subsequence converging a.e. (μ_X) and define

$$\tilde{Z}(B) = \tilde{Z}(B,\cdot) = \liminf_{k\to\infty} \phi_{n_k}(B,\cdot)\, 1_{\{x;\ \phi_{n_k}(B,x) \text{ converges}\}}(\cdot).$$

$\tilde{Z}(B,\cdot)$ is a $\mathcal{C}$-measurable function on X^T for each $B \in \mathcal{S}$. Hence $\tilde{Z} = (\tilde{Z}(B);\ B \in \mathcal{S})$ is a stochastic process on the probability space $(X^T,\mathcal{C},\mu_X)$, and from (3.7), $\tilde{Z}(B,X(\cdot,\omega)) = Z(B,\omega)$ a.s., so that $\tilde{Z}$ is equal in law to Z, i.e., $\tilde{Z}$ is an independently scattered SαS measure with control measure m.

i) $\Rightarrow$ ii). Let s be an admissible translate of X. From Proposition 2, $s \in \mathbf{F}$, and thus for $p \in (0,\alpha)$,

$$|\Sigma_{k=1}^{n} a_k s(t_k)| \le \|s\|_{\mathbf{F}} \|\Sigma_{k=1}^{n} a_k X(t_k)\|_{L_p(P)}.$$

Hence as in Proposition 4, $F[\Sigma_{k=1}^{n} a_k X(t_k)] = \Sigma_{k=1}^{n} a_k s(t_k)$ is a well defined continuous linear functional on $\mathcal{L}(X)$ and $s(t) = F[X(t)]$. Thus by (3.7),

$$\phi_n(B,s) = \sum_{k=1}^{N_n(B)} a_{n,k}(B)\, s(t_{n,k}(B))$$

$$= F[\sum_{k=1}^{N_n(B)} a_{n,k}(B)\, X(t_{n,k}(B)] \to F[Z(B)] \quad \text{as } n\to\infty.$$

Hence for all $B \in \mathcal{S}$,

$$\tilde{Z}(B,s) = F[Z(B)] \quad \text{and} \quad \tilde{Z}(B,s+x) = \tilde{Z}(B,s) + \tilde{Z}(B,x). \tag{3.8}$$

Now if $\tilde{Z}_d(B,\cdot) = \tilde{Z}(A^c \cap B,\cdot)$, then $\tilde{Z}_d = (\tilde{Z}_d(B,\cdot);\ B \in \mathcal{S})$ is an independently scattered SαS measure with control measure m_d and by (3.8) it has $\tilde{Z}_d(\cdot,s)$ as an admissible translate. But m_d is non-atomic, thus by Proposition 4, $\tilde{Z}_d(\cdot,s) = 0$, i.e. for all $B \in \mathfrak{I}$,

$$0 = \tilde{Z}_d(B,s) = \tilde{Z}(A^c \cap B,s) = F(Z(A^c \cap B)) = F(Z_d(B)),$$

and hence

$$s(t) = F[X(t)] = F[X_a(t)+X_d(t)] = F[X_a(t)]$$

(since X_d is obtained by a linear operation on Z_d which implies $F[X_d(t)] = 0$). Therefore

$$s(t) = F[X_a(t)] = F[\textstyle\sum_{n=1}^N f_n(t)\, Z_n] = \sum_{n=1}^N f_n(t)\, F(Z_n) \tag{3.9}$$

$$= \textstyle\sum_{n=1}^N f_n(t)\, s_n = \sum_{n=1}^N f_n(t,a_n)\, s_n',$$

where $s_n = F(Z_n)$ and $s_n' = m^{1/\alpha}(\{a_n\})\, s_n$. On the other hand $\tilde{X}_{at} = (\tilde{X}_{at}(t,x) = \Sigma_{n=1}^N f(t,a_n)\tilde{Z}(\{a_n\},x);\ t \in T)$ has distribution $\mu_{X_{at}}$ and by the linearity of the map $x \to \tilde{X}_{at}(\cdot,x)$, the function $\tilde{X}_{at}(\cdot,s)$ is an admissible translate of $\tilde{X}$ and hence of X_{at}. But

$$\tilde{X}_{at}(t,s) = \textstyle\sum_{n=1}^N f(t,a_n)\, F[Z(\{a_n\})] = \sum_{n=1}^N f(t,a_n)\, s_n' = \sum_{n=1}^N f_n(t)\, s_n = s(t),$$

i.e., s is an admissible translate of X_{at}.

ii) $\Rightarrow$ i). The proof is identical to that in Proposition 4.

ii) $\Rightarrow$ iii). The proof is as in Proposition 4, with $\psi_n(n,x) = \tilde{Z}(\{a_n\})/m^{1/\alpha}(\{a_n\})$, so that by (3.8),

$$\psi(n,s) = F[Z(\{a_n\})]/m^{1/\alpha}(\{a_n\}) = s_n'/m^{1/\alpha}(\{a_n\}) = s_n. \tag{3.10}$$

To prove that a translate which is not admissible is singular, it suffices to consider $s \in F(X)$, i.e., $s(t) = F[X(t)]$, as by Proposition 2, $s \notin F$ implies singularity. Suppose $F[X_d(t)] \neq 0$. Then there exists $B \in \mathfrak{I}$ such that $F[Z_d(B)] \neq 0$ and by (3.8), $\tilde{Z}_d(B,s) = \tilde{Z}(A^c \cap B,s) = F[Z(A^c \cap B)] = F[Z_d(B)] \neq 0$. It follows from Proposition 4 that $\mu_{s+X}\tilde{Z}_d^{-1} \perp \mu_X \tilde{Z}_d^{-1}$ and hence $\mu_{s+X} \perp \mu_X$. Therefore $s(t) = F[X_a(t)] = \Sigma_{n=1}^N f_n(t)\, s_n$ and as in the proof of Proposition 4, $\Sigma_{n=1}^N |s_n|^2 = \infty$ implies $\mu_{s+X} \perp \mu_X$. □

It follows from Proposition 5 that for an invertible SαS process with nonatomic control measure every (non-zero) translate is singular. In particular, this contains Corollary 10.1 of [28]. Applied to SαS processes with purely atomic control measure, Proposition 5 is a stochastic process version of a result proved in [26, Th. 4] for SαS measures with discrete spectral measures on separable Banach spaces. The proposition completes the result in [26], providing a dichotomy.

Proposition 5 also provides examples where the set of admissible translates is a nontrival proper subset of the function space F of the process X. E.g., if $X(t) = \Sigma_{n=1}^{\infty} f_n(t) Z_n$, $t \in \mathsf{T}$, where $Z_1, Z_2, \ldots$ are i.i.d. standard SαS random variables with $1 < \alpha \leq 2$ and $L_\alpha\text{-}\overline{sp}\{\{f_n(t)\}_n;\ t \in \mathsf{T}\} = l_\alpha$, then

$$\mathsf{F}_\alpha = \{ s;\ s(t) = \Sigma_{n=1}^{\infty} s_n f_n(t),\ \{s_n\} \in \ell_{\alpha^*} \},$$

while the set of admissible translates is the infinite dimensional subspace (since $\alpha^* \geq 2$) of F_α for which $\{s_n\} \in \ell_2$; hence we have equality only if $\alpha = 2$, and proper inclusion if $1 < \alpha < 2$.

Important examples of invertible SαS processes are presented in the following.

Harmonizable SαS processes (and sequences). Let

$$X(t) = \int_{\mathsf{I}} e^{i<t,u>} Z(du), \quad t \in \mathbf{R}^d \text{ or } \mathbf{Z}^d,$$

where $\mathsf{I} = \mathbf{R}^d$ and $[-\pi,\pi]^d$ respectively, d = 1,2,..., and Z is a complex SαS independently scattered random measure with finite control measure m, referred to as the spectral measure of the harmonizable process X. If the spectral measure m is nonatomic and $0 < \alpha < 2$ then it follows from Proposition 5 that X has no nontrivial admissible translate. When the stable distribution of Z is radially symmetric, i.e. when X is stationary, this result is in sharp contrast with the stationary Gaussian processes $\alpha = 2$, whose admissible translates are:

$$\{ s;\ s(t) = \int_{\mathsf{I}} e^{i<t,u>} \bar{z}(u) m(du),\ z \in L_2(m) \}.$$

If m is purely discrete there is no difference between the Gaussian and non-Gaussian stable cases. Namely, if X has a Fourier series representation

$$X(t) = \Sigma_{n=1}^{N} b_n e^{i<c_n,t>} Z_n, \quad N \leq \infty,$$

with Z_n's i.i.d. standard SαS random variables and $\Sigma_{n=1}^{N} |b_n|^\alpha < \infty$, the set of admissible translates is

$$\{ s;\ s(t) = \Sigma_{n=1}^{N} s_n e^{i<c_n,t>},\ \Sigma_{n=1}^{N} |s_n/b_n|^2 < \infty \},$$

and depends on α, $0 < \alpha \leq 2$, only via the α-summable sequence $(b_n;\ n \in \mathsf{N}) \in l_\alpha$. In other words if for fixed $\{b_n\} \in l_\beta$, $1 \leq \beta \leq 2$, we define $X_\alpha(t) = \Sigma_{n=1}^{\infty} b_n \exp\{i<c_n,t>\} Z_{n,\alpha}$, where the $Z_{n,\alpha}$'s are standard i.i.d. SαS with $\beta \leq \alpha \leq 2$ for $1 < \beta \leq 2$ and $1 < \alpha \leq 2$ for $\beta = 1$, then all these processes X_α have the same set of admissible translates.

Continuous time SαS moving averages. Let

$$X(t) = \int_{\mathbf{R}} f(t-u) Z(du), \quad t \in \mathbf{R},$$

where Z is independently scattered SαS with Lebesgue control measure and $f \in L_\alpha(\text{Leb})$.

When f vanishes on the negative line, the moving average X is nonanticipating; the stationary solutions of nth order linear stochastic differential equations with constant coefficients driven by stable motion Z are nonanticipating moving averages.

In the Gaussian case $\alpha = 2$ the admissible translates are

$$F_2 = \{ s;\ s(t) = \int_{\mathbf{R}} f(t-u)z(u)du,\ z \in L_2(Leb) \} = \{ s;\ s \in L_2(Leb),\ \hat{s}/\hat{f} \in L_2(Leb) \},$$

where $\hat{}$ denotes Fourier transform.

Examples of invertible SαS moving averages with $\alpha < 2$ and therefore with no admissible translates, can be obtained by taking i) f continuous and equal to zero on $(-\infty,0)$ and at $+\infty$ [2, Th. 2]; ii) $\alpha \in (1,2)$ and f the Fourier transform of some function F in $L_{\alpha^*}(Leb)$ with $F \neq 0$ a.e. (Leb) [27, Th. 75]. Case i) includes nonanticipating moving averages with continuous kernel f, while case ii) contains certain nonanticipating moving averages with discontinuous kernels f, namely the stationary solutions of nth order linear stochastic differential equations with constant coefficients. There f(t) is a linear combination of functions of the type $t^{k-1}e^{-at}1_{(0,\infty)}(t)$ with integer k and $a > 0$, which are Fourier transforms of the $L_{\alpha^*}(Leb)$ functions $\Gamma(k)/[2\pi(a+iu)]$ so that X is invertible. Thus solutions of nth order stochastic differential equations driven by SαS motion have no admissible translate for $1 < \alpha < 2$, in sharp contrast with the Gaussian case $\alpha = 2$. E.g., if $n = 1$, f(t) $= e^{-t}\, 1_{(0,\infty)}(t)$, the S$\alpha$S Ornstein-Uhlenbeck process

$$X(t) = \int_{-\infty}^{t} e^{-(t-u)} Z(du), \quad t \in \mathbf{R},$$

has no admissible translates for $1 < \alpha < 2$, while for $\alpha = 2$ its admissible translates are $s(t) = \int_{-\infty}^{t} e^{-(t-u)} z(u)du,\ t \in \mathbf{R},\ z \in L_2(Leb)$.

Invertible discrete time SαS processes (sequences) have similar admissible translates in the Gaussian and non-Gaussian cases, and of course nonadmissible translates are singular.

Mixed auto-regressive moving averages of order (p,q) (ARMA (p,q)). Let $X = (X_n;\ n = ..., -1,0,1,...)$ be defined by the difference equation

$$X_n - a_1X_{n-1} - \dots - a_pX_{n-p} = Z_n + b_1Z_{n-1} + \dots + b_qZ_{n-q}$$

where the Z_n's are i.i.d. standard SαS random variables with $0 < \alpha \leq 2$. If the polynomials $P(u) = 1 - a_1u - \dots - a_pu^p$ and $Q(u) = 1 + b_1u + \dots + b_qu^q$ satisfy the condition $P(u)Q(u) \neq 0$ for all complex u with $|u| \leq 1$, then the difference equation defining X has a unique stationary solution of the moving average form

$$X_n = \sum_{k=-\infty}^{n} g_{n-k} Z_k \qquad \text{and} \qquad Z_n = X_n - \sum_{j=1}^{\infty} h_j X_{n-j},$$

where the coefficients are determined by $Q(u)/P(u) = \sum_{j=0}^{\infty} g_j u^j$, $P(u)/Q(u) = 1-\sum_{j=1}^{\infty} h_j u^j$, $|u| \leq 1$, (see, e.g., [11]). Thus $\mathcal{L}(X) = \mathcal{L}(Z)$, i.e. X is invertible, and hence, by Proposition 5, the admissible translates of X are

$$\{ s = \{s_n\}_{-\infty}^{\infty};\ s_n = \sum_{k=-\infty}^{n} g_{n-k} z_k\ ,\ \sum_{k=-\infty}^{\infty} z_k^2 < \infty \}.$$

We should note the different behavior of non-Gaussian SαS moving averages in continuous and in discrete time. A continuous time moving average may have no admissible translates, whereas a discrete time ARMA sequence has a set of admissible translates identical to the Gaussian case.

REFERENCES

1. A. Araujo and E. Giné. *The Central Limit Theorem for Real and Banach Valued Random Variables.* Wiley, New York, 1980.
2. A. Atzman. Uniform approximation by linear combinations of translations and dilations of a function. *J. London Math. Soc.*, 2, 1983, 51-54.
3. P. Brockett and H. Tucker. A conditional dichotomy theorem for stochastic processes with independent increments. *J. Multivariate Anal.*, 7, 1977, 13-27.
4. S. Cambanis. The measurability of a stochastic process of second order and its linear space. *Proc. Amer. Math. Soc.*, 47, 1975, 467-475.
5. S. Cambanis. Complex symmetric stable variables and processes. In "*Contribution to Statistics: Essays in Honour of Norman L. Johnson*", P.K. Sen, Ed. North Holland, New York, 1982, 63-79.
6. S. Cambanis, C.D. Hardin Jr. and A. Weron. Innovations and Wold decompositions for stable sequences. *Probab. Th. Rel. Fields*, 1988, to appear.
7. S. Cambanis and A.G. Miamee. On prediction of harmonizable stable processes. *Sankya A*, 1988, to appear.
8. S. Cambanis and G. Miller. Linear problems in p^{th} order and stable processes. *SIAM J. Appl. Math.*, 41, 1981, 43-69.
9. S. Cambanis and A.R. Soltani. Prediction of stable processes: Spectral and moving average representations. *Z. Wahrsch. ver. Geb.*, 66, 1982, 593-612.
10. S. Chatterji and V. Mandrekar. Equivalence and singularity of Gaussian measures and applications. In "*Probabilistic Analysis and Related Topics* 1", A.T. Bharucha-Reid, Ed. Academic Press, New York, 1978, 163-167.
11. D.B. Cline and P.J. Brockwell. Linear prediction of ARMA processes with infinite variance. *Stochastic Proc. Appl.*, 19, 1985, 281-296.
12. R.M. Dudley, Singularity of measures on linear spaces, *Z. Wahrsch. verw. Geb.*, 6, 1966, 129-132.
13. W.H. DuMouchel. On the asymptomatic normality of the maximum likelihood estimate when sampling from a stable distribution. *Ann. Stat.*, 1, 1973, 948-957.
14. R. Fortet. Espaces à noyau reproduisant et lois de probabilités des fonctions aléatoires. *Ann. Inst. H. Poincaré* B, IX, 1973, 41-48.

15. I. Gihman and A. Skorohod. On densities of probability measures in function spaces. *Russ. Math. Surveys*, 21, 1966, 83-156.

16. P.R. Halmos. *Measure Theory*. Springer, New York, 1974.

17. D.C. Hardin. On the spectral representation of symmetric stable processes. *J. Multivariate Anal.*, 12, 1982, 385-401.

18. S.T. Huang and S. Cambanis. Spherically invariant processes: Their non-linear structure, discrimination and estimation. *J. Multivariate Anal.*, 9, 1979, 59-83.

19. J.F.C. Kingman and S.J. Taylor. *Introduction to Measure and Probability*, Cambridge University Press, London, 1973.

20. J. Kuelbs. A representation theorem for symmetric stable processes and stable measures on H. *Z. Wahrsch. verw. Geb.*, 26, 1973, 259-271.

21. R. LePage. Multidimensional infinitely divisible variables and processes. Part I: Stable case. *Tech. Rep. No.* 292, Statistics Department, Stanford University, Stanford, CA, 1980.

22. Y.,-M. Pang. Simple proofs of equivalence conditions for measures induced by Gaussian processes. *Sel. Transl. Math. Stat. Probab.*, 12, Amer. Math. Soc., Providence, RI, 1973, 109-118.

23. M. Schilder. Some structural theorems for symmetric stable laws. *Ann. Math. Stat.*, 41, 1970, 412-421.

24. L.A. Shepp. Distinguishing a sequence of random variables from a translate of itself. *Ann. Math. Stat.*, 36, 1965, 1107-1112.

25. I. Singer. *Best Approximation in Normed Linear Spaces by Elements of Linear Subspaces*, Springer, New York, 1970.

26. D. Thang and N. Tien. On symmetric stable measures with discrete spectral measure on Banach spaces. In "*Probability Theory and Vector Spaces II*", A. Weron, Ed. *Lectures Notes in Mathematics* No. 828. Springer, Berlin, 1979, 286-301.

27. E.C. Titchmarsch. *Fourier Integrals*. University Press, Oxford, 1928.

28. J. Zinn. Admissible translates of stable measures. *Studia Math.*, 54, 1975, 245-257.

Probability Theory on Vector Spaces IV
Lancut, June '87, Springer's LNM 1391

AN OUTLINE OF THE INTEGRATION THEORY OF BANACH SPACE VALUED MEASURES

by

P. Masani
Department of Mathematics
University of Pittsburgh
Pittsburgh, PA 15260, USA

and

H. Niemi
Department of Statistics
University of Helsinki
Helsinki, Finland

CONTENTS

1. INTRODUCTION

Our primary concern is with integrals of the form

$$x = \int_\Omega f(\omega)\xi(d\omega) \in X, \tag{1.1}$$

where ξ is a countably additive measure over a set Ω with values in a Banach space X over the field F (=R or C), and f is an F-valued function on Ω. Product measures $\xi \times \mu$, and iterated integrals such as

$$y = \int_\Lambda \{ \int_\Omega f(\omega, \lambda)\xi(d\omega)\}\mu(d\lambda) \in X,$$

where μ is a countably additive, F-valued or $[0, \infty]$-valued measure over a space Λ, and f is F-valued on $\Omega \times \Lambda$, will also concern us. This obliges us to attend also to the more familiar vector-integrals

$$y = \int_{\Lambda} \vec{g}(\lambda)\mu(d\lambda) \in X, \tag{1.2}$$

where $\vec{g}$ is an X-valued function on Λ.

To gain maximum generality we shall interpret the integral in (1.2) in the sense of Pettis [18], not in the narrower sense of Bochner; i.e. we shall define the y in (1.2) by the condition

$$\forall x' \in X', \quad x'(y) = \int_{\Lambda} x'\{\vec{g}(\lambda)\}\mu(d\lambda) \in \mathsf{F}, \tag{1.3}$$

X' being the dual of X. It is then only natural to adopt the same policy with regard to the integral in (1.1), and to define x by the condition analogous to (1.3), viz.

$$\forall x' \in X', \quad x'(x) = \int_{\Omega} f(\omega)(x' \circ \xi)(d\omega) \in \mathsf{F}. \tag{1.4}$$

Bartle [1] deals with general measures (vector and operator valued) finitely additive on algebras, but with countably additive measures on σ-algebras. Thus he deals only with bounded countably additive measures. The same applies to the treatments of Dunford & Schwartz [6] and Diestel & Uhl [4]. As for Dinculeanu [5] he confines his integration theory to measures of finite variation. Thus the literature on the integration of the type (1.1) with (1.4), free of boundedness and other restraints on ξ comprises two papers of D.R. Lewis [10], [11] and the cognate one by E.G.F. Thomas [20]. For Lewis, X is a locally convex topological vector space and Ω an arbitrary set. For Thomas, X is a Banach space, but Ω is a locally compact Hausdorff space and ξ is a so-called vectorial "Radon" measure. (There is also an unpublished work [21] by Thomas to which we shall refer later, cf. 4.15.) In the present paper we shall initially lean heavily on the Lewis papers, but soon obtain new results specific to X being a Banach space. We shall omit proofs. These will appear in fuller papers in the Advances in Mathematics.

With $\emptyset \neq \mathcal{F} \subseteq 2^{\Omega}$ and $\emptyset \neq \mathcal{G} \subseteq 2^{X}$ we shall adopt the notation:

$$\begin{cases} FA(\mathcal{F}, X) := \{\xi : \xi \textit{ is a finitely additive measure on } \mathcal{F} \textit{ to } X\}, [1] \\ CA(\mathcal{F}, X) := \{\xi : \xi \textit{ is a countably additive measure on } \mathcal{F} \textit{ to } X\}, [2] \\ \mathcal{M}(\mathcal{F}, \mathcal{G}) := \{f : f \in X^{\Omega} \ \& \ \forall G \in \mathcal{G},\ f^{-1}(G) \in \mathcal{F}\}, \textit{ i.e. } \mathcal{M}(\mathcal{F}, \mathcal{G}) \textit{ is the class of } \\ \quad \mathcal{F}, \mathcal{G} \textit{ measurable functions}, \\ \mathcal{F}^{\text{loc}} := \{A : A \subseteq \Omega \ \& \ \forall F \in \mathcal{F},\ A \cap F \in \mathcal{F}\}, \\ Bl(Y) := \textit{the smallest } \sigma\textit{-ring containing all open sets of a topological space } Y. \end{cases} \tag{1.5}$$

$\mathcal{F}^{\text{loc}}$ is the class of subsets of Ω that are "locally in $\mathcal{F}$". $\mathcal{F}^{\text{loc}}$ is an algebra or σ-algebra according as $\mathcal{F}$ is a ring or a δ-ring (i.e. a ring closed under countable intersections).

It is now recognized that the natural domain of a countably additive measure ξ with range other than $[0, \infty]$ is merely a δ-ring in general. Accordingly, we shall assume in the sequel that

---

[1] This requires $\xi(\emptyset) = 0$ when $\emptyset \in \mathcal{F}$, and for all $n \geq 1$, if $F_1, \ldots, F_n \in \mathcal{F}$ are disjoint, and if $\cup_1^n F_k \in \mathcal{F}$, then $\sum_1^n \xi(F_k) = \xi(\cup_1^n F_k)$.

[2] This requires $\xi(\emptyset) = 0$ when $\emptyset \in \mathcal{F}$, and if for $n \geq 1$, the $F_n \in \mathcal{F}$ are disjoint, and if $F = \cup_1^\infty F_n \in \mathcal{F}$, then $\sum_1^n \xi(F_k)$ converges unconditionally to $\xi(F)$ in the topology of X.

$$(1.6)\qquad \begin{cases} \mathcal{D} \textit{ is a } \delta\textit{-ring over a (non-void) set } \Omega \\ \xi \in CA(\mathcal{D}, X), \textit{ and so } \forall x' \in X', \quad x' \circ \xi \in CA(\mathcal{D}, \mathsf{F}). \end{cases}$$

In order that a function f on Ω to F be integrable with respect to ξ, it has to be appropriately "measurable". As is clear from Lewis' papers, it is best to assume that f is $\mathcal{D}^{\text{loc}}$, $Bl(\mathsf{F})$-measurable, i.e. in symbols:

$$(1.7)\qquad f \in \mathcal{M}(\mathcal{D}^{\text{loc}}, Bl(\mathsf{F})), \quad \text{cf. } (1.5).$$

Here $\mathcal{D}^{\text{loc}}$ is a σ-algebra, since $\mathcal{D}$ is a δ-ring. Obviously, $\mathcal{D} \subseteq \sigma\text{-alg}(\mathcal{D}) \subseteq \mathcal{D}^{\text{loc}}$.

It is also natural to take the larger σ-algebra $\mathcal{D}^{\text{loc}}$ as the domain of the various *variation measures* associated with ξ, the definitions of which we now recall:

1.8. Definition. For all $A \in \mathcal{D}^{\text{loc}}$, let

$$\Pi_A := \{\pi : \pi \text{ is a finite class of disjoint sets in } \mathcal{D} \cap 2^A\}.$$

(a) The *quasi-variation* $q_\xi(\cdot)$ *of* ξ is the function on $\mathcal{D}^{\text{loc}}$ defined by: $\forall A \in \mathcal{D}^{\text{loc}}$,

$$q_\xi(A) := \sup \{|\xi(\Delta)| : \Delta \in \mathcal{D} \cap 2^A\}.$$

(b) The *semi-variation* $s_\xi(\cdot)$ *of* ξ is the function on $\mathcal{D}^{\text{loc}}$ defined by: $\forall A \in \mathcal{D}^{\text{loc}}$,

$$s_\xi(A) := \sup\{|\sum_{\Delta \in \pi} \alpha(\Delta)\xi(\Delta)|_X : \pi \in \Pi_A, \ \alpha \in \mathsf{F}^\pi \ \& \ |\alpha(\cdot)| \le 1\}.$$

(c) The *variation measure* $|\xi|(\cdot)$ *of* ξ is the function on $\mathcal{D}^{\text{loc}}$ defined by: $\forall A \in \mathcal{D}^{\text{loc}}$,

$$|\xi|(A) := \sup \{\sum_{\Delta \in \pi} |\xi(\Delta)| : \pi \in \Pi_A\}.$$

The obvious relationship between these variations, and certain simple properties that they share are stated in:

1.9. Triviality. *Let* $\xi, \eta \in CA(\mathcal{D}, X)$. *Then*
(a) $\forall A \in \mathcal{D}^{\text{loc}}, \quad 0 \le q_\xi(A) \le s_\xi(A) \le |\xi|(A) \le \infty$;
(b) $\forall c \in \mathsf{F}, \quad q_{c\xi}(A) = |c| q_\xi(A), \quad s_{c\xi}(A) = |c| s_\xi(A), \quad |c\xi|(A) = |c| \cdot |\xi|(A)$;
(c) $q_{\xi+\eta} \le q_\xi + q_\eta, \quad s_{\xi+\eta} \le s_\eta + s_\eta, \quad |\xi + \eta| \le |\xi| + |\eta|$.

From the conditions (1.6) and (1.4), it is clear that a preliminary task confronting us is to study measures $\mu \in CA(\mathcal{D}, \mathsf{F})$, and the concept of integration with respect to such μ. This, without localizability-restraints on μ, is done in §2.

In §3 we introduce the classes of Gelfand integrable functions and show that $\mathcal{G}_{1,\xi} :=$ Gelfand integrable scalar functions w.r.t a given $\xi \in CA(\mathcal{D}, X)$ is a Banach space under a suitable norm. In §4 we go over to the Pettis integration and show that the class $\mathcal{P}_{1,\xi} :=$ Pettis integrable scalar functions w.r.t. a given $\xi \in CA(\mathcal{D}, X)$ is the closure of $\mathcal{D}$-simple functions in $\mathcal{G}_{1,\xi}$.

One of our main goals has been to obtain a vectorial extension of integration over product spaces for unbounded measures. In §§7–9 we first obtain an unrestricted version of the Product Measure Theorem as well as the Slicing Theorem and, finally, of the Tonelli and Fubini Theorems.

We now list some notation we will use in the interest of brevity:

1.10. Notation.

(a) The symbol ":=" means "equal by definition". The indicator function of a set A is denoted by χ_A. $\text{Rstr.}_A f$ stands for the restriction of the function f to a subset A of its domain. The symbol $\|$ means "disjoint".

(b) $\mathbf{F}$ refers to either the real or complex number fields $\mathbf{R}$ or $\mathbf{C}$, and $\mathbf{N}$ to the set of all integers. $\mathbf{R}_+, \mathbf{N}_+$ and $\mathbf{R}_{0+}, \mathbf{N}_{0+}$ denote the subsets of positive elements, and the subsets of non-negative elements of $\mathbf{R}$ and $\mathbf{N}$, respectively.

(c) For normed vector spaces X and Y, and $A \subseteq X$, $< A >$ is the linear manifold spanned by A in X, and $\mathfrak{S}(A) := \text{cls} < A >$, where cls stands for the closure in the norm-topology. $CL(X, Y)$ is the set of continuous linear operators on X to Y.

(d) For a ring $\mathcal{R}$ over Ω, and $\xi \in FA(\mathcal{R}, X)$, $\mathcal{M}_\xi := <\text{Range } \xi>$ and $\mathcal{S}_\xi = \mathfrak{S}\{\text{Range } \xi\}$. (Thus $\mathcal{M}_\xi \subseteq \mathcal{S}_\xi \subseteq X$.)

$$\mathcal{N}_\xi = \{A : A \in \mathcal{R}^{\text{loc}} \ \& \ \forall R \in \mathcal{R}, \ \ \xi(A \cap R) = 0\}.$$

For $\emptyset \neq \mathcal{F} \subseteq 2^\Omega$, σ-alg$(\mathcal{F})$ is the smallest σ-algebra over Ω containing $\mathcal{F}$; likewise for a σ-ring $(\mathcal{F})$, etc.

(e) The bold face letters $\mathbf{\cap}$ and $\mathbf{\cup}$, are the operations induced by $\cap$ and $\cup$ on the class of set-families, i.e.

$$\forall \mathcal{F}, \mathcal{G} \subseteq 2^\Omega, \quad \mathcal{F} \mathbf{\cap} \mathcal{G} := \{F \cap G : F \in \mathcal{F} \ \& \ G \in \mathcal{G}\}$$

$$\mathcal{F} \mathbf{\cup} \mathcal{G} := \{F \cup G : F \in \mathcal{F} \ \& \ G \in \mathcal{G}\}.$$

This convention will govern other operations as well. For instance, for $\emptyset \neq \mathcal{F} \subseteq 2^\Omega$ & $\emptyset \neq \mathcal{G} \subseteq 2^\Lambda$,

$$\mathcal{F} \mathbf{\times} \mathcal{G} = \{F \times G : F \in \mathcal{F} \ \& \ G \in \mathcal{G}\}.$$

2. PRELIMINARIES

We first record a result from the classical theory of non-negative measures on σ-algebras, which is not in the literature but is needed in the proof of 3.12 (b) below.

2.1. Fatou's Lemma (for sequences of measures). *Let*

(i) $\mathcal{A}$ *be a σ-algebra over Ω,*

(ii) $(\mu_n)_1^\infty$ *be a sequence in $CA(\mathcal{A}, [0, \infty])$, and*

(iii) $\mu \in CA(\mathcal{A}, [0, \infty])$ & $\mu(\cdot) \leq \underline{\lim}_{n \to \infty} \mu_n(\cdot)$ *on $\mathcal{A}$.*

Then $\forall f \in \mathcal{M}(\mathcal{A}, Bl(\mathbf{R}_{0+}))$,

$$0 \leq \int_\Omega f(\omega) \mu(d\omega) \leq \underline{\lim}_{n \to \infty} \int_\Omega f(\omega) \mu_n(d\omega) \leq \infty.$$

From here on it will be understood that

$$(2.2) \qquad \begin{cases} \mathcal{D} \text{ is a } \delta\text{-ring over } \Omega, \\ \mu \in CA(\mathcal{D}, \mathbf{F}), \\ f \in \mathcal{M}(\mathcal{D}^{\text{loc}}, Bl(\mathbf{F})). \end{cases}$$

Central to the question of the definition of integrability and integration with respect to μ is the basic result that

$$(2.3)\qquad \begin{cases} (a) & |\mu| \in CA(\mathcal{D}^{\mathrm{loc}}, [0,\infty]); \\ (b) & \forall D \in \mathcal{D}, \quad |\mu|(D) < \infty \quad \text{(cf. [5 : p. 48, Corollary 1])}. \end{cases}$$

The result (2.3) (a) allows us to introduce the $L_{1,\mu}$ norm of f by

$$(2.4)\qquad |f|_{1,\mu} := \int_\Omega |f(\omega)| \cdot |\mu|(d\omega) \in [0,\infty],$$

the last integral being defined as in elementary analysis (cf. e.g. Rudin [19]). We then define the class $L_{1,\mu} = L_1(\Omega, \mathcal{D}, \mu : \mathsf{F})$ à la Bochner, by

$$(2.5)\qquad \begin{cases} (a) & L_{1,\mu} := \{f : f \in \mathcal{M}(\mathcal{D}^{\mathrm{loc}}, Bl(\mathsf{F})) \ \& \ |f|_{1,\mu} < \infty\} \\ (b) & L_{1,\mu}^{\mathrm{loc}} := \{f : f \in \mathcal{M}(\mathcal{D}^{\mathrm{loc}}, Bl(\mathsf{F})) \ \& \ \forall A \in \mathcal{D},\ \chi_A \cdot f \in L_{1,\mu}\}. \end{cases}$$

It follows as in elementary analysis that

2.6. Proposition.
(a) $L_{1,\mu}$ *is a Banach space over* F *under the norm* $|\cdot|_{1,\mu}$, *when functions* f, g *for which* $\mathrm{supp}(f-g) \in \mathcal{N}_\mu$ *are identified.*
(b) *The set*

$$\mathcal{S}(\mathcal{D}, \mathsf{F}) := \{f : f \textit{ is a } \mathcal{D}\textit{-simple}^3 \textit{ function on } \Omega \textit{ to } X\}$$

is a linear manifold everywhere dense in $L_{1,\mu}$.

(c)
$$\begin{aligned} \mathcal{D}_\mu &:= \{A : A \in \mathcal{D}^{\mathrm{loc}} \ \& \ \chi_A \in L_{1,\mu}\} \\ &= \{A : A \in \mathcal{D}^{\mathrm{loc}} \ \& \ |\mu|(A) < \infty\}. \end{aligned}$$

Proposition 2.6 (b) allows us to define $\int_\Omega f(\omega)\mu(d\omega)$ for $f \in L_{1,\mu}$ as a Bochner integral. It is convenient to adopt the abbreviation

$$(2.7)\qquad E_\mu(f) = \int_\Omega f(\omega)\mu(d\omega).$$

Our definition then reads:

2.8. Definition. E_μ is defined on $L_{1,\mu}$ as follows:
(a) if $f = \sum_{k=1}^r a_k \chi_{D_k} \in \mathcal{S}(\mathcal{D}, \mathsf{F})$, then $E_\mu(f) = \sum_{k=1}^r a_k \mu(D_k)$; [4]
(b) if $f \in L_{1,\mu}$, then $E_\mu(f) := \lim_{n\to\infty} E_\mu(s_n)$, where $(s_n)_1^\infty$ in $\mathcal{S}(\mathcal{D}, \mathsf{F})$ is such that $|s_n - f|_{1,\mu} \to 0$, as $n \to \infty$. [5]

Obviously E_μ is a contractive linear functional on $L_{1,\mu}$ to F with $|E_\mu| = 1$. Also, it is easily checked that depending on whether $\mathsf{F} = \mathsf{R}$ or C, $L_{1,\mu}$ is the intersection of the L_1 classes we get with respect to the two or four measures in the Hahn-Jordan Decomposition of μ, cf. [5: pp.

---

[3] i.e. $f \in \mathsf{F}^\Omega$, Range f is finite, & $\forall y \in \text{Range } f \setminus \{0\}$, $f^{-1}(y) \in \mathcal{D}$.

[4] $E_\mu(f)$, so defined, is independent of the particular expression used to represent f. This is shown by applying the disjoint normal form to the family $\{D_1, \ldots, D_r\}$.

[5] $E_\mu(f)$, so defined, is independent of the particular sequence $(s_n)_1^\infty$, as is easily seen.

44–50], and that E_μ is the usual linear combination of the E's with respect to these component measures.

Next we associate with the pair μ, f the *indefinite integral* $\nu_{\mu,f}(\cdot)$ *of* f *with respect to* μ. The central result, parts of which seem to be new, is as follows:

2.9. Theorem. (The indefinite integral). *Let*
(i) $f \in \mathcal{M}(\mathcal{D}^{\text{loc}}, \text{Bl}(\mathsf{F}))$,
(ii) $\mathcal{D}_\mu(f) := \{A : A \in \mathcal{D}^{\text{loc}} \ \& \ \chi_A \cdot f \in L_{1,\mu}\}$,
(iii) $\forall A \in \mathcal{D}_\mu(f),\ \ \nu_{\mu,f}(A) := \int_A f(\omega)\mu(d\omega)$.
Then
(a) $\mathcal{D}_\mu(f) = a\ \delta\text{-ring} \subseteq \mathcal{D}^{\text{loc}} \subseteq [D_\mu(f)]^{\text{loc}}$;
(b) $\nu_{\mu,f} \in CA(\mathcal{D}_\mu(f), \mathsf{F})$;
(c) $\forall A \in \mathcal{D}^{\text{loc}},\ |\nu_{\mu,f}|(A) = \int_A |f(\omega)| \cdot |\mu|(d\omega)$, *both terms being finite for* $A \in \mathcal{D}_\mu(f)$ *and both being* ∞ *for* $A \in \mathcal{D}^{\text{loc}} \setminus \mathcal{D}_\mu(f)$;
(d) $\forall f \in L^{\text{loc}}_{1,\mu},\ \ \mathcal{D} \subseteq \mathcal{D}_\mu(f) \subseteq \mathcal{D}^{\text{loc}} = [\mathcal{D}_\mu(f)]^{\text{loc}}$.

The part (c), the best possible of its kind, is quite useful in the later sections. Its proof is quite long and rests on several lemmas. From Theorem 2.9, we get a good extension of the classical substitution principle:

2.10. Theorem. (Substitution principle). *Let* $f, g \in \mathcal{M}(\mathcal{D}^{\text{loc}}, Bl(\mathsf{F}))$. *Then*
(a) $\mathcal{D}^{\text{loc}} \cap \mathcal{D}_{\nu_{\mu,f}}(g) = \mathcal{D}_\mu(fg)$;
(b) $\forall A \in \mathcal{D}^{\text{loc}} \cap \mathcal{D}_{\nu_{\mu,f}}(g),\ \ \int_A g(\omega) \cdot \nu_{\mu,f}(d\omega) = \int_A g(\omega) f(\omega) \cdot \mu(d\omega)$.

3. THE VARIATIONS OF A VECTOR MEASURE AND GELFAND INTEGRABILITY

Central to the advanced theory of vector measures and vector integration is the Baire category theorem for complete metric spaces. Guided by our cognate work on Banach groups [15], we shall first enunciate a very general uniform boundedness principle that exhausts the appeal to the Baire category theorem made in this paper. This principle involves the three variation measures discussed in 1.8 and 1.9.

3.1. Theorem. (Uniform boundedness principle). *Let*
(i) $(Y, +, |\cdot|)$ *be a Banach group, i.e. an abelian group with a norm* $|\cdot|$ [6] *that yields a complete metric,*
(ii) $\forall \zeta \in FA(\mathcal{D}, Z), \|\zeta\|(\cdot)$ *stand for any one of* $|\zeta|(\cdot), s_\zeta(\cdot), q_\zeta(\cdot)$, *where* Z *is a Banach space over* F,
(iii) $\eta(\cdot)$ *be a homomorphism on the group* Y *to the group* $FA(\mathcal{D}, Z)$ *such that* $\forall A \in \mathcal{D}^{\text{loc}}, \|\eta(\cdot)\|(A)$ *is lower semi-continuous on* Y *to* $[0, \infty]$, *i.e.*

$$y_n \to y \text{ in } Y \Rightarrow \|\eta(y)\|(A) \le \underline{\lim}_{n\to\infty} \|\eta(y_n)\|(A) \le \infty.$$

Then
(a) $\forall A \in \mathcal{D}^{\text{loc}}$ *we have*

$$\|\eta(\cdot)\|(A) < \infty \text{ on } Y \Rightarrow \exists r > 0 \text{ such that } \sup_{|y| \le r} \|\eta(y)\|(A) < \infty,$$

[6] i.e. $|\cdot|$ is a function on Y to $\mathbf{R}_{0+}$ such that $\forall y_1, y_2, y \in Y,\ |y_1 + y_2| \le |y_1| + |y_2|,\ |-y| = |y| \ \&\ |y| = 0 \Leftrightarrow y = 0$.

(b) *when* Y *is a Banach space over* F, *and the* $\eta(\cdot)$ *in (iii) is a linear operator on* Y *to* $FA(\mathcal{D}, Z)$ *we have* $\forall A \in \mathcal{D}^{\text{loc}}$,

$$\|\eta(\cdot)\|(A) < \infty \text{ on } Y \Rightarrow \sup_{|y| \le 1} \|\eta(y)\|(A) < \infty.$$

We shall omit the proof. Applying Theorem 3.1(b) with $Y = X'$, $Z = \mathsf{F}$ and $\eta(x') = x' \circ \xi$, cf. (1.6), we get the following useful result on the semi-variation:

$$\text{(3.2)} \qquad \forall A \in \mathcal{D}^{\text{loc}},\ s_\xi(A) < \infty \text{ iff } \forall x' \in X',\ |x' \circ \xi|(A) < \infty.$$

We refer to Dinculeanu [5, pp. 51–57] for the standard properties of the semi-variation measure. Here we will only note that (3.2) entails both the results:

$$\text{(3.3)} \qquad \begin{cases} \text{(a)} & \forall D \in \mathcal{D},\ s_\xi(D) < \infty; \\ \text{(b)} & \mathcal{D} \text{ is a } \sigma\text{-ring } \Rightarrow \xi \text{ is bounded on } \mathcal{D}. \end{cases}$$

Another important property of s_ξ that we will need is contained in Lewis [10: 1.3] and [11: 2.4]:

$$\text{(3.4)} \qquad (D_n)_1^\infty \text{ is in } \mathcal{D} \ \&\ D_n \downarrow D, \text{ as } n \to \infty \Rightarrow s_\xi(D_n) \downarrow s_\xi(D), \text{ as } n \to \infty.$$

Next we turn to the concept of absolute continuity for X-valued measures. We adapt the definition given for non-negative measures by von Neumann [17: p. 197, 11.2.1]:

3.5. Definition. Let $\mathcal{R}$ be a ring over Ω, $\xi \in X^{\mathcal{R}}$ and $\mu \in FA(\mathcal{R}, [0, \infty])$. We say that ξ *is absolutely continuous with respect to* μ *on* $\mathcal{R}$ (in symbols $\xi \prec\prec \mu$ on $\mathcal{R}$) iff $\forall E \in \mathcal{R}$ & $\forall \varepsilon > 0$, $\exists \delta_{E,\varepsilon} > 0$ such that

$$R \in \mathcal{R} \cap 2^E \ \&\ 0 \le \mu(R) < \delta_{E,\varepsilon} \Rightarrow |\xi(R)| < \varepsilon.$$

We then have the following full-fledged extension of von Neumann's result for R_{0+}-valued measures on δ-rings [17; p. 199, 11.2.4]. The proof hinges on the Lewis result (3.4).

3.6. Proposition. *Let* $\mu \in CA(\mathcal{D}, [0, \infty])$ *and* $\xi \in CA(\mathcal{D}, X)$. *Then the following conditions are equivalent:*
(α) $D \in \mathcal{D}$ & $\mu(D) = 0 \Rightarrow \xi(D) = 0$,
(β) $\xi \prec\prec \mu$ *on* $\mathcal{D}$.

The development of the theory of the X-valued Pettis integrals (1.2), (1.3) and (1.1), (1.4) leads quite naturally to a related concept of integration due to Gelfand [7], in which the integrals fall not in X but in X'', the second dual of X. Logically, the Gelfand theory preceeds the Pettis, and we now turn to it.

3.7. Definition.
(a) $\forall \mu \in CA(\mathcal{D}, \mathsf{F}) \cup CA(\mathcal{D}, [0, \infty])$, we define $\mathcal{G}_{1,\mu} = \mathcal{G}_{1,\mu}(\Omega, \mathcal{D}, \mu; X)$ by

$$\begin{aligned} \mathcal{G}_{1,\mu} &:= \{f : f \in X^\Omega \ \&\ \forall x' \in X',\ x' \circ f \in L_1(\Omega, \mathcal{D}, \mu; \mathsf{F})\} \\ &= \cap_{x' \in X'} x'^{-1}\{L_1(\Omega, \mathcal{D}, \mu; \mathsf{F})\}. \end{aligned}$$

(b) $\forall \xi \in CA(\mathcal{D}, X)$ we define $\mathcal{G}_{1,\xi} = \mathcal{G}_{1,\xi}(\Omega, \mathcal{D}, \xi; \mathsf{F})$ by

$$\begin{aligned} \mathcal{G}_{1,\xi} &:= \{f : f \in \mathsf{F}^\Omega \ \&\ \forall x' \in X',\ f \in L_1(\Omega, \mathcal{D}, x' \circ \xi; \mathsf{F})\} \\ &= \cap_{x' \in X'} L_1(\Omega, \mathcal{D}, x' \circ \xi; \mathsf{F}). \end{aligned}$$

It follows at once from Definition 3.7 that $\mathcal{G}_{1,\mu}$ and $\mathcal{G}_{1,\xi}$ are vector spaces over F, and

$$(3.8)\qquad \begin{cases} \text{(a)} & \mathcal{G}_{1,\mu} \subseteq \mathcal{M}(\mathcal{D}^{\mathrm{loc}}, \mathcal{W}(X)) \\ \text{(b)} & \mathcal{G}_{1,\xi} \subseteq \mathcal{M}(\mathcal{D}^{\mathrm{loc}}, Bl(\mathsf{F})), \end{cases}$$

where $\mathcal{W}(X)$ is the σ-algebra generated by the base $\mathcal{N}_w$ of weak neighborhoods of X, cf. [12: 2.1 (e)].[7] Next, membership in these spaces is easily expressible in terms of the L_1-norm defined in (2.4); thus:

$$(3.9)\qquad \begin{cases} \text{(a)} & f \in \mathcal{G}_{1,\mu} \Leftrightarrow \forall x' \in X',\ |x' \circ f|_{1,\mu} < \infty; \\ \text{(b)} & f \in \mathcal{G}_{1,\xi} \Leftrightarrow \forall x' \in X',\ |f|_{1,x' \circ \xi} < \infty. \end{cases}$$

It is natural to introduce the following *x'-independent norms of f*[8]:

$$(3.10)\qquad \begin{cases} \text{(a)} & \forall f \in \mathcal{M}(\mathcal{D}^{\mathrm{loc}}, \mathcal{W}(X)),\ \|f\|_{1,\mu} = \sup\limits_{\substack{x' \in X' \\ |x'| \leq 1}} |x' \circ f|_{1,\mu} \leq \infty \\ \text{(b)} & \forall f \in \mathcal{M}(\mathcal{D}^{\mathrm{loc}}, Bl(\mathsf{F})),\ |f|_{1,\xi} = \sup\limits_{\substack{x' \in X' \\ |x'| \leq 1}} |f|_{1,x' \circ \xi} \leq \infty. \end{cases}$$

On applying Theorem 3.2 (b), taking $Y = X'$, $Z = \mathsf{F}$, and first letting

$$\eta(x')(D) := \int_D (x' \circ f)(\omega) \cdot \mu(d\omega),\ D \in \mathcal{D},$$

where $f \in \mathcal{M}(\mathcal{D}^{\mathrm{loc}}, \mathcal{W}(X))$ and $\|f\|_{1,\mu} < \infty$, and then letting

$$\eta(x')(D) := \int_D f(\omega) \cdot (x' \circ \xi)(d\omega),\ D \in \mathcal{D},$$

where $f \in \mathcal{M}(\mathcal{D}^{\mathrm{loc}}, Bl(\mathsf{F}))$ and $|f|_{1,\xi} < \infty$, we get the following important result:

3.11. Theorem. *Let $\mu \in CA(\mathcal{D}, \mathsf{F}) \cup CA(\mathcal{D}, [0, \infty])$ and $\xi \in CA(\mathcal{D}, X)$. Then*
(a) $\mathcal{G}_{1,\mu} = \{f : f \in \mathcal{M}(\mathcal{D}^{\mathrm{loc}}, \mathcal{W}(X))\ \&\ \|f\|_{1,\mu} < \infty\}$
(b) $\mathcal{G}_{1,\xi} = \{f : f \in \mathcal{M}(\mathcal{D}^{\mathrm{loc}}, Bl(\mathsf{F}))\ \&\ \|f\|_{1,\xi} < \infty\}$.

From (3.9) (b), we easily see that all F-valued $\mathcal{D}$-simple functions are in $\mathcal{G}_{1,\xi}$. Also, by (3.10) (b) and 3.11 (b), $|\cdot|_{1,\xi}$ is a norm on the vector space $\mathcal{G}_{1,\xi}$, when functions f, g differing on ξ-negligible sets are identified. We can show that $\mathcal{G}_{1,\xi}$ is actually a Banach space, and so sum up the information on this space as follows:

3.12. Proposition. *Let $\xi \in CA(\mathcal{D}, X)$. Then*

---

[7] $\mathcal{M}(\mathcal{D}^{\mathrm{loc}}, \mathcal{W}(X))$ comprises all f in X^Ω such that $\forall x' \in X'$, $x' \circ f \in \mathcal{M}(\mathcal{D}^{\mathrm{loc}}, Bl(\mathsf{F}))$. The different concepts of "Borel measurability" that exist for functions f in X^Ω are discussed in [12: §§2,3].

[8] We write $\| f \|_{1,\mu}$ instead of $|f|_{1,\mu}$, as it is natural to let

$$|f|_{1,\mu} := \int_\Omega |f(\omega)|_X |\nu|(d\omega).$$

Obviously, $\| f \|_{1,\mu} \leq |f|_{1,\mu} \leq \infty$.

(a) $\mathcal{G}_{1,\xi}$ *is a Banach space over* F *under the norm* $|\cdot|_{1,\xi}$, *when functions* f, g *in* $\mathcal{G}_{1,\xi}$ *such that* $\text{supp}(f-g) \in \mathcal{N}_\xi$ *are identified;*
(b) $\mathcal{S}(\mathcal{D}, \mathsf{F})$ *is a linear manifold in* $\mathcal{G}_{1,\xi}$;
(c) *For all Cauchy sequences* $(g_n)_{n=1}^\infty$ *in* $\mathcal{G}_{1,\xi}$, $\exists$ *a subsequence* $(g_{n_k})_{k=1}^\infty$ *and* $\exists\, g \in \mathcal{G}_{1,\xi}$ *such that* $\lim_{k\to\infty} g_{n_k}(\cdot) = g(\cdot)$, *a.e.* (ξ) *on* Ω;
(d) $\forall f \in \mathcal{G}_{1,\xi}$, $|f|_{1,\xi} = 0$ *iff* $\text{supp} f \in \mathcal{N}_\xi$;
(e) $\forall A \in \mathcal{D}^{\text{loc}}$, $|\chi_A|_{1,\xi} = s_\xi(A)$.

For the space $\mathcal{G}_{1,\mu}$, we have only the following weaker result:

3.13. Triviality. *Let* $\mu \in CA(\mathcal{D}, \mathsf{F}) \cup CA(\mathcal{D}, [0,\infty])$. *Then*
(a) $\mathcal{G}_{1,\mu}$ *is a pre-Banach space over* F *under the norm* $\|\cdot\|_{1,\mu}$, *when functions* f, g *in* $\mathcal{G}_{1,\mu}$ *such that* $\|f-g\|_{1,\mu} = 0$ *are identified*[9];
(b) $\mathcal{S}(\mathcal{D}, X)$ *is a linear manifold in* $\mathcal{G}_{1,\mu}$.

Better results on $\mathcal{G}_{1,\mu}$ can be had by imposing the requirements that X is separable, but these considerations are not germane to our work, nor is the consideration of the two Gelfand integrals that can be defined for $f \in \mathcal{G}_{1,\mu}$ and $g \in \mathcal{G}_{1,\xi}$ respectively, viz.

$$\int_\Omega f(\omega)\mu(d\omega) \in X'' \;\;\&\;\; \int_\Omega g(\omega)\xi(d\omega) \in X''.$$

4. THE TWO PETTIS INTEGRALS

In this section it is to be understood that

$$\mu \in CA(\mathcal{D}, \mathsf{F}), \quad \xi \in CA(\mathcal{D}, X). \tag{4.1}$$

The next two definitions introduce the two Pettis classes:

4.2. Definition. Let $f \in X^\Omega$ & $C \in \mathcal{D}^{\text{loc}}$. Then we say that
(a) f is *Pettis integrable on* C with respect to μ iff $\exists x_C \in X$ such that $\forall x' \in X'$,

$$x' \circ f \in L_1(\Omega, \mathcal{D}, \mu; \mathsf{F}) \;\;\&\;\; x'(x_C) = \int_C x'\{f(\omega)\}\mu(d\omega);$$

(b) f is *Pettis integrable over* $\mathcal{C} \subseteq \mathcal{D}^{\text{loc}}$ with respect to μ, iff $\forall C \in \mathcal{C}$, f is Pettis integrable on C with respect to μ;

(c)
$$\begin{aligned}\mathcal{P}_{1,\mu} &:= \mathcal{P}_1(\Omega, \mathcal{D}, \mu; X)\\ &:= \{f : f \in X^\Omega \;\&\; f \text{ is Pettis integrable over } \mathcal{D}^{\text{loc}} \text{ with respect to } \mu\};\end{aligned}$$

(d)
$$\forall C \in \mathcal{D}^{\text{loc}} \;\&\; \forall f \in \mathcal{P}_{1,\mu}, \quad \int_C f(\omega)\mu(d\omega) := \textit{the } x_C \textit{ in } (a);$$

$$\forall f \in \mathcal{P}_{1,\mu}, \quad E_\mu(f) := \int_\Omega f(\omega)\mu(d\omega).$$

---

[9] This condition is not in general replaceable by $\text{supp}(f-g) \in \mathcal{N}_\mu$.

4.3. Definition. Let $f \in \mathsf{F}^\Omega$ & $C \in \mathcal{D}^{\text{loc}}$. Then we say that

(a) f is *Pettis integrable on* C with respect to ξ, iff $\exists x_C \in X$ such that $\forall x' \in X'$,

$$f \in L_1(\Omega, \mathcal{D}^{\text{loc}}, |x' \circ \xi|; \mathsf{F}) \;\&\; x'(x_C) = \int_C f(\omega) \cdot x'\{\xi(d\omega)\};$$

(b) f is *Pettis integrable over* $\mathcal{C} \subseteq \mathcal{D}^{\text{loc}}$ with respect to ξ, iff $\forall C \in \mathcal{C}$, f is Pettis integrable on C with respect to ξ;

(c)
$$\mathcal{P}_{1,\xi} := \mathcal{P}_1(\Omega, \mathcal{D}, \xi; \mathsf{F})$$
$$:= \{f : f \in \mathsf{F}^\Omega \;\&\; f \quad \text{is Pettis integrable over } \mathcal{D}^{\text{loc}} \text{ with respect to } \xi\};$$

(d)
$$\forall C \in \mathcal{D}^{\text{loc}} \;\&\; \forall f \in \mathcal{P}_{1,\xi}, \quad \int_C f(\omega)\xi(d\omega) := \textit{the } x_C \textit{ in } (a);$$
$$\forall f \in \mathcal{P}_{1,\xi}, \quad E_\xi(f) := \int_C f(\omega)\xi(d\omega).$$

It follows at once from 4.2 (c),(a), 3.8 (a) and (3.9) (a) and from 4.3 (c), (a), 3.8 (b) and (3.9) (b), that

$$(4.4) \qquad \begin{cases} (a) \quad \mathcal{P}_{1,\mu} \subseteq \mathcal{G}_{1,\mu} \subseteq \mathcal{M}(\mathcal{D}^{\text{loc}}, \mathcal{W}(X)) \\ (b) \quad \mathcal{P}_{1,\xi} \subseteq \mathcal{G}_{1,\xi} \subseteq \mathcal{M}(\mathcal{D}^{\text{loc}}, Bl(\mathsf{F})), \end{cases}$$

where $\mathcal{W}(X)$ is the σ-algebra generated by the base $\mathcal{N}_w$ of weak neighborhoods of X.

The fundamental properties of $\mathcal{P}_{1,\xi}$ and E_ξ have been worked out by Lewis [11]. Central are those listed in the next two theorems. The parts (a), (c) of the first and parts (b), (c) of the second are due to Lewis [11]. The other parts are obvious.

4.5. Theorem. *Let* $f \in \mathcal{P}_{1,\xi}$. *Then*

(a) $\forall A \in \mathcal{D}^{\text{loc}}, \chi_A \cdot f \in \mathcal{P}_{1,\xi}$, & *with* $\eta(A) := \int_A f(\omega)\xi(d\omega) = E_\xi(\chi_A \cdot f)$, *we have* $\eta \in CA(\mathcal{D}^{\text{loc}}, X)$; *thus* η *is bounded on* $\mathcal{D}^{\text{loc}}$;

(b) $\forall A \in \mathcal{D}^{\text{loc}}$, $s_\eta(A) = \sup_{\substack{x' \in X' \\ |x'| \leq 1}} \int_A |f(\omega)| \cdot |x' \circ \xi|(d\omega) \in \mathbf{R}_{0+}$;

(c) $\exists$ *a monotone increasing sequence* $(D_n)_1^\infty$ *in* $\mathcal{D}$ *such that*

$$\forall n \geq 1, \; s_\eta(\Omega \setminus D_n) < 1/n, \;\&\; s_\eta(\Omega \setminus \bigcup_{k=1}^\infty D_k) = 0.$$

4.6. Theorem.

(a) $\mathcal{P}_{1,\xi}$ *is a linear manifold in the Banach space* $\mathcal{G}_{1,\xi}$, *and* E_ξ *is a linear contraction on* $\mathcal{P}_{1,\xi}$ *into* X, *i.e.*

$$\forall f \in \mathcal{P}_{1,\xi}, \; |E_\mu(f)|_X \leq |f|_{1,\xi};$$

(b) $\mathcal{S}(\mathcal{D}, \mathsf{F}) \subseteq \mathcal{P}_{1,\xi}$, *cf. 2.1 (f), and for a* $\mathcal{D}$*-simple,* F*-valued* $f = \sum_{k=1}^r a_k \chi_{D_k}$, *we have* $E_\xi(f) = \sum_{k=1}^r a_k \xi(D_k)$.

(c) *(Dominated Convergence). Let* (i) $(f_n)_1^\infty$ *be a sequence in* $\mathcal{P}_{1,\xi}$, (ii) $|f_n(\cdot)| \leq g(\cdot) \in \mathcal{P}_{1,\xi}$ *and* (iii) $f_n(\cdot) \to f(\cdot)$ *on* Ω, *as* $n \to \infty$. *Then*

$$f \in \mathcal{P}_{1,\xi} \;\&\; \forall A \in \mathcal{D}^{\text{loc}}, \; \lim_{n\to\infty} \int_A f_n(\omega)\xi(d\omega) = \int_A f(\omega)\xi(d\omega);$$

moreover, the convergence is uniform with respect to A in $\mathcal{D}^{\text{loc}}$.

(d) *(Simple function approximation.) Let $f \in \mathcal{P}_{1,\xi}$. Then $\forall n \geq 1$, $\exists$ a $\mathcal{D}$-simple function s_n on Ω to F such that*

$$\forall x' \in X', \int_\Omega |f(\omega) - s_n(\omega)||x' \circ \xi|(d\omega) \leq \frac{|x'|}{n},$$

&

$$\forall A \in \mathcal{D}^{\text{loc}}, \int_A f(\omega)\xi(d\omega) = \lim_{n\to\infty} \int_A s_n(\omega)\xi(d\omega).$$

Theorem 4.6 (d) yields the following useful result:

4.7. Corollary. w.cls Range E_ξ = cls Range E_ξ = S_ξ = w.cls < Range ξ >, *where* w.cls *stands for "weak closure", and $S_\xi := \mathfrak{S}\{\text{Range } \xi\}$, cf. 1.10 (c).*

Theorem 4.6 (d) tells us that

$$\mathcal{P}_{1,\xi} \subseteq \text{ cls } \mathcal{S}(\mathcal{D},\mathsf{F}) \subseteq \mathcal{G}_{1,\xi},$$

the closure being taken in the Banach space $\mathcal{G}_{1,\xi}$. Conversely, let $f \in$ cls $\mathcal{S}(\mathcal{D},\mathsf{F})$, $C \in \mathcal{D}^{\text{loc}}$, and $|f - s_n|_{1,\xi} < 1/n$. Then we can show that the sequence $(x_n)_1^\infty$, where $x_n := E_\xi(\chi_C \cdot s_n)$, is Cauchy in X, and that its limit y_C satisfies

$$x'(y_C) = E_{x' \circ \xi}(\chi_C \cdot f), \quad \forall x' \in X',$$

and therefore that $f \in \mathcal{P}_{1,\xi}$. We thus get the following important result which suggests that Pettis integration (unlike Gelfand integration) resembles Lebesgue integration.

4.8. Theorem.

(a) *$\mathcal{P}_{1,\xi}$ is the closure of the linear manifold $\mathcal{S}(\mathcal{D},\mathsf{F})$ in the Banach space $\mathcal{G}_{1,\xi}$, cf. 3.8 (b); briefly,* $\mathcal{P}_{1,\xi} =$ cls $\mathcal{S}(\mathcal{D},\mathsf{F})$.

(b) $\forall f \in \mathcal{P}_{1,\xi}$ & $\forall (s_n)_{n=1}^\infty$ *in* $\mathcal{S}(\mathcal{D},\mathsf{F})$ *for which* $|s_n - f|_{1,\xi} \to 0$, *as* $n \to \infty$, *we have*

$$\forall A \in \mathcal{D}^{\text{loc}}, \quad \int_A f(\omega)\xi(d\omega) = \lim_{n\to\infty} \int_A s_n(\omega)\xi(d\omega);$$

(c) *$\mathcal{P}_{1,\xi}$ is a Banach space under the norm $|\cdot|_{1,\xi}$ when functions f, g in $\mathcal{P}_{1,\xi}$, for which* supp $(f-g) \in \mathcal{N}_\xi$, *are identified.*

Since for $s_n \in \mathcal{S}(\mathcal{D},\mathsf{F})$, we have $|s_n(\cdot)| \in \mathcal{S}(\mathcal{D},\mathsf{R}_{0+})$, and

$$\big||s_n(\cdot)| - |f(\cdot)|\big|_{1,\xi} \leq \big||s_n(\cdot) - f(\cdot)|\big|_{1,\xi} = |s_n - f|_{1,\xi}$$

we see at once that whenever $f(\cdot)$ is in cls $\mathcal{S}(\mathcal{D},\mathsf{F})$, so in it is $|f(\cdot)|$. Thus from Theorem 4.8 (a), we may immediately conclude that

$$f \in \mathcal{P}_{1,\xi} \Rightarrow |f(\cdot)| \in \mathcal{P}_{1,\xi}. \tag{4.9}$$

We turn next to two important results, the first on the support of a ξ-integrable function and the second on a domination condition sufficient to ensure ξ-integrability.

4.10. Theorem. (Supp$f$, for $f \in \mathcal{P}_{1,\xi}$). *Let $f \in \mathcal{P}_{1,\xi}$, and $\forall A \in \mathcal{D}^{\text{loc}}$, $\eta(A) := E_\xi(\chi_A \cdot f)$. Then $\exists$ a $\uparrow$ sequence $(\Delta_k)_{k=1}^\infty$ in $\mathcal{D}$ and $\exists N \in \mathcal{N}_\eta \cap \mathcal{N}_\xi$ $(= \mathcal{N}_\xi)$ such that*

(a) $S :=$ supp $f = N \cup \bigcup_{k=1}^\infty \Delta_k$ & $N \parallel \bigcup_{k=1}^\infty \Delta_k$.

(b) $\exists B \in \sigma\text{-}ring(\mathcal{D})$ *such that* $B \subseteq S$ & $S \setminus B \in \mathcal{N}_\xi$,
(c) $\forall k \in \mathbb{N}_+$, $\Delta_k \subseteq S$ & $\lim_{n\to\infty} s_\eta(\Omega \setminus \Delta_n) = 0$.

4.11. Theorem. (Domination Principle). *Let*
(i) $f \in \mathcal{M}(\mathcal{D}^{\mathrm{loc}}, Bl(\mathsf{F}))$,
(ii) $|f(\cdot)| \leq \varphi(\cdot) \in \mathcal{P}_1(\Omega, \mathcal{D}, \xi; \mathbb{R}_{0+})$.
Then $f \in \mathcal{P}_{1,\xi}$.

Two corollaries of the Domination Principle 4.11 follow immediately, the first on taking $\varphi = r \cdot \chi_A$, where $r \in \mathbb{R}_+$ and $\chi_A \in \mathcal{P}_{1,\xi}$, the second on taking $\varphi(\cdot) = |f(\cdot)|$.

4.12. Corollary. *Let* $f \in \mathcal{M}(\mathcal{D}^{\mathrm{loc}}, Bl(\mathsf{F}))$ *and* $|f(\cdot)| \leq r \cdot \chi_A(\cdot)$, *where* $r \in \mathbb{R}_+$ *and* $A \in \mathcal{D}^{\mathrm{loc}}$ & $\chi_A \in \mathcal{P}_{1,\xi}$. *Then* $|f(\cdot)| \in \mathcal{P}_{1,\xi}$.

4.13. Proposition. (Equivalence). *Let* $f \in \mathcal{M}(\mathcal{D}^{\mathrm{loc}}, Bl(\mathsf{F}))$. *Then the membership in* $\mathcal{P}_{1,\xi}$ *of any one of the following blocks, separated by semi-colons, entails the membership in* $\mathcal{P}_{1,\xi}$ *of all of them:*

$$f;\ |f(\cdot)|;\ \overline{f};\ \mathrm{Re}\, f\ \&\ \mathrm{Im}\, f;\ (\mathrm{Re}\, f)_\pm\ \&\ (\mathrm{Im}\, f)_\pm.$$

It should be noted that for Banach spaces X such as c_0, there are measures $\xi \in CA(\mathcal{D}, X)$ for which $\mathcal{P}_{1,\xi}$ is a proper subset of $\mathcal{G}_{1,\xi}$. This is shown in the next example:

4.14. Fundamental Example. *Let*
(i) Ω *be an infinite set equipped with the discrete topology,*
(ii) $X := C_o(\Omega, \mathsf{F}) = \{f : f$ *is bounded on* Ω *to* F & $\lim_{\omega\to\infty} |f(\omega)| = 0\}$,
(iii) $\mathcal{D} = \{D : D \subseteq \Omega$ & D *is finite*$\}$,
(iv) $\forall D \in \mathcal{D}$, $\xi(D) := \chi_D$.
Then
(a) X *is a Banach space over* F *under the sup norm,* $\mathcal{D}$ *is a* δ*-ring over* Ω & $\mathcal{D}^{\mathrm{loc}} = 2^\Omega = Bl(\Omega)$;
(b) $\xi \in CA(\mathcal{D}, X)$;
(c) $\forall \mu \in CA(2^\Omega, \mathsf{F})$, $E_\mu \circ \xi = \mu$ *on* $\mathcal{D}$ & $|E_\mu \circ \xi|(\cdot) = |\mu|(\cdot)$ *on* 2^Ω;
(d) $\forall f \in \mathcal{M}(\mathcal{D}^{\mathrm{loc}}, Bl(\mathsf{F}))$, $|f|_{1,\xi} = |f|_\infty$ *and* $B(\Omega, \mathsf{F}) := \{f : f$ *is bounded on* Ω *to* $\mathsf{F}\} = \mathcal{G}_{1,\xi}$;
(e) $\mathcal{P}_{1,\xi} = X$, $E_\xi = I_X$ & $s_\xi(\cdot) = 1$ *on* $\mathcal{D}^{\mathrm{loc}} \setminus \{\emptyset\}$.

4.15. Remarks. In the light of Theorem 4.8, the question arises as to whether the Pettis integral E_ξ, for $\xi \in CA(\mathcal{D}, X)$ is a Lebesgue integral. The answer will of course depend on the interpretation we give to the term "Lebesgue integral". The classical definition of the Lebesgue class, involving the total variation measure $|\xi|(\cdot)$ of ξ, cf. (2.4) and (2.5) (a), viz.,

$$L_{1,\xi} = \{f : f \in \mathcal{M}(\mathcal{D}^{\mathrm{loc}}, Bl(\mathsf{F}))\ \&\ |f|_{1,|\xi|} < \infty\},$$

where $|\xi|$ is the total variation of ξ and

$$|f|_{1,|\xi|} := \int_\Omega |f(\omega)| \cdot |\xi|(d\omega),$$

is inadequate, since there are many important measures ξ in $CA(\mathcal{D}, X)$ for which

$$\forall D \in \mathcal{D},\ |\xi|(D) = 0 \text{ or } |\xi|(D) = \infty.$$

For such ξ, the last $L_{1,\xi}$ will collapse to the unit-set comprising exclusively the equivalence class [0] of the zero-function.

A more adequate procedure would be to take instead of $|\cdot|_{1,|\xi|}$, the norm $|\cdot|_{1,\xi}$ defined in (3.10) (b). This would make $L_{1,\xi} = \mathcal{G}_{1,\xi}$. This, however, would deprive the Lebesgue integral from generally having the Pettis property, since in general $\mathcal{P}_{1,\xi}$ is a proper subspace of $\mathcal{G}_{1,\xi}$, cf. Example 4.14 (d), (e).

An option open to us is to define

$$(*) \qquad L_{1,\xi} := \text{cls } \mathcal{S}(\mathcal{D}, \mathsf{F}), \text{ (closure in the Banach space } \mathcal{G}_{1,\xi})$$

and then define $\int_A f(\omega)\xi(d\omega)$ to be the limit of the integrals of simple functions, as in 4.8 (b). Then, cf. Theorem 4.8, we get all the properties expected of the Lebesgue integral, including the Pettis property. From a structural standpoint, however, the definition (*) has the shortcoming that it gives the L_1-norm $|\cdot|_{1,\xi}$, a definiens involving the entire unit ball of the dual space X' of X, and is thus heavily "Pettis", cf. (3.10). In this approach the L_1 theory has to await the development of a good deal of the Pettis theory. It would be much better to define the L_1-norm, ab initio, in an entirely dual-free manner.

From this standpoint Professor Thomas's paper [21] (unpublished) is of considerable interest. It develops the integration of F-valued functions with respect to measures $\xi \in CA(\mathcal{D}, X)$, where X is not necessarily a Banach space. His definition of the integral makes sense even for so-called F-spaces for which $X' = \{0\}$. Professor Thomas shows moreover that his integral, which by necessity has a dual-free definition, is also Pettis when the F-space X happens to be a Banach space. It seems clear that his concepts of integrability and integration coincide for Banach spaces X with our $\mathcal{P}_{1,\xi}$ and E_ξ, but the nexus between his norm and the norms (3.10) remain to be found.

In the present paper we shall not adopt the definition (*), and continue to write $\mathcal{P}_{1,\xi}$ instead of $L_{1,\xi}$. This is both because of the shortcomings of the norm $|\cdot|_{1,\xi}$ of (3.10) from the Lebesgue standpoint, and because to prove the Tonelli and Fubini theorems we are obliged to hit with functionals in X', and deal with scalars.

We turn finally to the effects of a linear transformation of the class $\mathcal{P}_{1,\xi}$ and on ξ-integration. The following transformation law is routinely proved.

4.16. Lemma. (Transformation Rule). *Let* (i) Y *be a Banach space over* F, *and* (ii) $T \in CL(X,Y)$. *Then*

(a) $T \circ \xi \in CA(\mathcal{D}, Y)$ & $\mathcal{P}_{1,\xi} \subseteq \mathcal{P}_{1,T\circ\xi}$;

(b) *for bijective* T, $\mathcal{P}_{1,\xi} = \mathcal{P}_{1,T\circ\xi}$;

(c) $\forall f \in \mathcal{P}_{1,\xi}$ & $\forall A \in \mathcal{D}^{\text{loc}}$, $T\{\int_A f(\omega)\xi(d\omega)\} = \int_A f(\omega)T\{\xi(d\omega)\}$.

5. THE δ-RING $\mathcal{D}_\xi$ OF SETS WITH ξ-INTEGRABLE INDICATORS AND THE δ-RING $\overline{\mathcal{D}}_\xi$ OF SETS WITH FINITE SEMI-VARIATIONS

To complete the study of Pettis integration, we have to consider along with the measure ξ on the δ-ring $\mathcal{D}$, the class $\mathcal{D}_\xi$ of so-called "ξ-integrable sets", as well as the extension $\overline{\xi}$ of ξ from $\mathcal{D}$ to $\mathcal{D}_\xi$. The novel questions of measurability that arise have to be addressed.

We now need to augment the notation introduced in (4.1), to which we shall adhere, by stipulating that

$$(5.1) \qquad \begin{cases} \text{(a)} \quad \mathcal{D}_\xi := \{A : A \in \mathcal{D}^{\text{loc}} \ \& \ \chi_A \in \mathcal{P}_{1,\xi}\}, \\ \text{(b)} \quad \forall A \in \mathcal{D}_\xi, \ \overline{\xi}(A) = E_\xi(\chi_A). \end{cases}$$

The notation $\mathcal{D}_\xi$ extends the notation $\mathcal{D}_\mu$ introduced in 2.6 (c). The equality asserted in 2.6 (c) does not, of course, prevail. It is the semi-variation s_ξ, not the total variation $|\xi|$, that is now germane. We have to consider the family

$$(5.2) \qquad \overline{\mathcal{D}}_\xi := \{C : C \in \mathcal{D}^{\text{loc}} \ \& \ s_\xi(C) < \infty\}.$$

The equality $\mathcal{D}_\xi = \overline{\mathcal{D}}^{\mathrm{loc}}$, corresponding to the one asserted in 2.6 (c), is valid for a very important subclass of Banach spaces X (cf. §10), but not for all as we shall see in Example 5.8.

We now turn to the precise relation between $\mathcal{D}, \mathcal{D}_\xi, \overline{\mathcal{D}}_\xi, \mathcal{D}^{\mathrm{loc}}, \mathcal{D}_\xi^{\mathrm{loc}}, \overline{\mathcal{D}}_\xi^{\mathrm{loc}}$ and between s_ξ and $s_{\overline{\xi}}, \mathcal{G}_{1,\xi}$ and $\mathcal{G}_{1,\overline{\xi}}, \mathcal{P}_{1,\xi}$ and $\mathcal{P}_{1,\overline{\xi}}, E_\xi$ and $E_{\overline{\xi}}$. We shall deal with $\overline{\mathcal{D}}_\xi$ last.

5.3. Main Lemma.
(a) $\mathcal{D} \subseteq \mathcal{D}_\xi \subseteq \sigma\text{-ring}(\mathcal{D}_\xi) \subseteq \sigma\text{-alg}(\mathcal{D}_\xi) \subseteq \mathcal{D}^{\mathrm{loc}} = \mathcal{D}_\xi^{\mathrm{loc}}$;
(b) $\mathcal{D}_\xi$ *is a* δ*-ring* $\subseteq \mathcal{D}^{\mathrm{loc}}$;
(c) $\xi \subseteq \overline{\xi} \in CA(\mathcal{D}_\xi, X)$;
(d) $\forall x' \in X'$, $|x' \circ \xi| = |x' \circ \overline{\xi}|$ *on* $\mathcal{D}^{\mathrm{loc}}$;
(e) $s_\xi = s_{\overline{\xi}}$ *on* $\mathcal{D}^{\mathrm{loc}}$.

In this the crucial assertion is that of the last equality in (a). A corollary of Lemma 5.3 and Theorem 4.10 on supports is the result that every set in $\mathcal{D}_\xi$ is, so-to-speak, "ξ-essentially" in σ-ring($\mathcal{D}$). More precisely, we have:

5.4. Proposition. *Let* $A \in \mathcal{D}_\xi$. *Then* $\exists$ *a* $\uparrow$ *sequence* $(\Delta_k)_{k=1}^\infty$ *in* $\mathcal{D}$ *and* $\exists N \in \mathcal{N}_\xi$ *such that*
(a) $A = N \cup \bigcup_{k=1}^\infty \Delta_k \;\&\; N \parallel \bigcup_{k=1}^\infty \Delta_k \in \sigma\text{-ring}(\mathcal{D})$.
(b) $N \in \mathcal{D}_\xi,\ \overline{\xi}(N) = 0 \;\&\; \overline{\xi}(A) = \lim_{n\to\infty} \xi(\Delta_n)$.

The converse of 5.4 (b) that suggests itself, viz.

$$(\Delta_k)_{k=1}^\infty \textit{ is } \uparrow \textit{ sequence in } \mathcal{D} \;\&\; x := \lim_{n\to\infty} \xi(\Delta_n) \textit{ exists } \in \mathcal{D}$$

$$\Rightarrow \bigcup_{k=1}^\infty \Delta_k \in \mathcal{D}_\xi \;\&\; \overline{\xi}\Big(\bigcup_{k=1}^\infty \Delta_k\Big) = x$$

fails even for $X = \mathbf{R}$ as the following simple example shows:

5.5. Example. *Let* $\Omega := \mathbf{N}_+$,

$$\mathcal{D} := \{D : D \subseteq \mathbf{N}_+ \;\&\; D \textit{ is finite}\}$$

$$\forall D \in \mathcal{D},\ \xi(D) = \sum\nolimits_{k\in D} (-1)^{k-1}/k$$

$$\forall n \in \mathbf{N}_+,\ D_n = \{1, 2, \ldots, n\}.$$

Then $D_n \uparrow \Omega \;\&\; \xi(D_n) \to \log 2$, *but* $\Omega \notin \mathcal{D}_\xi$.

Integrability and integration also turn out to be the same for ξ and $\overline{\xi}$. We have:

5.6. Theorem.
(a) $\forall f \in \mathcal{M}(\mathcal{D}^{\mathrm{loc}}, Bl(\mathbf{F}))$, $|f|_{1,\xi} = |f|_{1,\overline{\xi}} \leq \infty$.
(b) $\mathcal{G}_{1,\xi} = \mathcal{G}_{1,\overline{\xi}}$ & $\forall f \in \mathcal{G}_{1,\xi}$, $|f|_{1,\xi} = |f|_{1,\overline{\xi}} < \infty$.
(c) $\mathcal{S}(\mathcal{D}_\xi, \mathbf{F}) \subseteq \mathcal{P}_{1,\xi}$, *and for a* $\mathbf{F}$*-valued*, $\mathcal{D}_\xi$*-simple* $f = \sum_{k=1}^r a_k \chi_{A_k}$ *in standard form*, $E_\xi(f) = \sum_{k=1}^r a_k \overline{\xi}(A_k)$.
(d) $\mathcal{P}_{1,\xi} = \mathcal{P}_{1,\overline{\xi}}$ *and* $E_\xi(\cdot) = E_{\overline{\xi}}(\cdot)$.
(e) $\mathcal{D}_{\overline{\xi}} = \mathcal{D}_\xi, \overline{\overline{\xi}} = \overline{\xi}$.
(f) $\forall x' \in X'$, $\overline{x' \circ \overline{\xi}} = \overline{x' \circ \xi}$.

Turning to the family $\overline{\mathcal{D}}_\xi$ defined in (5.2), we have in analogy with 5.3 (a), (b).

5.7. Lemma.
(a) $\mathcal{D}_\xi \subseteq \overline{\mathcal{D}}_\xi \subseteq \sigma$-ring $(\overline{\mathcal{D}}_\xi) \subseteq \sigma$-alg $(\overline{\mathcal{D}}_\xi) \subseteq \mathcal{D}^{\mathrm{loc}} = \mathcal{D}_\xi^{\mathrm{loc}} = \overline{\mathcal{D}}_\xi^{\mathrm{loc}}$;
(b) $\overline{\mathcal{D}}_\xi =$ *a σ-ring* $\subseteq \mathcal{D}_\xi^{\mathrm{loc}}$.

In this, as in 5.3, the crucial assertion is that of the last equality in (a). That the inclusion $\mathcal{D}_\xi \subseteq \overline{\mathcal{D}}_\xi$ asserted in 5.7 (a) is in general proper is shown by the following example:

5.8. Example.
(a) *Let $\Omega, X, \mathcal{D}, \xi$ be as in Example 4.14. Then*

$$\mathcal{D}_\xi = \mathcal{D}, \text{ but } \overline{\mathcal{D}}_\xi = 2^\Omega.$$

(b) *In* (a) *let Ω have the cardinality of* R. *Then*

$$\sigma\text{-ring } (\mathcal{D}_\xi) \subsetneq \sigma\text{-ring}(\overline{\mathcal{D}}_\xi).$$

Obviously σ-ring$(\mathcal{D}) \subseteq \sigma$-ring$(\mathcal{D}_\xi)$, but even for $X = R$, equality may fail as is easy to check. The best we are able to say on this issue is:

$$\sigma\text{-ring}(\mathcal{D}_\xi) = \sigma\text{-ring}(\mathcal{D}) \cup [\mathcal{N}_\xi \cap \sigma\text{-ring}(\mathcal{D}_\xi)]. \tag{5.9}$$

Roughly, every set in σ-ring$(\mathcal{D}_\xi)$ is "ξ-essentially" contained in the smaller σ-ring$(\mathcal{D})$.

As a final remark in this section, we should point out that care must be excerised in applying the bar to the inclusion relations of measures in $CA(\mathcal{D}, X)$. The implication

$$\xi_1 \subseteq \xi_2 \Rightarrow \overline{\xi}_1 \subseteq \overline{\xi}_2$$

does *not* prevail in general. Indeed we may have $\xi_1 \subsetneq \xi_2$ and $\overline{\xi}_2 \subsetneq \overline{\xi}_1$. This happens, since for $\xi_1 \in CA(\mathcal{D}_1, X), \xi_2 \in CA(\mathcal{D}_2, X)$, the definitions of $\overline{\xi}_1, \overline{\xi}_2$ involve $\mathcal{D}_1^{\mathrm{loc}}, \mathcal{D}_2^{\mathrm{loc}}$, cf. 5.1 (a), and we may have $\mathcal{D}_2^{\mathrm{loc}} \subsetneq \mathcal{D}_1^{\mathrm{loc}}$ even when $\mathcal{D}_1 \subsetneq \mathcal{D}_2$. When in addition to $\xi_1 \subseteq \xi_2$, we are given that $\mathcal{D}_2 \subseteq \mathcal{D}_1^{\mathrm{loc}}$, we have the following result:

5.10. Triviality. *Let* (i) $\xi_1 \in CA(\mathcal{D}_1, X)$, $\xi_2 \in CA(\mathcal{D}_2, X)$, (ii) $\xi_1 \subseteq \xi_2$ *and* (iii) $\mathcal{D}_1 \subseteq \mathcal{D}_2 \subseteq \mathcal{D}_1^{\mathrm{loc}}$. *Then*
(a) $\mathcal{D}_2^{\mathrm{loc}} \subseteq \mathcal{D}_1^{\mathrm{loc}}$;
(b) $\forall A \in \mathcal{D}_2^{\mathrm{loc}}$ & $\forall x' \in X'$, $|x' \circ \xi_1|(A) \leq |x' \circ \xi_2|(A)$;
(c) $\forall A \in \mathcal{D}_2^{\mathrm{loc}}$, $s_{\xi_1}(A) \leq s_{\xi_2}(A)$;
(d) $\forall f \in \mathcal{M}(\mathcal{D}_2^{\mathrm{loc}}, Bl(\mathsf{F}))$, $|f|_{1,\xi_1} \leq |f|_{1,\xi_2}$;
(e) $\mathcal{G}_{1,\xi_2} \subseteq \mathcal{G}_{1,\xi_1}$.

An interesting question is as to what happens when the premiss $\xi_1 \subseteq \xi_2$, $\mathcal{D}_2 \subseteq \mathcal{D}_1^{\mathrm{loc}}$ is strengthened to $\xi_1 \subseteq \xi_2 \subseteq \overline{\xi}_1$. In this case it turns out that $\overline{\xi}_2 \subseteq \overline{\xi}_1$, as the following useful proposition reveals.

5.11. Proposition. *Let* (i) $\xi_1 \in CA(\mathcal{D}_1, X)$, $\xi_2 \in CA(\mathcal{D}_2, X)$, (ii) $\xi_1 \subseteq \xi_2 \subseteq \overline{\xi}_1$ & *therefore* $\mathcal{D}_1 \subseteq \mathcal{D}_2 \subseteq \mathcal{D}_{\xi_1}$. *Then*
(a) $\mathcal{D}_2^{\mathrm{loc}} \subseteq \mathcal{D}_1^{\mathrm{loc}}$;
(b) $\mathcal{P}_{1,\xi_2} = \mathcal{P}_{1,\xi_1} \cap \mathcal{M}(\mathcal{D}_2^{\mathrm{loc}}, Bl(\mathsf{F})) \subseteq \mathcal{P}_{1,\xi_1}$,
(c) $E_{\xi_2} \subseteq E_{\xi_1}$,
(d) $(\mathcal{D}_2)_{\xi_2} = (\mathcal{D}_1)_{\xi_1} \cap \mathcal{D}_2^{\mathrm{loc}} \subseteq (\mathcal{D}_1)_{\xi_1}$, & $\overline{\xi}_2 = \mathrm{Rstr.}(\mathcal{D}_2)_{\xi_2}\overline{\xi}_1 \subseteq \overline{\xi}_1$,
(e) $\forall A \in \mathcal{D}_2^{\mathrm{loc}}$, $|\xi_1|(A) \leq |\xi_2|(A)$ & $s_{\xi_1}(A) \leq s_{\xi_2}(A)$.

6. SET FAMILIES OVER CARTESIAN PRODUCTS

In order to turn to the theory of product measures, we now have to consider set families over two given non-void sets, which in this section will be denoted by Ω_1, Ω_2. A well-known result concerns pre-rings (called semi-clans in [5, p. 7]):

$$(6.1) \qquad \begin{cases} \mathcal{P}, \mathcal{Q} \text{ are pre-rings over } \Omega_1, \Omega_2 \text{ respectively} \\ \Rightarrow \mathcal{P} \times \mathcal{Q} \text{ is pre-ring over } \Omega_1 \times \Omega_2. \end{cases}$$

The following lemma on disjoint sequences of sets in $\mathcal{P}, \mathcal{Q}$ is useful. Its proof and that of other results in this section are left to the reader.

6.2. Lemma. *Let*
(i) $\mathcal{P}, \mathcal{Q}$ *be pre-rings over* Ω_1, Ω_2 *respectively,*
(ii) $(R_k)_1^\infty$ *be a* $\|$ *sequence in* $\mathcal{P} \times \mathcal{Q}$ *such that* $R := \cup_1^\infty R_j \in \mathcal{P} \times \mathcal{Q}$,
(iii) $\forall k$ *in* $\mathbb{N}_+$, $R_k := P_k \times Q_k$, $R := P \times Q$, P_k, $P \in \mathcal{P}$ & Q_k, $Q \in \mathcal{Q}$,
(iv) $\forall \omega \in \Omega_1$, $\mathbb{N}_\omega := \{n : n \in \mathbb{N}_+ \ \&\ \omega \in P_n\}$.
Then $\forall \omega \in P$, *the sets* Q_n, $n \in \mathbb{N}_\omega$, *are* $\|$ & $\bigcup_{n \in \mathbb{N}_\omega} Q_n = Q$.

6.3. Triviality. *Let* $\mathcal{R}_1, \mathcal{R}_2$ *be rings over* Ω_1, Ω_2 *and* $\mathcal{D} = \delta\text{-ring}(\mathcal{R}_1 \times \mathcal{R}_2)$. *Then* $\forall D \in \mathcal{D}$, $\exists R_1 \in \mathcal{R}_1$ *and* $\exists R_2 \in \mathcal{R}_2$ *such that* $D \subseteq R_1 \times R_2$.

For products of δ-rings we have:

6.4. Lemma. *Let* $\mathcal{D}_1, \mathcal{D}_2$ *be* δ*-rings over* Ω_1, Ω_2 *respectively. Then*

$$\sigma\text{-alg}(\mathcal{D}_1^{\text{loc}} \times \mathcal{D}_2^{\text{loc}}) \subseteq [\delta\text{-ring}(\mathcal{D}_1 \times \mathcal{D}_2)]^{\text{loc}}.$$

The lemma 6.4 serves to prove the following proposition pertaining to pairs of δ-rings which are related in the manner of $\mathcal{D}, \mathcal{D}_\xi$ or $\mathcal{D}_\xi, \overline{\mathcal{D}}_\xi$, cf. 5.3 (a), 5.7 (a):

6.5. Proposition. *Let*
(i) *for* $i = 1, 2$, $\mathcal{D}_i, \tilde{\mathcal{D}}_i$ *be* δ*-rings over* Ω_i *such that* $\mathcal{D}_i \subseteq \tilde{\mathcal{D}}_i$ & $\mathcal{D}_i^{\text{loc}} = \tilde{\mathcal{D}}_i^{\text{loc}}$,
(ii) $\mathcal{D} := \delta\text{-ring}(\mathcal{D}_1 \times \mathcal{D}_2)$, $\tilde{\mathcal{D}} := \delta\text{-ring}(\tilde{\mathcal{D}}_1 \times \tilde{\mathcal{D}}_2)$.

Then

$$\mathcal{D} \subseteq \tilde{\mathcal{D}} \subseteq \sigma\text{-alg}(\mathcal{D}_1^{\text{loc}} \times \mathcal{D}_2^{\text{loc}}) \subseteq \tilde{\mathcal{D}}^{\text{loc}} \subseteq \mathcal{D}^{\text{loc}}.$$

The equalities $\tilde{\mathcal{D}}_1^{\text{loc}} = \mathcal{D}_1^{\text{loc}}$ & $\tilde{\mathcal{D}}_2^{\text{loc}} = \mathcal{D}_2^{\text{loc}}$ in 6.5 (i) do not in general entail $\tilde{\mathcal{D}}^{\text{loc}} = \mathcal{D}^{\text{loc}}$, i.e. the inclusion $\tilde{\mathcal{D}}^{\text{loc}} \subseteq \mathcal{D}^{\text{loc}}$ in Proposition 6.5 is proper in general. An example to this effect is deferred, cf. 6.10, since it utilizes the idea of sections of sets in $\Omega_2 \times \Omega_2$, which we shall now introduce.

6.6. Definition. Let Ω_1, Ω_2 be any non-void sets, $E \subseteq \Omega_1 \times \Omega_2$, and $\mathcal{D}_1, \mathcal{D}_2$ be δ-rings over Ω_1, Ω_2, respectively. Then
(a) $\forall \omega_1 \in \Omega_1$, $E_{\omega_1} := \{\omega_2 : \omega_2 \in \Omega_2 \ \&\ (\omega_1, \omega_2) \in E\}$,
$\forall \omega_2 \in \Omega_2$, $E^{\omega_2} := \{\omega_1 : \omega_1 \in \Omega_1 \ \&\ (\omega_1, \omega_2) \in E\}$.
(b) $[\mathcal{D}_1, \mathcal{D}_2] := \{E : E \subseteq \Omega_1 \times \Omega_2 \ \&\ \forall (\omega_1, \omega_2) \in \Omega_1 \times \Omega_2, E_{\omega_1} \in \mathcal{D}_2 \ \&\ E^{\omega_2} \in \mathcal{D}_1\}$.

The following results on sections of sets are trivial:

6.7. Triviality.

(a) $\forall A_1 \subseteq \Omega_1,\ \forall A_2 \subseteq \Omega_2\ \&\ \forall \omega_1 \in \Omega_1,\ (A_1 \times A_2)_{\omega_1} = \begin{cases} A_2, & \omega_1 \in A_1 \\ \emptyset, & \omega_1 \in \Omega_1 \setminus A_1, \end{cases}$

(b) $\forall E \subseteq \Omega_1 \times \Omega_2\ \&\ \forall (\omega_1, \omega_2) \in \Omega_1 \times \Omega_2,$

$$\chi_{E_{\omega_1}}(\omega_2) = \chi_E(\omega_1, \omega_2) = \chi_{E^{\omega_2}}(\omega_1),$$

(c) $\forall E_j \subseteq \Omega_1 \times \Omega_2,\ j \in J\ \&\ \forall \omega_1 \in \Omega_1,$

$$(\bigcup_{j \in J} E_j)_{\omega_1} = \bigcup_{j \in J} (E_j)_{\omega_1}\ \&\ (\bigcap_{j \in J} E_j)_{\omega_1} = \bigcap_{j \in J} (E_j)_{\omega_1},$$

(d) $\forall E, F \subseteq \Omega_1 \times \Omega_2\ \&\ \forall \omega_1 \in \Omega_1,$

$$(E \setminus F)_{\omega_1} = E_{\omega_1} \setminus F_{\omega_1}\ \&\ (\Omega_1 \times \Omega_2 \setminus F)_{\omega_1} = \Omega_2 \setminus F_{\omega_1},$$

(e) $E \subseteq F \subseteq \Omega_1 \times \Omega_2 \Rightarrow \forall \omega_1 \in \Omega_1,\ E_{\omega_1} \subseteq F_{\omega_1},$

(f) $E, F \subseteq \Omega_1 \times \Omega_2\ \&\ E \parallel F \Rightarrow \forall \omega_1 \in \Omega_1,\ E_{\omega_1} \parallel F_{\omega_1}.$

Analogous results hold for ω_2-sections in Ω_1.

The following theorem is needed in settling the measurability questions that arise in the Slicing and Tonelli, Fubini Theorems (§§8,9):

6.8. Theorem. *Let $\mathcal{D}_1, \mathcal{D}_2$ be δ-rings over Ω_1, Ω_2, respectively. Then*
(a) $\delta\text{-ring}(\mathcal{D}_1 \times \mathcal{D}_2) \subseteq [\mathcal{D}_1, \mathcal{D}_2] = \delta\text{-ring} \subseteq 2^{\Omega_1 \times \Omega_2}$,
(b) *when* $\Omega_1 = \bigcup_{D \in \mathcal{D}_1} D\ \&\ \Omega_2 = \bigcup_{D \in \mathcal{D}_2} D$, *we have*

$$\{\delta\text{-ring}(\mathcal{D}_1 \times \mathcal{D}_2)\}^{\mathrm{loc}} \subseteq [\mathcal{D}_1^{\mathrm{loc}}, \mathcal{D}_2^{\mathrm{loc}}].$$

In case the spaces Ω_1, Ω_2 are countable unions of sets belonging to the basic δ-rings (associated with the component measures), we get the following useful result governing the product δ-ring and its localization algebra:

6.9. Lemma. *Let*
(i) $\mathcal{D}_1, \mathcal{D}_2$ *be δ-rings over* Ω_1, Ω_2, *respectively,*
(ii) $\Omega_1 \in \sigma\text{-ring}(\mathcal{D}_1)$, $\Omega_2 \in \sigma\text{-ring}(\mathcal{D}_2)$,
(iii) $\mathcal{D}_{12} := \delta\text{-ring}(\mathcal{D}_1 \times \mathcal{D}_2)$.
Then
(a) $\mathcal{D}_1^{\mathrm{loc}} = \sigma\text{-ring}(\mathcal{D}_1)\ \&\ \mathcal{D}_2^{\mathrm{loc}} = \sigma\text{-ring}(\mathcal{D}_2)$,
(b) $\Omega_1 \times \Omega_2 \in \sigma\text{-ring}(\mathcal{D}_{12}) = \mathcal{D}_{12}^{\mathrm{loc}}$,
(c) $\sigma\text{-ring}(\mathcal{D}_1^{\mathrm{loc}} \times \mathcal{D}_2^{\mathrm{loc}}) = \sigma\text{-ring}(\mathcal{D}_{12}) = \mathcal{D}_{12}^{\mathrm{loc}}$.

We now give an example to show that the last inclusion in Proposition 6.5 can be proper:

6.10. Example. *Let*
(i) Ω *be any set such that* $\operatorname{card} \Omega > c$,
(ii) $\mathcal{D}_1 = \{D : D \subseteq \Omega\ \&\ D \text{ is finite}\}, \tilde{\mathcal{D}}_1 = 2^\Omega = \mathcal{D}_2 = \tilde{\mathcal{D}}_2$
(iii) $\mathcal{D} = \delta\text{-ring}(\mathcal{D}_1 \times \mathcal{D}_2)$, $\tilde{\mathcal{D}} = \delta\text{-ring}(\tilde{\mathcal{D}}_1 \times \tilde{\mathcal{D}}_2)$.
Then
(a) $\mathcal{D}_1 \subseteq \tilde{\mathcal{D}}_i\ \&\ \mathcal{D}_1^{\mathrm{loc}} = \tilde{\mathcal{D}}_i^{\mathrm{loc}} = 2^\Omega$, *for* $i = 1, 2$,
(b) $\tilde{\mathcal{D}}^{\mathrm{loc}} = \sigma\text{-alg}(2^\Omega \times 2^\Omega)\ \&\ \mathcal{D}^{\mathrm{loc}} = 2^{\Omega \times \Omega}$,
(c) *the diagonal* $\Delta = \{(\omega, \omega) : \omega \in \Omega\} \in \mathcal{D}^{\mathrm{loc}} \setminus \tilde{\mathcal{D}}^{\mathrm{loc}}$.

7. PRODUCT MEASURE

In this section it will be understood that

$$(7.1)\qquad \begin{cases} X \textit{ is a Banach space over } \mathsf{F}, \\ \mathcal{D}, \mathcal{C} \textit{ are } \delta\textit{-rings over sets } \Omega, \Lambda, \textit{ respectively}, \\ \xi \in CA(\mathcal{D}, X), \quad \mu \in CA(\mathcal{C}, \mathsf{F}), \\ \hat{\mathcal{D}} := \delta\text{-ring}(\mathcal{D} \times \mathcal{C}),\ \overline{\mathcal{D}} := \delta\text{-ring}(\mathcal{D}_\xi \times \mathcal{C}_\mu), \\ \overline{\overline{\mathcal{D}}} := \delta\text{-ring}(\overline{\mathcal{D}}_\xi \times \mathcal{C}_\mu),\ \overline{\overline{\overline{\mathcal{D}}}} := \delta\text{-ring}(\mathcal{D}^{\mathrm{loc}} \times \mathcal{C}^{\mathrm{loc}}). \end{cases}$$

We use the letter $\mathcal{C}$ for a δ-ring for notational reasons, to distinguish sets $D \in \mathcal{D}$ from sets $C \in \mathcal{C}$. By 5.3 (a), 5.7 (a),

$$\mathcal{D} \subseteq \mathcal{D}_\xi \subseteq \overline{\mathcal{D}}_\xi,\ \mathcal{C} \subseteq \mathcal{C}_\mu,\ \mathcal{D}^{\mathrm{loc}} = \mathcal{D}_\xi^{\mathrm{loc}} = \overline{\mathcal{D}}_\xi^{\mathrm{loc}},\ \mathcal{C}^{\mathrm{loc}} = \mathcal{C}_\mu^{\mathrm{loc}}.$$

Hence by two applications of Proposition 6.5, we get

$$(7.2)\qquad \hat{\mathcal{D}} \subseteq \overline{\mathcal{D}} \subseteq \overline{\overline{\mathcal{D}}} \subseteq \overline{\overline{\overline{\mathcal{D}}}} = \overline{\overline{\overline{\mathcal{D}}}}^{\mathrm{loc}} \subseteq \overline{\overline{\mathcal{D}}}^{\mathrm{loc}} \subseteq \overline{\mathcal{D}}^{\mathrm{loc}} \subseteq \hat{\mathcal{D}}^{\mathrm{loc}}.$$

The last two inclusions in (7.2) are proper as can be shown by building on the Example 4.14.

Our objective is to show that there is a unique product $\pi = \xi \times \mu$ of the measures ξ and μ such that

$$(1)\qquad \begin{cases} \pi \in CA(\hat{\mathcal{D}}, X), \\ \forall D \times C \in \mathcal{D} \times \mathcal{C},\ \pi(D \times C) = \xi(D) \cdot \mu(C), \end{cases}$$

and then study its properties. For this we shall lean primarily on an extension theorem, due essentially to Brooks&Dinculeanu [3], which rests on Brooks' theorem on the existence of a control measure [2], central to which is the following concept:

7.3. Definition. Let $\mathcal{R}$ be a ring over Ω and $\xi \in FA(\mathcal{R}, X)$. We say that ξ is *locally strongly bounded* on $\mathcal{R}$, iff

$$R \in \mathcal{R} \ \&\ (R_k)_1^\infty \text{ is a } \| \text{ sequence in } \mathcal{R} \cap 2^R \Rightarrow \lim_{k \to \infty} \xi(R_k) = 0.$$

The extension theorem in question then reads as follows:

7.4. Extension Theorem. *Let $\mathcal{R}$ be a ring over Ω and $\xi \in CA(\mathcal{R}, X)$ and ξ be locally strongly bounded on $\mathcal{R}$. Then*

$$\exists_1 \overline{\xi} \textit{ such that } \xi \subseteq \overline{\xi} \in CA(\delta\text{-ring}(\mathcal{R}), X).$$

This enunciation differs from that of Brooks &Dinculeanu [3] in that our condition *local strong boundedness* replaces their *local strong additivity*. But the two notions are equivalent, as is easy to see.

To be able to apply Theorem 7.4, we must show that a measure defined on the pre-ring $\mathcal{D} \times \mathcal{C}$ by the equality in (1) is countably additive on $\mathcal{D} \times \mathcal{C}$, and so extends trivially to a countably additive measure on $\hat{\mathcal{R}} := \text{ring}(\mathcal{D} \times \mathcal{C})$, and then show that this extension is locally strongly bounded on $\hat{\mathcal{R}}$. This is done in the next three lemmas, in the proof of the first of which Lemma 6.2 is crucial.

7.5. Lemma. *Let* $\forall E = D \times C \in \mathcal{D} \times \mathcal{C}$, $\zeta(E) := \xi(D) \cdot \mu(C)$. *Then*

$$\zeta \in CA(\mathcal{D} \times \mathcal{C}, X).$$

We know, cf. Dinculeanu [4, p. 57], that every countably additive measure ν on a pre-ring $\mathcal{P}$ extends to a countably additive measure on ring($\mathcal{P}$). Applying this to $\mathcal{P} = \mathcal{D} \times \mathcal{C}$ and $\nu = \zeta$, we get:

7.6. Lemma. *Let* $\hat{\mathcal{R}} = \text{ring}(\mathcal{D} \times \mathcal{C})$. *Then*

$$\exists_1 \zeta \in CA(\hat{\mathcal{R}}, X) \textit{ such that } \forall R := D \times C \in \mathcal{D} \times \mathcal{C}, \ \zeta(R) = \xi(D) \cdot \mu(C).$$

7.7. Fundamental Lemma. *The measure* ζ *in 7.6 is locally strongly bounded on* $\hat{\mathcal{R}}$.

The proof of 7.7 rests on showing that

$$\forall R \in \hat{\mathcal{R}}, \ \zeta(R) = \int_\Omega \mu(R_\omega) \cdot \xi(d\omega).$$

It follows from Lemmas 7.6, 7.7 and Theorem 7.4 that ζ has a countably additive extension to $\hat{\mathcal{D}}$. We thus arrive at the following theorem:

7.8. Theorem. (Product-measure.) $\exists_1 \pi \in CA(\hat{\mathcal{D}}, X)$ *such that*

$$\forall R = D \times C \in \mathcal{D} \times \mathcal{C}, \ \pi(R) = \xi(D) \cdot \mu(C).$$

7.9. Definition. The unique measure π obtained in the last theorem is called *the product of* ξ *and* μ, and is denoted by $\xi \times \mu$.

Let ξ, μ be as in (7.1). Then since

$$\overline{\xi} \in CA(\mathcal{D}_\xi, X) \ \& \ \overline{\mu} \in CA(\mathcal{C}_\mu, \mathsf{F}), \text{ cf. } (5.1) \ \& \ 5.3(\text{c}),$$

therefore by Theorem 7.8 applied to these measures, we see that $\overline{\xi} \times \overline{\mu} \in CA(\overline{\mathcal{D}}, X)$. The connections between $\xi \times \mu$ and $\overline{\xi} \times \overline{\mu}$ are studied in the next lemma, the proof of part (b) of which hinges on Triviality 5.10.

7.10. Lemma. *With the notation of (7.1), we have*

(a) $\xi \times \mu \subseteq \overline{\xi} \times \overline{\mu} \in CA(\overline{\mathcal{D}}, X)$;

(b) $\forall E \in \overline{\mathcal{D}}^{\text{loc}}$, $|x' \circ (\xi \times \mu)|(E) \le |x' \circ (\overline{\xi} \times \overline{\mu})|(E)$ & $s_{\xi \times \mu}(E) \le s_{\overline{\xi} \times \overline{\mu}}(E)$;

(c) $\forall f \in \mathcal{M}(\overline{\mathcal{D}}^{\text{loc}}, Bl(\mathsf{F}))$, $|f|_{1,\xi \times \mu} \le |f|_{1,\overline{\xi} \times \overline{\mu}}$;

(d) $\mathcal{G}_{1,\overline{\xi} \times \overline{\mu}} \subseteq \mathcal{G}_{1,\xi \times \mu} = \mathcal{G}_{1,\overline{\xi \times \mu}}$.

7.11. Remark. For the important case of Banach spaces X for which for any ξ in $CA(\mathcal{D}, X)$ we have $\mathcal{G}_{1,\xi} = \mathcal{P}_{1,\xi}$, the relations 7.10 (d) reduce to

$$\mathcal{P}_{1,\overline{\xi} \times \overline{\mu}} \subseteq \mathcal{P}_{1,\xi \times \mu} = \mathcal{P}_{1,\overline{\xi \times \mu}}.$$

Actually these relations prevail for *all* Banach spaces X. But it is best to defer the proof until after the establishment of the Slicing Theorem 8.6.

A fundamental fact, easily established, is that for any $x' \in X'$, $x' \circ (\xi \times \mu)$ is itself an F-valued product measure; more fully:

7.12. Corollary. *With the notation of (7.1), we have*

$$\forall x' \in X', \ x' \circ (\xi \times \mu) = (x' \circ \xi) \times \mu \in CA(\hat{\mathcal{D}}, \mathsf{F}).$$

The equality is easily checked for sets in $\mathcal{D} \times \mathcal{C}$; the Identity Theorem [5, p. 24, Proposition 5] then shows that it holds on $\hat{\mathcal{D}}$.

8. THE SLICING THEOREM

Our goal now is to prove the "Slicing Theorem" to the effect that $\forall E \in \hat{\mathcal{D}}$,

$$\int_{\Omega} \mu(E_\omega)\xi(d\omega) = (\xi \times \mu)(E) = \int_{\Lambda} \xi(E^\lambda)\mu(d\lambda).$$

By virtue of 6.8 (a),

$$\hat{\mathcal{D}} := \delta\text{-ring}(\mathcal{D} \times \mathcal{C}) \subseteq [\mathcal{D}, \mathcal{C}] = \text{a } \delta\text{-ring}.$$

Hence the two integrands are meaningful. Write

$$\text{(8.1)} \qquad \begin{cases} \forall E \in [\mathcal{D}, \mathcal{C}] \ \& \ \forall (\omega, \lambda) \in \Omega \times \Lambda, \\ \varphi_E(\omega) := \mu(E_\omega) \in \mathsf{F}, \ \psi_E(\lambda) := \xi(E^\lambda) \in X. \end{cases}$$

Our objective is then to show that

$$\text{(1)} \qquad \int_{\Omega} \varphi_E(\omega)\xi(d\omega) = (\xi \times \mu)(E) = \int_{\Lambda} \psi_E(\lambda)\mu(d\lambda).$$

Since the first of these Pettis integrals is essentially a Lebesgue integral (cf. Remark 4.15), whereas the second is not, the first equality is more directly accessible than the second. When $X = \mathsf{F}$, both ξ and μ are F-valued, the two equalities in (1) are on a par. This fact has to be exploited in establishing the second equality for arbitrary X. We thus arrive at the following scheme of lemmas for proving (1).

8.2. Lemma. (Measurability of φ_E, ψ_E).
(a) *Let $\mathcal{D}, \mathcal{C}, \hat{\mathcal{D}}, \xi$ be as in (7.1) and $\forall E \in [\mathcal{D}, \mathcal{C}]$ & $\forall \lambda \in \Lambda$, $\psi_E(\lambda) := \xi(E^\lambda)$. Then*

$$\forall E \in \hat{\mathcal{D}}, \ \psi_E \in \mathcal{M}(\mathcal{C}^{\text{loc}}, \mathcal{W}(X)). \ (\textit{For } \mathcal{W}(X), \ \textit{cf. (3.8) et seq.})$$

(b) *Let $\mathcal{D}, \mathcal{C}, \hat{\mathcal{D}}, \mu$ be as in (7.1) and $\forall E \in [\mathcal{D}, \mathcal{C}]$ & $\forall \omega \in \Omega$, $\varphi_E(\omega) := \mu(E_\omega)$. Then*

$$\forall E \in \hat{\mathcal{D}}, \ \varphi_E \in \mathcal{M}(\mathcal{D}^{\text{loc}}, Bl(\mathsf{F})).$$

(c) *Let the premisses of (a) hold. Then*

$$\forall E \in \hat{\mathcal{D}} \ \& \ \forall B \in \mathcal{C}^{\text{loc}}, \ E \cap (\Omega \times B) \in \hat{\mathcal{D}} \ \& \ \psi_{E \cap (\Omega \times B)} = \psi_E \cdot \chi_B \ \textit{on } \Lambda.$$

The proof of 8.2(a) requires use of the result [5, p. 11. Proposition 16], on δ-monotone classes. Part (b) is a special case of (a) for $X = \mathsf{F}$, since obviously $\mathcal{W}(\mathsf{F}) = Bl(\mathsf{F})$. Part (c) is obvious.

8.3. Lemma. (1st part of slicing equality). *Let $\mathcal{D}, \mathcal{C}, \hat{\mathcal{D}}, \xi, \mu$ be as in (7.1) and φ_E be as in (8.1). Then*

$$\forall E \in \hat{\mathcal{D}},\ \varphi_E \in \mathcal{P}_{1,\xi}\ \&\ \int_{\Omega} \varphi_E(\omega)\xi(d\omega) = (\xi \times \mu)(E).$$

The result $\varphi_E \in \mathcal{P}_{1,\xi}$ in 8.3 is easily checked for $E \in \mathcal{D} \times \mathcal{C}$, and thence for $E \in \hat{\mathcal{R}} :=$ ring$(\mathcal{D} \times \mathcal{C})$. The Domination Principle 4.11 then yields it for all $E \in \hat{\mathcal{D}}$. The equality is proved by letting $\zeta(E)$ be the left hand side, showing that (1) $\zeta = \xi \times \mu$ on $\mathcal{D} \times \mathcal{C}$, (2) $\zeta \in CA(\hat{\mathcal{D}}, X)$, by use of the Dominated Convergence Theorem 4.6 (c), and (3) appealing to the Identity Theorem [5, p. 24, Proposition 5].

For $X = \mathsf{F}$, both ξ and μ are F-valued countably additive measures on the δ-rings $\mathcal{D}$ and $\mathcal{C}$. Thus we may apply 8.3 not only to $\xi \in CA(\mathcal{D}, \mathsf{F})$ but also to $\mu \in CA(\mathcal{C}, \mathsf{F})$. This gives

8.4. Lemma. (Slicing equality for F-valued measures). *Let*
(i) $\mathcal{D}, \mathcal{C}, \hat{\mathcal{D}}$ *be as in (7.1)*
(ii) $\nu \in CA(\mathcal{D}, \mathsf{F})$, $\mu \in CA(\mathcal{C}, \mathsf{F})$,
(iii) $\forall E \in [\mathcal{D}, \mathcal{C}]$ & $\forall (\omega, \lambda) \in \Omega \times \Lambda$,

$$\varphi_E(\omega) := \mu(E_\omega),\ \psi_E(\lambda) := \nu(E^\lambda).$$

Then $\forall D \in \hat{\mathcal{D}}$, $\varphi_E \in L_1(\Omega, \mathcal{D}, \nu; \mathsf{F})$, $\psi_E \in L_1(\Lambda, \mathcal{C}, \mu; \mathsf{F})$ *and*

$$\int_{\Omega} \varphi_E(\omega)\nu(d\omega) = (\nu \times \mu)(E) = \int_{\Lambda} \psi_E(\lambda)\mu(d\lambda)$$

Finally to get the general slicing equality, we apply 8.4 with $\nu = x' \circ \xi$, where $x' \in X'$, and with the μ in (7.1). By Corollary 7.11

$$\nu \times \mu = x' \circ (\xi \times \mu).$$

Working on these facts we get our main result:

8.5. Slicing Theorem. *Let*
(i) $\mathcal{D}, \mathcal{C}, \hat{\mathcal{D}}, \xi, \mu$ *be as in (7.1),*
(ii) $\forall E \in [\mathcal{D}, \mathcal{C}]$, φ_E, ψ_E *be as in (8.1).*
Then $\forall E \in \hat{\mathcal{D}}, \varphi_E \in \mathcal{P}_1(\Omega, \mathcal{D}, \xi; \mathsf{F})$, $\psi_E \in \mathcal{P}_1(\Lambda, \mathcal{C}, \mu; X)$ *and*

$$\int_{\Omega} \varphi_E(\omega)\xi(d\omega) = (\xi \times \mu)(E) = \int_{\Lambda} \psi_E(\lambda)\mu(d\lambda).$$

The next theorem supercedes all known results on the relationship between $|\nu \times \mu|$ and $|\nu| \times |\mu|$, where μ, ν are F-valued countably additive measures on δ-rings. Part (a) recapitulates (7.2) for $\xi = \nu$. Use of the Slicing Theorem 8.5 yields a proof of part (b) that is appreciably shorter than its ab initio proof. The other parts follow readily from (b).

8.6. Theorem. *Let*
(i) $\mathcal{D}, \mathcal{C}, \hat{\mathcal{D}}$ *be as in (7.1),*
(ii) $\nu \in CA(\mathcal{D}, \mathsf{F})$, $\mu \in CA(\mathcal{C}, \mathsf{F})$, $\overline{\mathcal{D}} := \delta\text{-}ring(\mathcal{D}_\nu \times \mathcal{C}_\mu)$.
Then
(a) $\hat{\mathcal{D}} \subseteq \overline{\mathcal{D}} \subseteq \overline{\mathcal{D}}^{\mathrm{loc}} \subseteq \hat{\mathcal{D}}^{\mathrm{loc}}$;
(b) $\overline{\mathcal{D}} \subseteq (\hat{\mathcal{D}})_{\nu \times \mu}$ & $\forall E \in \overline{\mathcal{D}}$, $|\nu \times \mu|(E) = (|\nu| \times |\mu|)(E) < \infty$,
(c) $\forall E \in \overline{\mathcal{D}}^{\mathrm{loc}}$, $|\nu \times \mu|(E) = \left||\nu| \times |\mu|\right|(E) \le \infty$, *i.e.* $\left||\nu| \times |\mu|\right| = \mathrm{Rstr.}_{\overline{\mathcal{D}}^{\mathrm{loc}}} |\mu \times \nu|$;

(d) $\forall E \in (\overline{\mathcal{D}})_{|\nu|\times|\mu|}$, $|\nu \times \mu|(E) = \Big||\nu| \times |\mu|\Big|(E) < \infty$;
(e) $(\overline{\mathcal{D}})_{|\nu|\times|\mu|} = \overline{\mathcal{D}}^{\mathrm{loc}} \cap (\hat{\mathcal{D}})_{\nu\times\mu}$.

A useful corollary of 8.6 (b) concerns the semi-variation $s_{\xi\times\mu}$:

$$(8.7)\qquad \begin{cases} \text{With } \mathcal{D}, \mathcal{C}, \xi, \mu, \text{ as in (7.1),} \\ \forall B \times G \in \mathcal{D}^{\mathrm{loc}} \times \mathcal{C}^{\mathrm{loc}},\ s_{\xi\times\mu}(B \times G) = s_\xi(B)|\mu|(G). \end{cases}$$

With the aid of 8.6 and (8.7), we can address the issue raised in Remark 7.11. Here again result (a) is the difficult part. The remainder follows easily.

8.8. Theorem. *Let $\mathcal{D}, \mathcal{C}, \hat{\mathcal{D}}, \xi, \mu, \overline{\mathcal{D}}$ be as in (7.1). Then*
(a) $\overline{\mathcal{D}} \subseteq (\hat{\mathcal{D}})_{\xi\times\mu}$ & $\bar{\xi} \times \bar{\mu} \subseteq \overline{\xi \times \mu}$;
(b) $\bar{\bar{\xi}} \times \bar{\mu} \subseteq \overline{\xi \times \mu}$;
(c) $(\overline{\mathcal{D}})_{\bar{\xi}\times\bar{\mu}} = \overline{\mathcal{D}}^{\mathrm{loc}} \cap (\hat{\mathcal{D}})_{\xi\times\mu} \subseteq (\hat{\mathcal{D}})_{\xi\times\mu}$;
(d) $\mathcal{P}_{1,\bar{\xi}\times\bar{\mu}} = \mathcal{M}(\overline{\mathcal{D}}^{\mathrm{loc}}, Bl(\mathsf{F})) \cap \mathcal{P}_{1,\xi\times\mu} \subseteq \mathcal{P}_{1,\xi\times\mu} = \mathcal{P}_{1,\overline{\xi\times\mu}}$;
(e) $E_{\bar{\xi}\times\bar{\mu}} \subseteq E_{\xi\times\mu} = E_{\overline{\xi\times\mu}}$.

We turn finally to the δ-ring $\overline{\overline{\mathcal{D}}}$ in (7.1), left out so far, and assert

$$(8.9)\qquad \overline{\overline{\mathcal{D}}} \subseteq \overline{[(\hat{\mathcal{D}})_{\xi\times\mu}]} \qquad \text{(cf. (5.2)).}$$

9. THE FUBINI AND TONELLI THEOREMS

Adhering to the notation of (7.1), we shall assume in the sequel that

$$(9.1)\qquad \Omega \in \sigma\text{-ring}(\mathcal{D}_\xi)\ \&\ \Lambda \in \sigma\text{-ring}(\mathcal{C}_\mu).$$

It then follows from Lemma 6.9 that

$$(9.2)\qquad \begin{cases} \mathcal{D}^{\mathrm{loc}} = \sigma\text{-ring}(\mathcal{D}_\xi)\ \&\ \mathcal{C}^{\mathrm{loc}} = \sigma\text{-ring}(\mathcal{C}_\mu), \\ \overline{\mathcal{D}}^{\mathrm{loc}} = \sigma\text{-ring}(\overline{\mathcal{D}}) = \sigma\text{-alg}(\mathcal{D}^{\mathrm{loc}} \times \mathcal{C}^{\mathrm{loc}}). \end{cases}$$

The following theorem then settles the question of the measurability of partial integrals.

9.3. Lemma. (a) *Let $f \in \mathcal{M}(\overline{\mathcal{D}}^{\mathrm{loc}}, Bl(\mathsf{F}))$ and*

$$\forall \in \lambda,\ f(\cdot, \lambda) \in \mathcal{P}_1(\Omega, \mathcal{D}, \xi; \mathsf{F})\ \&\ \psi_f(\lambda) := \int_\Omega f(\omega, \lambda)\xi(d\omega) \in X.$$

Then $\psi_f \in \mathcal{M}(\mathcal{C}^{\mathrm{loc}}, \mathcal{W}(X))$, where $\mathcal{W}(X)$ is as in (3.8) *et seq.*
(b) *Let $f \in \mathcal{M}(\overline{\mathcal{D}}^{\mathrm{loc}}, Bl(\mathsf{F}))$ and*

$$\forall \omega \in \Omega,\ f(\omega, \cdot) \in \mathcal{P}_1(\Lambda, \mathcal{C}, \mu; \mathsf{F})\ \&\ \varphi_f(\omega) := \int_\Lambda f(\omega, \lambda)\mu(d\lambda) \in \mathsf{F}.$$

Then $\varphi_f \in \mathcal{M}(\mathcal{D}^{\text{loc}}, Bl(\mathsf{F}))$.

Our vectorical version of the classical Fubini Theorem reads as follows:

9.4. Fubini's Theorem. *Let*
(i) $f \in \mathcal{M}(\overline{\mathcal{D}}^{\text{loc}}, Bl(\mathsf{F}))$ *be Pettis integrable over* $\overline{\mathcal{D}}^{\text{loc}}$ *w.r.t.* $\xi \times \mu$
(ii) $\exists$ *a carrier* Λ_0 *of* μ *such that* $\forall \lambda \in \Lambda_0$, $f(\cdot, \lambda) \in \mathcal{P}_1(\Omega, \mathcal{D}, \xi; \mathsf{F})$
(iii) $\exists$ *a carrier* Ω_0 *of* ξ *such that* $\forall \omega \in \Omega_0$, $f(\omega, \cdot) \in L_1(\Lambda, \mathcal{C}, \mu; \mathbf{R})$.
Then

(a) $$\int_{\Lambda} f(\cdot, \lambda)\mu(d\lambda) \in \mathcal{P}_1(\Omega, \mathcal{D}, \xi; \mathsf{F}), \quad \int_{\Omega} f(\omega, \cdot)\xi(d\omega) \in L_1(\Lambda, \mathcal{C}, \mu; X);$$

(b) $$\int_{\Omega}[\int_{\Lambda} f(\omega, \lambda)\mu(d\lambda)]\xi(d\omega) = \int_{\Omega\times\Lambda} f(\omega, \lambda) \cdot (\xi \times \mu)\{d(\omega, \lambda)\} = \int_{\Lambda}[\int_{\Omega} f(\omega, \lambda)\xi(d\omega)]\mu(d\lambda).$$

Our vectorical version of the classical Tonelli theorem cf. [6: p. 194, Theorem 14 & Corollary 15] reads as follows:

9.5. Tonelli's Theorem. *Let*
(i) $f \in \mathcal{M}(\overline{\mathcal{D}}^{\text{loc}}, Bl(\mathsf{F}))$,
(ii) $\forall \lambda \in \Lambda$, $f(\cdot, \lambda) \in \mathcal{P}_1(\Omega, \mathcal{D}, \xi; \mathsf{F})$,
(iii) $\forall \omega \in \Omega$, $f(\omega, \cdot) \in L_1(\Lambda, \mathcal{C}, \mu; \mathsf{F}) =: L_{1,\mu}$, &

$$\Phi(\cdot) := \int_{\Lambda} |f(\cdot, \lambda)| \cdot |\mu|(d\lambda) \in \mathcal{P}_1(\Omega, \mathcal{D}, \xi; \mathbf{R}_{0+}),$$

(iv) $\forall C \in \overline{\mathcal{D}}^{\text{loc}}$, $\forall \omega \in \Omega$ & $\forall \lambda \in \Lambda, f_C(\cdot, \cdot) := \chi_C(\cdot, \cdot) \cdot f(\cdot, \cdot)$,

$$\varphi_C(\omega) := \int_{\Lambda} f_C(\omega, \lambda)\mu(d\lambda) \in \mathsf{F}, \quad \psi_C(\lambda) := \int_{\Omega} f_C(\omega, \lambda)\xi(d\omega) \in X.$$

Then
(a) $\varphi_C \in \mathcal{P}_{1,\xi}(\Omega, \mathcal{D}, \xi; \mathsf{F})$, $\psi_C \in \mathcal{P}_{1,\mu}(\Lambda, \mathcal{C}, \mu; X)$;
(b) f *is Pettis integrable over* $\overline{\mathcal{D}}^{\text{loc}}$ *w.r.t.* $\xi \times \mu$, *and* $\forall C \in \overline{\mathcal{D}}^{\text{loc}}$;

$$\int_{C} f(\omega, \lambda)(\xi \times \mu)\{d(\omega, \lambda)\} = \int_{\Omega}[\int_{\Lambda} f_C(\omega, \lambda)\mu(d\lambda)]\xi(d\omega)$$

$$= \int_{\Lambda}[\int_{\Omega} f_C(\omega, \lambda)\xi(d\omega)]\mu(d\lambda);$$

(c) $\forall B \times G \in \mathcal{D}^{\text{loc}} \times \mathcal{C}^{\text{loc}}$,

$$\int_{B\times G} f(\omega, \lambda)(\xi \times \mu)\{d(\omega, \lambda)\} = \int_{B}[\int_{G} f(\omega, \lambda)\mu(d\lambda)]\xi(d\omega)$$

$$= \int_{G}[\int_{B} f(\omega, \lambda)\xi(d\omega)]\mu(d\lambda).$$

These equalities hold in particular for $B = \Omega$ *and* $G = \Lambda$.

Consideration of the arbitrary set C in 9.5 (iv) is required to get the general form of the equality in (c), and more importantly to justify the Pettis integrability asserted in (b), by an appeal to the Definition 4.3.

9.6. Remarks. In our versions of the Fubini and Tonelli theorems, the measurability requirement

$$(1) \qquad f \in \mathcal{M}(\overline{\mathcal{D}}^{\text{loc}}, Bl(\mathsf{F}))$$

is stronger than the usual one, viz.

$$(1') \qquad f \in \mathcal{M}(\hat{\mathcal{D}}^{\text{loc}}, Bl(\mathsf{F}));$$

and the conclusion in 9.5 (b), viz.

$$(2) \qquad f \text{ is Pettis integrable over } \overline{\mathcal{D}}^{\text{loc}} \text{ w.r.t. } \mu$$

is weaker than the more natural condition

$$(2') \qquad f \in \mathcal{P}_{1,\xi\times\mu},$$

i.e. that f is Pettis integrable over $\hat{\mathcal{D}}^{\text{loc}}$ w.r.t. $\xi \times \mu$. This happens because the "σ-finiteness" constraints imposed in (9.1) are weaker than the ones suggested by the classical theory, viz.

$$(9.1') \qquad \Omega \in \sigma\text{-ring}(\mathcal{D}) \;\&\; \Lambda \in \sigma\text{-ring}(\mathcal{C}).$$

Under the stronger assumption (9.1′), these blemishes disappear, and we get more natural versions of the Fubini and Tonelli theorems in which $\hat{\mathcal{D}}^{\text{loc}}$ appears in place of $\overline{\mathcal{D}}^{\text{loc}}$, and Pettis integrability is over $\hat{\mathcal{D}}^{\text{loc}}$, not $\overline{\mathcal{D}}^{\text{loc}}$.

10. THE THEORY FOR WEAKLY Σ-COMPLETE BANACH SPACES

We now consider the simplifications which accrue in the theory when the Banach space X is weakly Σ-complete, in the following sense:

10.1. Definition. We say that a Banach space X over F is *weakly Σ-complete*[11], iff for any sequence $(x_n)_1^\infty$ in X, the following conditions are equivalent:

(α) $\forall x' \in X'$, $\sum_{n=1}^{\infty} |x'(x_n)| < \infty$,

(β) $\exists x \in X$ such that $\forall x' \in X'$, $\sum_{n=1}^{\infty} x'(x_n)$ converges unconditionally to $x'(x)$ in F.

By the Orlicz-Pettis Theorem [9, Theorem 3.2.3] we can replace (β) by its more profound equivalent:

$$(\beta') \qquad \sum_{n=1}^{\infty} x_n \text{ converges unconditionally in } X \text{ in the norm topology.}$$

[11] This nomenclature, due to E.G.F. Thomas, is suggested by the fact that the definition reduces to that of *weak sequential completeness* when the series in it is replaced by a sequence. Accordingly we may call our Banach space *weakly summationally complete*, and abbreviate "summationally" by "Σ".

It is known that the class of all weakly Σ-complete Banach spaces X contains the weakly sequential ones, in particular all reflexive spaces, as well as all dual spaces that are separable. Bessaga and Pelczynski have shown that the class comprises precisely the Banach spaces that contain no "copies" of the Banach space c_o.

The major simplication in the theory of Pettis integration that accrues from the weak Σ-completeness of the Banach space X is that the first inclusions in (4.4) now become equalities, cf. [4, p. 54, # 7] and [11, p. 301, # 5.1]:

$$(10.2) \quad \begin{cases} \text{(a)} \quad \mathcal{P}_{1,\mu} = \mathcal{G}_{1,\mu} \subseteq \mathcal{M}(\mathcal{C}^{\mathrm{loc}}, \mathcal{W}(X)) \qquad (X \text{ separable})^{12} \\ \text{(b)} \quad \mathcal{P}_{1,\xi} = \mathcal{G}_{1,\xi} \subseteq \mathcal{M}(\mathcal{D}^{\mathrm{loc}}, Bl(\mathsf{F})). \end{cases}$$

Another improvement comes from the fact that the inclusion $\mathcal{D}_\xi \subseteq \overline{\mathcal{D}}_\xi$, noted in Lemma 5.7 and Example 5.8, now becomes an equality:

$$(10.3) \qquad \overline{\mathcal{D}}_\xi = \mathcal{D}_\xi.$$

Other simplifications result. For instance, for measurable f, the equivalence between the ξ-integrability of $f(\cdot)$ and that of $|f(\cdot)|$, which is hard to establish in general, cf. 4.8, 4.11–4.13, is now trivial.

But while several proofs are considerably shorthened and the notation is reduced by the imposition of the weakly Σ-complete condition, the theorems do not change in essence.

Hilbert spaces are of course weakly Σ-complete, as are the L_p spaces for $p \in (1, \infty)$. Hence for the kind of applications of vector measures made in the papers [13], [14], [16], it is the simplified theory that is most often germane.

References

[1] Bartle, R.G.: A general bilinear integral, Studia Math. 15 (1956), 337–352.

[2] Brooks, J.K.: On the existence of a control measure for strongly bounded vector measures, Bull. Amer. Math. Soc. 77 (1971), 999–1001.

[3] Brooks, J.K., and Dinculenu, N.: Strong additivity, absolute continuity, and compactness in spaces of measures. J. Math. Anal. Appl. 45 (1974), 156–175.

[4] Diestel, J., and Uhl, J.J.: Vector measures, Amer. Math. Soc., Providence, R.I., 1977.

[5] Dinculeanu, N.: Vector measures, Pergamon Press, Oxford, 1953.

[6] Dunford, N., and Schwartz, J.: Linear operators, I, Interscience, New York, 1958, 1963.

[7] Gelfand, I.M.: Abstrakte funktionen und lineare operatoren, Mat. Sb. (N.S.) 4 (46) (1938), 235–286.

[8] Halmos, P.R.: Measure theory, Van Nostrand, Princeton, N.J., 1950.

[9] Hille, E., and Phillips, R.S.: Functional analysis and semigroups, Amer. Math. Soc., Providence, R.I., 1957.

[10] Lewis, D.R.: Integration with respect to vector measures, Pacific J. Math. 23 (1970), 157–165.

[11] Lewis, D.R.: On integrability and summability in vector spaces, Illinois J. Math. 16 (1972), 294–307.

[12] Masani, P.: Measurability and Pettis integration in Hilbert spaces, J. Reine Angew. Math. 297 (1978), 92–135.

[13] Masani, P.: The theory of stationary vector-valued measures over **R**, J. Reine Angew. Math. 339 (1983), 105–132.

[12] For reflexive X the separability is not needed, cf. [19: p. 78].

[14] Masani, P.: An outline of the Fourier transformation as integration with respect to a vector measure, in "Anniversary Volume on Approximation Theory and Functional Analysis". (P.L. Butzer, B.Sz.-Nagy and R.L. Stens, Eds.), Birkhauser, Basel, 1984, pp. 463–479.

[15] Masani. P.R.: On conditional homomorphisms and semi-norms in Banach groups and the unification of operator theory and measure theory, Oper. Theory: Adv. Appl. 28 (1988), 173–196, Birkhauser, Basel.

[16] Niemi, H.: Stationary vector measures and positive definite translation invariant bimeasures, Ann. Acad. Sci. Fenn. Ser. A I Math. 4 (1978), 209–226.

[17] von Neumann, J.: Functional operators, I, Princeton Univ. Press, Princeton, N.J., 1950.

[18] Pettis, B.J.: On integration in vector spaces, Trans. Amer. Math. Soc. 44 (1938), 277–304.

[19] Rudin, W.: Real and complex analysis, McGraw Hill, New York, 1966.

[20] Thomas, E.G.F.: L'integration par rapport a une measure de Radon vectorielle, Ann. Inst. Fourier (Grenoble), 20:2, (1970), 55–191.

[21] Thomas, E.G.F.: Vector integration (unpublished).

Probability Theory on Vector Spaces IV
Lancut, June'87, Springer's LNM 1391

NORM DEPENDENT POSITIVE DEFINITE FUNCTIONS AND MEASURES ON VECTOR SPACES

J.K.Misiewicz
Cz. Ryll-Nardzewski

1.Introduction.

Distributions for which all one dimensional projections are the same up to a scale parameter play particular role in statistics and probability theory. For example, symmetric Gaussian measures, symmetric stable measures, and rotationally invariant measures have this property. This paper is devoted to an investigation of some properties of the class of such measures.

Let E be a Banach space and E^* be its dual. By P we denote the set of all symmetric probability measures on the real line, and by P^+ the set of all probability measures on $[0,\infty)$. By a nondegenerate measure on E we will understand a measure whose linear support is the whole E

Definition. A nondegenerate measure μ on E is called an E-valued version of a measure $\mu_1 \in P$ if there exists a function $c : E^* \to [0,\infty)$ such that the characteristic function of μ is of the form :

$$\hat{\mu}(x^*t) = \hat{\mu}_1(c(x^*)|t|) , \qquad t \in R,\ x^* \in E^*.$$

It is easy to see that the function c has the following properties:

1. $c(x^*) = 0$ if and only if $x^* = 0$,
2. $c(tx^*) = |t|c(x^*)$, for $t \in R$, $x^* \in E^*$,
3. c is continuous on every finite dimensional subspace $F \in E^*$,
4. for every finite dimensional subspace $F \in E^*$ there exist positive numbers K, M and m such that for every $x^*, y^* \in F$ we have
   $c(x^* + y^*) \le K(c(x^*) + c(y^*))$ and $m\|x^*\| \le c(x^*) \le M\|x^*\|$,
   where $\| \ \|$ is the norm in E .

The first two properties immediately follow from Definition, the third one is easy, and the last one follows from the continuity of c on the unit sphere in F. For details see [14].

We do not know whether for each such a function c there exists a nontrivial measure $\mu_1 \in P$ such that $\hat{\mu}_1(c(x^*))$ is a characteristic function. It is also an open question, how large is the class $m(c,E)$ of measures $\mu_1 \in P$ (or equivalently the class $M(c,E)$ of cylindrical measures μ on E) for which $\hat{\mu}_1(c(x^*))$ is the characteristic function. It is, however, easy to see that :

(i) if $\mu_1, \mu_2 \in m(c,E)$ then $\mu_1 * \mu_2 \in m(c,E)$,

(ii) if $\mu_1, \mu_2 \in m(c,E)$, $0 \le a \le 1$, then $a\mu_1 + (1-a)\mu_2 \in m(c,E)$,

(iii) if $\mu_n \in m(c,E)$, and μ_n weakly converges to μ_0, then $\mu_0 \in m(c,E)$,

(iv) if $\mu_1 \in m(c,E)$, $\lambda \in P^+$, then $\mu_1 \circ \lambda \in m(c,E)$, where

$$\mu_1 \circ \lambda \ (A) = \int_0^\infty \mu_1(At^{-1})\ \lambda(dt)\ , \qquad \text{for every } A \in \mathcal{B}(R)\ .$$

The class $M(c,E)$ has the same properties.

If we assume that a measure $\mu_1 \in P$ has (at least two)-dimensional nondegenerate version then $\operatorname{supp} \mu_1 = R$, i.e. $\mu_1(U) > 0$ for every open set $U \subset R$ (for detailes see [14]).

The classes $m(c,E)$ and $M(c,E)$ were studied for some special c's and E's but only in one case we have the full characterization.

EXAMPLE 1.

If we take $c(x^*)^2 = (x^*, Qx^*)$, where $Q : E^* \to E$ is a Gaussian covariance operator then the set $M(c,E)$ is called the class of elliptically contoured measures and we have the following characterization :

<u>If $\dim E = n < \infty$</u> then $\mu \in M(c,E)$ if and only if $\mu = \mathcal{L}(AU\Theta)$, where $A : R^n \to R^n$ is a linear nondegenerate operator, U is a random vector uniformly distributed on the unit sphere $S^{n-1} \subset R^n$, and $\Theta > 0$ is a random variable such that U and Θ are independent.

<u>If $\dim E = \infty$</u> then $\mu \in M(c,E)$ if and only if $\mu = \mathcal{L}(X\sqrt{\Theta})$, where X is a Gaussian symmetric random vector, $\Theta > 0$ is a random variable, X and Θ are independent; or equivalently, if and only if $\hat{\mu}(x^*) = \hat{\mu}_1(c(x^*))$, where

$$\hat{\mu}_1(t) = \int e^{-t^2 s}\ \lambda(ds) \qquad \text{for some measure } \lambda \in P^+.$$

For detailes see [5] , [8] , [9] , [13] , [16] .

Remark.

It is easy to see that if μ is an E-valued version of the measure $\mu_1 \in P$ such that $\int |x|^2 \mu_1(dx) < \infty$ then μ is an elliptically contoured measure. Equivalently, we can say that the following equivalence holds : E is a Hilbert space if and only if there exists $\mu_1 \in P$ such that $\int |x|^2 \mu_1(dx) < \infty$ and such that $\hat{\mu}_1(\|x^*\|)$ is a positive definite function on E^* (see [4]).

EXAMPLE 2.

Let $c(x) = \Sigma |x_i|$ for $x = (x_1,\ldots,x_n) \in R^n$. Cambanis, Keener and Simons [3] gave a full characterization of the sets $m(c,R^n)$ and $M(c,R^n)$ in this case, namely

$\mu \in M(c,R^2)$ if and only if $\hat{\mu}(x^*) = \int \varphi_o(sc(x^*))\lambda(ds)$ for some measure $\lambda \in P^+$, where

$$\varphi_o(r) = \frac{2}{\pi} \int_r^\infty \frac{\sin t}{t}\, dt \qquad \text{for } r > 0,$$

or equivalently, if and only if $\mu = \mathcal{L}((U/\sqrt{B}, V/\sqrt{1-B})\,\Theta)$, where (U,V) is a random vector uniformly distributed on the unit sphere S^1, B is distributed Beta(1/2,1/2) , $\Theta > 0$ is a random variable, (U,V) , B and Θ are independent.

$\mu \in M(c,R^n)$ if and only if $\mu = \mathcal{L}((U_1/\sqrt{D_1},\ldots,U_n/\sqrt{D_n})\Theta)$, where $U=(U_1,\ldots,U_n)$ is uniformly distributed on S^{n-1}, $D = (D_1,\ldots,D_n)$ has the Dirichlet distribution with parameters $(1/2,\ldots,1/2)$, $\Theta > 0$ is a random variable, U, D and Θ are independent.

And finally (see [16]) $\hat{\mu}_1(\|x^*\|)$ is a positive definite function on l^p, $0 < p \le 2$, if and only if

$$\hat{\mu}_1(t) = \int e^{-|t|^p s}\lambda(ds) \qquad \text{for some measure } \lambda \in P^+.$$

2. Norm dependent positive definite functions.

Let us assume that $c(x) = \|x\|$, where $\|\ \|$ is the norm on a Banach space E or the p-norm on an L^p-space. We want to know for what kind of functions $f:R^+ \to R$ the function $f(\|x\|)$ is positive definite on E . Bretagnolle et al. (see [2]) gave the following characterization:

If $L^p(S,\beta,P)$ is an infinite dimensional space, then for every function $f : R^+ \to R$ with $f(0) = 1$, the function $f(\|\ \|_p)$ is positive definite on $L^p(S,\beta,P)$ if and only if

$$f \equiv 1 , \qquad p > 2$$

and

$$f(t) = \int e^{-|t|^p s}\lambda(ds) , \qquad \text{for some measure } \lambda \in P^+ \text{ if } 0 < p \le 2.$$

Using the argument (Dvoretzki's theorem) that every infinite dimensional Banach space contains l_n^2 uniformly, Christensen and Ressel proved (see [4]) that if $f(\| \|)$ is a positive definite function on an infinite dimensional Banach space E, and $f(0) = 1$, then there exists a measure $\lambda \in P^+$ such that

$$f(t) = \int e^{-t^2 s}\lambda(ds) , \qquad t > 0 .$$

Using the same arguments (replacing only "2" by "p" in the proof of the above characterization) we can prove the following:

Proposition 1.

Let E be an infinite dimensional space endowed with a norm (or a p-norm) and let $f : R^+ \to R$, $f(0) = 1$ be such that $f(\| \|)$ is positive definite on E. If E contains l_n^q uniformly, then there exists a measure $\lambda \in P^+$ such that

$$f(t) = \int e^{-t^q s}\lambda(ds) , \qquad t > 0.$$

This Proposition has an interesting consequence.

Proposition 2.

Let E be an infinite dimensional Banach space (or an infinite dimensional closed subspace of an L^p-space) such that $\exp(-\|x\|^q)$ is a positive definite function on E, $p \le q \le 2$. Then for each $0 < r < q$ E does not contain l_n^r uniformly (so E has a stable cotype r, see [11]).

Proof.

Suppose that there exists $0 < r < q$ such that E contains l_n^r uniformly. According to Proposition 1 we see that there exists a measure $\lambda \in P^+$ such that

$$\exp(-|t|^q) = \int e^{-|t|^q s}\lambda(ds) .$$

Let $g(t) := \exp(-t^{q/r}) = \int \exp(-ts)\lambda(ds)$, $t > 0$. It is easy to see that the function g is an absolutely monotonic function on R^+ as the Laplace transform of the measure λ. On the other hand we have

$$g''(t) = e^{-t^a} a^2 t^{a-2} (t^a - (a-1)/a) , \qquad \text{where } a = q/r > 1 ,$$

and it is easy to see that g'' changes its sign on R^+ and so g can not be an absolutely monotonic function. A contradiction.

So far it is known from the result of Bretagnolle et al. that the only function f , $f(0) = 1$, such that $f(\| \|)$ is positive definite on infinite dimensional L^p--space , $p > 2$, is the function $f \equiv 1$. In [6] the same result was proved for $C([0,1])$ (this fact follows also immediately from the universality of the space $C([0,1])$). To characterize all spaces with this property we can prove (as the simple consequence of Prop. VI2.3 and Theorem VI5.1 of [17]) the following:

Proposition 3.

Let E be a Banach space, and let a function $f: R^+ \to R$ with $f(0) = 1$ and $f \not\equiv 1$ be such that $f(\| \|)$ is a positive definite function on E . Then:

a) there exists a probability space (S,β,P) such that E is isomorphic to a closed subspace of the space $L_o(S,\beta,P)$, so E has the cotype 2 ,

b) for every $0 < p < 1$ there exists a probability space (S,β,P) such that E is isomorphic to a closed subspace of the space $L^p(S,\beta,P)$.

Recently L. Mezrag (see [12]) has proved that if a space E is isomorphic to a closed subspace of L^p-space , $0 < p < 1$, and the space E^* is of the finite Rademacher cotype then E is isomorphic to a closed subspace of an L^q-space for some $q \in (1,2]$. Consequently, every Banach space E imbedded into an L^p-space such that E^* is of cotype 2 , is isomorphic to a Hilbert space.

In 1984 Kuritsyn and Shestakov (see [10]) proved that $\exp(-(|x|^p + |y|^p)^{1/p})$ is a positive definite function on R^2 for every $p > 2$. It is easy to see that the same function is positive definite on R^2 for every $1 \leq p \leq 2$, and it is not positive definite for any $0 < p < 1$. It turns out that their result is a simple consequence of the following:

Proposition 4. (see also [7] , [1]).

A function $f(t,s) = \exp(-c(t,s))$, $t,s \in R$, is the characteristic function of a symmetric 1-stable distribution on R^2 if and only if the function $c(t,s)$ is anorm in R^2.

Proof.

If f is the characteristic function of a symmetric 1-stable distribution on R^2 then (see [11])

$$c(t,s) = \int |tx + sy| \, \lambda(dx,dy)$$

for some finite measure λ on the unit sphere $S^1 \subset R^2$ so c is an L^1-norm on R^2.

Now we need only to prove that for every norm $c(t,s)$ on R^2 there exists a finite measure λ on $(0,2\pi]$ such that

$$c(t,s) = \int_0^{2\pi} |t \cos\alpha + s \sin\alpha| \, \lambda(d\alpha) \quad .$$

Let us define a function q on $(0,2\pi]$ as follows:

$$q(\alpha) = c(\cos\alpha, \sin\alpha) \quad ,$$

and assume for a while that q has continuous second derivative (it means that the norm c is smooth enough) . In this case the convexity of the set $\{(t,s) : c(t,s) \le 1\}$ is equivalent to the inequality $q'' + q \ge 0$. Now it is easy to check that

$$4c(t,s) = \int_0^{2\pi} |t \cos\alpha + s \sin\alpha| (q''(\alpha + \pi/2) + q(\alpha + \pi/2)) \, d\alpha \; ,$$

and we obtain an explicit formula for the density of the measure λ . Less smooth norms can always be approximated by ones which are smooth enough.

It is interesting that Proposition 4 does not hold in higher dimensional spaces. For example, the function $\exp(-\max(|t|,|s|,|u|))$ is not positive definite on R^3.

As we will see in chapter 3 the problem of positive definitness of a function $\exp(-\|x\|)$ on E is equivalent to the negative definitness of the norm $\| \;\|$ on E and it is equivalent to the isometric embeddability of E into an L^1-space. For a long time it was an open question whether or not there exists an n-dimensional normed space which is not embeddable into L^1, but whose every k-dimensional subspace $(k \le n)$ is isometrically embeddable in L^1. Of course, for $k = 2$ the answer is yes. Recently Neyman (see [15]) solved this problem in the following way:

For every $1 \le p < \infty$, $p \ne 2$, and every positive integers k,n with $2 < k \le n$ there exists a norm $\| \;\|$ on R^n which restricted to every $(k-1)$-dimensional subspace is an L^p-norm, but there is a k-dimensional subspace for which $\| \;\|$ is not an L^p-norm.

3. E-valued versions of symmetric measures and stable measures.

Let us return to E-valued versions of the measures from P , and let, as before, $M(c,E)$ denote the set of all cylindrical measures μ on E for which the characteristic function is of the form $\hat{\mu}(x^*t) = f(c(x^*)|t|)$, $t \in R$, $x^* \in E^*$. Of course, f is the characteristic function of some measure $\mu_1 \in P$, and the set of all such measures is denoted by $m(c,E)$.

It is well known (see for example [4] and the Remark after Example 1) that there exists $\mu_1 \in m(c,E)$ such that $\mu_1 \neq \delta_0$ and $\int x^2 \mu_1(dx) < \infty$ if and only if c is a Hilbert norm on E^*. So, if the function c is not a Hilbert norm on E^* , then for every $\delta_0 \neq \mu_1 \in m(c,E)$ $\int x^2 \mu_1(dx) = \infty$. Now if we define a number

$$p(c,E) := \sup \{p \in [0,2] : \exists\, \mu_1 \in m(c,E),\ \mu_1 \neq \delta_0 ,\ \int |x|^p \mu_1(dx) < \infty \}$$

then we can formulate the following:

Theorem.

Let E be a Banach space and let $p > 0$. Then the following conditions are equivalent:

1) $p \leq p(c,E)$,

2) E^* with c as a quasinorm is isometrically embeddable into an L^p-space,

3) $\exp(- c(x^*)^p)$ is a positive definite function on E^*,

4) for every measure $\lambda \in P^+$ the function $f(x^*) = \int \exp(- c(x^*)^p s)\, \lambda(ds)$ is positive definite on E^*.

Proof. 1) ⇒ 2). Let $\delta_0 \neq \mu_1 \in m(c,E)$ be such that $\int |x|^p \mu_1(dx) < \infty$, and let X be a cylindrical random vector on E with the characteristic function $\hat{\mu}(x^*) = \hat{\mu}_1(c(x^*))$. We know that $\hat{\mu}(tx^*) = \hat{\mu}_1(|t|c(x^*))$ so the distribution of the random variable $x^*(X)$ is equal to the distribution of $c(x^*)X_1$, where $\mathcal{L}(X_1) = \mu_1$. So we have:

$$\int |x^*(x)|^p \mu(dx) = \int c(x^*)^p |x|^p \mu_1(dx) = c(x^*)^p \int |x|^p \mu_1(dx) .$$

Without loss of generality we can assume that $\int |x|^p \mu_1(dx) = 1$ and we obtain that c is an L^p-norm on E^*, and (E^*,c) is isometrically embeddable into an L^p-space.

If $p = p(c,E)$ we can use the above arguments for a sequence of measures

$\mu_{1,q}$ for which $\int |x|^q \mu_{1,q}(dx) < \infty$, $q < p$, and we obtain negative definitness of c^p on E^* from the negative definitness of every c^q , $q < p$.

<u>2) ⇒ 3).</u> If (E^*,c) is isometrically embeddableinto an L^p-space then c^p is a negative definite function on E^* (see for example [11]), hence $\exp(-c(x^*)^p)$ is positive definite.

<u>3) ⇒ 4).</u> It is an immediate consequence of the property (iv) of the set $M(c,E)$.

<u>4) ⇒ 1).</u> It is enough to take $\lambda = \delta_1$. We conclude that $\exp(-c(x^*)^p)$ is positive definite on E^*, and, because this function corresponds to the symmetric p-stable measure μ_1 (which has every moment less than p) , $p(c,E) \geq p$.

This Theorem tells us that in the case of $p(c,E) > 0$ the only homogeneous functions c such that $\hat{\mu}_1(c(x^*))$ is positive definite on E^* for some $\mu_1 \in P$, $\mu_1 \neq \delta_0$, are L^p-norms on E^* for some $p \in (0,2]$. Thus we obtain the same functions which appear in the characteristic functions of symmetric stable measures. But we still do not know what happens when $p(c,E) = 0$.

Let us notice that the sets

$$A_p := \{f : f(t) = \int \exp(-|t|^p s)\, \lambda(ds),\ \lambda \in P^+\}, \qquad p \in (0,2]$$

satisfy $A_2 \supset A_p \supset A_q$ for $2 \leq p \leq q$ and $\bigcap A_p = \{1\}$. According to this observation we conjecture that if $p(c,E) = 0$ then $m(c,E) = \{\delta_0\}$ at least in large enough spaces E . The only fact we know is that if $\dim E \geq 2$ then a measure $\mu_1 \in m(c,E)$, $\mu_1 \neq \delta_0$ has no finite moments of any order, $\operatorname{supp} \mu_1 = R$, and μ_1 has no other atoms than possibly $\{0\}$.

<u>References.</u>

[1] P.Assouad, Caracterizations des suous-espaces normes de L^1 de dimension finie, Seminaire d'Analyse Fonctionelle 1979-1980, preprint.

[2] J.Bretagnolle, D. Dacunha-Castelle, J.L. Krivine, Lois stables et espaces L^p, Lecture Notes in Math., Springer-Verlag 31(1967).

[3] S. Cambanis, R. Keener, G. Simons, On α -symmetric distributions, Journal of Multivariate Analysis 13(1983), 213-233.

[4] J.P.R. Christensen, P. Ressel, Norm dependent positive definite functions on B-spaces, Probab. in Banach Spaces, Oberwolfach 1982, Springer-Verlag 990 (1983), 47-53.

[5] J.J. Crawford, Elliptically contoured measures on infinite dimensional Banach spaces, Studia Math. 60(1977), 15-32.

[6] S.J. Einhorn, Functions positive definite in C([0,1]), Proc. Amer. Math. Soc. 22(1969), 702-703.

[7] T.S. Ferguson, A representation of the symmetric bivariate Cauchy distribution, Ann. Math. Stat. 33(1962), 1256-1266.

[8] S. Das Gupta, M.L. Eaton, I. Olkin, M. Perlman, L.J. Savage, M. Sobel,

Inequalities on the probability content of convex regions for elliptically contoured distributions, Proc. Sixth Berkeley Symp. 2(1977), 241-265.

[9] C.D. Hardin, On the linearity of regression, Zeit. Wahr. verw. Gebiete(1982)61.

[10] Yu.G. Kuritsyn, A.V. Shestakov, On α -symmetric laws, Teoria Verojatn. i Primen. XXIX(1984)4, 769-771.

[11] W. Linde, Infinitely divisible and stable measures on Banach spaces, Teubner-Texte zur Mathematik, Bd.58, Leipzig 1983.

[12] L. Mezrag, Theoremes de factorisation et de prolongement pour les operateurs a valeurs dans les espaces L^p, pour $p < 1$, C.R.Acad; Sc. Paris, t.300, Serie I, n 10(1985), 299-302.

[13] J.K. Misiewicz, Characterization of the elliptically contoured measures on infinite dimensional Banach spaces, Probab. and Math. Stat.4(1984)1, 47-56.

[14] J.K. Misiewicz, Some remarks on measures with n-dimensional version, Probab. and Math. Stat., to appear.

[15] A. Neyman, Representations of L^p-norms and isometric embedding in L^p-spaces, Israel Journ. of Math. 48(1984)2-3, 129-138.

[16] J. Schoenberg, Metric spaces and completely monotonic functions, Annals Math., 38(1938), 811-841.

[17] N.N. Vakhania, V.I. Tarieladze, S.A. Chobanian, Wierojatnostnyje raspredelenia w Banachowych prostranstwach, Moskva, Nauka 1985

Probability Theory on Vector Spaces IV
Lancut, June'87, Springer's LNM 1391

ON THE CONVERGENCE OF WEIGHTED SUMS OF MARTINGALE DIFFERENCES

by

NGUYEN DUY TIEN

Department of Mathematics

University of Hanoi

Viet Nam

and

NGUYEN VAN HUNG

Institute of Computer Sciences and Cybernetics

Nghia Do - Tu Liem - Hanoi

Viet Nam

1. INTRODUCTION

Let (Ω,F,P) be a probability space and $(F_n\ ,\ n \geq 1)$ an increasing sequence of sub-σ-fields of F. Let $X_n : \Omega \to R$ be an F_n-measurable random variable.

A sequence $(X_n\ ,\ n \geq 1)$ is said to be uniformly integrable if

(1-1) $$\sup_{n\in \mathbb{N}} \{ \int_{|X_n|>a} |X_n| dP\} \to 0 \quad \text{as} \quad a \uparrow \infty .$$

Note that (1-1) implies

(1-2) $$\sup_{n\in \mathbb{N}} \{aP(|X_n| > a)\} \to 0 \quad \text{as} \quad a \uparrow \infty .$$

A sequence $(X_n, F_n,\ n \geq 1)$ is said to be a martingale difference if $E(X_n/F_{n-1}) = 0$ for all $n \geq 1$.

Let (a_{nk}) be an array of real numbers such that

$$(1-3)\qquad \begin{aligned} &\text{(i)} \quad \lim_{n\to\infty} a_{nk} = 0 \quad \text{for each} \quad k \ge 1 \ , \\ &\text{(ii)} \quad \sum_k |a_{nk}|^p \le M \quad \text{for each} \quad n \ge 1 \ , \end{aligned}$$

and some $M < +\infty$. Such sequence is said to be ℓ_p - Toeplitz matrix.

The stochastic convergence of $A_n^{-1} \sum_{k=1}^{n} a_k X_k$ or $\sum_{k=1}^{n} a_{nk} X_k$ has been extensively studied where (X_k) are independent random variables, (a_k) is a sequence of positive real numbers and $A_n = (\sum_{k=1}^{n} a_k) \uparrow \infty$, (a_{nk}) is a Toeplitz matrix, i.e., (1-3) is satisfied for $p = 1$ ([2], [4], [7]). The purpose of this paper is to extend these results to martingale differences.

In Section 2, we shall show that $\sum_{k=1}^{n} a_{nk} X_k \to 0$ in probability, where (a_{nk}) is a Toeplitz matrix, for any uniformly integrable martingale difference $(X_k, F_k, \ k \ge 1)$ if and only if $\max_{k \le n} |a_{nk}| \to 0$ as $n \to \infty$. The same result is also obtained if the tail probabilities of (X_n) are uniformly bounded by the tail probabilities of a random variable $X \in L^1$. In Section 3, we shall study the almost sure convergence of $A_n^{-1} \sum_{k=1}^{n} a_k X_k$ where $(X_n, F_n, \ n \ge 1)$ is a martingale difference and (a_k) is a sequence of positive real numbers and $A_n = (\sum_{k=1}^{n} a_k^p)^{1/p} \uparrow \infty$, $1 \le p < 2$. It is easy to check that the sequence (a_{nk}), where $a_{nk} = \frac{a_k}{A_n}$ for $k \le n$ and $a_{nk} = 0$ for $k > 0$, satisfies (1-3). In this section, we also study the a.s. convergence of $\sum_{k=1}^{n} a_{nk} X_k$, where (a_{nk}) is a Toeplitz matrix such that $\max_{1 \le k \le n} |a_{nk}| \downarrow 0$ (monotone decreasing sequence) as $n \to \infty$. In Section 4, we shall study the convergence of $\sum_{k=1}^{n} a_{nk} X_k$ in L^p , $1 \le p < 2$, (a_{nk}) is a Toeplitz matrix such that $\max_{k \le n} |a_{nk}| \to 0$ as $n \to \infty$.

Recall that a sequence $(X_n \ , \ n \ge 1)$ of random variables has uniformly bounded tail probabilities by the tail probabilities of a random variable $X \in L^p$, $p > 0$ (for short $(X_n) \prec X$) if there exists a positive constant C such that for all $x > 0$ and $n=1,2,\dots,$ we have

$$P(|X_n| > x) \le CP(|X| > x) .$$

2. THE CONVERGENCE IN PROBABILITY

Thoughout this section (a_{nk}) is a Toeplitz matrix, i.e., (1-3) is satisfied for $p = 1$. Let $S_n = \sum_{k=1}^{n} a_{nk}X_k$.

Theorem 1. $(S_n) \to 0$ *in probability as* $n \to \infty$ *for any uniformly integrable martingale difference sequence* $(X_n, F_n,\ n \ge 1)$ *if and only if*

$$\max_{k\le n} |a_{nk}| \to 0 \quad \text{as} \quad n \to \infty . \tag{2-1}$$

Proof.

(a) The sufficiency :

We need the following elementary fact: if $f_n : R \to R^+$ is such that $0 \le f_n \le 1$ for all $n \ge 1$ and $\sup_{n\in N} (xf_n(x)) \to 0$ as $x \to \infty$, then

$$\sup_{n\in \mathbb{N}} \left(\frac{1}{y}\int_0^y xf_n(x)dx\right) \to 0 \quad \text{as} \quad y \to \infty .$$

Put $f^*(x) = \sup_{n\in \mathbb{N}} (xf_n(x))$, $0 \le f^*(x) \to 0$ as $x \to \infty$.

Note that

$$\sup_{n\in \mathbb{N}} \left(\frac{1}{y}\int_0^y xf_n(x)dx\right) \le \frac{1}{y}\int_0^y f^*(x)dx \quad \text{for all} \quad y > 0 ,$$

it suffices to show that

$$\frac{1}{y}\int_0^y f^*(x)dx \to 0 \quad \text{as} \quad y \to \infty .$$

Since $f^*(x) \to 0$ as $x \to \infty$, hence for all $\varepsilon > 0$, there exists an $x_o(\varepsilon)$ such that $x > x_o(\varepsilon)$, $y > x_o(\varepsilon)$:

$$0 \le f^*(x) < \varepsilon ,$$

$$\frac{1}{y}\int_0^y f^*(x)dx = \frac{1}{y}\left\{\int_0^{x_0(\varepsilon)} f^*(x)dx + \int_{x_0(\varepsilon)}^y f^*(x)dx\right\} .$$

Now, we have

$$\text{(i)} \qquad \frac{1}{y}\int_0^{x_0(\varepsilon)} f^*(x)dx \le \frac{x_0^2(\varepsilon)}{2y} \to 0 \quad \text{as} \quad y \to \infty ,$$

$$\text{(ii)} \qquad \frac{1}{y}\int_{x_0(\varepsilon)}^{y} f^*(x)dx \le \frac{\varepsilon}{y}\int_{x_0(\varepsilon)}^{y} dx = \frac{\varepsilon}{y}(y - x_0(\varepsilon)) < \varepsilon .$$

Using the avove result together with (1-2), we have

$$\text{(2-2)} \qquad \sup_{n\in\mathbb{N}} \left(\frac{1}{a}\int_0^a xP(|X_n| > x)dx\right) \to 0 \quad \text{as} \quad a \to \infty .$$

Next, put

$$X_{n,k} = a_{nk}X_kI(|X_k| \le |a_{nk}|^{-1}) ,$$

where $I(A)$ denotes the indicator function of the set A, and

$$Z_n = \sum_{k=1}^n X_{nk} .$$

Note that (2-1) implies that $\min_{k\le n} (|a_{nk}|^{-1}) \to \infty$ as $n \to \infty$ which together with (1-2) shows that

$$\begin{aligned} P(S_n \ne Z_n) &\le \sum_{k=1}^n P(X_{nk} \ne a_{nk}X_k) \\ &\le \sum_{k=1}^n |a_{nk}|\{|a_{nk}|^{-1}P(|X_k| > |a_{nk}|^{-1})\} \\ &\le \sum_{k=1}^n |a_{nk}| \cdot \varepsilon \le \varepsilon \cdot M , \end{aligned}$$

for n large enough.

Hence, it suffices to show that $Z_n \to 0$ in probability as $n \to \infty$.

Note again, for n large enough

$$\begin{aligned} \sum_{k=1}^n E|X_{nk} - E(X_{nk}|F_{k-1})|^2 &\le \sum_{k=1}^n E|X_{nk}|^2 \\ &\le 2\sum_{k=1}^n |a_{nk}|^2 \int_{\{0<x\le|a_{nk}|^{-1}\}} xP(|X_k| > x)dx \end{aligned}$$

$$\leq 2\sum_{k=1}^{n} |a_{nk}|\{\frac{1}{|a_{nk}|^{-1}}\int_{\{0<x\leq|a_{nk}|^{-1}\}} xP(|X_k| > x)dx\}$$
$$\leq 2\sum_{k=1}^{n} |a_{nk}| \cdot \varepsilon \leq \varepsilon \cdot M ,$$

by (2-2).

This implies by the Chebyshev inequality that

$$\sum_{k=1}^{n} (X_{nk} - E(X_{nk}|F_{k-1})) \to 0 \quad \text{in probability.} \tag{2-3}$$

On the other hand, since $E(X_n|F_{n-1}) = 0$ for all $n \geq 1$, we get that

$$E(X_{nk}|F_{k-1}) = -E(a_{nk}X_k I(|X_k| > |a_{nk}|^{-1})|F_{k-1}) .$$

Next, for n large enough, we have

$$E|\sum_{k=1}^{n} E(X_{nk}|F_{k-1})| \leq \sum_{k=1}^{n} |a_{nk}|E(|X_k|I(|X_k| > |a_{nk}|^{-1}))$$
$$\leq \sum_{k=1}^{n} |a_{nk}|\{\int_{x>|a_{nk}|^{-1}} P(|X_k| > x)dx\}$$
$$\leq \sum_{k=1}^{n} |a_{nk}| \cdot \varepsilon \leq M \cdot \varepsilon .$$

Hence $\sum_{k=1}^{n} E(X_k|F_{k-1}) \to 0$ in probability. This together with (2-3) completes the proof.

(b) <u>The necessity</u> :

Suppose that $S_n = \sum_{k=1}^{n} a_{nk}X_k \to 0$ in probability for any uniformly integrable martingale difference sequence $(X_n, F_n, n \geq 1)$. We must show that $\max_{k\leq n} |a_{nk}| \to 0$ as $n \to \infty$. To do this, it suffices to show that $(X_n, n \geq 1)$ is a sequence of independent random variables with $EX_n = 0$, $EX_n^2 < +\infty$ and $P(X_n \neq 0) > 0$.

Put $X_{nk} = a_{nk}X_k$, $1 \leq k \leq n$ and $S_n = \sum_{k=1}^{n} X_{nk}$. Let f_{nk}, g_n denote the characteristic functions of X_{nk} and S_n, respectively. We get $S_n \to 0$ in probability $\Leftrightarrow$ $g_n(u) = \prod_{k=1}^{n} f_{nk}(u) \to 1$ uniformly on each finite interval $[-c,c]$.

Using the inequality of the truncation, ([5], page 209), i.e., for all $\varepsilon > 0$

$$\int_{\{|x|\geq\varepsilon\}} dF(x) \leq -7\varepsilon \int_0^{1/\varepsilon} (\log \operatorname{Re} f(u)du ,$$

we obtain that

$$\sum_{k=1}^{n} P(|X_{nk}| \geq \varepsilon) \leq -7\varepsilon \int_0^{1/\varepsilon} \log|g_n(u)|du \to 0 .$$

On the other hand, we have

$$S_n \to 0 \quad \text{in probability} \Rightarrow L(S_n) \to L(0) .$$

Because of the independence of $(X_{nk}\ ,\ 1 \leq k \leq n)$, we get that

$$\sum_{k=1}^{n} P(|X_{nk}| \geq \varepsilon) \to 0 \Leftrightarrow \max_{k\leq n} |X_{nk}| \to 0$$

in probability, ([5], page 330 or [2], page 53). Since $(X_n\ ,\ n \geq 1)$ are arbitrary, $\max_{k\leq n}|a_{nk}| \to 0$ as $n \to \infty$. □

Now, we present some corollaries of Theorem 1.

<u>Corollary 1</u>. *Let* $(X_k,\ F_k,\ k \geq 1)$ *be a martingale difference such that* $(X_k) \prec X \in L^1$. *Then* $(S_n) \to 0$ *in probability if and only if* $\max_{k\leq n} |a_{nk}| \to 0$ *as* $n \to \infty$.

<u>Proof.</u> It is clear that $E|X| < \infty$ implies (1-2), i.e.

$$\sup_{k\in\mathbb{N}} \{aP(|X_k| > a)\} \to 0 \quad \text{as} \quad a \to \infty ,$$

since

$$\sup_{k\in\mathbb{N}} \{aP(|X_k| > a)\} \leq C\, aP(|X| > a) \to 0 \quad \text{as} \quad a \to \infty .$$

And so the corollary is obtained from Theorem 1. □

<u>Corollary 2</u>. *Let* $(M_n,F_n,\ n \geq 1)$ *be a martingale with the increments* $D_n = M_n - M_{n-1}$, $D_0 = 0$ *such that* $(D_n) \prec X \in L^1$ *or* (D_n) *is a uniformly integrable family. Then*

$$M_n = o(n^{-1}) \quad \textit{in probability.}$$

<u>Proof</u>. We obtain immediately the corollary by putting $a_{nk} = 1/n$ for $1 \le k \le n$, $a_{nk} = 0$ for $k > n$, and using Theorem 1. □

<u>Remark 1</u>. J. Elton ([1]) has proved that $M_n = o(n^{-1})$ almost surely if (D_n) are identically distributed and $D_1 \in L\log^+ L$. He has also constructed an interesting example which showed that if $f \in L^1$, $Ef = 0$ and $f \notin L\log^+ L$, then there exists a martingale difference sequence $(D_n, F_n, n \ge 1)$ with the same distribution as f but $n^{-1} \sum_{k=1}^{n} D_k$ diverges almost surely.

3. THE ALMOST SURE CONVERGENCE

Throughout this section the following assumptions are made: $(X_n, F_n, n \ge 1)$ is a martingale difference, $a_k > 0$, $A_n = (\sum_{k=1}^{n} a_k^p)^{1/p}$, $1 \le p < 2$, $A_n \uparrow \infty$, $a_n/A_n \to 0$. $S_n = \sum_{k=1}^{n} a_k X_k$ denotes the partial weighted sums.

It is known that if (X_n) are independent identically distributed random variables, $a_k > 0$, $A_n = \sum_{k=1}^{n} a_k \uparrow \infty$ and $a_n/A_n \to 0$, $E|X_1| < +\infty$, $EN(|X_1|) < +\infty$ and

$$\int_0^{\infty} x^2 \int_{y \ge x} \frac{N(y)}{y^3}\, dy\, dF(x) < +\infty , \tag{3-1}$$

where $F(\cdot)$ is the distribution of X_1 and

$$N(x) = \operatorname{card}\{n : A_n/a_n \le x\}, \quad x > 0 ,$$

then $S_n/A_n \to 0$ a.s. ([4], Theorem 2 or [1], Theorem 1). If the $(X_n, F_n, n \ge 1)$ is a martingale difference, $a_{nk} = 1/n$ for $n = 1,2,\ldots$ and $k = 1,2,\ldots,n$, the sequence $S_n/n \to 0$ a.s. if

$$(X_n) \prec X \in L\operatorname{Log}^+ L = \{f \in L^1 : E|f|\operatorname{Log}^+|f| < +\infty\} ,$$

([1], Theorem 2 or [2], Theorem 2.19).

The purpose of this section is to study the a.s. convergence of S_n/A_n where $A_n = (\sum_{k=1}^{n} a_k^p)^{1/p}$, $1 \le p < 2$. As a result we obtain $n^{-1/p}\sum_{k=1}^{n} X_k \to 0$ a.s., $1 < p < 2$, if $(X_k) \prec X \in L^p$. The last result of this section deals with the case when (a_{nk}) is a Toeplitz matrix such that $\max_{k \le n} |a_{nk}| \to 0$ as $n \to \infty$.

Now, we give a lemma which be often applied in this section.

<u>Lemma 1</u>. (Kronecker's lemma)

Let $\{x_n, n \ge 1\}$ be a sequence of real numbers such that $\sum_n x_n$ converges, and let $\{b_n, n \ge 1\}$ be a monotone sequence of positive constants with $b_n \uparrow \infty$. Then

$$b_n^{-1} \sum_{k=1}^{n} b_k x_k \to 0 .$$

<u>Theorem 2</u>. *Let* $a_k > 0$, $A_n = (\sum_{k=1}^{n} a_k^p)^{1/p}$, $1 \le p < 2$, $A_n \uparrow \infty$, $a_n/A_n \to 0$, $(X_n, F_n, n \ge 1)$ *be a martingale difference such that* $(X_n) \prec X$, $EN(|X|) < +\infty$ *and*

$$\text{(3-2)} \qquad \int_0^\infty xP(|X| > x) \int_{y \ge x} \frac{N(y)}{y^3}\, dydx < +\infty ,$$

$$\text{(3-3)} \qquad \int_1^\infty P(|X| > x) \int_{1 \le y \le x} \frac{N(y)}{y^2}\, dydx < +\infty .$$

Then S_n/A_n *converges almost surely to zero.*

<u>Proof</u>. Put $Y_n = X_n I(|X_n| \le A_n/a_n)$ and

$$T_n = \sum_{k=1}^{n} a_k X_k .$$

Clearly,

$$\sum_{k=1}^{\infty} P(X_k \ne Y_k) = \sum_{k=1}^{\infty} P(|X_k| > A_k/a_k)$$
$$\le c\sum_{k=1}^{\infty} P(|X| > A_k/a_k) = c\sum_{k=1}^{\infty} \int_{\{x > A_k/a_k\}} dP(|X| \le x)$$
$$= \int_0^\infty N(x)\,dP(|X| \le x) = EN(|X|) < +\infty ,$$

so that the sequences (T_n/A_n) and (S_n/A_n) converge on the same set and to the same limit. We shall show that the series T_n/A_n converges a.s. to zero.

Now, by the same way as the proof of Theorem 2.1 [3], we have

$$\sum_{k=1}^{n} (a_k/A_k)^2 E[Y_k - E(Y_k|F_{k-1})]^2 \le 2\sum_{k=1}^{\infty} (a_k/A_k)^2 E|Y_k|^2$$

$$= 4\sum_{k=1}^{\infty} (a_k/A_k)^2 \int_{\{0<x\le A_k/a_k\}} xP(|X_k| > x)dx$$

$$\le 4C\sum_{k=1}^{\infty} (a_k/A_k)^2 \int_{\{0<x\le A_k/a_k\}} xP(|X| > x)dx$$

$$\le 4C\int_0^{\infty} xP(|X| > x) \sum_{\{k:A_k/a_k\ge x\}} (a_k/A_k)^2 dx$$

$$\le 8C\int_0^{\infty} x\,P(|X| > x) \int_x^{\infty} \frac{N(y)}{y^3}\,dydx < +\infty .$$

The last inequality follows from the fact that

$$\sum_{\{k:A_k/a_k\ge x\}} (a_k/A_k)^2 = \lim_{u\to\infty} \sum_{\{k:x\le A_k/a_k<u\}} (a_k/A_k)^2 = \lim_{u\to\infty} \int_x^u \frac{dN(y)}{y^2}$$

$$= \lim_{u\to\infty}\left(\frac{N(u)}{u^2} - \frac{N(x)}{x^2} + 2\int_x^u \frac{N(y)}{y^3}\,dy\right)$$

and because

$$\frac{N(u)}{u^2} \le 2\int_u^{\infty} \frac{N(y)}{y^3}\,dy \to 0 \quad \text{as} \quad u \to \infty .$$

Hence, in view of the martingale convergence theorem, we get

$$\text{(3-4)} \qquad \frac{1}{A_n}\sum_{k=1}^{n} a_k(Y_k - E(Y_k|F_{k-1})) \to 0 \quad \text{a.s.}.$$

Note that

$$0 = E(X_n|F_{n-1}) = E(Y_n|F_{n-1}) + E(X_n I(|X_n| > A_n/a_n)|F_{n-1}) ,$$

and

$$\sum_{k=1}^{\infty} (a_k/A_k)E(|X_k|I(|X_k| > A_k/a_k))$$

$$= \sum_{k=1}^{\infty} (a_k/A_k) \int_{\{x>A_k/a_k\}} P(|X_k| > x)dx$$

$$\le c \sum_{k=1}^{\infty} (a_k/A_k) \int_{\{x>A_k/a_k\}} P(|X| > x)dx$$

$$= c \int_1^{\infty} P(|X| > x) \sum_{\{k: 1\le A_k/a_k\}} (a_k/A_k)dx$$

$$c \int_1^{\infty} P(|X| > x) \int_1^{x} \frac{N(y)}{y^2} dydx < +\infty .$$

Hence, by Kronecker's lemma we obtain

$$\frac{1}{A_n} \sum_{k=1}^{n} a_k E(Y_k|F_{k-1}) \to 0 \quad \text{a.s.} \tag{3-5}$$

which, together with (3-4), completes the proof. □

Remark 2.

(i) If $p = 1$, $A_n = \sum_{k=1}^{n} a_k$, $A_n \uparrow \infty$ and $a_n/A_n \to 0$, i.e.,

$$a_{nk} = \begin{cases} a_k/A_n & \text{for} \quad k \le n \\ 0 & \text{for} \quad k > n , \end{cases}$$

then is easy to see that (a_{nk}) is a Toeplitz matrix.

If the X_n are independent with $(X_n) \prec X \in L^1$, we can see that $E(X_nI(|X_n| > A_n/a_n)|F_{n-1}) = E(X_nI(|X_n| > A_n/a_n)) = c_n \to 0$ as $n \to \infty$. Note that $A_n \uparrow \infty$ and in view of Toeplitz lemma we have $A_n^{-1} \sum_{k=1}^{n} a_kc_k \to 0$, i.e., (3-5) holds without the assumption (3-3). In this case, we obtain Theorem 2 of [4], because (3-2) and (3-1) are equivalent.

If the independence of (X_n) is omitted, one must use the assumption (3-3), for example, consider

$$a_{nk} = \begin{cases} 1/n & \text{for} \quad k \le n , \\ 0 & \text{for} \quad k > n , \end{cases}$$

i.e. $S_n/n = \frac{1}{n} \sum_{k=1}^{n} X_k$. In this case, if $E|X| < +\infty$, then $EN(|X|) < +\infty$ and (3-2) holds. On the other hand,

$$\int_1^{\infty} P(|X| > x) \int_1^{x} N(y)/y^2 \, dydx = \int_1^{\infty} \log x P(|X| > x)dx = E|X|\log|X| .$$

So $S_n \to 0$ a.s. only if $X \in L\log L$, as the Remark 1.

(ii) If $1 < p < 2$, $a_n \equiv 1$, $A_n = n^{1/p}$ we have

Corollary 3. *If* $1 < p < 2$, $a_n \equiv 1$, $A_n = n^{1/p}$, $(X_n, F_n, n \geq 1)$ *is a martingale difference such that* $(X_n) \prec X \in L^p$, *then* $S_n/n^{1/p} \to 0$ a.s. *as* $n \to \infty$.

Proof. Note that $a_{nk} = n^{-1/p}$ for $n = 1,2,\ldots$, $k = 1,2,\ldots,n$ satisfies (1-3). It is easy to check that $N(x) \sim x^p$ so (3-2) and (3-3) hold if $X \in L^p$. In view of Theorem 2, this completes the proof. □

The following result deals with the case when the weights (a_n) are bounded and (A_n) are p-norms of $a_1, a_2, \ldots, a_n$. At first, we give a lemma which is used to obtain this result.

Lemma 2. ([3], Lemma 2.1)

Let $A_n = \sum_{k=1}^{n} a_k^p)^{1/p}$, $n = 1,2,\ldots$, $0 < p \leq 2$, $(a_n) \in \ell_\infty$, $a_n > 0$ and $A_n \uparrow \infty$. There exists a constant C such that

$$N(x) \leq C\, x^p \log x\,,$$

for $x \in R^+$ large enough.

Theorem 3. *Let* $1 \leq p < 2$, $a_n > 0$, $(a_n) \in \ell_\infty$, $A_n = (\sum_{k=1}^{n} a_k^p)^{1/p} \uparrow \infty$. *If* $(X_n, F_n, n \geq 1)$ *is a martingale difference such that* $(X_n) \prec X \in L^p \log L$, *then* $S_n/A_n \to 0$ a.s. *as* $n \to \infty$.

Proof. It is an application of Theorem 2 and Lemma 2. It suffices to check the assumptions (3-2) and (3-3).

$$\int_0^\infty xP(|X| > x) \int_x^\infty N(y)/y^3\, dy\, dx \leq C \int_0^\infty xP(|X| > x) \int_x^\infty \frac{y^p \log y}{y^3}\, dy\, dx$$

$$\leq C_1 \int_0^\infty x^{p-1} \log x P(|X| > x)dx$$

$$= C_1 E|X|^p \log|X| < +\infty\,.$$

By the same way, we obtain (3-3) when $X \in L^p \log L$. □

The last result of this section deals with the case when (a_{nk}) is a Toeplitz matrix (i.e., (1-3) holds with $p = 1$) and $\max_{k\le n} |a_{nk}|$ is a monotone sequence, $b_n = \max_{k\le n} |a_{nk}| \downarrow 0$. Note that $N(x) = \operatorname{card}\{n : b_n^{-1} \le x\}$ and $S_n = \sum_{k=1}^{n} a_{nk}X_k$.

<u>Theorem 4</u>. *Suppose* $b_n = \max_{k\le n} |a_{nk}| \downarrow 0$ *as* $n \to \infty$ *and let* $(X_n , F_n , n \ge 1)$ *be a martingale difference such that* $(X_n) \prec X$, $EN(|X|) < +\infty$, $N(x) = \operatorname{card}\{n : b_n^{-1} \le x\}$, $x \in \mathbb{R}^+$. *If* (3-2) *and* (3-3) *are satisfied, then* $S_n \to 0$ a.s. *as* $n \to \infty$

Proof. It is easy to see that

$$
\text{(3-5)} \qquad \Big|\sum_{k=1}^{n} a_{nk}X_k\Big| \le \sum_{k=1}^{n} |a_{nk}||X_k| \le \max_{k\le n} |a_{nk}| \sum_{k=1}^{n} |X_k|
$$

$$
= \frac{1}{b_n^{-1}} \sum_{k=1}^{n} |X_k| ,
$$

$b_n^{-1} \uparrow \infty$ as $n \to \infty$.

Now, put

$$R_n = b_n \sum_{k=1}^{n} |X_k| ,$$

$$Y_n = X_n I(|X_n| \le b_n^{-1})$$

and

$$T_n = b_n \sum_{k=1}^{n} |Y_k| .$$

Clearly

$$
\begin{aligned}
\sum_{k=1}^{\infty} P(|X_k| \ne |Y_k|) &\le \sum_{k=1}^{\infty} P(|X_k| > b_k^{-1}) \\
&\le C \sum_{k=1}^{\infty} P(|X| > b_n^{-1}) \\
&= C \sum_{k=1}^{\infty} \int_{\{x > b_n^{-1}\}} dP(|X| \le x) \\
&= C \int_0^{\infty} N(x)\, dP(|X| \le x) \\
&= C\, EN(|X|) < +\infty ,
\end{aligned}
$$

Therefore the sequences (T_n) and (R_n) converge on the same set and to the same limit. It suffices to show that $T_n \to 0$ a.s. as $n \to \infty$.

The rest of the proof is proceeded as the proof Theorem 2 .

4. THE CONVERGENCE IN L^r $(1 \leq r < 2)$

Throughout this section, (a_{nk}) denotes a Toeplitz matrix (i.e. $\lim_{n\to\infty} a_{nk} = 0$ for each $k \geq 1$, $\sum_k |a_{nk}| \leq M$ for all n and some constant M) and $\max_{k\leq n} |a_{nk}| \to 0$ as $n \to \infty$.

<u>Theorem 5</u>. *Let* $(X_n, F_n, n \geq 1)$ *be a martingale difference such that* $(X_n) \prec X \in L^r$, $1 \leq r < 2$. *Then* $E(|S_n|^r) \to 0$.

<u>Proof</u>. Suppose first that $1 < r < 2$. By Burkholder's inequality ([2], page 23) for the martingale array $(S_{nj} = \sum_{k=1}^{j} a_{nk}X_k$, F_j , $1 \leq j \leq n)$

$$E|S_n|^r = E|\sum_{k=1}^{n} a_{nk}X_k|^r \leq B(r)E\{(\sum_{k=1}^{n} a_{nk}^2X_k^2)^{r/2}\} \tag{4-1}$$

where $B(r)$ is a positive constant depending on r .

Now, put

$$Y_{nk} = a_{nk}X_kI(|X_k| \leq |a_{nk}|^{-1}) ,$$

and

$$Z_{nk} = a_{nk}X_k - Y_{nk} .$$

By C_p- inequality, i.e.,

$$E|X + Y|^p \leq C_p(E|X|^p + E|Y|^p) ,$$

$C_p = 1$ if $0 < p \leq 1$ and $C_p = 2^{p-1}$ if $p > 1$, and by $(a+b)^2 \leq 2(a^2+b^2)$, from (4-1) we have

$$\begin{aligned} E(|S_n|)^r &\leq B(r)E\{[\sum_{k=1}^{n} (Y_{nk} + Z_{nk})^2]^{r/2}\} \\ &\leq B(r)E\{[2\sum_{k=1}^{n} (Y_{nk}^2 + Z_{nk}^2)]^{r/2}\} \end{aligned} \tag{4-2}$$

$$\leq 2^{r/2}B(r)E\{[\sum_{k=1}^{n} (Y_{nk}^2 + Z_{nk}^2)]^{r/2}\}$$

$$\leq 2^{r/2}B(r)\{E(\sum_{k=1}^{n} Y_{nk}^2)^{r/2} + E(\sum_{k=1}^{n} Z_{nk}^2)^{r/2}\} .$$

Next, by C_p- inequality again, $0 < r/2 < 1$, we have

$$(4\text{-}3) \qquad E(\sum_{k=1}^{n} Y_{nk}^2)^{r/2} \leq \sum_{k=1}^{n} E|Y_{nk}|^r$$

$$= \sum_{k=1}^{n} |a_{nk}|^r \int_{\{0<x\leq |a_{nk}|^{-1}\}} x^{r-1}P(|X_k| > x)dx$$

$$\leq C.r \sum_{k=1}^{n} |a_{nk}|\{\frac{1}{|a_{nk}|^{-1}} \int_{\{0<x\leq |a_{nk}|^{-1}\}} x^{r-1}P(|X|>x)dx\}$$

$$\leq r.C.M.\varepsilon ,$$

because for n large enough and $(X_n) \prec X \in L^r$, $1 < r < 2$

$$\frac{1}{\lambda^{p-1}} \int_{\{0<x\leq\lambda\}} x^{r-1}P(|X| > x)dx \to 0 \quad \text{as} \quad \lambda \to \infty .$$

By the same way, we obtain

$$(4\text{-}4) \qquad E(\sum_{k=1}^{n} Z_{nk}^2)^{r/2} \to 0 \quad \text{as} \quad n \to \infty ,$$

which, together with (4-3) and (4-2), completes the proof.

Now, suppose that $r = 1$. Then

$$E|\sum_{k=1}^{n}[Y_{nk} - E(Y_{nk}|F_{k-1})]|^2 \leq \sum_{k=1}^{n} E|Y_{nk}|^2 \to 0 ,$$

and

$$E|\sum_{k=1}^{n} [Z_{nk} - E(Z_{nk}|F_{k-1})]| \leq 2\sum_{k=1}^{n} E|Z_{nk}| \to 0 ,$$

in the proof of Theorem 1. It implies that

$$E|S_n| \leq E|\sum_{k=1}^{n}[Y_{nk} - E(Y_{nk}|F_{k-1})]| + E|\sum_{k=1}^{n}[Z_{nk} - E(Z_{nk}|F_{k-1})]| \to 0$$

$$\text{as} \quad n \to \infty . \quad \square$$

<u>Remark 3</u>. The condition $(X_n) \prec X \in L^p$ in Theorem 5 can be replaced

by the uniformly integrable condition of the family $(|X_n|^p,\ n \geq 1)$.

Acknowledgments. We wish to thank Dang Hung Thang and Ho Dang Phuc for their help to complete this paper.

REFERENCES

[1] J. Elton, A law of large numbers for identically distributed martingale differences, Annals of Probability, 9 (1981), 405-412.

[2] P. Hall, C.C. Hayde, Martingale limit theory and its application, Academic Press, New York, (1980).

[3] J. Howell, R.L. Taylor, W.A. Woyczynski, Stability of linear forms in independent random variables in Banach spaces. Lecture Notes in Math., 860 (1980).

[4] B. Jamison, S. Oray, W. Pruitt, Convergence of weighted Averages of independent random variables, Z.W., 4 (1965), 40-44.

[5] M. Loeve, Probability Theory, 1962 (in Russian).

[6] J. Neveu, Martingales a Temps discret, (1972).

[7] A. Stout, Almost sure convergence, Academic Press, New York, (1974).

Probability Theory on Vector Spaces IV
Lancut, June '87, Springer's LNM 1391

Density of the norm of translated stable vectors in Banach spaces

Gyula Pap
Department of Mathematics
University of Debrecen, Hungary

Summary. Let Y be a stable random vector in a real, separable Banach space B. Suppose that the distribution of Y has infinite dimensional support and the space B has smooth enough norm and B is uniformly convex with power order (see in the paper). We prove that the density $p_b(x)$ of the distribution function $F_b(x) = P\{\|Y+b\| < x\}$, $b \in B$ can be estimated in the form $p_b(x) \le c(1+\|b\|^{\hat{m}})$, where c is a constant, depending on Y, and $\hat{m}$ depends on the geometry of the space B.

Introduction

Let $(B, \|\cdot\|)$ be a real, separable Banach space. Suppose that the norm $\|\cdot\|$ is twice differentiable in the Fréchet sense. Denote $\nabla^{(k)}(b)[b_1,...,b_k]$ the k-th derivative of the norm at the point $b \in B$ in the directions $b_1,...,b_k \in B$. Denote $\|\nabla^{(k)}(b)\|$ the norm of the derivative considered as a multilinear form. We shall use the following assumptions:

Assumption 1. $\|\nabla^{(1)}(.)\|$ and $\|\nabla^{(2)}(.)\|$ are bounded on the unit sphere S of the space B.

Assumption 2. The norm $\|.\|$ is uniformly convex with power order m, that is there exist $\sigma > 0, c > 0$ such that

$$\nabla^{(1)}(b + r\gamma)[\gamma] \ge c \cdot r^m \tag{1}$$

for all $b, \gamma \in S$, $\nabla^{(1)}(b)[\gamma] \ge 0$, and $0 \le r \le \sigma$. (This condition guarantees that the unit sphere S does not have flat or almost flat pieces.)

It is known that for example the spaces $L_p, p \ge 2$ satisfy assumptions 1 and 2 with $m = p - 1$.

Lifschitz and Smorodina [7] have shown that if Y is a stable random vector in a Banach space B satisfying assumptions 1 and 2, and the distribution of Y has infinite dimensional support then the distribution of $\|Y\|$ is absolutely continuous with respect to the Lebesgue measure and has bounded density.

For symmetric Y the absolute continuity was investigated first by Paulauskas [10], then by Sztencel [14],[15], Zak [16], Byczkowski and Samotij [3]. The boundedness of the density for symmetric α-stable, $\alpha > 1$, vectors in a Hilbert space was established by Pap [8],[9]; for not necessary symmetric, α-stable, $\alpha > 1$, vectors in a Hilbert space that was proved by Bentkus and Pap [2]. For symmetric α-stable, $\alpha < 1$, vectors see Lewandowski and Zak [6], Zak [17], Ryznar [12].

My aim is to investigate the dependence of the density of the norm on translation of Y. For $b \in B$ denote $p_b(x)$ the density of $\|Y + b\|$. We shall prove the following result:

Theorem. Let Y be a stable random vector in a real, separable Banach space B, satisfying assumptions 1 and 2. Suppose that the distribution of Y has infinite dimensional support. Then the density $p_b(x)$ of $\|Y + b\|, b \in B$ admits the following estimation for arbitrary $\delta > 0$:

$$\sup_x p_b(x) \le c(Y)(1 + \|b\|^{\hat{m}}), \qquad b \in B \tag{2}$$

where $\hat{m} = 2m - 1 + \delta$, m is the power order of the uniform convexity of the norm.

Remark. In [2] the estimation (2) was proved in the Hilbert space case with $\hat{m} = 1$, and moreover it was shown that (2) is not valid with $\hat{m} < 1$. In an analogous way one can show that in L_p spaces (2) is not valid with $\hat{m} < p - 1$, see [11].

The method of the proof is a Malliavin-type calculus, used by Lifschitz and Smorodina [7]. Denote C_0^∞ the space of infinitely many times differentiable functions with compact support. To prove estimation for the density of a probability measure we use the following lemma.

Lemma 1. Let ν be a probability measure on R and suppose that there exists a constant $K > 0$ such that for all $\phi \in C_0^\infty$ we have

$$\left| \int_R \phi'(x)\nu(dx) \right| \le K\|\phi\|_\infty \tag{3}$$

Then ν has density, bounded with K (See Ikeda, Watanabe [4], p.334).

Roughly speaking, for the computation of the integral $\int \phi'(x)\nu(dx)$ we shall use the formula of partial integration, so we get an integral $\int \phi(x)\nu'(dx)$ in some sense. Differentiation of ν will be understood in the direction of a vector field. But first we need an appropriate representation of the stable vector Y on a suitable probability space. Such a construction will be given in §1. The formula of integration by parts and related definitions can be found in §2. In §3 we build up a vector field and calculate the derivatives with respect to it. The estimation of the integral $\int \phi'(x)\nu(dx)$ for ν being the distribution of $\|Y + b\|$ is given in §4.

§1. Representation of stable vectors on the space of configurations

Let Y be an α-stable, $0 < \alpha < 2$, random vector in a real, separable Banach space B. Denote μ the distribution of Y. It is well-known that there exists a process $\{\xi(t), t \in [0,1]\}$ with independent, stationary increments such that $\xi_0 = 0$ and

$$P\{\xi(t_2) - \xi(t_1) \in A\} = \mu\Big(\frac{1}{(t_2 - t_1)^{1/\alpha}} A\Big)$$

for $A \in \mathbf{B}(B)$, where $\mathbf{B}(B)$ is the Borel σ-field of B. Let $U_\varepsilon = \{x \in B : \|x\| \geq \varepsilon\}$ for $\varepsilon > 0$, and $\mathbf{B}_\varepsilon = \mathbf{B}(U_\varepsilon)$ be the Borel σ-field of U_ε. We can take a right-continuous version of the process $\xi(t)$, so if we consider

$$\nu(t, A) = card\{s \in [0, t] : \xi(s) - \xi(s - 0) \in A\}, \qquad A \in \mathbf{B}_\varepsilon,$$

(the number of jumps, falling into the set A), then we have $P\{\nu(t, A) < \infty\} = 1$, and $\{\nu(t, A), t \in [0, 1]\}$ is a Poisson process. Let $E\nu(t, A) = \Pi(t, A)$. Taking into account that the process $\xi(t)$ has no Gaussian component we have the following integral representation:

$$\xi(t) = b_0 t + \int_{\|x\|\leq 1} x(\nu(t, dx) - \Pi(t, dx)) + \int_{\|x\|>1} x\nu(t, dx), \tag{4}$$

where $b_0 \in B$.

Consider polar co-ordinates in B : for $x \in B$ let $x = (\theta(x), r(x))$, where $r(x) = \|x\|$, $\theta(x) = x/\|x\|$. Making transformation $r \to r^{-1/\alpha}$ in (4) we get

$$\xi(t) = b_0 t + \frac{1}{\alpha}\Big(\int_{r\geq 1} \frac{\theta(x)}{(r(x))^{1/\alpha}}(\nu - \Pi)(dx) + \int_{r<1} \frac{\theta(x)}{(r(x))^{1/\alpha}}\nu(dx)\Big). \tag{5}$$

Now the jumps of the process in the interval [0,1] determine the value of the process in $t = 1$, and the distribution of $\xi(1)$ is exactly μ. Moreover, the number of jumps falling into U_ε is finite with probability 1.

The subset $X \subset B$ is called **locally finite**, if the intersection of X with any bounded set is finite, that is $card(X \cap \{x \in B : \|x\| < K\}) < \infty$ for all $K > 0$. Locally finite subsets are also called **configurations**.

Let $\mathbf{X}$ be the space of all locally finite subsets of B, endowed with the natural σ-field Σ (Σ is the least σ-field such that the mappings $X \mapsto card(X \cap V)$, $X \in \mathbf{X}$ are measurable for all bounded Borel sets $V \subset B$).

After transformation $r \to r^{-1/\alpha}$ the set of jumps of the original process $\{\xi(t), t \in [0, 1]\}$ belongs to $\mathbf{X}$ with probability 1, so from (5) with $t = 1$, taking into account that $\xi(t)$ is a stable process, we get the following representation for the α-stable vector Y denoting its spectral measure by π:

$$Y(X) = b_0 + \frac{1}{\alpha}\Big(\sum_{x\in X} \frac{\theta(x)}{r^{1/\alpha}(x)} - \int_S \int_1^\infty r^{-1/\alpha}\theta\lambda(dr)\pi(d\theta)\Big), \qquad X \in \mathbf{X}, \tag{6}$$

where λ is the Lebesgue-measure.

§2. Differential operators related to vector fields and the formula of integration by parts

Consider now $\mathbf{X}$, the space of all locally finite subset of the Banach space B with polar coordinates $x = (\theta(x), r(x)) \in B$. Denote

$$H(X) = \Big\{h : X \to R \mid \sum_{x\in X} h^2(x) < \infty\Big\}$$

the tangent Hilbert space at the point $X \in \mathbf{X}$.

For $x = (\theta, r) \in B$ and $v \in R$ let $x + v = (\theta, r + v) \in B$. Analogously for $X \in \mathbf{X}$ and $h \in H(X)$ let $X + h = \{x + h(x) | x \in X\} \in \mathbf{X}$. Moreover for $X \in \mathbf{X}$ and $h \in H(X)$ define translation of the tangent space $T(X,h) : H(X+h) \mapsto H(X)$ by

$$T(X,h)(g)(x) = g(x + h(x)), \qquad g \in H(X+h), \quad x \in X.$$

We shall use differential operators with respect to a vector field

$$\mathbf{K} = \{\mathbf{K}(X) \, | X \in \mathbf{X}, \; \mathbf{K}(X) \in H(X)\}.$$

If we have a function $f : \mathbf{X} \mapsto R$ then it has a natural extension $\tilde{f}(X,.) : H(X) \mapsto R$ to the tangent space:

$$\tilde{f}(X,h) = f(X+h) \qquad \text{for } h \in H(X).$$

Analogously, the vector field $\mathbf{K}$ has the natural extension $\tilde{\mathbf{K}}(X,.) : H(X) \mapsto H(X)$ to the tangent space:

$$\tilde{\mathbf{K}}(X,h) = T(X,h) \circ \mathbf{K}(X+h) \qquad \text{for } h \in H(X).$$

The derivative of a function $f : \mathbf{X} \mapsto R$ with respect to $\mathbf{K}$ at the point $X \in \mathbf{X}$ is

$$(D^{(1)}_{\mathbf{K}} f)(X) = < \mathbf{K}(X), \tilde{f}^{(1)}(X,.)|_0 >_{H(X)}, \tag{7}$$

where $\tilde{f}^{(1)}(X,.)|_0$ is the derivative of the function $\tilde{f}(X,.)$ on the tangent space $H(X)$ taking at the point $h = 0 \in H(X)$, and $< .,. >_{H(X)}$ is the scalar product of the tangent space $H(X)$. It can be considered also as the derivative in the direction $\mathbf{K}(X)$, because

$$(D^{(1)}_{\mathbf{K}} f)(X) = \lim_{\varepsilon \to 0} \frac{1}{\varepsilon}[f(X + \varepsilon \mathbf{K}(X)) - f(X)].$$

The definition (7) can also be written as

$$D^{(1)}_{\mathbf{K}} f = \tilde{f}^{(1)}[\tilde{\mathbf{K}}], \tag{8}$$

so the second derivative of the function $f : \mathbf{X} \mapsto R$ with respect to $\mathbf{K}$ at the point $X \in \mathbf{X}$ can be calculated with the help of the chain rule:

$$D^{(2)}_{\mathbf{K}} f = \tilde{f}^{(2)}[\tilde{\mathbf{K}}, \tilde{\mathbf{K}}] + \tilde{f}^{(1)}[\tilde{\mathbf{K}}^{(1)}[\tilde{\mathbf{K}}]], \tag{9}$$

taking all derivatives at the point $h = 0 \in H(X)$.

Now it can be proved (see Smorodina [13]), that for $\phi \in C_0^\infty$ we have

$$\int_{\mathbf{X}} \phi'(f) dP = \int_{\mathbf{X}} \phi(f) \Big(\frac{D^{(2)}_{\mathbf{K}} f}{(D^{(1)}_{\mathbf{K}} f)^2} - \frac{tr \tilde{\mathbf{K}}^{(1)}}{D^{(1)}_{\mathbf{K}} f} \Big) dP. \tag{10}$$

It follows from Lemma 1 that in order to prove our Theorem it is enough to show that choosing an appropriate vector field $\mathbf{K}$ we have the following estimation for the function $f_b(X) = \|Y(X) + b\|$:

$$\int_{\mathbf{X}} \Big(\frac{|D^{(2)}_{\mathbf{K}} f_b|}{|D^{(1)}_{\mathbf{K}} f_b|^2} + \frac{|tr \tilde{\mathbf{K}}^{(1)}|}{|D^{(1)}_{\mathbf{K}} f_b|} \Big) dP \le c(Y)(1 + \|b\|^{m}). \tag{11}$$

§3. Construction of the suitable vector field

First we cut the function $Y(X)$ into two parts:

$$Y_1(X) = \frac{1}{\alpha} \sum_{\{x\in X, r(x)<1\}} r^{-1/\alpha}(x)\theta(x), \qquad Y_0(X) = Y(X) - Y_1(X).$$

Moreover, define the function

$$V(X) = 1 + \frac{1}{\alpha} \sum_{\{x\in X, 1\le r(x)\le 2\}} r^{-1/\alpha}(x).$$

Let us fix an infinitely differentiable function $\kappa : R \mapsto R$ such that $0 \le \kappa(r) \le 1$ and

$$\kappa(r) = \begin{cases} 0, & \text{if } r \le 0; \\ 1, & \text{if } r \ge 1. \end{cases}$$

Define the functions

$$\chi(r) = \kappa(2-r), \qquad \varphi(r) = \kappa(r-2), \qquad \psi(r) = \kappa(r-100).$$

Let us consider the radial vector field $\varrho(X)(x) = r(x)\chi(r(x)),\ x \in X$, and the tangential vector field

$$g(X)(x) = \left|\nabla^{(1)}(Y(X)+b)[\theta(x)]\right| \cdot r^{-1-1/\alpha}(x)\varphi(r(x))1_{\{\theta(x)\in\cup_{i=1}^k U_i\}},$$

where the sets $U_1, \dots, U_k \subset S$ are chosen so that they have positive spectral measure and they are uniformly linearly independent in the sense

$$\inf\Big\{\|\sum_{i=1}^k c_i\theta_i\| : c_i \in R, \sum_{i=1}^k c_i^2 = 1, \theta_i \in U_i, i = 1,\dots,k\Big\} > 0.$$

Now take the following mixture of the radial and tangential vector fields:

$$\mathbf{K}(X) = \psi_1(X)\varrho(X) + \psi_0(X)g(X),$$

where

$$\psi_1(X) = \psi\Big(\frac{\|Y_1(X)\|}{\|Y_0(X)+b\|}\Big)\psi\Big(\frac{\|Y_1(X)\|}{V(X)}\Big), \qquad \psi_0(X) = 1 - \psi_1(X).$$

We can easily calculate the derivatives of the function $f(x) = \|Y(X)+b\|$ with respect to $\mathbf{K}$. First compute the derivative of the extended function $\tilde{f}(X,.) : H(X) \mapsto R$ at the point $h = 0 \in H(X)$. Consider the orthonormal system $\{h_y, y \in X\}$, where

$$h_y(x) = \begin{cases} 1, & \text{if } x = y; \\ 0, & \text{if } x \ne y. \end{cases}$$

The derivative of the function $\tilde{Y}(X,.) : H(X) \mapsto B$ at the point $h = 0 \in H(X)$ in the direction h_y is

$$\tilde{Y}^{(1)}(X,.)|_0(h_y) = \lim_{\varepsilon \to 0} \frac{Y(X+\varepsilon h_y) - Y(X)}{\varepsilon} =$$
$$= \lim_{\varepsilon \to 0} \frac{1}{\varepsilon} \frac{(r(y)+\varepsilon)^{-1/\alpha}\theta(y) - r^{-1/\alpha}(y)\theta(y)}{\alpha} = -\frac{1}{\alpha^2}\theta(y) r^{-1-1/\alpha}(y),$$

so using the chain rule the derivative of the extended function $\tilde{f}(X,.) : H(X) \mapsto R$ in the direction h_y will be

$$\tilde{f}^{(1)}(X,.)|_0(h_y) = -\frac{1}{\alpha^2} r^{-1-1/\alpha}(y) \nabla^{(1)}(Y(X)+b)[\theta(y)]. \tag{12}$$

So the first derivative of f with respect to K is

$$(D^{(1)}_{\mathbf{K}} f)(X) = < \mathbf{K}(X), \tilde{f}^{(1)}(X,.)|_0 >_{H(X)} = \sum_{y \in X} \tilde{f}^{(1)}(X,.)|_0(h_y) \cdot \mathbf{K}(X)(y) =$$
$$= -\frac{\psi_0(X)}{\alpha^2} \sum_{\{\theta(y) \in \cup_{i=1}^{k} U_i\}} s\Big(\nabla^{(1)}(Y(X)+b)[\theta(y)]\Big)^2 r^{-2(1+1/\alpha)}(y)\varphi(r(y)) -$$
$$- \frac{\psi_1(X)}{\alpha^2} \sum_{y \in X} \nabla^{(1)}(Y(X)+b)[\theta(y)] r^{-1/\alpha}(y)\chi(r(y)),$$

where

$$s = \text{sign}\Big(\nabla^{(1)}(Y(X)+b)[\theta(y)]\Big).$$

The second derivative of f with respect to $\mathbf{K}$ can be calculated with the help of the formula (9). So first we have to determine the second derivative of the extended function $\tilde{f}(X,.) : H(X) \mapsto R$ at the point $h = 0$ in the directions $h_y, h_z \in H(X)$:

$$\tilde{f}^{(1)}(X,.)|_0(h_y,h_z) = \nabla^{(2)}(Y(X)+b)\Big[-\frac{1}{\alpha^2}\theta(y)r^{-1-1/\alpha}, -\frac{1}{\alpha^2}\theta(y)r^{-1-1/\alpha}\Big] +$$
$$+ \nabla^{(1)}(Y(X)+b)\Big[\frac{\delta_{y,z}}{\alpha^2}(1+\frac{1}{\alpha})\theta(y)r^{-2-1/\alpha}(y)\Big],$$

because the second derivative of $\tilde{Y}(X,.) : H(X) \mapsto B$ in the directions h_y, h_z is

$$\tilde{Y}^{(2)}(X,.)|_0(h_y,h_z) = \lim_{\varepsilon \to 0} \frac{-\theta(y)\big(r(y)+\varepsilon h_z(y)\big)^{-1-1/\alpha} + \theta(y)r^{-1-1/\alpha}(y)}{\varepsilon\alpha^2} =$$
$$= \frac{\delta_{y,z}}{\alpha^2}(1+\frac{1}{\alpha})\theta(y)r^{-2-1/\alpha}(y),$$

where

$$\delta_{y,z} = \begin{cases} 1, & \text{if } y = z; \\ 0, & \text{if } y \neq z \end{cases}$$

is the Kronecker symbol.

Returning now to the first term of (9) we have

$$\tilde{f}^{(2)}[\tilde{\mathbf{K}},\tilde{\mathbf{K}}] = \sum_{y\in X}\sum_{z\in X}\tilde{f}^{(1)}(X,.)|_0(h_y,h_z)\cdot K(X)(y)\cdot K(X)(z) =$$
$$= \frac{1}{\alpha^4}\sum_{y\in X}\sum_{z\in X}K(X)(y)K(X)(z)\nabla^{(2)}(Y(X)+b)[\theta(y),\theta(z)]\cdot(r(y)r(z))+$$
$$+\frac{1}{\alpha^2}\left(1+\frac{1}{\alpha}\right)\sum_{y\in X}K^2(X)(y)\nabla^{(1)}(Y(X)+b)[\theta(y)]r^{-2-1/\alpha}.$$

For the second term of (9) we have to compute the derivative of the extended vector field $\tilde{\mathbf{K}}(X,.) : H(X) \mapsto H(X)$ at the point $h = 0 \in H(X)$. The derivative of the radial vector field $\varrho(X)(x) = r(x)\chi(r(x))$, $x \in X$ in the direction h_y will be

$$\tilde{\varrho}^{(1)}(X,.)|_0(h_y)(x) = \lim_{\varepsilon\to 0}\frac{(r(x)+\varepsilon h_y(x))\chi(r(x)+\varepsilon h_y(x)) - r(x)\chi(r(x))}{\varepsilon} =$$
$$= \delta_{x,y}(\chi(r(x)) + r(x)\chi'(r(x))).$$

So it does not vanish only if $r(x) < 2$ and $y = x$.

The derivative of the tangential vector field

$$g(X)(x) = \left|\nabla^{(1)}(Y(X)+b)[\theta(x)]\right| r^{-1-1/\alpha}(x)\varphi(r(x))1_{\{\theta(x)\in\cup_{i=1}^{k}U_i\}}$$

in the direction h_y will be

$$\tilde{g}^{(1)}(X,.)|_0(h_y)(x) =$$
$$= s\nabla^{(2)}(Y(X)+b)\left[-\frac{1}{\alpha^2}\theta(y)r^{-1-1/\alpha}(y),\theta(x)\right]\cdot r^{-1-1/\alpha}(x)\varphi(r(x))1_{\{\theta(x)\in\cup_{i=1}^{k}U_i\}}$$
$$+\left|\nabla^{(1)}(Y(X)+b)[\theta(x)]\right|\left(-1-\frac{1}{\alpha}\right)r^{-2-1/\alpha}(x)h_y(x)\varphi(r(x))1_{\{\theta(x)\in\cup_{i=1}^{k}U_i\}}$$
$$+\left|\nabla^{(1)}(Y(X)+b)[\theta(x)]\right|r^{-1-1/\alpha}(x)\varphi'(r(x))h_y(x)1_{\{\theta(x)\in\cup_{i=1}^{k}U_i\}}.$$

So it does not vanish only if $r(x) > 2$, and in the case $y \neq x$ only the first term differs from 0.

For calculating of the derivative of the real valued weight functions $\psi_0(X)$ and $\psi_1(X)$ we compute derivatives of Y_0, Y_1 and V as follows:

$$\tilde{Y}_0^{(1)}(X+b,.)|_0(h_y) = -\frac{1}{\alpha^2}\theta(y)r^{-1-1/\alpha}(y)1_{\{r(y)\geq 1\}},$$
$$\tilde{Y}_1^{(1)}(X,.)|_0(h_y) = -\frac{1}{\alpha^2}\theta(y)r^{-1-1/\alpha}(y)1_{\{r(y)<1\}},$$
$$\tilde{V}_0^{(1)}(X,.)|_0(h_y) = -\frac{1}{\alpha^2}r^{-1-1/\alpha}(y)1_{\{1\leq r(y)\leq 2\}}.$$

From this with the help of the chain rule we get

$$\|\tilde{Y}_0(X+b,.)\|^{(1)}|_0(h_y) = -\frac{1}{\alpha^2}\nabla^{(1)}(Y_0(X)+b)[\theta(y)]r^{-1-1/\alpha}(y)1_{\{r(y)\geq 1\}},$$

$$\|\tilde{Y}_1(X,.)\|^{(1)}|_0(h_y) = -\frac{1}{\alpha^2}\nabla^{(1)}(Y_1(X))[\theta(y)]r^{-1-1/\alpha}(y)1_{\{r(y)\geq 1\}}.$$

$$\left(\tilde{\psi}\Big(\frac{\|Y_1(X,.)\|}{\|Y_0(X,.)+b\|}\Big)\right)^{(1)}\Bigg|_0(h_y) = \psi'\Big(\frac{\|Y_1(X)\|}{\|Y_0(X)+b\|}\Big)\frac{r^{-1-1/\alpha}(y)}{\|Y_0(X)+b\|^2\alpha^2}\cdot$$
$$\left\{\|Y_1(X)\|\nabla^{(1)}(Y_0(X)+b)[\theta(y)]1_{\{r(y)\geq 1\}} - \|Y_0(X)+b\|\nabla^{(1)}(Y_1(X))[\theta(y)]1_{\{r(y)<1\}}\right\},$$

$$\left(\tilde{\psi}\Big(\frac{\|Y_1(X,.)\|}{V(X,.)}\Big)\right)^{(1)}\Bigg|_0(h_y) = \psi'\Big(\frac{\|Y_1(X)\|}{V(X)}\Big)\frac{r^{-1-1/\alpha}(y)}{V(X)^2\alpha^2}\cdot$$
$$\cdot\left\{\|Y_1(X)\|1_{\{1\leq r(y)\leq 2\}} - V(X)\nabla^{(1)}(Y_1(X))[\theta(y)]1_{\{r(y)<1\}}\right\}.$$

So the derivatives of the weight functions are

$$\begin{aligned}\tilde{\psi}_1^{(1)}(X,.)|_0(h_y) =& \tilde{\psi}\Big(\frac{\|Y_1(X,.)\|}{\|Y_0(X,.)+b\|}\Big)\Bigg)^{(1)}\Bigg|_0(h_y)\psi\Big(\frac{\|Y_1(X)\|}{V(X)}\Big)+\\ &\psi\Big(\frac{\|Y_1(X)\|}{\|Y_0(X)+b\|}\Big)\tilde{\psi}\Big(\frac{\|Y_1(X,.)\|}{V(X,.)}\Big)\Bigg)^{(1)}\Bigg|_0(h_y),\end{aligned}$$

$$\tilde{\psi}_0^{(1)}(X,.)|_0(h_y) = -\tilde{\psi}_1^{(1)}(X,.)|_0(h_y).$$

Finally, we have

$$\begin{aligned}\tilde{\mathbf{K}}^{(1)}(X,.)|_0(h_y)(x) =& \tilde{\psi}_1^{(1)}(X,.)|_0(h_y)\varrho(X)(x) + \psi_1(X)\tilde{\varrho}^{(1)}(X,.)|_0(h_y)(x)+\\ &+\tilde{\psi}_0^{(1)}(X,.)|_0(h_y)g(X)(x) + \psi_0(X)\tilde{g}^{(1)}(X,.)|_0(h_y)(x).\end{aligned} \tag{13}$$

From (12) and (13) we can combine the second term of (9):

$$\begin{aligned}&\tilde{f}^{(1)}[\tilde{\mathbf{K}}^{(1)}[\tilde{\mathbf{K}}]] = \sum_{z\in X}\tilde{f}^{(1)}(X,.)|_0(h_z)\tilde{\mathbf{K}}^{(1)}[\tilde{\mathbf{K}}](z) =\\ &= -\frac{1}{\alpha^2}\sum_{z\in X}\nabla^{(1)}(Y(X)+b)[\theta(z)]r^{-1-1/\alpha}(z)\sum_{y\in X}\tilde{\mathbf{K}}^{(1)}(X,.)|_0(h_y)(z)\mathbf{K}(X)(y).\end{aligned}$$

Thus we can compute the second derivative of the function f with respect to $\mathbf{K}$:

$$D_{\mathbf{K}}^{(2)}f = \tilde{f}^{(2)}[\tilde{\mathbf{K}},\tilde{\mathbf{K}}] + \tilde{f}^{(1)}[\tilde{\mathbf{K}}^{(1)}[\tilde{\mathbf{K}}]].$$

In the integral in (11) the trace of $\tilde{\mathbf{K}}^{(1)}$ appears:

$$tr\tilde{\mathbf{K}}^{(1)} = \sum_{y\in X}\tilde{\mathbf{K}}^{(1)}(X,.)|_0(h_y)(y).$$

§4 Proof of the estimation (11)

We shall frequently use the following statements (see [7]):

Lemma 2. Consider the following random variables on the space of configurations:

a) $\sum_{\{r(x)>a\}} r^{-1-c}(x), \qquad \sum_{\{r(x)<a\}} 1, \qquad V(X) = 1 + \frac{1}{\alpha} \sum_{\{1 \le r(x) \le 2\}} r^{-1/\alpha}(x),$

where $a, c > 0$;

b) $\|Y_0(X) + b\|, \qquad b \in B.$

c) $\|Y(X) + b\|^{-1}, \qquad \|Y_0(X) + b\|^{-1}, \qquad b \in B.$

Then all moments of these variables are finite, and in the case c) the moments can be estimated uniformly in $b \in B$.

Proof of Lemma 2. The first variable in a) has infinitely divisible distribution and its spectral measure is concentrated in a bounded set, so the statement is obvious. The second variable in a) has also infinitely divisible distribution, and the statement follows from straightforward computation. The random vector $Y_0(X) + b$ has infinitely divisible distribution, and its Levy measure satisfies the condition of [5], so the variable in b) has finite exponential moments. The random vector $Y(X) + b$ is stable and not concentrated in any finite dimensional subspaces, so the statement for the first variable in c) follows from [1]. Now if we assume that one of the moments of $\|Y_0(X) + b\|^{-1}$ is not finite, then it would be so for $\|Y_{0,\varepsilon}(X) + b\|^{-1}$, where

$$Y_{1,\varepsilon}(X) = \frac{1}{\alpha} \sum_{\{x \in X, r(x) < \varepsilon\}} r^{-1/\alpha}(x)\theta(x), \qquad Y_{0,\varepsilon}(X) = Y(X) - Y_{1,\varepsilon}(X).$$

But if ε tends to 0, it would be in contradiction with the previous statement. #

Everywhere in this part of the paper c will stand for a positive constant, not depending on $b \in B$, and $M(X)$ will stand for a random variable on the space of configurations, not depending on $b \in B$, too.

Lemma 3.

$$|(D_{\mathbf{K}}^{(j)} f)(X)| \le M(X) \cdot (1 + \|Y_1\|) \qquad j = 1, 2$$
$$|tr\tilde{\mathbf{K}}^{(1)}| \le M(X) \cdot (1 + \|Y_1\|)$$

Proof of Lemma 3. From assumption 1 the following inequalities can be obtained:

$$|\nabla^{(1)}(b)[b_1]| \le c\|b_1\|, \qquad |\nabla^{(2)}(b)[b_1, b_2]| \le c\|b_0\|^{-1}\|b_1\| \cdot \|b_2\|,$$

where $b \ne 0, b_1, b_2 \in B$. Using these inequalities, $tr\tilde{\mathbf{K}}^{(1)}$ for example can be estimated as follows.

From the first term of (13):

$$c\|Y_1\| \sum_{\{1\le r(y)\le 2\}} r^{-1/\alpha}(y)\Big(\frac{\nabla^{(1)}(Y_0(X)+b)[\theta(y)]}{\|Y_0+b\|^2}+\frac{1}{V^2}\Big)+$$
$$+c\sum_{\{r(y)<1\}} r^{-1/\alpha}(y)\Big(\frac{1}{V}+\frac{1}{\|Y_0+b\|}\Big)\nabla^{(1)}(Y_1(X))[\theta(y)] \le$$
$$\le c\frac{\|Y_1\|}{V^2}\sum_{\{1\le r(y)\le 2\}} r^{-1/\alpha}(y)+c\frac{\|Y_1\|}{\|Y_0+b\|^2}\nabla^{(1)}(Y_0(X)+b)[Y_0(X)]+$$
$$+c\Big(\frac{1}{V}+\frac{1}{\|Y_0+b\|}\Big)\nabla^{(1)}(Y_1(X))[Y_1(X)]\le M(X)(1+\|Y_1\|).$$

Second term:

$$c\sum_{\{r(y)<1\}}1+c\sum_{\{1\le r(y)\le 2\}}(1+r(y))\le M(X).$$

Third term:

$$c\frac{\|Y_1\|}{\|Y_0+b\|^2}\sum_{\{r(y)\ge 2\}} r^{-2-2/\alpha}(y)\big(\nabla^{(1)}(Y_0(X)+b)[\theta(y)]\big)^2\le M(X)\|Y_1\|.$$

Fourth term:

$$c\sum_{\{r(y)\ge 2\}}\nabla^2(Y(X)+b)[\theta(y),\theta(y)]r^{-2-2/\alpha}(y)+$$
$$+c\sum_{\{r(y)\ge 2\}}\nabla^1(Y(X)+b)[\theta(y)]r^{-1-1/\alpha}(y)\le M(X).$$

Analogously can we prove the other statements. #

As in [7], first we shall estimate the integral in (11) over the domain $\{X\in\mathbf{X}:\psi_1(X)\ge\frac{1}{2}\}$. For the estimation of the denominators from below we use the second term of $(D^{(1)}_{\mathbf{K}}f)(X)$:

$$\big|(D^{(1)}_{\mathbf{K}}f)(X)\big|\ge c\sum_{y\in X}\nabla^1(Y(X)+b)[\theta(y)]r^{-1/\alpha}(y)\chi(r(y))=$$

$$=c\nabla^1(Y(X)+b)\Big[Y_1(X)+\frac{1}{\alpha}\sum_{\{1\le r(y)<2\}}\theta(y)r^{-1/\alpha}(y)\chi(r(y))\Big]\ge$$

$$\ge c\Big(\nabla^{(1)}(Y(X)+b)[Y_1(X)]-V(X)\Big)\ge c\Big(\nabla^{(1)}(Y(X)+b)[Y(X)+b]-\|Y_0(X)\|-V(X)\Big)=$$
$$=c(\|Y(X)+b\|-\|Y_0(X)\|-V(X))\ge c\|Y(X)+b\|.$$

Here we used linearity of $\nabla^{(1)}$, Assumption 1 and that in this domain we have

$$\|Y(X)+b\|-\|Y_0(X)+b\|-V(X)\ge\frac{\|Y(X)+b\|}{2}.$$

For the estimation of the numerators from above we use Lemma 3. So for the integral in (11) over the domain $\{X\in\mathbf{X}:\psi_1(X)\ge\frac{1}{2}\}$ we get the estimation

$$\int_{\{X\in\mathbf{X}:\psi_1(X)\ge\frac{1}{2}\}}\frac{M(X)(1+\|Y_1(X)\|)}{\|Y(X)+b\|^2}P(dX).$$

But in this domain $\|Y_1(X)\| \le c\|Y(X)+b\|$ and $\|Y(X)+b\| \ge c$, so we get estimation $c(1+\|b\|^{1+\delta}), \delta > 0$.

Now we have to estimate the integral in (11) over the domain $\left\{X \in \mathbf{X} : \psi_1(X) \le \frac{1}{2}\right\}$. We follow again the method of [7]. For the estimation of the denominators from below we use the first term of $(D^{(1)}_{\mathbf{K}} f)(X)$:

$$|(D^{(1)}_{\mathbf{K}} f)(X)| \ge c \sum_{\{r(y)\ge 3, \theta(y) \in \cup_{i=1}^k U_i\}} \left(\nabla^{(1)}(Y(X)+b)[\theta(y)]\right)^2 r^{-2-2/\alpha}(y) = cZ(X).$$

Thus we have to estimate the integrals

$$\int_{\{\psi_1(X)\le\frac{1}{2}\}} \frac{|(D^{(2)}_{\mathbf{K}} f)(X)|}{Z^2(X)} P(dX), \qquad \int_{\{\psi_1(X)\le\frac{1}{2}\}} \frac{|tr\tilde{\mathbf{K}}^{(1)}|}{Z(X)} P(dX).$$

We shall deal only with the first term of the first integral after replacing $(D^{(2)}_{\mathbf{K}} f)(X)$ from (9). (The other term and the second integral can be handled in the same way.) So we estimate the integral

$$\int_{\{\psi_1(X)\le\frac{1}{2}\}} \frac{P(dX)}{Z^2(X)} \sum_y \sum_z \nabla^{(2)}(Y(X)+b)[\theta(y),\theta(z)]\nabla^{(1)}(Y(X)+b)[\theta(y)]\cdot$$

$$\cdot\nabla^{(1)}(Y(X)+b)[\theta(z)](r(y)r(z))^{-2-2/\alpha}.$$

We compute the conditional expectation with respect the following partition of the probability space $(\mathbf{X}, \Sigma, \mathbf{P})$. First we cut the space into the parts

$$\mathbf{X}_{n_1,\dots,n_k} = \left\{X \in \mathbf{X} : \min\{r \ge 3 | (\theta, r) \in X, \theta \in U_i\} \in [n_i, n_i+1)\right\},$$

where $n_1, \dots, n_k$ are natural numbers. Then divide $\mathbf{X}_{n_1,\dots,n_k}$ into k-dimensional layers fixing the points of the configurations outside of the set $\cup_{i=1}^k U_i \times [n_i, n_i+1)$, and fixing all but one point in each set $U_i \times [n_i, n_i+1)$. Lastly fix the first co-ordinate $\theta_i \in U_i \cap S$ of each non-fixed point $(\theta_i, r_i) \in U_i \times [n_i, n_i+1)$. Now only the second co-ordinates of r_i of these points left non-fixed. The conditional expectation can be estimated from above with the integral

$$\int_{\Delta_{n_1}\times\dots\times\Delta_{n_k}} \frac{\sum_{i=1}^k \sum_{j=1}^k (r_i r_j)^{-2-2/\alpha} \nabla^{(1)}(Y(X)+b)[\theta_i]\nabla^{(1)}(Y(X)+b)[\theta_j]}{\|Y(X)+b\| \left(\sum_{i=1}^k \left(\nabla^{(1)}(Y(X)+b)[\theta_i]\right)^2 r_i^{-2-2/\alpha}\right)^2} dr_1 \dots dr_k,$$

where $\Delta_{n_i} = [n_i, n_i+1)$, and

$$Y(X) = Y(r_1, \dots, r_k) = \hat{b} + \frac{1}{\alpha}\sum_{i=1}^k r_i^{-1/\alpha}\theta_i,$$

where $\hat{b} \in B$ is constant for each element of the partition. After the transformation $r_i \to (\alpha s_i)^{-\alpha}$ we get

$$\int_{\Delta_{n_1}\times\dots\times\Delta_{n_k}} \frac{\left(\sum_{i=1}^k \nabla^{(1)}(\hat{b}+b+\sum_{l=1}^k s_l\theta_l)[\theta_i]s_i^{2+2\alpha}\right)^2}{\left(\sum_{i=1}^k \left(\nabla^{(1)}(\hat{b}+b+\sum_{l=1}^k s_l\theta_l)[\theta_i]\right)^2 s_i^{2+2\alpha}\right)^2} \frac{\prod_{i=1}^k s_i^{1+\alpha}}{\|Y(X)+b\|} ds_1 \dots ds_k.$$

Taking into consideration that the distribution of $\min\{r|(\theta,r)\in X\}$ is stable, so its negative powers has finite moments, and extending the domain of integration to the cube $[0,c]^k$ we have to estimate the following integral :

$$\int_{[0,c]^k} \frac{\left(\sum_{i=1}^k \nabla^{(1)}(\hat{b}+b+\sum_{l=1}^k s_l\theta_l)[\theta_i]\right)^{2+\delta}}{\left(\sum_{i=1}^k(\nabla^{(1)}(\hat{b}+b+\sum_{l=1}^k s_l\theta_l)[\theta_i])^2\right)^{2+\delta}} \frac{ds_1...ds_k}{\|Y(X)+b\|},$$

where $\delta>0$. Let $\hat{s}_i = s_i(\sum_{j=1}^k s_j^2)^{-1/2}$. Then applying Hölder inequality :

$$(\nabla^{(1)}(\hat{b}+b+\sum_{l=1}^k s_l\theta_l)[\sum_{i=1}^k \hat{s}_i\theta_i])^2 = (\sum_{i=1}^k \hat{s}_i\nabla^{(1)}(\hat{b}+b+\sum_{l=1}^k s_l\theta_l)[\theta_i])^2 \le$$

$$\le \sum_{i=1}^k(\nabla^{(1)}(\hat{b}+b+\sum_{l=1}^k s_l\theta_l)[\theta_i])^2.$$

Changing the vector $\hat{b}$ we can assume that the convex function $\|\hat{b}+b+\sum_{i=1}^k s_i\theta_i\|$ takes its minimum value for $s_1 = ... = s_k = 0$. Considering polar co-ordinates with the origin in $\hat{b}$ and denoting $\hat{\gamma} = \sum_{i=1}^k \hat{s}_i\theta_i$, $\gamma = \hat{\gamma}/\|\hat{\gamma}\|$ we obtain

$$\int ds \int du \|\hat{b}+b\|^{-1}(\nabla^{(1)}(\hat{b}+b+u\gamma)[\gamma])^{-2-\delta}u^{k-1}1_{\{\nabla^{(1)}(\hat{b}+b)[\gamma]\ge 0\}}.$$

Here we used that the vectors $\theta_1,...,\theta_k$ were chosen so that $0<c_1<\|\hat{\gamma}\|<c_2$. Now from the Assumption 2 it follows that if we chose $k>(2+\delta)m$ then the last integral can be estimated by $c(1+\|\hat{b}+b\|^{(2+\delta)m-1})$. Returning to the unconditional expectation we get the estimation

$$\int_{\{\psi_1(X)\le\frac{1}{2}\}}(1+\|Y(X)+b\|^{(2+\delta)m-1})P(dX).$$

But in this domain $\|Y_1(X)\|\le c(\|Y_0(X)+b\|+V(X))$, so choosing δ appropriately we get the required estimation $c(1+\|b\|^{2m-1+\delta})$.

Remark. Let us fix $x_0>0$. Whithout the Assumption 2 we can show that on the domain $\{X\in\mathbf{X}: f(X)\ge x_0\}$ the integral in (11) is finite (exactly as for $\{X\in\mathbf{X}:\psi_1(X)\le\frac{1}{2}\}$). It is easy to notice that if the inequality (3) holds for all $\phi\in C_0^\infty$ with support in (x_0,∞), then the density of ν is bounded on (x_0,∞). So we have the following result.

Statement. Let Y be a stable random vector in a real, separable Banach space B, satisfying assumption 1. Suppose that the distribution of Y has infinite dimensional support. Then the density $p(x)$ of $\|Y\|$ is bounded on (x_0,∞) for every $x_0>0$.

Acknowledgements. I am very grateful to V.Bentkus for supporting my work.

References

[1] Acosta de A. *Stable measures and seminorms.* — Ann. Prob., 1975, vol.3, No.5, 865-875.

[2] Bentkus V., Pap Gy. *On the distribution of the norm of stable random vectors taking values in Hilbert space* (in Russian). — Lithuanian Math. J., 1986, vol.XXVI, No.2, 211-220.

[3] Byczkowski T., Samotij K. *Absolute continuity of stable seminorms.* — Ann. Prob., 1986, vol.14, 299-312.

[4] Ikeda N., Watanabe S. *Stochastic differential equations and diffusion processes.* — Amsterdam: North Holland / Kondasha, 1981, 464 p.

[5] Kruglov V.M., Antonov S.I. *Again on the asymptotic behavior of infinitely divisible laws in Banach spaces* (in Russian). — Theory of Prob. and its Appl., 1984, vol.XXIX, No.4, 735-742.

[6] Lewandowski M., Zak T. *On the density of the distribution of p-stable seminorms,* $0 < p < 1$. — Proc. American Math. Soc., 1987, vol.100, No.2, 345-351.

[7] Lifschitz M.A., Smorodina N.V. *On the distribution of stable vectors* (in Russian). — Theory of Prob. and its Appl., 1987, to appear.

[8] Pap Gy. *Boundedness of the density of the norm of stable random vectors in a Hilbert space* (in Russian). — Theory of Prob. and Math. Stat. (Kiev), 1987, vol.36, 102-105.

[9] Pap Gy. *Density of the norm of stable vectors in Hilbert space.* — Abstracts of Communications of the IV. International Vilnius Conference on Probab. Theory and Math. Statistics, 1985, vol.4, 223.

[10] Paulauskas V. *The rates of convergence to stable laws and the law of iterated logarithm in Hilbert space.* — Göteborgs University preprint, 1977, No.5.

[11] Paulauskas V., Račkauskas A. *On accuracy of approximation in the central limit theorem in Banach spaces* (in Russian; to appear in English). — Vilnius: Mokslas, 1987, 188 p.

[12] Ryznar M. *Density of stable seminorms.* — Bull. Pol. Ac.:Math., 1985, vol.33, 431-440.

[13] Smorodina N.V. *Differential calculus on the space of configurations and stable measures* (in Russian). — Theory of Prob. and its Appl., to appear.

[14] Sztencel R. *On the lower tail of stable seminorm.* — Bull. Pol. Ac.:Math., 1984, vol.32, 715-719.

[15] Sztencel R. *Absolute continuity of the lower tail of stable seminorms.* — Bull. Pol. Ac.:Math., 1986, vol.34, 231-234.

[16] Zak T. *On the continuity of the distribution function of a seminorm of stable random vectors.* — Bull. Pol. Ac.:Math., 1984, vol.32, 519-521.

[17] Zak T. *A formula for the density of the norm of stable random vector in Hilbert spaces,* to appear in Probab. and Math. Stat., 1988.

Probability Theory on Vector Spaces
Lancut, June'87, Springer's LNM 1391

MARTINGALE APPROACH TO BOLTZMANN'S ENTROPY AND EXACT TRANSFORMATIONS

Krzysztof Podgórski

Institute of Mathematics, Technical University of Wrocław
50-370 Wrocław, Poland

ABSTRACT. Using the martingale convergence theorem we present a simple proof of the fact that a measure preserving transformation is exact if and only if entropy of iteration of the Frobenius-Perron operator corresponding to this transformation tends to zero for all densities.

1. PRELIMINARES.

In this paper we will consider a fixed probability space $<X,\Sigma,\mu>$. A transformation $S : X \to X$ is called measure-preserving if for every set $A \in \Sigma$ we have $S^{-1}A \in \Sigma$ and $\mu(A) = \mu(S^{-1}A)$. By D_1 we denote the set of all densities defined on $<X,\Sigma,\mu>$ and by $D_2 \subseteq D_1$ the set of densities belonging to $L_2 = L_2<X,\Sigma,\mu>$.

Let f be a measureable, nonnegative function on X . We define a function $h : <0,\infty) \to (-\infty,e^{-1}>$ in the following way

$$h(x) = \begin{cases} -x \lg(x) & : x > 0 \\ 0 & : x = 0 \end{cases} .$$

Then the number $H(f) = \int h \circ f d\mu$ is called entropy of a function f.

A positive linear contraction on $L_1 = L_1\langle X, \Sigma, \mu\rangle$ is called a Markov operator on L_1.

For a measure preserving transformation S we can define a Markov operator which is called the Frobenius-Perron operator corresponding to S in the following way. For $f \in L_1$ we define a finite, σ-additive signed measure ν on Σ by the formula

$$\nu(A) = \int_{S^{-1}A} f \, d\mu \, , \quad \text{for each} \quad A \in \Sigma \, .$$

Since ν is absolutely continuous with respect to μ, by the Radon--Nikodym theorem, there exists an unique function $g \in L_1$ such that

$$\nu(A) = \int_A g d\mu \, , \quad \text{for each} \quad A \in \Sigma \, .$$

Let us define a map $P : L_1 \to L_1$ by the equality $Pf = g$. It is easy to check (see [3]) that P is a linear operator on L_1 and

$\|P\| \le 1$,

if $f \ge 0$, $f \in L_1$ then $Pf \ge 0$ and $\|Pf\| = \|f\|$,

$P1 = 1$.

We will need the following property of the Frobenius-Perron operator.

<u>Lemma 1</u>. Let S be a measure-preserving transformation and P the Frobenius -Perron operator corresponding to S. Then for every $f \in L_1$

$$Pf \circ S = E(f|S^{-1}\Sigma) \quad \text{a.e.}$$

<u>Proof.</u> Let $g = Pf$. Since $g \circ S$ is a $S^{-1}\Sigma$ measureable function so it is enough to show that for every $B \in S^{-1}\Sigma$

$$\int_B g \circ S \, d\mu = \int_B f \, d\mu \, .$$

If $B \in S^{-1}\Sigma$ then there exists a set $A \in \Sigma$ such that $B = S^{-1}A$ and we have

$$\int_B g\circ S\, d\mu = \int 1_A g\circ S\, d\mu = \int 1_A g\, d\mu = \int_A g\, d\mu = \int_{S^{-1}A} f d\mu = \int_B f d\mu .$$

The second equality follows from the fact that S preserves measure μ . □

Collorary 1. For every $n \in N$, P^n is the Frobenius -Perron operator corresponding to S^n , and $P^n f\circ S^n = E(f|S^{-n}\Sigma)$ a.e.

Proof. In the proof we will employ the Principle of Induction. If $n = 1$ then $Pf\circ S = E(f|S^{-1})$ by Lemma 1. Let us assume that the thesis is satisfied for fixed $n \in N$ then

$$\int_A P^{n+1} f\, d\mu = \int_{S^{-1}A} P^n f\, d\mu = \int_{S^{-n}(S^{-1}A)} f d\mu = \int_{S^{-(n+1)}A} f d\mu ,$$

which proves that the thesis is satisfied for $n+1$. □

We will say that a measure preserving transformation S is exact if σ-field $\bigcap_{n\in N} S^{-n}\Sigma$ is trivial. We have the following characterization exact transformations (see [5]).

Proposition 1. A transformations S is exact if and only if for every $A \in \Sigma$ with $\mu(A) > 0$ such that for every $n \in N$ $S^n A \in \Sigma$ we have that $\lim_{n\to\infty} \mu(S^n A) = 1$.

Now one can see at once that if a σ-field Σ is nontrivial then exact transformations can not be invertible.

The following characterization of exact transformations by behaviour of iterations of Frobeniusz-Perron operator is well known (see [2], [3], [4]).

Proposition 2. Let S be a measure-preserving transformation and P the Frobenius -Perron operator corresponding to S . Then the following conditions are equivalent

(i) S is exact,

(ii) for every $f \in L_1$ $\lim_{n\to\infty} \|P^n f - \int f d\mu\|_1 = 0$.

Now we extend the concept of exactness to the case of an arbitrary Markov operators. We will say that a Markov operator P is exact if for every $f \in L_1$ $\lim_{n\to\infty} \|P^n f - \int f d\mu\|_1 = 0$.

It is easy to check that for every $f \in D_1$ $H(f) \le 0$ and $H(f)=0$ if and only if $f = 1$ a.e.

The following property is the simple consequence of the Jensen inequality for Markov operators (see [2]).

Proposition 3. If P is a Markov operator and $P1 = 1$ then for every $n \in N$ and for every $f \in D_1$ $H(P^n f) \ge H(P^{n-1} f)$.

Thus if P is the Frobenius-Perron operator corresponding to the transformation S then there exists the limit $\lim_{n\to\infty} \uparrow H(P^n f)$.

It was proved in [3] that if S is exact then this limit is equal to zero for all $f \in D_1$ such that $H(f) > -\infty$ and moreover if the limit is equal to zero for all bounded densities then S is exact. Using another techniques we obtain a slightly stronger result with the simpler proof (see Theorem 2).

2. THE MAIN RESULTS.

The following theorem allows to identify the limit $\lim_{n\to\infty} H(P^n f)$ when P is the Frobenius-Perron operator.

Theorem 1. Let S be a measure preserving transformation and P the Frobenius-Perron operator corresponding to S. Then for every function $f \in D_1$

$$\lim_{n\to\infty} H(P^n f) = H(E(f | \bigcap_{k=1}^{\infty} S^{-k}\Sigma)) .$$

Proof. It follows from the Jensen inequality for conditional expectations that for $\Sigma_o = \bigcap_{k=1}^{\infty} S^{-k}\Sigma$ and fixed $x \in N$

$$h(E(f|\Sigma_o)) = h(E(E(f|S^{-n}\Sigma)|\Sigma_o)) \geq E(h(E(f|S^{-n}\Sigma))|\Sigma_o) .$$

By Lemma 1

$$H(E(f|\Sigma_o)) \geq \int E(h(E(f|S^{-n}\Sigma))|\Sigma_o)d\mu = \int h(E(f|S^{-n}\Sigma))d\mu = H(P^n f)$$

and so

$$\lim_{n\to\infty} H(P^n f) \leq H(E(f|\Sigma_o)) .$$

By the martingale convergence theorem we have that

$$\lim_{n\to\infty} E(f|S^{-n}\Sigma) = E(f|\bigcap_{n=1}^{\infty} S^{-n}\Sigma) \quad \text{a.e.}$$

Thus for $g_n = h(E(f|S^{-n}\Sigma))$

$$\lim_{n\to\infty} g_n = h(E(f|\bigcap_{n=1}^{\infty} S^{-n}\Sigma)) \quad \text{a.e.}$$

Moreover since $g_n \leq 1/e$ we have by the Fatou lemma

$$H(E(f|\bigcap_{k=1}^{\infty} S^{-k}\Sigma)) = \int \lim_{n\to\infty} g_n \, d\mu \leq \lim_{n\to\infty} \int g_n \, d\mu ,$$

so

$$\lim_{n\to\infty} H(P^n f) > H(E(f|\Sigma_o)) . \qquad \square$$

Theorem 2. Let S be a measure-preserving transformation and P the Frobenius- Perron operator corresponding to S . The following conditions are equivalent

(i) S is exact,

(ii) for every $f \in D_1$ $\lim_{n\to\infty} H(P^n f) = 0$,

(iii) for every $A \in \Sigma$ $\lim_{n\to\infty} H(P^n((1/\mu(A))1_A)) = 0$.

Proof. The implication (i) $\Rightarrow$ (ii) follows from Theorem 1 if we notice that for an exact transformation $E(f|\bigcap_{n=1}^{\infty} S^{-n}\Sigma) = \int f d\mu$.

Since the implication (ii) $\Rightarrow$ (iii) is obvious it remains to prove the implication (iii) $\Rightarrow$ (i). By Theorem 1 it is enough to demonstrate that if for any $A \in \Sigma$ with $\mu(A) > 0$ $H(E((1/\mu(A))1_A|\bigcap_{n=1}^{\infty} S^{-n}\Sigma)) = 0$ then σ-field $\bigcap_{n=1}^{\infty} S^{-n}\Sigma$ consists of sets of measure zero or one. Let $A \in \bigcap_{n=1}^{\infty} S^{-n}\Sigma$ and $\mu(A) > 0$. Then

$$H(E((1/\mu(A))1_A|\bigcap_{n=1}^{\infty} S^{-n}\Sigma)) = H((1/\mu(A))1_A) =$$

$$= \int_A -(1/\mu(A))\lg(1/\mu(A))d\mu = \lg \mu(A) .$$

Thus $\lg \mu(A) = 0$ and consequently $\mu(A) = 1$. □

3. REMARKS

The second law of thermodynamics excludes existence of reversible physical processes on the macroscopic level. There exist at least two aproaches to find suitable mathematical model for such processes. First aproach assumes apriori that a transformation describing evolution of a system is not reversible, see [3]. In this case the evolution of the system is described by iteration of the Frobenius -Perron operator. If $f \in D_1$ is treated as a state of the system at the begining of observation then a state at the time n is given by $P^n f$. Because of the Boltzmann H-theorem one can only consider such transformation that

$$\lim_{n\to\infty} \uparrow H(P^n f) = 0 .$$

By Theorem 2 this condition is satisfied if and only if S is exact.

The basic idea of the second aproach presented in the Prigogine theory of irreversibility is to find equivalence between an irreversible system represented by a stochastic process and a highly ustable but classical dynamical system represented by a reversible transformation (see [1], [6], [7]). More precisely having a group $\{U^n\}_{n\in Z}$ of unitary operators on L_2 defined by formula

$$Uf = f \circ S^{-1}, \quad \text{for every} \quad f \in L_2 ,$$

where S is a K-authomorphism, we can construct a bounded operator Λ on L_2 such that family $\{W^{*n}\}_{n\in N}$ defined for every $n \in N$ by the equation

$$W^{*n} = \Lambda U^n \Lambda^{-1}$$

is a well defined semigroup of Markov operators for which Boltzmann H-theorem is satisfied.

Namely, we have the following property

Proposition 4. If W^* is defined as above, then for every $f \in D_2$ $\lim_{n\to\infty} H(W^{*n}f) = 0$.

Proof. It follows from the construction of Λ that

$$\|W^{*n}f - 1\|_2 \to 0 \quad \text{for every} \quad f \in D_2$$

(see [6], [7]). This implies that W^* is exact (see Proposition 2). In [3] one can find the proof of the fact that exactness of a Markov operator implies the convergence of the Boltzmann entropy to zero although the theorem is formulated only for Frobenius -Perron operators there. □

REFERENCES

[1] M. Courbage and I. Prigogine, "Intrinsic randomnes and intrinsic irreversibility in classical dynamical system." Proc. Natl. Acad. Sci. USA, 8 (1983), 2412-2416.

[2] S. Horowitz, "Strong ergodic theorems for Markov processes." Proc. Amer. Math. Soc. 23 (1969), 328-334.

[3] A. Lasota, M. Mackey, "Probabilistic properties of deterministic systems!" Cambridge Univ. Press 1985.

[4] M. Lin, "Mixing for Markov Operators." Z. Wahr. 19 (1971), 231-242.

[5] V.A. Rochlin, "Exact endomorphisms of Lebesgue spaces." Amer. Math. Soc. Transl. 39 (1964), 1-36.

[6] M. Rybaczuk, Z. Suchanecki, A. Weron and K. Weron, "An explicit approach to the Λ operator and the H-theorem in the Prigogine theory of irreversibility." Physica A (to apear).

[7] Z. Suchanecki and A. Weron, "Applications of an operator stochastic integral in Prigogine's theory of irreversible dynamical systems." Exp. Math. (to apear).

Probability Theory on Vector Spaces IV
Lancut, June '87, Springer's LNM 1391

ON THE MARTINGALE CENTRAL LIMIT THEOREM IN BANACH SPACES

A. J. Račkauskas

Department of Mathematics
Vilnius, V. Kapsukas University
Vilnius, USSR

1. Introduction. Throughout this paper E, F are real Banach spaces with norms $\|\cdot\|_E$ and $\|\cdot\|_F$ respectively, L(E, F) is the space of baunded linear operators from E to F, I_E — identity map, E' — the dual space of E, u' — the dual operator of $u \in L(E, F)$.

Let X_{nk}, $1 \leq k \leq k_n$, $n \in N$ be an array of random elements of the space E, defined on a probability space (Ω, F, P) and let F_{nk}, $0 \leq k \leq k_n$, $n \in N$ be a set of sub-σ-fields of F with $F_{nk-1} \subset F_{nk}$ and $X_{nk} \in F_{nk}$ (F_{nk}-measurable). We always suppose that $E\|X_{nk}\|_E < \infty$ for all $k=1, \ldots, k_n$, $n \in N$ and that (X_{nk}, F_{nk}) constitute a martingale difference array (m.d.a. for short).

The main aim of the present paper is to get sufficient conditions for asymptotic normality of row sums of m.d.a.'s of the space C(S), where (S, ρ) is a compact metric space and to estimate the speed of convergence.

In the case E is 2-smoothable Banach space, sufficient conditions have been given by Rosinski [7]. Moreover, in [7] 2-smoothability is characterized via central limit theorem for m.d.a. The speed of convergence in martingale central limit theorem was studed by Butzer, Hahn, Roeckerath [1] and by Rhee and Talagrand [8].

In section 2 we study p-smoothing operators, $p > 1$, $u \in L(E, C(S))$. In terms of metric entropy, sufficient conditions for identical embedding of Lip_d (S) into C(S) to be p-smoothing, $p \geq 2$, are given. This allows us to prove analogue of Rosinski's result in C(S) case, wich is stated in section 3. Finally, in section 4 we investigate the speed of convergence in the martingale central limit theorem in the space C(S).

2. Smoothing operators. Let $D_k(E)$, $k \in N$ be a space of k-time Frechet differentiable functions $f: E \to R^1$ and $D_{k,\alpha}(E) = (f \in D_k(E): \sup_{x \neq y} |f^{(k)}(x) - f^{(k)}(y)| \, \|x-y\|^{-\alpha} < \infty)$, $0 < \alpha \leq 1$. A function $f: E \to R^1$ is said to have u-bounded support if the set u(supp (f)) is bounded, where $u \in L(E, F)$ and supp (f) denotes the support of the function f.

For a number $p > 0$ let k(p) be an integer such that $p-1 \leq k(p) < p$ and $\alpha(p) := p - k(p)$.

Definition 2.1. Let $p > 0$. An operator $u \in L(E, F)$ is said to be p-smoothing, if the set $D_{k(p),\alpha(p)}(E)$ contains a nontrivial function with u-bounded support.

The set of p-smoothing operators we shall denote by S_p. In the following proposition we collect a few facts about the set S_p.

Proposition 2.1. a) S_p, $p > 1$ is injective and regular operator ideal;

b) A Banach space E is p-smoothable iff I_E is p-smoothing, $1 < p \leq 2$;

c) If Π_r denotes the ideal of r-summing operators, then $\Pi_r \subset S_p$ for all $r > 0$ and $p > 1$.

Proof. a) It is obvious that $I_R \in S_p$. Let u_1, $u_2 \in S_p(E, F)$. There are functions $f_i \in D_{k(p),\alpha(p)}(E)$ such that $f_i \not\equiv 0$, $0 \leqslant f_i \leqslant 1$ and $f_i(x)=0$ if $\|u_i(x)\| > 1$, $i=1,2$. Let $f\colon E \to R^1$ be the product function $f_1 f_2$. By the mean value theorem one immediatly checks, that there is a constant $C > 0$ such that

$$\max_{i=1,2} (\sup_x \| f_i^{(j)}(x) \|) < C \text{ for all } j=1, \ldots, k(p).$$

Now, it is easy to prove, that $f \in D_{k(p),\alpha(p)}(E)$. If $\|(u_1+u_2)(x)\| > 2$ then $\|u_1(x)\| > 1$ or $\|u_2(x)\| > 1$, consequently $f(x)=0$ if $\|(u_1+u_2)(x)\| > 2$ and we have proved that $u_1+u_2 \in S_p(E, F)$.

Now let $u \in S_p(E, F)$, $v \in L(E_1, E)$, $w \in L(F, F_1)$ and let $f \in D_{k(p),\alpha(p)}(E)$ be a function with u-bounded support. Then the function $g\colon E_1 \to R^1$, $g=fv$ has $w \circ u \circ v$-bounded support and obviously $g \in D_{k(p),\alpha(p)}(E_1)$. Hence $w \circ u \circ v \in S_p$

To prove regularity of the operator ideal S_p, let $J \in L(F_0, F)$ be any injective operator, $u \in L(F, F_0)$ be such that $J \circ u \in S_p(E, F)$. Hence, there is a function $f \in D_{k(p),\alpha(p)}(E)$ such that $f(x)=0$ if $\|J \circ u(x)\| > 1$. Let $0 < \tau < j(J)$, where $j(J)$ denotes the modulus of injection J. Then $\|J(x)\| \geqslant \tau \|x\|$ for all $x \in F_0$. Therefore $f(x)=0$ if $\|u(x)\| > 1/\tau$ and $u \in S_p(E, E_0)$. Regularity of S_p is trivial.

b) Let us recall that a Banach space E is said to be p-smoothable, $1 < p \leqslant 2$, if it possible to renorm E in such a way, that $\sup_{0<\tau\leqslant 1} \rho_E(\tau)\tau^{-p} < \infty$, where $\rho_E(\tau) = \sup(2^{-1}(\|x+y\| + \|x-y\|) - 1 : \|x\|=1, \|y\| = \tau)$.

To prove the „if" part one can use the arguments of the proof of theorem 3.2 in [2] to construct the corresponding equivalent norm in E. The „only if" part is standart.

c) If $u \in \Pi_r(E, F)$ then by known theorem of Pietsch, there is a probality m on $S_{E'} = (a \in E' : \|a\| = 1)$ and a positive number C such that for all $h \in E$

$$\|u(h)\|_F \leqslant C(\int_{S_{E'}} (h, a)^q m(da))^{1/q}$$

for every even number $q > r$. Let $\mathcal{K}$ be a function of the class C^∞ such that $0 \leqslant \mathcal{K} \leqslant 1$, $\mathcal{K}(t)=1$ if $t<1$ and $\mathcal{K}(t)=0$ if $t>2$. Then the function $f\colon E \to R^1$ defined by $f(h) = \mathcal{K}((\int_{S_{E'}} (h, a)^q\, m(da))^{1/q})$ is in the class $D_{k,\alpha}(E)$ for every $k > 1$, $0 < \alpha \leqslant 1$. Hence $u \in S_p$ for every $p > 1$.

The next proposition shows smoothing properties of identity $I_{l_\infty^m}$ of the space l_∞^m. The norm of l_∞^m we shall denote by $|\cdot|_m$, $|x|_m = \sup_{1\leqslant i\leqslant m} |x_i|$.

Proposition 2.2. Let an integer number $m > 1$. There existes a function $f \in D_\infty(l_\infty^m)$ with the following properties:

a) $f(0)=1$ and $f(x)=0$ if $|x|_m > 1$;

b) for every $k \in N$ there is a constant C_k depending on k only, such that

$$\|f^{(k)}\|_\infty = \sup_x \|f^{(k)}(x)\| \leqslant C_k \log^{k-1} m. \tag{2.1}$$

Proof. Let $q_s(x) = \sum_{j=1}^m |x_j|^s$ where s is an even number.

The function $q_s \in D_\infty(l_\infty^m)$ and

$$q_s^{(k)}(x)(h^1 \ldots h^k) = \sum_{j=1}^m s(s-1)\ldots(s-k+1)|x_j|^{s-k}(\operatorname{sgn} x_j)^k h_j^1 \ldots h_j^k,$$

where $x \in l_\infty^m$, $h^1, \ldots, h^k \in l_\infty^m$, $k \geqslant 1$. One immediatly checks that

$$|q_s^{(k)}(x)(h^1, \ldots, h^k)| \leqslant s(s-1)(s-2)\ldots(s-k+1)(q_s(x))^{(s-k)/s}|h^1|_m \ldots |h^k|_m\, m^{k/s}. \tag{2.2}$$

Now consider a function $g_s(x)=(q_s(x))^{1/s}$ wich is infinitely many time differentiable away from zero. Using (2.2) we get the following estimate:

$$|g_s^{(k)}(x)(h^1,\ldots,h^k)|\leqslant C_k s^{k-1} m^{k/s} g_s^{1-k}(x)\,|h^1|_m\cdots|h^k|_m, \tag{2.3}$$

where constant C_k dependes on k only.

Let $\mathcal{K}\in D_\infty(R^1)$ be a function such that $0\leqslant\mathcal{K}\leqslant 1$, $\mathcal{K}(t)=0$ if $t>0$. Define $f:l_\infty^m\to R^1$ by $f(x)=\mathcal{K}(g_s(x)-2^{-1})$. Then $f\in D_\infty(l_\infty^m)$. Moreover, $f(0)=1$ and $f(x)=0$ if $|x|_m>1$. From (2.3) we get

$$\|f^{(k)}(x)\|\leqslant C_k\, s^{k-1}\, m^{k/s}$$

for all $x\in l_\infty^m$. To finish the proof we take s such that $\log m\leqslant s\leqslant 2\log m$.

Given a compact metric space (S,ρ), $C(S)$ denotes the space of continuouse functions on S, equipped with the sup-norm. For any continuous with respect to ρ a semmi-metric d on S, the space of d-Hölder functions is defined as

$$C_d(S)=(x\in C(S):\|x\|_{C_d}=\max(\sup_{s\in S}|x(s)|,\ \sup_{s,t\in S,s\neq t}|x(s)-x(t)|/d(s,t))<\infty).$$

For any continuous with respect to ρ semi-metric d, as usual $N_d(S,x)$ denotes the minimal number of d-balls of radius x, wich cover S and $H_d(S,x)=\log_2 N_d(S,x)$. If $u\in L(E,C(S))$ then a semi-metric d_u on S is defined as $d_u(s,t)=\|u'(\delta_s-\delta_t)\|_E$.

Theorem 2.1. Let $u\in L((E,C(S))$, $m\in N, m>1$. If there exists continuous with respect to ρ semi-metric $d\geqslant d_u$ such that

$$\int_0^a H_d^{(m-1)/m}(S,x)\log^{2/m} H_d(S,x)\,dx<\infty \tag{2.4}$$

for some $a>0$, then $u\in S_m(E,C(S))$.

Proof. Let $N_n=N_d(S,2^{-n})$, $n\geqslant 1$. There is a set $V_n=(s_i^n,\ i=1,\ldots,N_n)\subset S$ such that $S=\bigcup_{j\leqslant N_n} B_d(s_j^n,2^{-n})$, where $B_d(s,x)$ denotes the d-ball with center s and radius x. Put $m_1=N_1$ and let m_n, $n>1$, denotes the number of pairs (s, t) of the set $V_n U V_{n+1}$ such that $0<d(s,t)\leqslant 3\cdot 2^{-n}$. Define operators $T_n:E\to l_\infty^{m_n}$ by

$$T_1x=(u(x)(s),\ s\in V_1);$$

$$T_nx=(ux(s)-ux(t):s,t\in V_n\cup V_{n+1},\ 0<d(s,t)\leqslant 3\cdot 2^{-n}).$$

Let $s\in S$. Then for every $n\geqslant 1$ we can find $s_n\in V_n$ such that $d(s,s_n)\leqslant 2^{-n}$. Hence, if $x\in E$

$$ux(s)=\sum_{k=1}^{\infty} ux(s_{k+1})-ux(s_k)+ux(s_1).$$

Since $d(s_{k+1},s_k)\leqslant 3\cdot 2^{-k}$ we have

$$\|u(x)\|_{C(S)}\leqslant\sum_{k=1}^{\infty}\|T_kx\|_{m_k}, \tag{2.5}$$

where $\|\cdot\|_{m_k}$ denotes the norm of the space $l_\infty^{m_k}$. Now let the functions $f_k:l_\infty^{m_k}\to R^1$ are as constructed in proposition 2.2. $f_k(x)=0$ if $\|x\|_{m_k}>1$, $f_k(0)=1$, $\|f_k^{(j)}\|_\infty\leqslant C\log^{j-1}m_k$, $j=1,\ldots,m$. Let $b_k>0$ be a sequence such that $\sum_{k=1}^{\infty} b_k=1$, and define a function $g:E\to R^1$ as $g(x)=\prod_{k=1}^{\infty} f_k(T_kx\,b_k^{-1})$. Because of (2.5), $g(x)=0$ if $\|u(x)\|_{C(S)}>1$. Moreover, according to the construction of the functions f_k, one can see, that $g_k(x)=1$ if $k\geqslant 1$ and $\|T_kx\|_{m_k} b_k^{-1}\leqslant 1/20$. Note, that $\|T_kx\|_{m_k}\leqslant 3\cdot 2^{-k}\|x\|_E$ if $k>1$ and $\|T_1x\|_{m_1}\leqslant\|x\|_E\cdot\|u\|$.

Therefore, for every $x\in E$, the product in the definition of the function g is finite and only differentiability of g is to verify. Since $g_k^{(j)}(x)(h)^j = f_k^{(j)}(T_k x\, b_k^{-1})(T_k h)^j\, b_k^{-j}$ and $m_k \leq N_k^2$, therefore $\| g_k^{(j)}\|_\infty \leq C H_k^{j-1}\, 2^{-jk} b_k^{-j}$, $j=1,\ldots,m$ and easy computations show, that $g\in D_{m-1,1}(E)$ if $\sum_{k=1}^{\infty} H_k^{j-1}\; 2^{-jk}\, b_k^{-j} < \infty$ for every $j=1,\ldots,m$. Choose $a_k = \log^{2/m}(H_k) H_k^{(m-1)/m}\cdot\; 2^{-k}$ if $H_k > 2^k\, k^{-4}$ and $a_k = k^{-2}$ otherwise. Because of the condition (2.4) $A=\sum_{k=1}^{\infty} a_k < \infty$. Now we can put $b_k = A^{-1} a_k$ and we have finished to construct the non-trivial function $g\in D_{m-1,1}(E)$ with u-bounded support.

Remark. As it is shown in [5], if instead the condition (2.4) we have

$$\int_0^a H_d^q (S,x)\, dx < \infty$$

for some $0<q\leq 1/2$, then u is p-smoothing for every $1<p<1/(1-q)$.

For the embedding $I_d : C_d(S)\to C(S)$ we have the following corollary.

Corollary 2.1. If the metric d satisfies the condition (2.4) then I_d is m-smoothing operator.

Recall, that the n-th entropy number of an operator $u\in L(E,F)$ is given by

$e_n(u) = \inf(\varepsilon>0$: There exists an ε-net for $u(U_E)$ in F of cardinality not exceeding 2^{n-1})

where U_E denotes the closed unit ball in E.

Theorem 2.2. Let $u\in L(E,F)$, a natural number $m>1$. If

$$\sum_{k=1}^{\infty} k^{-1/m}\, e_k(u')\log^{2/m} k < \infty$$

then $u\in S_m(E,F)$.

Proof. We use the fact, that the set $S=U_{F'}$, equipped with the distance $d_u(f,g) = \| u'(f) - u'(g)\|_E$ is compact. Consider an operator $J: F\to C(S)$ given by $Jx(f) = (f,x)$. Since $d_{J\circ u}(f,g) = d_u(f,g)$ and

$$\int_0^{e_1(u')} H_{d_u}^{(m-1)/m}(S,x)\log^{2/m} H_{d_u}(S,x)dx =$$

$$= \sum_{k=1}^{\infty}\int_{e_{k-1}(u')}^{e_k(u')} H_{d_u}^{(m-1)/m}(S,x)\log^{2/m} H_{d_u}(S,x)\,dx \leq C\sum_{k=1}^{\infty} e_k(u')k^{-1/m}\log^{2/m} k$$

therefore $J\circ u\in S_m(E, C(S))$ and according to the injectivity of the operator ideal S_m, $u\in S_m(E,F)$ as well.

3. Central limit theorem. For a martingale $(X_n)_{n\geq 0}$ as usual $dX_n = X_n - X_{n-1}$ if $n\geq 1$, $dX_0 = X_0$. We denote the set $\{-1,+1\}^N$ by Ω' and by $\varepsilon_n : \Omega' \to \{-1,+1\}$ the n-th coordinate. Let A' be the Borel σ-algebra on Ω' and P' — the Haar measure on Ω'. Let A_n, $n\geq 1$, denotes the σ-algebra on Ω' generated by the first n-coodinates $(\varepsilon_1,\ldots,\varepsilon_n)$, $A_0 = (\phi, \Omega')$. Every martingale $(X_n)_{n\geq 0}$ on the probebility space (Ω', A', P') relative to $(A_n)_{n\geq 0}$ is called a Walsh-Paley martingale.

The main step to the consideration of the martingale central limit theorem is the following martingale characterization of p-smoothing operators, $1<p\leq 2$.

Theorem 3.1. Let $u\in L(E,F)$, $1<p\leq 2$. The following statments are equivalent:

a) $u\in S_p(E,F)$;

b) There exists a constant $C>0$ such that for any E-valued martingale $(X_n)_{n\geq 0}$ with $X_0=0$

$$\sup_n E\| u(X_n)\|_F^p \leq C\sum_{n\geq 1} E\| dx_n\|_E^p; \tag{3.1}$$

c) There is a constant $C>0$ such that for every E-valued Walsh-Paley martingale $(X_n)_{n\geq 0}$ with $X_0=0$ the inequality (3.1) is satisfied.

Proof. We give the sketch of the proof only. Details can be found in [5]. Let $u \in S_p(E, F)$, $\|u\| \leq 1$ and let $(X_n)_{n \geq 0}$ be a Walsh-Paley martingale such that $X_0 = 0$. From the definition of p-smoothing operator it follows, that there exists a function $f \in D_1(E)$ such that $f(0)=1$, $f(x)=0$ if $\|u(x)\|_F > 1$ and $C := \sup_{x \neq y} \|f'(x) - f'(y)\| \cdot \|x-y\|_E^{1-p} < \infty$. Denote $g = 1-f$. Then

$$P(\|u(X_n)\|_F > 1) \leq Eg(X_n) = I_n. \tag{3.2}$$

Since $E\, g'(\Sigma_{i=1}^{n-1} dX_i)(dX_n) = 0$, using Taylor's formula, we get

$$I_n = E \int_0^1 f'(\Sigma_{i=1}^{n-1} dX_i + t dX_n) - f'(\Sigma_{i=1}^{n-1} dX_i)(dX_n)\, dt + I_{n-1} \leq C\, E\, \|dX_n\|_E^p + I_{n-1}. \tag{3.3}$$

Noting that $g(0)=0$, in the same way

$$I_1 = E\, f(dX_1) \leq E\|dX_1\|_E^p. \tag{3.4}$$

From (3.2)—(3.4) we get

$$P(\|u(X_n)\|_F > 1) \leq C\, \Sigma_{i=1}^{n} E\| dX_i\|_E^p.$$

for all $n \geq 1$ and the inequality (3.1) follows by standart stopping time technique.

To prove b) ⇐ c) and c) ⇒ a) one can construct a semi-norm q on E such that for all $h, x \in E$

i) $\|u(x)\|_F \leq q(x) \leq C_1 \|h\|_E$;

ii) $q^p(x+h) + q^p(x-h) \leq 2q^p(x) + C_2 \|h\|_E^p$.

For this purpose one can use the arguments of the proof of theorem 3.1 in [4]. Now b) follows from the inequalities

$$E\, q^p(X_{n+1}) + E\, q^p(X_n - dX_{n+1}) \leq 2E\, q^p(X_n) + CE\|dX_{n+1}\|_E^p,$$

$$E\, q^p(X_n - dX_{n-1}) \geq E\, q^p(X_n).$$

Taking the function $\mathcal{K} \in D_\infty(R^1)$ such that $\mathcal{K}(t)=0$ if $t>2$, $\mathcal{K}(t)=1$ if $t<1$ we have the function $g(x)=\mathcal{K}(q(x))$ wich is in the class $D_{1,p-1}(E)$ and has u-bounded support.

Now we can pass to the central limit theorem. Let (X_{nk}, F_{nk}) be a m.d.a. of the space C(S) where (S, ρ) is compact metric space and let d be a continuouse with respect to ρ semi-metric on S.

Theorem 3.2. Suppose that m.d.a. (X_{nk}, F_{nk}) satisfies the following conditions:

a) There exists a function $f: S \to R_+^1$ such that for every $s \in S$

$$\Sigma_{k=1}^{k_n} E\, X_{nk}^2(s)/F_{nk-1} \xrightarrow{P} f(s) \quad \text{as } n \to \infty;$$

b) For all $s \in S$ and $\varepsilon > 0$

$$\Sigma_{k=1}^{k_n} E\, X_{nk}^2(s)\, I(|X_{nk}(s)| > \varepsilon)/F_{nk-1} \xrightarrow{P} 0 \text{ as } n \to \infty;$$

c) $\sup_n \Sigma_{k=1}^{k_n} E\|X_{nk}\|_{C_d}^2 < \infty$

where d is such that

$$\int_0^a H_d^{1/2}(S, x) \log H_d(S, x)\, dx < \infty \text{ for some } a > 0,$$

then there exists a Gaussian measure μ such that

$$L(\Sigma_{k=1}^{k_n} X_{nk}) \Rightarrow \mu \text{ as } n\to\infty.$$

Proof. The conditions a) and b) gives us the central limit theorem for m.d.a. $X_{nk}(s)$, for every $s\in S$, while metric entropy condition together with theorem 3.1 prove tightness of the distributions $\{L(\Sigma_{k=1}^{k_n} X_{nk}), n\geq 1\}$.

4. The speed of convergence. Let $(F_i)_{i\leq n}$ be an increasing sequence of σ-fields with $F_i\subset F$ and $F_0=(\phi,\Omega)$. We shall consider a martingale difference sequence X_k, $k=1, \dots, n$ of the space $C(S)$, relative to F_k, $0\leq k\leq n$. We always suppose that there exist a Gaussian $C(S)$-valued random element Y and a sequence $\sigma_1, \dots, \sigma_n$ of random variables such that

$$E\, X_k(s)X_k(t)/F_{k-1}=\sigma_k^2\, E\, Y(s)Y(t)$$

for all $s,t\in S$ and $1\leq k\leq n$. Denote $S_n=\Sigma_{k=1}^{n} X_k$.

Theorem 4.1. Suppose the following conditions to be satisfied.

a) There exists continuouse with respect to ρ semi-metric d on S such that

$$\int_0^a H_d^{2/3}(S,x)\log^{2/3} H_d(S,x)\,dx<\infty \tag{4.1}$$

for some $a>0$ and $E\|X_k\|_{C_d}^3<\infty$ for all $1\leq k\leq n$;

b) The distribution function of the random variable $\|Y\|_{C(S)}$ has a bounded density.

Then there exists a constant $C>0$ such that

$$\Delta_n:=\sup_{r\geq 0}|P(\|S_n\|_{C(S)}<r)-P(\|Y\|_{C(S)}<r)|\leq$$

$$C\inf_{\varepsilon\leq 0}[\varepsilon^{1/3}+(W_n+L_n)^{1/4}H_d^{3/4}(S,\varepsilon)\log^{9/4}H_d(S,\varepsilon)], \tag{4.2}$$

where $W_n=\Sigma_{k=1}^{n}E\|X_k\|C_d$, $L_n=E|\Sigma_{i=1}^{n}\sigma_1^2-1|^{3/2}$.

If $N_d(S,x)\leq C\exp(x^{-\beta})$ with $0<\beta<3/2$, then we have the estimate

$$\Delta_n\leq C(W_n+L_n)^{1/4(1-3\beta)}.$$

In the case $N_d(S,x)\leq Cx^{-\gamma}$, $\gamma>0$ the speed of convergence is better

$$\Delta_n\leq C(W_n+L_n)^{1/4}\log(W_n+L_n).$$

If $S=[0,1]$ and $d(s,t)=\rho(s,t)=|s-t|^{\alpha}$ with $\alpha>1/2$,

the logarithmic factor one can omit (see[6]).

Proof. We shall use the same notations as in the proof of Th. 2.1, and in addition we define operators $\tilde{T}_k$: $C(S)\to l_\infty^{N_k}$ by $\tilde{T}_k x=(x(s), s\in V_k)$, $k\geq 1$. Assume that $\Sigma_{k=1}^{n}\sigma_k^2=1$ a.s. and $W_n<1$ (the case $W_n\geq 1$ is trivial). It is easy to prove, that for all $K\geq 1$, $\varepsilon>0$

$$\Delta_n\leq\Delta_{n,K}+I_1+I_2+I_3,$$

where

$$\Delta_{n,K} = \sup_{r\geq 0} |P(\|\widetilde{T}_K S_n\|_{N_K}<r) - P(\|\widetilde{T}_K Y\|_{N_K}<r)|,$$

$$I_1 = P(\sup_{d(s,t)\leq 2^{-K}} |S_n(s) - S_n(t)| > \varepsilon),$$

$$I_2 = P(\sup_{d(s,t)\leq 2^{-K}} |Y(s) - Y(t)| > \varepsilon),$$

$$I_3 = \sup_{r\geq 0} P(r-\varepsilon \leq \|Y\|_{C(S)} \leq r+\varepsilon).$$

Because of the condition b) $I_3 \leq C\varepsilon$. Let $b_k > 0$ be a seguence such that $\Sigma_{k=K}^{\infty} b_k \leq 1$. Then

$$P(\sup_{d(s,t)\leq 2^{-K}} |S_n(s) - S_n(t)| > \varepsilon) \leq$$

$$\Sigma_{k=K}^{\infty} P(\|T_k S_n\|_{m_k} > \varepsilon b_k).$$

Using the functions, constructed in the proposition 2.2, similar as in the proof of lemma 4 in [6], we get

$$I_1 \leq C(\varepsilon^{-3} W_n \Sigma_{k=K}^{\infty} b_k^{-3} H_k^2 2^{-3k} + \varepsilon^{-2} \Sigma_{k=K}^{\infty} b_k^{-2} H_k 2^{-2k}).$$

By the same arguments

$$I_2 \leq C\varepsilon^{-2} \Sigma_{k=K}^{\infty} b_k^{-2} H_k 2^{-2k}.$$

From the proof of the lemma 6 in [6]

$$\Delta_{n,K} \leq C(W_n^{1/4} H_K^{3/4} \log^{9/4} H_K + \varepsilon + \varepsilon^{-2} \Sigma_{k=K}^{\infty} b_k^{-2} H_k 2^{-2k}).$$

Let $a_k = H_k^{2/3} 2^{-k} \log^{2/3} H_k$ if $H_k 2^{-3/2k} > k^{-4}$ and $a_k = k^{-2}$ otherwise. Put $b_k = A^{-1} a_k$, where $A = \Sigma_{k=1}^{\infty} a_k$.

Collecting the estimates we get

$$\Delta_n \leq C(W_n^{1/4} H_K^{3/4} \log^{9/4} H_K + \varepsilon + \varepsilon^{-2} 2^{-K} + \varepsilon^{-3} W_n K^{-1}).$$

Choosing ε in an appropriate way, we have

$$\Delta_n \leq C(2^{-1/3K} + W_n^{1/4} H_K^{3/4} \log^{9/4} H_K)$$

and (4.2) is proved in the case $\Sigma_{k=1}^{n} \sigma_k^2 = 1$ a.s.

To overcome this assumption we follow the paper [3] and use the following result.

Lemma 4.1. If the condition a) of Th. 4.1 is fulfiled, then there exists a constant $C > 0$ such that

$$E\|S_n\|_{C(S)}^3 \leq C(\Sigma_{i=1}^{n} E\|X_i\|_{C_d}^3 + E(\Sigma_{i=1}^{n} \sigma_i^2)^{3/2}). \tag{4.3}$$

Under stronger assumption on the semi-metric d, this inequality is proved in [6]. But only a litle changes are needed to get (4.3) under the condition (4.1).

Let us continue the proof of (4.2) in the case $\Sigma_{k=1}^{n} \sigma_k^2 \neq 1$. Define a stooping time $\tau = \max(k=0, \ldots, n: \Sigma_{i=1}^{k} \sigma_i^2 \leq 1)$, and let $\widetilde{X}_i = X_i I(i \leq \tau)$ for $i=1, \ldots, n$, $\widetilde{X}_{n+1} = (1-\Sigma_{i=1}^{\tau} \sigma_i^2)^{1/2} \widetilde{Y}$, where $\widetilde{Y}$ is independent copy of Y, independent with the completion $\hat{F}_k$ of F_k for all $1 \leqslant k \leqslant n$. Then $\tilde{X}_1, \ldots, \tilde{X}_{n+1}$ is martingale difference sequence relative to $F_0, \ldots, F_n, F$ and

$$E\, \tilde{X}_i(s)\tilde{X}_i(t)/F_{i-1} = \tilde{\sigma}_i^2 E\, Y(s)Y(t)$$

where $\tilde{\sigma}_j = \sigma_i I(i \leqslant \tau), i=1, \ldots, n, \tilde{\sigma}_{n+1} = (1 - \Sigma_{i=1}^{\tau} \sigma_i^2)^{1/2}$.

Hence $\Sigma_{i=1}^{n+1} \tilde{\sigma}_i^2 = 1$, therefore

$$\tilde{\Delta}_n = \sup_{r \geqslant 0} |P(\| \Sigma_{i=1}^{n+1} \tilde{X}_i \|_{C(S)} < r) - P(\| Y \|_{C(S)} < r)| \leqslant$$

$$C \inf_{0 < \varepsilon < 1} (\varepsilon^{1/3} + \tilde{W}_n^{1/4} H_d^{3/4}(S, \varepsilon) \log^{9/4} H_d(S, \varepsilon)), \tag{4.4}$$

where $\tilde{W}_n = \Sigma_{k=1}^{n+1} E \| \tilde{X}_k \|_{C_d}^3$.

On the other hand, we have

$$\Delta_n \leqslant \tilde{\Delta}_n + P(\| \Sigma_{i=1}^{n} X_i - \Sigma_{i=1}^{n+1} \tilde{X}_i \|_{C(S)} > \varepsilon) +$$

$$\sup_{r \geqslant 0} P(r - \varepsilon \leqslant \| Y \|_{C(S)} \leqslant r + \varepsilon) \leqslant$$

$$\Delta_n + \varepsilon^{-3} E \| \Sigma_{i=1}^{n} X_i - \Sigma_{i=1}^{n+1} \tilde{X}_i \|_{C(S)}^3 + C\varepsilon \tag{4.5}$$

fore every $\varepsilon > 0$. Using lemma 4.1 we get the following estimate

$$E \| \Sigma_{i=1}^{n} X_i - \Sigma_{i=1}^{n+1} \tilde{X}_i \|_{C(S)}^3 \leqslant C(W_n + L_n)$$

and (4.2) follows from (4.4), (4.5), since $\tilde{W}_n \leqslant C(W_n + L_n)$.

References

1. Butzer P. L., Hahn L., Roeckerath M. Th. (1983) Central limit theorem and weak law of large numbers with rates for martingales in Banach spaces J. Mulivar. Anal., 13, 287—301.
2. Fabian M., Whitfield J. H. M., Ziezler V. (1983) Norms with locally lipschitzian derivatives. Jsrael J. Math., 44, 3, 262—276.
3. Haeusler E. (1984) A note on the rate of convergence in the martingale central limit theorem. Ann. Probab., 12, 635—639.
4. Pisier G. (1975) Martingales with values in uniformly convex spaces. Israel J. Math., 20, 3—4, 326—350.
5. Račkauskas A. (1985) On the smoothing of operators. F. Schiller University Jena, preprint N/85/17.
6. Račkauskas A. (1987) On convergence rate in the martingale central limit theorem in the space C(S). Liet. matem. rink., 27, 3, 523—534 (in Russian).
7. Rosinski J. (1981) Central limit theorems for dependent random vectors in Banach spaces. Lecture Notes in Math., 939, 157—180.
8. Wan Soo Rhee, Talagrand M. (1986) Uniform bound in the central limit theorem for Banach space valued dependent random variables. J. Multivar. Anal., 20, 303—320.

Probability Theory on Vector Spaces IV
Lancut, June'87, Springer's LNM 1391

ON SERIES EXPANSIONS OF STOCHASTIC INTEGRALS

by

Andrzej Russek

Department of Mathematics
Louisiana State University
Baton Rouge, Louisiana 70803

1. Introduction.

Let (E,H,μ) be an abstract Wiener space; E is a separable real Banach space, H a separable Hilbert space densely imbedded in E and μ is the Gaussian measure on Borel subsets of E with the characteristic functional given by

$$\int_E \exp i(x,h)\mu(dx) = \exp\{-\frac{1}{2}\|h\|_H^2\} ,$$

$h \in E' \subset H'$ where E' is the dual of E and (,) denotes the pairing between E and E'.

For the dual spaces we have the inclusion $E' \subset H' \simeq H$ and we can identify E' with a subspace of H. Thus we have the triple $E' \subset H \subset E$.

For h in E' the random variable $\xi(x) = (h,x)$ is Gaussian with mean zero and variance $\|h\|^2$. If $h \in H$ then there exists a sequence $\{u_n: n = 1,2,\dots\} \subset E'$ such that $h = \lim u_n$. The $L^2(E,\mu)$-limit, $\lim_{n\to\infty}(u_n,\cdot)$ exists independently of the choice of (u_n) and is Gaussian with mean zero and variance $\|h\|^2$. We will use the same symbol $(h,\cdot)$ to denote this random variable. In particular if $H = L^2(I)$, $I = [0,1]$ then $b_t = (1_{(0,t)},\cdot)$, $t \in I$ is a Brownian motion and $(h,\cdot)$ is the Wiener integral of h i.e.

$$(h,\cdot) = \int_0^1 h(t)db_t.$$

Let $\phi: E \to H$ be a measurable functional and let $(h_n)_{n=1}^\infty$ be an orthonormal basis in H. In this paper we consider the series

$$\sum_{n=1}^\infty (\phi,h_n)_H(h_n,\cdot) \tag{1.1}$$

where $(,)_H$ denotes the scalar product in H.

If $H = L^2(I)$ and $\phi = \phi_t(x)$, $t \in I$, $x \in E$ is a square integrable stochastic process, (1.1) has the form

$$\sum_{n=1}^{\infty} a_n \int_0^1 h_n(t)db_t \tag{1.2}$$

where $a_n(x) = \int_0^1 \phi_t(x)h_n(t)dt$.

This series was studied by several authors (cf. [2, 6, 7, 17]. In particular in [2] sufficient conditions for $L^2(E,\mu)$ convergence of (1.2) were given and the sum was identified as the Stratonovich integral of ϕ_t.

In this paper we give a characterization of convergence of the series (1.1) generalizing the one given in [7] and evaluate its sum to be trace $D\phi + D^*\phi$. Here D is the weak derivative and D^* its adjoint defined below. We also give the Fock space representation of (1.1). In section 4 we consider the case $H = L^2(I)$, i.e. the series (1.2), and we give some conditions for convergence of (1.2) to the Stratonovich integral. Our conditions are formulated in a different way than those in [2] and [6]. The methods we use are those of calculus on the white noise space and are different from the ones used in [2, 7, 17].

Sections 3 and 4 are independent of one another.

2. The series expansion.

Let K be a Hilbert space. $L^2(E;K)$ denotes the space of K-valued functions square integrable with respect to μ. The space $L^2(E;\mathbb{R})$ will be denoted by $L^2(E)$. Following [11] we define the Sobolev spaces on Wiener space as follows. Let $\mathcal{P}(K)$ denote the set of K-valued functions on E of the form

$$\phi(x) = \sum_{i=1}^{n} f_i(x)k_i \tag{2.1}$$

where $k_i \in K$, $i = 1,\dots,n$ and each f_i is of the form

$$f(x) = w((h_1,x),\dots,(h_k,x)) \tag{2.2}$$

where w is a polynomial and $h_1,\dots,h_k \in E'$. The space $\mathcal{P}(\mathbb{R})$ is denoted by $\mathcal{P}$.

Any function ϕ of the form (2.1) is arbitrarily many times differentiable at every point x in E. The derivative $D^n\phi(x)$ is a continuous n-linear form on E

$$D^n\phi(x): E \times \dots \times E \rightarrow K.$$

Its restriction to $H \times \dots \times H$ is a Hilbert-Schmidt form (see [3]). Let $|D^n\phi(x)|_{H-S}$ denote the Hilbert-Schmidt norm of this restriction. Define

$$\|\phi\|^p_{n,p} = \int_E |D^n\phi(x)|^p_{H-S}\mu(dx).$$

The Sobolev space $W^k_p(E;K)$ is now defined as the completion of $\mathcal{P}(K)$ with respect to the norm

$$\|\phi\|_{W^k_p} = \sum_{n=0}^{k} \|\phi\|_{n,p} \; . \tag{2.3}$$

The extension of the differentiation operator D from $\mathcal{P}$ to $W^1_2(E;\mathbb{R}) = W^1_2(E)$ is denoted by the same symbol D and is called the weak derivative. We have

$$D: W^1_2(E) \subset L^2(E) \longrightarrow L^2(E;H)$$

The adjoint operator D^*: domain$(D^*) \subset L^2(E;H) \longrightarrow L^2(E)$ is defined through

$$\int_E D^*\phi\cdot\xi \; d\mu = \int_E (\phi, D\xi)_H \; d\mu, \quad \xi \in W^1_2(E) \; . \tag{2.4}$$

Proposition 2.1. For h in H and ξ in $W^1_2(E)$ we have

$$D^*\xi\cdot h + D\xi[h] = (h,\cdot)\xi \; . \tag{2.5}$$

Proof. Let $\xi,\eta \in \mathcal{P}$. By the well known translation formula for Gaussian measure we have

$$\int_E \xi(x)\cdot\eta(x + \varepsilon h)\mu(dx) = \int_E \xi(x - \varepsilon h)\cdot\eta(x) \; e^{(x,\varepsilon h) - \frac{1}{2}\|\varepsilon h\|^2} \mu(dx) \; .$$

Differentiating both sides of this equality $d/d\varepsilon$ at $\varepsilon = 0$ we get

$$\int_E \xi\cdot D\eta[h] \; d\mu = -\int_E D\xi[h]\cdot\eta \; d\mu + \int_E \xi\cdot\eta\cdot(\cdot,h) \; d\mu$$

which proves (2.5) for ξ in $\mathcal{P}$. The operators D and D^* are continuous from $W^1_2(E;H)$ to $L^2(E)$ (see (3.9) below or [9] p. 44). The operator of multiplication by $(h,\cdot)$ is closed on $L^2(E)$ thus the proposition follows by an obvious approximation argument.

Let A: H → H be a linear operator. We say that A has a finite trace if for any orthonormal basis $(h_n)_{n=1}^{\infty}$, $\sum_{n=1}^{\infty} (Ah_n,h_n)_H$ is convergent and the limit does not depend on $(h_n)_{n=1}^{\infty}$. The limit is denoted by trace A. Note that since any permutation of (h_n) is also an orthonormal basis in H, the convergence of $\sum(Ah_n,h_n)_H$ is unconditional.

Theorem 2.2. Let ϕ be in $W_2^1(E;H)$. The series

$$\sum_{n=1}^{\infty} (\phi,h_n)_H(h_n,\cdot)$$

converges in $L^2(E)$ for every orthonormal basis $(h_n)_{n=1}^{\infty}$ if and only if for every ξ in $L^2(E)$ the operator $A_\xi = \int_E D\phi(x)\xi(x)\mu(dx)$ has a finite trace.

Proof. From Proposition 2.1 we have

$$\sum_{n\leq N} (\phi,h_n)_H(\cdot,h_n) = \sum_{n\leq N} D^*(\phi,h_n)_H h_n + \sum_{n\leq N} D(\phi,h_n)[h_n]. \tag{2.6}$$

Note that

$$\phi = \lim_{N\to\infty} \sum_{n\leq N} (\phi,h_n)_H h_n \tag{2.7}$$

in the space $W_2^1(E;H)$. Indeed, (2.7) holds in the sense of $L^2(E;H)$ convergence. Now $D\phi$ is in $L^2(E;K)$ where K is the space of Hilbert-Schmidt operators on H; hence, for any orthonormal basis $(e_j)_{j\geq 1}$ of K, $D\phi = \sum_j (D\phi,e_j)_{H-S}e_j$ in the $L^2(E;K)$ norm. But the operators $h_i \otimes h_k = (h_i,\cdot)h_k$, $1 \leq i, k < \infty$, form an orthonormal basis in K thus

$$D\phi = \lim \sum_{n\leq N} D((\phi,h_n)_H h_n)$$

in the space $L^2(E;K)$ i.e. $\phi = \sum_n (\phi,h_n)_H h_n$ in $W_2^1(E;H)$.
The operator D^* is continuous from $W_2^1(E;H)$ to $L^2(E)$ (see [9] p. 44 or (3.9) below) thus from (2.7) we have

$$\lim_{N\to\infty} \sum_{n\leq N} D^*(\phi,h_n)_H h_n = D^*\phi .$$

It follows that the convergence of the series (1.1) is equivalent to the convergence of $\sum_n D(\phi,h_n)_H[h_n] = \sum_n (D\phi[h_n],h_n)_H$.

Let $c_n(x) = (D\phi(x)h_n,h_n)_H$. For every ξ in $L^2(E)$, $(c_n,\xi)_{L^2(E)} = (A_\xi h_n,h_n)_H$ and

$$\sum_n (c_n,\xi)_{L^2(E)} = \text{trace } A_\xi. \tag{2.8}$$

Thus, the condition that A_ξ has finite trace for all $\xi \in L^2(E)$ is equivalent to the condition

$$\sum_n |(c_n,\xi)_{L^2(E)}| < \infty \quad \text{for all} \quad \xi \in L^2(E).$$

But in $L^2(E)$ this condition is equivalent to the strong unconditional convergence of $\sum c_n$ ([1] p. 93). Since $c_n = D(\phi,h_n)_H[h_n]$ we have the equivalence claimed in the theorem.

Corollary 2.3. Let ϕ be in $W_2^1(E;H)$ and suppose that for almost every x in E

$$\text{trace } D\phi(x) < \infty .$$

Then the series (1.1) converges in probability and we have

$$\sum_n (\phi,h_n)_H(\cdot,h_n) = D^*\phi + \text{trace } D\phi .$$

Proof. From (2.5) we have

$$\sum_{n\leq N} (\phi,h_n)_H(h_n,\cdot) = \sum_{n\leq N} D^*(\phi,h_n)_H h_n + \sum_{n\leq N} (D\phi[h_n],h_n)_H .$$

By (2.7) and the continuity of D^*, the first term converges to $D^*\phi$ in $L^2(E)$. The second term converges to trace $D\phi(x)$ for μ-almost every x.

3. Homogeneous chaos expansion.

The space $L^2(E)$ has the well known Wiener-Ito orthogonal decomposition

$$L^2(E) = \sum_{n=0}^{\infty} \oplus K_n$$

where K_n is the n-th homogeneous chaos. The space K_n is spanned by the functions of the form

$$\phi = \mathcal{H}_{\alpha_1}((h_{k_1},\ )) \ldots \mathcal{H}_{\alpha_r}((h_{k_r},\ )), \quad \alpha_1 + \ldots + \alpha_r = n . \tag{3.1}$$

Here $(h_k)_{k=1}^{\infty}$ is an orthonormal basis in H and $\mathcal{H}_k$ is the Hermite polynomial of order k:

$$\mathcal{H}_k = (-1)^n(n!)^{-\frac{1}{2}} \exp(x^2/2)D^n \exp(-x^2/2).$$

Let ϕ be in $L^2(E)$. The S-transform of ϕ is the function on H defined by the formula

$$S\phi(h) = \int_E \phi(x + h)\mu(dx), \quad h \in H.$$

For ϕ of the form (3.1) we have

$$S\phi(h) = (h_{k_1},h)^{\alpha_1} \ldots (h_{k_r},h)^{\alpha_r}$$

thus for any ϕ in K_n, $S\phi$ is a homogeneous form of degree n on H. Hence, there exists a unique symmetric n-linear form f on H such that

$$S\phi(h) = f(h,\dots,h).$$

Let H_n denote the space of symmetric n-linear forms on H with the scalar product

(3.2) $$(f,g)_n = \sum_{i_1\dots i_n} f(h_{i_1},\dots,h_{i_n})g(h_{i_1},\dots,h_{i_n})$$

where $(h_k)_{k=1}^{\infty}$ is an orthonormal basis in H. The inverse of S restricted to K_n is denoted by I_n i.e. I_n: $H_n \to K_n$ and $SI_n(f) = f$.

It is well known that

$$(I_n(f),I_n(g))_{L^2(E)} = n!(f,g)_n .$$

Thus every ϕ in $L^2(E)$ has an expansion $\phi = \sum_{n=0}^{\infty} I_n(f_n)$ with f_n in H_n, and we have

(3.3) $$\|\phi\|^2_{L^2(E)} = \sum_{n=0}^{\infty} n!\|f_n\|_n^2$$

where $\|f\|_n = (f,f)_n$.

If $H = L^2(I)$ then H_n is isometric to $\hat{L}_2(I^n)$ - the space of symmetric functions in $L^2(I^n)$ and $I_n(f)$ is the multiple Wiener integral or order n [10]

$$I_n(f) = \int \underset{I^n}{\dots} \int f(t_1,\dots,t_n)db_{t_1} \dots db_{t_n} .$$

Lemma 3.1. Let f be in H_n. Then $I_n(f)$ is in the space $W_2^1(E)$ and for every u in H, we have

(3.4) $$DI_n(f)[u] = I_{n-1}(nf(u,\cdot)) .$$

Proof. $I_n(f)$ is the $L^2(E)$ limit of a sequence of functionals $I_n(f^{(k)})$ of the form (3.1). For such $f^{(k)}$, (3.4) is verified by direct computation. Moreover we have

(3.5) $$\|DI_n(f^{(k)})\|_H = \sum_{i=1}^{\infty} |I_{n-1}(nf^{(k)}(h_i,\cdot))|^2$$

where $(h_i)_{i\geq 1}$ is the orthonormal basis in H. Now it is not hard to check that the sequences $\sum_i nf^{(k)}(h_i,\cdot)$ and $f^{(k)}(u,\cdot)$ converge in H_n as $k \to \infty$. It follows that $\|DI_n(f^{(k)})\|$ converge in $L^2(E)$ thus $I_n(f)$ is in $W_2^1(E)$. Moreover $DI_n(f^{(k)})[u]$ $= I_{n-1}(f^{(k)})(u,\cdot))$ converges in $L^2(E)$ thus we have (3.4).

Lemma 3.2. Let f be in H_{n+1} and let $\phi: E \to H$ be such that for each u in H

$$(\phi,u)_H = I_n(f(u,\cdot)).$$

Then ϕ is in the domain of D^* and we have

$$D^*\phi = I_{n+1}(f).$$

Proof. Let ξ be in $W_2^1(E)$ and suppose that ξ has the expansion $\xi = \sum_{j=0}^{\infty} I_j(g_j)$. Let $(h_k)_{k\geq 1}$ be an orthonormal basis in H. We have

$$\begin{aligned}(I_{n+1}(f),\xi)_{L^2(E)} &= (n+1)!(f,g_{n+1})_{n+1} \\ &= \sum_{k=1}^{\infty} (n+1)\big(I_n(f(h_k,\cdot)),I_n(g_{n+1}(h_k,\cdot))\big)_{L^2(E)} \\ &= \sum_{k=1}^{\infty} \big((\phi,h_k)_H,D\xi[h_k]\big)_{L^2(E)}.\end{aligned}$$

The last equality follows from Lemma 3.1. Thus we have

$$(I_{n+1}(f),\xi)_{L^2(E)} = \int_E (\phi,D\xi)_H d\mu$$

which proves the lemma. □

Let ϕ be in $L^2(E;H)$. For each u in H, $(\phi,u)_H$ has the expansion

$$(\phi,u)_H = \sum_{n=1}^{\infty} I_n(f_n(u,\cdot))$$

where $f_n(u,\cdot) \in H_n$. Let $\tilde{f}_n$ denote the symmetrization of f_n i.e.

$$\tilde{f}_n(u_1,\dots,u_{n+1}) = \frac{1}{(n+1)!}\sum_{\sigma} f_n(u_{\sigma(1)},\dots,u_{\sigma(n+1)})$$

where the sum is taken over all permutations σ of the integers $\{1,\dots,n+1\}$. If for $n > 0$, $\tilde{f}_n \in H_{n+1}$ and

$$\sum_{n=1}^{\infty} (n+1)!\,\|\tilde{f}_n\|_{n+1}^2 < \infty$$

then by Lemma 3.2, $D^*\phi$ is well defined and

$$D^*\phi = \sum_{n=0}^{\infty} I_{n+1}(f_n). \tag{3.7}$$

Moreover, we have

$$\|D^*\phi\|^2_{L^2(E)} = \int_E \|\phi\|^2_H \, d\mu + \sum_{k=1}^{\infty} \int_E \|D\phi[h_n]\|^2_H \, d\mu \; . \tag{3.8}$$

The last equality is obtained in the following way: since for fixed u in H, $f_n(u,\cdot)$ is symmetric we have

$$\tilde{f}_n(u_1,\ldots,u_{n+1}) = \frac{1}{n+1} \sum_{i=1}^{n} f_n(u_i,\tilde{u}_i)$$

where $\tilde{u}_i$ denotes the variables $u_1,\ldots,u_{n+1}$ with u_i deleted. Using this it is not hard to see that

$$\|\tilde{f}_n\|^2_{n+1} = \frac{1}{n+1} \sum_{k+1}^{\infty} \|f_n(h_k,\cdot)\|^2_n + \frac{n}{n+1} \sum_{k,l=1}^{\infty} \|f_n(h_k,h_l,\cdot)\|^2_{n-1}$$

where $(h_k)_{k=1}^{\infty}$ is an orthonormal basis in H. Now using (3.3), (3.4) and (3.7) we get (3.8).

As a corollary from (3.8) we get

$$\|D^*\phi\|_{L^2(E)} < \|\phi\|_{W_2^1(E;H)} \; . \tag{3.9}$$

From Lemma 3.1 we have

$$DI_n(f_n(u,\cdot))[h] = I_{n-1}(nf_n(u,h,\cdot))$$

and thus

$$\sum_{k=1}^{\infty} DI_n(f_n(h_k,\cdot))[h_k] = I_{n-1}(\mathrm{tr}\, T_n) \tag{3.10}$$

where we have denoted $\mathrm{tr}\, T_n = \sum_{k=1}^{\infty} nf_n(h_k,h_k,\ )$. Combining (2.6), Lemma 3.1, Lemma 3.2 and (3.10) we get the following theorem.

Theorem 3.3. Suppose ϕ is in $L^2(E;H)$ and that for each u in H, $(\phi,u)_H$ has the expansion

$$(\phi,u)_H = \sum_{n=0}^{\infty} I_n(f_n(u,\cdot))$$

with $\tilde{f}_n$ in H_{n+1}. Assume that

$$\sum_{n=0}^{\infty} (n+1)! \, \|\tilde{f}_n\|^2_{n+1} + \sum_{n=1}^{\infty} (n-1)! \, \|\mathrm{tr}\, T_n^2\|_{n-1} < \infty \; .$$

Then for every orthonormal basis $(h_n)_{n\geq 1}$ the series (1.1) is convergent in L^2 and

$$\sum_{k=1}^{\infty} (\phi,h_k)_H(h_k,\cdot) = \sum_{n=0}^{\infty} I_{n+1}(\tilde{f}_n) + \sum_{n=1}^{\infty} I_{n-1}(\mathrm{tr}\, T_n).$$

Remark. Lemma 3.1 implies that the weak derivative $D\phi$ is equivalent to the stochastic gradient of Skorokhod [16] p. 224 which by Theorem 1 of [18] is equivalent to the Malliavin derivative. Thus (3.7) coincides with Theorem 2 of [18]. The formula (3.8) for the norm of $D^*\phi$ appears in [16] p. 227.

4. The Stratonovich integral.

In this section we consider the case $H = L^2(I^d)$, $I = [0,1]$.

Let $\phi = \phi_t(x)$, $t \in I^d$, $x \in E$ be a stochastic process such that for every t in I^d, $\phi_t \in W_2^1(E)$ and $\int_{I^d} \|\phi_t\|^2_{W_2^1(E)} dt < \infty$.

By the Riesz representation theorem, $D\phi_t$ can be represented as follows

$$D\phi_t(x)h = \int_{I^d} k_\phi(t,s,x)h(s)ds \tag{4.1}$$

and we have

$$\int_{I^d} \|\phi_t\|^2_{W_2^1(E)} = \|\phi\|^2_{L^2(I^d)\otimes L^2(E)} + \|k_\phi\|^2_{L^2(I^{2d})\otimes L^2(E)} .$$

We will denote $\partial_s\phi_t = k_\phi(t,s,\cdot)$.

Suppose that for μ-almost every x, the operator $D\phi(x)$: $L^2(I^d) \to L^2(I^d)$ given by $D\phi(x)h(t) = \int_{I^d} \partial_s\phi_t(x)h(s)ds$ is a trace class operator and that $\partial_s\phi_t(x)$ is continuous in (t,s). Then by a well known theorem on the trace of an integral operator we have

$$\mathrm{trace}\ D\phi(x) = \int_{I^d} \partial_t\phi_t(x)\ dt$$

and thus

$$\sum_{n=1}^{\infty} (\phi,h_n)_H(h_n,\cdot) = D^*\phi + \int_{I^d} \partial_t\phi_t\ dt . \tag{4.2}$$

The last equality allows us to identify the series as a kind of Stratonovich integral on I^d. Namely for a partition $P = (\Delta_1,\Delta_2,\ldots,\Delta_n)$ of I^d into cubes let ϕ_P denote

$$\phi_P = \sum_{i=1}^{n} \frac{1}{|\Delta_i|} \int_{\Delta_i} \phi_t dt\cdot(\chi_{\Delta_i},\cdot)$$

where $|\Delta_i|$ is the volume of Δ_i. Let $\mathrm{mesh}(P) = \max\{|\Delta_i|: 1 \leq i \leq n\}$. The limit in probability of ϕ_P as $\mathrm{mesh}(P) \to 0$ if it exists is denoted by $\int_{I^d} \phi_t\circ db_t$.

Theorem 4.1. Let $\phi = \phi_t(x)$, $t \in I^d$, $x \in E$ be an $L^2(E)$ continuous stochastic process such that

$$\int_{I^d} \|\phi_t\|^2_{W_2^1(E)} dt < \infty .$$

Suppose that $t \to \partial_s \phi_t$ is $L^2(E;L^2(I^d))$ continuous and that for μ-almost every x, $\partial_s \phi_t(x)$ is continuous in (t,s) in a neighbourhood of the diagonal $\{(t,s) \in I^{2d};\ t = s\}$. Then the integral $\int_{I^d} \phi_t \circ db_t$ exists and we have

$$\int_{I^d} \phi_t \circ db_t = D^* \phi + \int_{I^d} \partial_t \phi_t \, dt . \tag{4.3}$$

Proof. Let us denote $\phi_{\Delta_i} = \frac{1}{|\Delta_i|} \int_{\Delta_i} \phi_t dt$. From Proposition 2.1 we have

$$\sum_{i=1}^n \phi_{\Delta_i} \cdot (\chi_{\Delta_i}, \cdot) = D^* \sum_{i=1}^n \phi_{\Delta_i} \cdot \chi_{\Delta_i} + \sum_{i=1}^n D\phi_{\Delta_i}[\chi_{\Delta_i}] .$$

By (3.9) and the $L^2(E)$ continuity of ϕ_t and $D\phi_t$, the first term converges in $L^2(E)$ to $D^*\phi$. The second term is equal to

$$\sum_{i=1}^n \frac{1}{|\Delta_i|} \int_{\Delta_i} \int_{\Delta_i} \partial_s \phi_t \, dtds$$

and by continuity in (t,s) converges to $\int_{I^d} \partial_t \phi_t(x)\, dt$.

The next theorem identifies $D^*\phi$ for nonanticipating ϕ with the Cairoli-Walsh integral [14, 15].

For $t = (t_1,\ldots,t_d)$ in I^d, let $(0,t)$ denote the interval $(0,t) = \{(s_1,\ldots,s_d) \in I^d;\ 0 \le s_i < t_i,\ 1 \le i \le d\}$ and let $\mathcal{F}_t$ denote the completion of the σ-field generated by $(\chi_{(0,u)}, \cdot)$ for all $(0,u) \subset (0,t)$. The process $\phi = \phi_t(x)$, $t \in I^d$, $x \in E$ is called nonanticipating if ϕ is $\mathcal{F}_{(1,\ldots,1)} \times \mathrm{Borel}(I^d)$-measurable and for each t in I^d, ϕ_t is measurable with respect to $\mathcal{F}_t$.

For a partition $\pi = \{0 = t_0 < t_1 < \ldots < t_n = 1\}$ of I, let P denote the partition of I^d into rectangles of the form $\Delta = (t_{i_1}, t_{i_{1+1}}) \times \ldots \times (t_{i_d}, t_{i_{d+1}})$ $0 \le i_k < n$. Let t_Δ denote the "lower left end corner", $t_\Delta = (t_{i_1},\ldots,t_{i_d})$. Define

$$\phi_P = \sum_\Delta \phi_{t_\Delta} \cdot (\chi_{t_\Delta}, \cdot) .$$

Theorem 4.2. Let $\phi = \phi_t(x)$, $t \in I^d$, $x \in E$ be a nonanticipating stochastic process such that

$$\int_{I^d} \|\phi_t\|^2_{W_2^1(E)} dt < \infty$$

and suppose that ϕ and $D\phi$ are $L^2(E)$ and $L^2(E;L^2(I^d))$ continuous respectively. Then the following $L^2(E)$-limit exists as mesh(P) $\to 0$ and we have the equality

$$\lim \sum_{\Delta\in P} \phi_{t_\Delta}(\chi_\Delta,\cdot) = D^*\phi .$$

Lemma 4.3. Suppose that $\psi = \psi(x)$, $x \in E$ is in $W_2^1(E)$ and that ψ is $\mathcal{F}_t$-measurable. Then for every h in $L^2(I^d)$ such that $h = 0$ on $(0,t)$ we have

$$D\psi[h] = 0.$$

Proof. Denote $E_t = E(\cdot|\mathcal{F}_t)$. From the isometric properties of $f \to I_n(f)$ we have

$$E_t(I_n(f) = I_n(P_t f) \tag{4.4}$$

where P_t is the projection

$$P_t f(s) = \begin{cases} f(s), & s \in (0,t) \\ 0 & \text{otherwise} \end{cases}$$

From the definition of $W_2^1(E)$, ψ is a limit of cylindrical functions of the form $\mathcal{H}_{k_1}(h_1, \ ) \ldots \mathcal{H}_{k_r}(h_r, \ ) = I_n(h_1^{\otimes k_1} \otimes \ldots \otimes h_r^{\otimes k_r})$ where $\mathcal{H}_k$ are Hermite polynomials. Now from (4.4) and (3.4) we get that $\psi = E_t\psi$ is the limit in $W_2^1(E)$ of cylindrical functions (2.2) with $h_i = 0$ on $(0,t)$. But for such ψ, $D\psi[h] = 0$ as claimed in the lemma.

Proof of Theorem 4.2. From Proposition 2.1 we have

$$\sum_\Delta \phi_{t_\Delta}\cdot(\chi_{t_\Delta},\cdot) = D^*\sum_\Delta \phi_{t_\Delta}\cdot\chi_\Delta + \sum_\Delta D\phi_{t_\Delta}[\chi_\Delta] .$$

By Lemma 4.3 the second term vanishes. By the $L^2(E)$ continuity of ϕ and $D\phi$, the expression $\sum\phi_{t_\Delta} \cdot \chi_\Delta$ converges to ϕ in the W_2^1 norm thus the first term converges in $L^2(E)$ to $D^*\phi$. □

For $d = 1$ the usual definition of the Stratonovich integral is the limit in probability of

$$\phi_P = \sum_{i=1}^{n-1} \frac{1}{2}(\phi_{t_i} + \phi_{t_{i+1}})(\chi_{(t_i,t_{i+1})},\cdot)$$

as mesh(P) $= \max\{t_{i+1} - t_i;\ 0 \le i \le n-1\} \to 0$ where $P = \{0 = t_0 < t_1 \ldots < t_n = 1\}$ is a partition of I.

Under the assumptions of Theorem 4.1 this integral exists and (4.2) holds. The continuity assumption of $\partial_s\phi_t$ in (t,s) however is not satisfied in many important cases e.g. if $\phi_t = (\chi_{(0,t)},\cdot)$ is a Brownian motion then $\partial_s\phi_t = \chi_{(0,t)}(s)$.

One can extend Theorem 4.1 as follows. Let $I_+ = \{(t,s) \in I^2;\ t < s\}$ and let $I_- = \{(t,s) \in I^2;\ t > s\}$ and let $\partial_s^+\phi_t$, $\partial_s^-\phi_t$ denote the symmetrizations of $\partial_s\phi_t \cdot \chi_{I_+}$ and $\partial_s\phi_t \cdot \chi_{I_-}$ respectively. In the same way as in Theorem 4.1 one can prove the following theorem.

Theorem 4.3. Let $\phi = \phi_t(x)$, $t \in I$, $x \in E$ be an L^2 continuous stochastic process such that $\int_I \|\phi_t\|^2_{W_2^1}\, dt < \infty$. suppose that for μ-almost every x, $\partial_s^+\phi_t$ and $\partial_s^-\phi_t$ are continuous in (t,s). Then the integral $\int_0^1 \phi_t \circ db_t$ exists and we have

$$\int_0^1 \phi_t \circ db_t = D^*\phi + \frac{1}{2}\int_0^1 (\partial_t^+\phi_t + \partial_t^-\phi_t)dt.$$

We now give some Sobolev type conditions for convergence of the series (1.2) to the Stratonovich integral.

We will use the following simple fact. Suppose that A is an operator on a real Hilbert space H such that the symmetrization $\tilde{A} = \frac{1}{2}(A + A^*)$ is a trace class operator. Then A has a finite trace and trace A = trace $\tilde{A}$.

Example. Let $\phi_t = (1_{(0,t)}, \cdot)$ be a Brownian motion. Then ϕ is in $W(E;L^2(I))$ and $D\phi(x)h = \int_0^{\cdot} h(s)ds$ is the operator of integration. Thus $D\phi(x)$ is not a trace class operator. $\tilde{D}\phi(x)$ is the projection onto the one dimensional subspace spanned by constants; thus, $D\phi(x)$ has a finite trace.

For a Hilbert space B let $H^\alpha(I;B)$ denote the Sobolev space of order α of B-valued functions on I. Let $e_n(t) = \sqrt{2}\sin n\pi t$, $n \geq 1$, be the trigonometric system. For u in $H^\alpha(I;B)$ we have

$$\|u\|^2_{L^2(I;B)} + \sum_n (n^\alpha \|a_n\|_B)^2 < \infty \tag{4.4}$$

where $a_n = \int_0^1 u(t)e_n(t)\, dt$.

Let $\partial_t^\sim\phi_s = \frac{1}{2}(\partial_t\phi_s + \partial_s\phi_t)$ denote the symmetrization of $\partial_t\phi_s$. Then $\tilde{D}\phi(x)h(t) = \int_I \partial_t^\sim\phi_s(x)h(s)ds$ is the symmetrization of the operator $D\phi(x)\colon L^2(I) \to L^2(I)$.

Theorem 4.5. Let $\phi = \phi_t(x)$, $t \in I$, $x \in E$ be a stochastic process such that $\int_0^1 \|\phi_t\|^2_{W_2^1}\, dt < \infty$. Suppose that for μ-almost every x, the kernel $t \to \tilde{\partial}_t\phi_\cdot$ is in

$H^{\alpha}(I;L^2(I))$ with $\alpha > \frac{1}{2}$. Then the series (1.2) converges in probability and we have

$$\sum_{n=1}^{\infty} (\phi,h_n)_{L^2(I)} \int_0^1 h_n(t)db_t = D^*\phi + \text{trace } D\phi.$$

Proof. The kernel $\tilde{\partial}_t\phi_s(x)$ has the expansion

$$\tilde{\partial}_t\phi_s(x) = \sum_n a_n(s,x)e_n(t)$$

where $a_n(s,x) = (\tilde{\partial}_\cdot\phi_s(x),e_n)_{L^2(I)}$.

Now $\tilde{D}\phi(x)$ can be expanded as follows

$$\tilde{D}\phi(x)h = \sum_n (a_n(\cdot,x),h)_{L^2(I)} e_n$$

whence

$$\text{trace } \tilde{D}\phi(x) < \sum_n \|a_n(\cdot,x)\|_{L^2(I)} .$$

But condition (4.4) with $\alpha > \frac{1}{2}$ implies $\sum_n \|a_n(\cdot,x)\|_{L^2(I)} < \infty$. It follows that $\tilde{D}\phi(x)$ is a trace class operator; thus $D\phi(x)$ has a finite trace. Thus the theorem follows from Corollary 2.3.

In a similar way we can prove a result on the $L^2(E)$ convergence of the series (1.2). Let Trace $D\phi$ denote the $L^2(E)$ limit

$$\text{Trace } D\phi = \sum_{n=1}^{\infty} (D\phi(\cdot)h_n,h_n)_{L^2(I)}$$

provided the series converges and the limit does not depend on the choice of an orthonormal basis (h_n).

Theorem 4.6. Let $\phi = \phi_t(x)$, $t \in I$, $x \in E$ satisfy $\int_0^1 \|\phi_t\|^2_{W_2^1} dt < \infty$ and assume that $\partial_\cdot\phi_\cdot$ is in $H^{\alpha}(I;L^2(I) \otimes L^2(E))$ with $\alpha > \frac{1}{2}$. Then the series (1.2) converges in $L^2(E)$ to $D^*\phi$ + Trace $D\phi$.

Proof. The theorem follows from Theorem 2.2. Indeed, for every ξ in $L^2(E)$, the function $\tilde{\partial}(t,s) = \int_E \tilde{\partial}_t\phi_s(x)\xi(x)\mu(dx)$ is in $H^{\alpha}(I;L^2(I))$. Now by the same reasoning as in the proof of Theorem 4.5 the operator $\tilde{A}_\xi$ with the kernel $\tilde{\partial}(t,s)$ is a trace class operator; thus A_ξ has a finite trace.

Remark. For $0 < \alpha < 1$, the space $H^{\alpha}(I;B)$ has an equivalent characterization as the space of B-valued functions $k(t)$ for which

$$\int_I \|k(t)\|_B^2\, dt + \iint_{I^2} \frac{\|k(t) - k(s)\|_B^2}{|t - s|^{1+2\alpha}}\, dtds < \infty$$

(see e.g. [4] p. 52). Conditions similar to the above appear in [7].

Theorem 4.7. Let $\phi = \phi_t(x)$, $t \in I$, $x \in E$ be an L^2-continuous stochastic process satisfying the condition $\int_0^1 \|\phi_t\|^2_{W_2^1} dt < \infty$. Assume that for μ-almost every x, the functions $t \to \partial_t^+\phi_\cdot(x)$ and $t \to \partial_t^-\phi_\cdot(x)$ are in $H^\alpha(I;L^2(I))$ with $\alpha > 3/2$. Then the integral $\int_0^1 \phi_t \circ db_t$ exists and we have

$$\int_0^1 \phi_t \circ db_t = D^*\phi + \text{trace } D\phi.$$

Proof. Set $\Delta_i = (t_i, t_{i+1})$ and $\| \ \| = \| \ \|_{L^2(I)}$. From Proposition 2.1 we have

$$\phi_P = \frac{1}{2} \sum_{i=1}^{n-1} D^*(\phi_{t_i} + \phi_{t_{i+1}})\chi_{\Delta_i} + \frac{1}{2} \sum_{i=1}^{n-1} D(\phi_{t_i} + \phi_{t_{i+1}})[\chi_{\Delta_i}] \ . \tag{4.5}$$

The condition (4.4) with $\alpha > \frac{1}{2}$ implies that the $L^2(I)$-valued functions $t \to \partial_t^{\pm}\phi_\cdot(x)$ have absolutely continuous Fourier series. Thus $t \to \partial_t^-\phi_\cdot(x)$ and $t \to \partial_t^+\phi_\cdot(x)$ are continuous. It follows that the expressions $\sum \phi_{t_i}\chi_{\Delta_i}$ and $\sum \phi_{t_{i+1}}\chi_{\Delta_i}$ converge to ϕ in the norm of $W_2^1(E;L^2(I))$. By (3.9) the first term in (4.5) converges in $L^2(E)$ to $D^*\phi$.

The second term in (4.5) can be evaluated as follows. Note that

$$D\phi_{t_i}[\chi_{\Delta_i}] = \int_0^1 \partial_{s_i}^+ \phi_{t_i}\chi_{\Delta_i}(s)\, ds$$

$$D\phi_{i+1}[\chi_{\Delta_i}] = \int_0^1 \partial_s^- \phi_{t_{i+1}}\chi_{\Delta_i}(s)\, ds$$

Let as before $e_n(t) = \sqrt{2} \sin \pi nt$ and set

$$a_n^{\mp}(s,x) = (\partial_s^{\mp}\phi_\cdot(x), e_n)_{L^2(I)} \ .$$

Let D_+ and D_- denote the operators on $L^2(I)$ with kernels $\partial_t^+\phi_s(x)$ and $\partial_t^-\phi_s(x)$ respectively i.e.

$$D_{\pm}h(t) = \int_0^1 \partial_t^{\pm}\phi_s(x)h(s)\, ds.$$

We have

$$\Big|\sum_{i=0}^{n-1} D\phi_{t_i}(x)[\chi_{\Delta_i}] - \operatorname{trace} D_+\Big|$$

$$\leq \sum_n \|a_n^+(\cdot,x)\| \, \Big\|\sum_{i=0}^{n-1} e_n(t_i)\chi_{\Delta_i} - e_n\Big\|$$

$$\leq \sum_n \|a_n^+(\cdot,x)\|\sqrt{2\pi}\ \mathrm{mesh}(P)\cdot n \ .$$

But if $\alpha > 3/2$, then

$$\sum_n \|a_n^+(\cdot,x)\|\cdot n \leq \mathrm{const}\Big(\sum_n \|n^\alpha a_n^+(\cdot,x)\|^2\Big)^{\frac{1}{2}}$$

$$\leq \mathrm{const}\|\partial_\cdot\phi_\cdot(x)\|_{H^\alpha(I;L^2(I))} \ .$$

Thus,

$$\Big|\sum_i D\phi_{t_i}(x)[\chi_{\Delta_i}] - \operatorname{trace} D_+\Big| \longrightarrow 0$$

as $\mathrm{mesh}(P) \to 0$.

The term $\sum_i D\phi_{t_{i+1}}[\chi_{\Delta_i}]$ can be estimated in the same way with D_- in place of D_+. In this way we obtain

$$\lim \frac{1}{2}\sum_{i=0}^{n-1} D(\phi_{t_i} + \phi_{t_{i+1}})[\chi_{\Delta_i}] = \operatorname{trace}(D_+ + D_-) = \operatorname{trace}\tilde{D}\phi$$

which concludes the proof of the theorem.

Acknowledgement. We would like to thank M. Beśka for useful comments.

Note: After this paper was completed the papers [19] and [20] appeared which contain several of the results in this paper.

References

1. N. Dunford, J. Schwartz, Linear Operators, Part 1, Interscience, New York.

2. H. H. Kuo, A. Russek, White noise approach to stochastic integration, J. Multivariate Anal. 24 (1988), 218-236.

3. H. H. Kuo, Gaussian measures on Banach spaces, Lecture Notes in Mathematics 463 (1975), Springer, Berlin.

4. J. L. Lions, E. Magenes, Nonhomogeneous boundary value problems and applications, vol. 1, Springer, Berlin, 1972.

5. P. A. Mever, Quelqes resultats analytiques sur le semi-groupe d'Orenstein-Uhlenbeck en dimension infinie, Lecture Notes in Control and Inf. Sci., vol. 49 (1983).

6. S. Ogawa, Une remarque sur l'approximation de l'integrale stochastique du type noncausal par une suite des integrales de Stieltjes, Tohoku Math. Journal, 36 (1984), 41-48.

7. J. Rosinski, On stochastic integration by series of Wiener integrals. Applied Mathematics and Optimization, to appear.

8. I. Shigekawa, Derivatives of Wiener functionals and absolute continuity of induced measures, J. Math. Kyoto Univ., 20 (1980), 263-289.

9. H. Sugita, Sobolev spaces of Wiener functionals and Malliavin calculus, J. Math. Kyoto Univ., 25 (1985), 31-48.

10. K. Ito, Multiple Wiener integrals, J. Math. Soc. Japan, 3 (1951), 157-169.

11. S. Wantanabe, Lectures on stochastic differential equations and Malliavin calculus, Tata Institute for Fundamental Research, Springer Verlag, 1984.

12. I. Shigekava, De Rham-Hodge-Kodaira decomposition on an abstract Wiener space, J. Math. Kyoto Univ.

13. I. Kubo, S. Takenaka, Calculus on Gaussian white noise, Proc. Japan Acad., 56, Ser. A (1980), 376-380.

14. R. Cairoli, J. B. Walsh, Stochastic integrals in the plane, Acta Math., 134 (1975), 111-183.

15. M. Yor, Representation des martingeles de carre integrable relative aux processus de Wiener et de Poisson a n parametrtes, Z. Wahrscheinlichkeit. verw. Gebiete 35 (1976), 121-129.

16. A. V. Skorokhod, On a generalization of a stochastic integral, Theory Probab. Apl. 20 (1975), 219-233.

17. D. Nualart, M. Zakai, Generalized stochastic integrals and the Malliavin calculus, Probab. Th. Rel. Fields 73 (1986), 255-280.

18. B. Gaveau, P. Trauber, L'integrale stochastique comme operateur de divergence dans l'espace fonctionnel, J. Funct. Anal. 46 (1982), 230-238.

19. D. Nualart, Noncausal stochastic integral and calculus, Lecture Notes in Math. 1316 (1988), 80-129.

20. D. Nualart, E. Pardoux, Stochastic calculus with anticipating integrals, Prob. Th. Rel. Fields 78 (1988), 535-583.

Probability Theory on Vector Spaces IV
Lancut, June'87, Springer's LNM 1391

MASTER EQUATION FOR QUANTUM OSCILLATOR WITH INFINITELY DIVISIBLE NOISE

M. Rybaczuk
Institute of Material Science and Applied Mechanics
Technical University of Wrocław

K. Weron

Institute of Physics
Technical University of Wrocław

ABSTRACT

It is shown how infinitely divisible and, in particular, stable distributions arise in some problems of statistical physics related to quantum oscillators.

1. INTRODUCTION

In theoretical physics, because of importance in technical applications (e.g., microelectronics), it can be observed growing interest in the study of classical and quantum systems subjected to stochastic perturbations. In the considerations of a system coupled to an external statistically described environment, almost always, the stochastic perturbations are assumed to have the characteristics of a Gaussian white noise [1]-[3]. It is well known that the Gaussian distribution is not unique limiting distribution, there exists more general class of infinitely divisible distributions which are completely described as a limit of the normalized sums of independent summands. Therefore, it is some interest in investigating which features of the behaviour of a system subjected to Gaussian white noise survive in the theory where the noise is characterized by more general distribution.

2. QUANTUM OSCILLATOR IN THE POINT-SOURCES RANDOM FIELD

A typical physical problem is the investigation of scattering on a point center system (e.g., Brownian motion). Mathematically, such systems can be represented by random fields. For finite systems there are no mathematical difficulties in calculations their statistical characteristics. However, for infinite systems the limit random field exists only under special assumptions [5].

Let consider a one-dimensional quantum oscillator in the infinite system of scattering point centers. It can be described by the Lagrangian in the following form

$$L[x(t)] = \frac{p^2}{2m} - \frac{1}{2} m\omega^2 x^2 + xf(t) \ , \qquad (1)$$

where $[\frac{p^2}{2m} - \frac{1}{2} m\omega^2 x^2]$ denotes the Lagrangian for a harmonic oscillator in the standard notation, and $xf(t)$ - the external forcing term which depends on the characteristics of the external noise given by the scattering system. The time evolution of the density operator is given by the equation. cf. [7],

$$\langle x|\rho(t)|y\rangle \equiv \rho(x,y;t) = \int_{-\infty}^{\infty} dx'dy' J(x,y,t;x'y',0)\rho(x',y',0) \ , \qquad (2)$$

where

$$J(x,y,t;x',y',0) = \iint D[x]\, D[y] \exp[\frac{i}{h} S_f[x] - \frac{i}{h} S_f[y]] \ , \qquad (3)$$

is the influence functional, and

$$S_f[x] \equiv S_f(x,x') = \int_{x(0)=x'}^{x(t)=x} L[x(t)]dt$$ is the classical action.

We assume that the initially prepared state $\rho(x,y,0)$ is statistically independent of the noise and that the effect of the noise is see through an averaged influence functional $\langle J(x,y,t;x',y',0)\rangle_{f(t)}$. Hence, the time development of the averaged density operator is given by

$$\bar{\rho}(x,y,t) = \int_{-\infty}^{\infty} dx'dy' \langle J(x,y,t;x',y',0)\rangle_{f(t)} \rho(x',y',0) , \qquad (4)$$

where

$$\langle J(x,y,t;x',y',0)\rangle_{f(t)} = \iint D[x]D[y] \langle \exp[\frac{i}{\hbar} S_f[x] - \frac{i}{\hbar} S_f[y]]\rangle_{f(t)} \qquad (5)$$

and

$$\langle \exp[\frac{i}{\hbar} S_f[x] - \frac{i}{\hbar} S_f[y]]\rangle_{f(t)} = \langle \exp\{-\frac{i}{\hbar} \int_0^T (x-y)(t) f(t) dt\}\rangle_{f(t)}$$
$$\times \exp[\frac{i}{\hbar} S_o[x] - \frac{i}{\hbar} S_o[y]] , \qquad (6)$$

and $S_o[x]$ is the noise independent part of the action $S_f[x]$.

In order to perform the averaging in (6) let define the functional $F[z]$ by

$$F[z] = \exp\{-i \int_0^T z(t) f(t) dt\} , \qquad (7)$$

and assume that the force $f(t)$ consists of independent point impulses with magnitudes described by some probability measure. Therefore $f(t)$ has the following form

$$f(t) = \sum_i f_i \delta(t_i - t) = \frac{d}{dt} \xi(t) , \qquad (8)$$

where $\xi(t)$ is the suitable jump process.

From (7) and (8) we have

$$F[z] = \exp\{-i \int_0^T z(t) \frac{d}{dt} \xi(t)\} . \qquad (9)$$

<u>THEOREM</u>.

Let $\xi(t)$ be the homogeneous process with independent increments. The magnitude of the first jump $\Delta\xi$ is given by

$$\Delta\xi = \Delta\Xi - \bar{\xi} , \qquad (10)$$

where the probability measure $\Pi^A(d\xi)$ corresponds to $\Delta\Xi$ and $\bar{\xi}$ denotes the mean jump in a time interval Δt , i.e.,

$$\bar{\xi} = \Delta t \int_{-\infty}^{\infty} \Pi^A(d\eta) \; . \tag{11}$$

If the support of the measure $\Pi^A(d\xi)$ is a bounded subset A of $\mathbb{R}^1$ then the mean value of the functional $F[z]$ equals to

$$\langle F[z]\rangle_{f(t)} = \exp\{\int_0^T \int_{\mathbb{R}^1} [e^{-i\eta z(t)} - 1 + i\eta z(t)]\Pi^A(d\eta)dt\} \; . \tag{12}$$

P r o o f .

The expression (7) can be replaced by the limit of the integral sums

$$F[z]= \lim_{\max(t_j-t_{j-1})\to 0} \exp\{-i \sum_{j=1}^{N} z[\frac{t_{j-1}+t_j}{2}][\xi(t_j) - \xi(t_{j-1})]\} \; . \tag{13}$$

For the jump process with independent increments all jumps

$$\xi(t_i) - \xi(t_{j-1}) = \Delta_j\xi = \Delta_j\Xi - \bar{\xi}_j \; , \tag{14}$$

are independent random variables. Averaging over all time intervals $(t_j - t_{j-1})$ independently, we obtain

$$\exp\{-i \sum_{j=1}^{N} z[\frac{t_{j-1}+t_j}{2}][\xi(t_j) - \xi(t_{j-1})]\}$$

$$= \exp\{ \sum_{j=1}^{N} \int_{t_{j-1}}^{t_j} \int_{\mathbb{R}^1} [\exp[-i\eta z[\frac{t_{j-1}+t_j}{2}]]- 1]\Pi^A(d\eta)ds$$

$$\times \exp\{i \sum_{j=1}^{N} z[\frac{t_{j-1}+t_j}{2}][t_j - t_{j-1}] \int_{-\infty}^{\infty} \eta \; \Pi^A(d\eta)\} \; , \tag{15}$$

where we applied the Theorem 2 from [8] for the homogeneous stochastic process. Now taking the fimit $\lim_{\max(t_j-t_{j-1})\to 0}$ of (15) we get (12).

Note, that for the homogeneous jump process with independent increments the jump moments are Poisson distributed in time, but in our notation the characteristic constant of the Poisson distribution is included in the measure $\Pi^A(d\xi)$.

Taking the family of measures $\Pi^A(d\xi)$ weakly convergent to the measure $\Pi(d\xi)$ we obtain

$$\lim_{A} \exp\{\int_0^T \int_{\mathbb{R}^1} [e^{i\eta z(t)} - 1 + i\eta z(t)]\Pi^A(d\eta)dt$$

$$= \exp\{\int_0^T \int_{\mathbb{R}^1} [e^{-i\eta z(t)} - 1 + i\eta z(t)]\Pi(d\eta)dt\} . \qquad (16)$$

3. MASTER EQUATION FOR QUANTUM OSCILLATOR WITH INFINITELY DIVISIBLE NOISE

Applying our Theorem for $z(t) = \frac{1}{\hbar}(x - y)(t)$ we obtain the averaged influence functional (5) in the following form

$$\langle J(x,y,t;x',y',0)\rangle_{f(t)} = \iint D[x]D[y]$$

$$\times \exp\{\frac{i}{\hbar} S_o[x] - \frac{i}{\hbar} S_o[y] + \int_0^T \Phi[\frac{1}{\hbar}(x - y)(t)]dt\} , \qquad (17)$$

where

$$\Phi(z) = \int_{\mathbb{R}^1} (e^{-i\eta z} - 1 + i\eta z)\Pi(d\eta) . \qquad (18)$$

The differential form of the evolution equation (4) follows from (17). The standard calculations, [7], give the master equation for the averaged density operator $\bar{\rho}(t)$

$$\frac{\partial\bar{\rho}}{\partial t} = \frac{i\hbar}{2m}\left(\frac{\partial^2\bar{\rho}}{\partial x^2} - \frac{\partial^2\bar{\rho}}{\partial^2 y}\right) - \frac{i}{\hbar}\left(\frac{1}{2} m\omega^2 x^2 - \frac{1}{2} m\omega^2 y^2\right)\bar{\rho} + \Phi\left(\frac{1}{\hbar}(x - y)(t)\right)\bar{\rho}. \qquad (19)$$

Note, that for certain measures $\Pi(d\xi)$ the expression (18) can be identified as the logarithm of the characteristic function of a symmetric α-stable distribution [6]. Thus, we can write

$$\Phi(z) = \lambda|z|^\alpha , \qquad (20)$$

where $0 < \alpha \leq 2$ and $\lambda \in \mathbb{R}_+$.

The form of the averaged influence functional (17) with $\Phi(z)$ given by (20) was used in paper [9] considering the external noise effect on quantum oscillator. Unfortunately, the assumptions in [9] do not yield easily the form (18) of the noise dependent part of the averaged influence functional (17). However, this can be achieved rigorously by employing our Theorem.

REFERENCES

[1] A.O.Caldeira and A.J.Leggett, Physica 121A, 587 (1983).

[2] E.Ott, T.M.Ansonen, Jr., and J.D.Hanson, Phys. Rev. Lett. 53, 2187 (1984).

[3] T.Schneider, M.Zannetti, R.Badir, and H.R.Jauslin, Phys. Rev. Lett. 53, 2191 (1984).

[4] W.Feller, "An Introduction to Probability Theory and Its Applications" Vol. 2, J. Wiley, New York 1966.

[5] V.M.Zolatariev, "One-dimensional Stable Distributions", Nauka, Moscow 1983, (in Russian).

[6] R.G.Laha and V.K.Rohatgi, "Probability Theory", J. Wiley & Sons, New York 1979.

[7] R.P.Feynman and A.R.Hibbs, "Quantum Mechanics and Path Integrals", Mc Graw-Hill Book Company, New York 1965.

[8] J.J.Gichman and A.W.Skorochod, "Introduction to the Theory of Stochastic Processes", chapter 6.2, Nauka, Moscow 1977, (in Russian).

[9] S.K.Bose, K.Datta, and F.Feinsilver, Phys. Rev. A32, 3547 (1985).

Probability Theory on Vector Spaces IV
Lancut, June '87, Springer's LNM 1391

On the convergence of the series $\sum K_n(U^n-U^{-n})$

by Krzysztof Samotij

Let H be a Hilbert space and let U be a unitary operator on H. By B(H) we denote the class of all bounded linear operators from H to H. Recently J. Woś [W] showed that if operators K_n, n = 1,2,.., from B(H) commute with U and if the sequence (K_n) satisfies the Hörmander's condition:

$$\|K_n\| \le \frac{A_1}{n} \quad , \quad \sum_{m>\alpha n} \|K_{m-n}-K_m\| \le A_2 \ , \quad n=1,2,\ldots$$

with some constants A_1, A_2 and $\alpha\ge 1$ then the series $\sum_{n=1}^{\infty} K_n(U^n-U^{-n})$ is convergent in the strong operator topology. In the present paper we will show that the commutativity assumption in that result can not be deleted except for the trivial case when 1 is not a cluster point of the spectrum of U.

First we introduce some notation. Let

$$(1) \qquad \tilde{D}_n(e^{it}) = -i \sum_{k=1}^{n} (e^{ikt}-e^{-ikt}) =$$

$$= \sin nt + \operatorname{ctg}\frac{t}{2}\,(1-\cos nt) \ , \ n=1,2,\ldots \ , \ \tilde{D}_0(e^{it}) \equiv 0.$$

Analogously, let $\tilde{D}_n(U) = -i \sum_{k=1}^{n} (U^k-U^{-k})$.

If E(dt) is the spectral measure from the representation

$$U = \int_0^{2\pi} e^{it}\, E(dt)$$

then we have $\tilde{D}_n(U) = \int_0^{2\pi} \tilde{D}_n(e^{it})\, E(dt)$.

We will consider first the case when 1 is not a cluster point of the spectrum of U. To this end we will need the following simple lemma.

Lemma 1. *If 1 is not a cluster point of the spectrum of U then the sequence* $(\tilde{D}_n(U))$ *is bounded in the operator norm.*

Proof. There is an ε, $0<\varepsilon<\pi$, such that for any $x \in H$ we have

$$|\tilde{D}_n(U)x|^2 = \int_{\varepsilon}^{2\pi-\varepsilon} |\tilde{D}_n(e^{it})|^2 \, (E(dt)x,x)$$

$$\leq \sup_{\varepsilon\leq t\leq 2\pi-\varepsilon} |D_n(e^{it})|^2 |x|^2 \leq (2|\mathrm{ctg}\,\tfrac{\varepsilon}{2}| + 1)^2 \, |x|^2.$$

Now we will show that in the considered case very little is needed to assure the convergence of the series $\sum K_n(U^n-U^{-n})$ and $\sum(U^n-U^{-n})K_n$.

Proposition. *If* (K_n) *is a sequence of elements of* B(H) *such that* $\lim_n \|K_n\| = 0$ *and* $\sum\|K_{n+1}-K_n\|<+\infty$, *and if 1 is not a cluster point of the spectrum of* U *then the both series* $\sum K_n(U^n-U^{-n})$ *and* $\sum (U^n-U^{-n})K_n$ *are convergent in the operator norm.*

Proof. Using the summation by parts we have, with arbitrary N_1 and N_2, $1\leq N_1\leq N_2$,

$$\|\sum_{k=N_1}^{N_2} K_k(U^k-U^{-k})\| = \|K_{N_2}\tilde{D}_{N_2}(U) - K_{N_1}\tilde{D}_{N_1-1}(U)$$

$$+ \sum_{k=N_1}^{N_2-1} (K_k-K_{k+1})\tilde{D}_k(U)\|$$

$$\leq \|K_{N_2}\| \, \|\tilde{D}_{N_2}(U)\| + \|K_{N_1}\| \|\tilde{D}_{N_1-1}(U)\| + \sum_{k=N_1}^{N_2-1} \|K_k-K_{k+1}\| \, \|\tilde{D}_k(U)\|.$$

By Lemma 1 and the assumptions of this proposition, each summand in the last sum converges to 0 when N_1 tends to infinity. Therefore the series $\sum K_n(U^n-U^{-n})$ converges in the operator norm. The case of the other series can be treated in a similar way.

Definition. *We say that a sequence* (x_n) *of elements of a normed space* $(X, \|\cdot\|)$ *satisfies the condition (K) if*

$$\|x_n\|\leq\frac{1}{n} \quad \text{and} \quad \|x_n - x_{n+1}\| \leq \frac{1}{n(n+1)}, \qquad n=1,2,\ldots$$

We will show that, except for the trivial case considered above, when the commutativity assumption is omited in the Woś's theorem nothing of his assertion remains true even if the Hörmander's condition is replaced by the, much stronger, condition (K).

Theorem. *Assume that* 1 *is a cluster point of the spectrum of* U . *Then there exists*

a) a sequence $(h_n^*)\subset H$ *satisfying* (K) *and an element* $h\in H$ *such that*

$$\lim_N \left| \sum_{n=1}^{N} (h_n^*,(U^n-U^{-n})h)\right| = +\infty ,$$

b) a sequence $(h_n)\subset H$ *satisfying* (K) *such that*

$$\lim_N \left| \sum_{n=1}^{N} (U^n-U^{-n})h_n\right| = +\infty ,$$

c) a sequence $(K_n')\subset B(H)$ *satisfying* (K) *and an element* $h'\in H$ *with*

$$\lim_N \left| \sum_{n=1}^{N} K_n'(U^n-U^{-n})h'\right| = +\infty ,$$

d) a sequence $(K_n'')\subset B(H)$ *satisfying* (K) *and an element* $h''\in H$ *such that*

$$\lim_N \left| \sum_{n=1}^{N} (U^n-U^{-n})K_n''h''\right| = +\infty .$$

To prove this theorem we need two more lemmas.

Lemma 2.*There are constants* C_1 *and* C_2 *such that for each non-negative integer* N *and each real* t *we have* $|\tilde{D}_N(e^{it})|\le C_1N$ *and*

$$\sup_{k\ge N} \left| \sum_{n=N}^{N'} \frac{1}{n(n+1)} \tilde{D}_n(e^{it})\right| \le C_2.$$

Also, there is a positive constant C_3 *and a non-negative integer* N_1 *such that the inequality*

$$\left| \sum_{n=N}^{N'} \frac{1}{n(n+1)} \tilde{D}_n(e^{it})\right| \ge C_3$$

holds for all $t\in[\frac{1}{N+1},\frac{1}{N})$ *if only* $N\ge N_1$ *and* N' *is sufficiently large.*

Proof. Follows from the formula (1) by simple calculations.

Lemma 3. *If* U *is as in the assumtion of Theorem then there exists a sequence* $(\delta_n^*)\subset H$ *with* $|\delta_n^*|\le\frac{1}{n(n+1)}$, $n=1,2,\ldots$, *and an element*

$g \in H$ *such that*

$$\lim_N \left| \sum_{n=1}^{N} (\delta_n^*, \tilde{D}_n(U)g) \right| = +\infty .$$

Proof. Without any loss of generality we can and do assume that $1 \in \overline{\sigma(U) \cap \{z: \operatorname{Im} z > 0\}}$. Otherwise we can replace U with U^{-1}. Let $\Delta_n = [\frac{1}{n+1}, \frac{1}{n})$, n=1,2,... Let N_1 be the constant from Lemma 2. Let (n_k) be an increasing sequence of positive integers chosen so that: $n_1 \geq N_1$, $E(\Delta_{n_k}) \neq \{0\}$, and

$$(2) \quad \left| \sum_{n=n_k}^{n_{k+1}-1} \frac{1}{n(n+1)} \tilde{D}_n(e^{it}) \right| \geq C_3, \qquad t \in \Delta_{n_k}$$

Such a choice is made possible by Lemma 2. Let (x_k) be a sequence of non-zero vectors of H chosen so that $x_k \in E(\Delta_{n_k})$, k=1,2,..., $\sum |x_k|^2$ +∞ and $\sum |x_k| = +\infty$. Observe that, by (2) we have

$$(3) \quad \left| \sum_{n=n_k}^{n_{k+1}-1} \frac{1}{n(n+1)} \tilde{D}_n(U)\, x_k \right|^2$$

$$= \int_{\Delta_{n_k}} \left| \sum_{n=n_k}^{n_{k+1}-1} \frac{1}{n(n+1)} \tilde{D}_n(e^{it}) \right|^2 (E(dt)x_k, x_k)$$

$$\geq C_3^2 \int_{\Delta_{n_k}} (E(dt)x_k, x_k) = C_3^2 \, |x_k|^2 .$$

A similar calculation and Lemma 2 give also

$$(4) \quad \left| \sum_{n=n_k}^{N} \frac{1}{n(n+1)} \tilde{D}_n(U)x_k \right|^2 \leq C_2^2 \, |x_k|^2 ,$$

for any $N \geq n_k$.

Let $g = \sum_{k=1}^{\infty} x_k$. The series is norm convergent since $\sum_{k=1}^{\infty} |x_k|^2 < +\infty$ and x_k's are mutually orthogonal. Let us denote

$$v_k = \sum_{n=n_k}^{n_{k+1}-1} \frac{1}{n(n+1)} \tilde{D}_n(U)x_k , \quad k=1,2,...$$

Let $\delta_n^*=0$ if $n<n_1$, and $\delta_n^* = \frac{1}{n(n+1)} v_k/|v_k|$ if $n_k \leq n < n_{k+1}$, k=1,2,...

Observe that for $j \geq 2$ we have, by (3),

$$\sum_{n=1}^{n_j-1} (\delta_n^*, \tilde{D}_n(U)g) = \sum_{k=1}^{j-1} \sum_{n=n_k}^{n_{k+1}-1} \frac{1}{n(n+1)} (v_k/|v_k|, \tilde{D}_n(U)x_k)$$

$$= \sum_{k=1}^{j-1} |v_k| \geq C_3 \sum_{k=1}^{j-1} |x_k|.$$

On the other hand if $n_j < N \leq n_{j+1}$ we have

$$\sum_{n=1}^{N} (\delta_n^*, \tilde{D}_n(U)g) = \left(\sum_{n=1}^{n_j-1} + \sum_{n=n_j}^{N} \right)(\delta_n^*, \tilde{D}_n(U)g).$$

But

$$\left|\sum_{n=n_j}^{N} (\delta_n^*, \tilde{D}_n(U)g)\right| = \left|(v_j/|v_j|, \sum_{n=n_j}^{N} \frac{1}{n(n+1)} \tilde{D}_n(U)x_j)\right|$$

$$\leq \left|\sum_{n=n_k}^{N} \frac{1}{n(n+1)} \tilde{D}_n(U)x_j\right| \leq C_2|x_j|,$$

where the last inequality follows by (4). Hence

$$\lim_N \left|\sum_{n=1}^{N} (\delta_n^*, \tilde{D}_n(U)g)\right| = +\infty.$$

Proof of Theorem. To prove *a)* let us put $h_n^* = \sum_{j=n} \delta_j^*$, $n=1,2,\ldots,$ and $h=g$, where (δ_j^*) and g are the same as in Lemma 3. Clearly, the sequence (h_n^*) satisfies (K). On the other hand

$$-i \sum_{n=1}^{N} (h_n^*, (U^n - U^{-n})h) = \sum_{n=1}^{N-1} (\delta_n^*, \tilde{D}_n(U)h) + (h_N^*, \tilde{D}_n(U)h).$$

But

$$|(h_N^*, D_N(U)h)|^2 \leq |h_N^*|^2 |\tilde{D}_N(U)h|^2$$

$$\leq (\textstyle\sum_{j=n} |\delta_N^*|)^2 \int_0^{2\pi} |\tilde{D}_N(e^{it})|^2 (E(dt)h,h)$$

$$\leq N^{-2} C_1^2 N^2 |h|^2 = C_1^2 |h|^2,$$

and *a)* follows by Lemma 3.

b) follows directly from *a)* with $h_n = h_n^*$.

To see that *a)* implies *c)* it is enough to take $h' = h$ and to define, for eachelement $f \in H$, $K_n'(f) = (h_n^*, f)e$, where e isan arbitrary fixed element of H of norm one.

Finally, we will derive *d)* from Lemma 3. Lemma 3 implies the existence of a sequence (G_n) of elements of $B(H)$ with $\|G_n\| \leq \frac{1}{n(n+1)}$ and

$$\lim_N \|\sum_{n=1}^{N} G_n \tilde{D}_n(U)\| = +\infty.$$

Hence

$$\lim_N \|\sum_{n=1}^{N} \tilde{D}_n(U)^* G_n^*\| = +\infty.$$

By the uniform boundedness theorem there isan $h'' \in H$ such that

$$\overline{\lim_N} |\sum_{n=1}^{N} \tilde{D}_n(U)^* G_n^* h''| = +\infty.$$

But $\tilde{D}_n(U)^* = \tilde{D}_n(U)$, so

$$\overline{\lim_N} |\sum_{n=1}^{N} \tilde{D}_n(U) G_n^* h''| = +\infty.$$

It is not hard to see that there is a sequence (ε_n), where $\varepsilon_n = 1$ or -1, $n=1,2,\ldots$, such that

$$\lim_N |\sum_{n=1}^{N} \varepsilon_n \tilde{D}_n(U) G_n^* h''| = +\infty.$$

Now, *d)* follows with $K_n'' = \sum_{j=n}^{\infty} \varepsilon_n G_n^*$ by the summation by parts.

Reference.

[W] Woś J., Singular integrals and second order stationary sequences, This Proceedings.

Institute of Mathematics
Technical University
Wybrzeze Wyspianskiego 27
50-370 Wroclaw
Poland

Probability Theory on Vector Spaces IV
Lancut, June'87, Springer's LNM 1391

ON ADMISSIBLE TRANSLATES OF SUBGAUSSIAN STABLE MEASURES

WŁODZIMIERZ SMOLEŃSKI AND RAFAŁ SZTENCEL

Let E be a separable Banach space and let $0 < p < 2$. The aim of this short note is to prove

THEOREM 1. *There exists a symmetric p-stable measure μ on E such that*
(i) the set A_μ of admissible translates of μ is dense in E;
(ii) the (symmetric) spectral measure σ of μ (on the unit sphere) does not charge finite-dimensional sets.

The theorem above follows immediately from Propositions 2 and 3 below. It extends an example of Żak [4] which in turn was a negative answer to a question of Zinn [3] who asked whether for a p-stable measure μ (ii) implied that $A_\mu = \{0\}$. We recall that $A_\mu = \{x \in E: \mu * \delta_x \ll \mu\}$.

We refer to Linde [2] for definitions and properties of stable measures which are used here without explanation.

In order to state our next result we need the following

DEFINITION. *A probability measure λ on the unit sphere S of E is said to be the radial projection of a probability measure ρ on $E \setminus \{0\}$ if, for $A \in \mathcal{B}(S)$*

$$\lambda(A) = \rho(\mathrm{cone}(A)),$$

where

$$\mathrm{cone}(A) = \{x \in E : \frac{x}{\|x\|} \in A\}.$$

Note that if ρ does not charge finite-dimensional sets, then neither does λ.

Let now γ be the distribution of an E-valued Gaussian random element X, $X \not\equiv 0$. Let ν be the distribution of a positive, stable random variable Y, independent of X, with the Laplace transform

$$\mathrm{E}\exp(-sY) = \exp(-s^{p/2}),\ s > 0.$$

It is known that the law μ of $Y^{1/2}X$ is a symmetric p-stable distribution on E (cf. [1], ch. VI, § 2).

PROPOSITION 2. *The (symmetric) spectral measure σ of μ is equivalent to the radial projection of γ (i.e. they are mutually absolutely continuous).*

PROOF. It is (more than) enough to show that for $A \in \mathcal{B}(S)$

$$(*) \qquad \sigma(A) = \frac{\sqrt{\pi}}{2^p\Gamma(\frac{p+1}{2})} \int_{\mathrm{cone}(A)} \|x\|^p \gamma(dx).$$

To get $(*)$ it is enough to write the characteristic functional $\hat{\mu}$ of μ in a proper way. Indeed, for $f \in E'$

$$\hat{\mu}(f) = \exp(i\langle f, Y^{1/2}X\rangle) = \exp(i\langle Y^{1/2}f, X\rangle)$$
$$= \mathrm{E}\hat{\gamma}(Y^{1/2}f) = \int_{R_+} \exp\left(-\frac{t}{2}\int_E \langle f, x\rangle^2 \gamma(dx)\right) \nu(dt).$$

But the Laplace transform of ν is $\exp(-s^{p/2})$, therefore

$$\hat{\mu}(f) = \exp\left(-\left(\frac{1}{2}\int_E \langle f, x\rangle^2 \gamma(dx)\right)^{p/2}\right) = \exp\left(-\frac{\sqrt{\pi}.}{2^p\Gamma(\frac{p+1}{2})}\int_E |\langle f, x\rangle|^p \gamma(dx)\right)$$
$$= \exp\left(-\frac{\sqrt{\pi}}{2^p\Gamma(\frac{p+1}{2})}\int_E |\langle f, \frac{x}{\|x\|}\rangle|^p \|x\|^p \gamma(dx)\right).$$

The desired formula $(*)$ follows immediately.

PROPOSITION 3. $A_\mu = A_\gamma$.

PROOF. Let $x \in A_\gamma$. Then $t^{-1}x \in A_\gamma$ for every positive real t. Take $A \in \mathcal{B}(E)$ such that $\mu(A) = 0$. Since $\mu(A) = \int_{R_+} \gamma(t^{-1}A)\nu(dt)$, it follows that $\gamma(t^{-1}A) = 0$ for ν-almost all t. Consequently $\gamma(t^{-1}(A + x)) = 0$ for ν-a.a. t. Hence $\mu(A + x) = 0$. This shows that $x \in A_\mu$.

Conversely, let $x \notin A_\gamma$. Then there exists a linear subspace $L \subset E$ such that $\gamma(L) = 1$ and $x \notin L$. Of course $\mu(L) = 1$. Thus $x \notin A_\mu$.

REFERENCES

[1] W. Feller, *An introduction to probability theory and its applications*, vol. II, 2nd ed., Wiley, New York 1971.

[2] W. Linde, *Probability on Banach spaces — infinitely divisible and stable measures* , Wiley, New York 1986.

[3] J. Zinn, *Admissible translates of stable measures*, Studia Math. **54** (1976), 245–257.

[4] T. Żak, *Admissible translates for subgaussian measures*, to appear.

INSTITUTE OF MATHEMATICS, TECHNICAL UNIVERSITY OF WARSAW, PL. JEDNOŚCI ROBOTNICZEJ 1, 00 661 WARSAW, POLAND

INSTITUTE OF MATHEMATICS, UNIVERSITY OF WARSAW, PKIN, 00 901 WARSAW, POLAND

Probability Theory on Vector Spaces IV
Lancut, June'87, Springer's LNM 1391

AN L^1 EXTENTION OF STOCHASTIC DYNAMICS FOR IRREVERSIBLE SYSTEMS

Zdzisław Suchanecki

Institute of Mathematics, Technical University of Wrocław
50-370 Wrocław, Poland

ABSTRACT. This is a continuation of our earlier work [7] on a martingale approach to the theory of intrinsically random dynamical systems.

1. INTRODUCTION.

Consider a dynamical system $(\Gamma, A, \mu, \{S_t\})$ where (Γ, A, μ) is a probability space and $\{S_t\}$, $t \in \mathbb{R}$, is a group of one-to-one measure preserving transformations of Γ. For example, in classical mechanics, such a system describes dynamical evolution of a finite number of interacting particles moving in some finite volume. Precisely, Γ is here an n-dimensional manifold, μ is an invariant measure on Γ and the evolution $\{S_t\}$ is described by the Hamilton equations ($\{S_t\}$ are invariant by Liouville's theorem). In statistical mechanics, one does not consider a given system but, so called, "statistical ensembles" which are represented by density functions on the space Γ. Therefore, for a given density function ρ its time evolution $\{\rho_t\}$ proceedes according to the formula

$$\rho_t(\omega) = (S_{-t}\omega) , \quad \omega \in \Gamma .$$

However, the above equation does not describe the thermodynamical evolution. In thermodynamics the evolution goes in the direction of increasing entropy and the system tends to equilibrium. Since the transformations S_t preserve the measure μ, it is obvious that the entropy functional

$$\Omega(\rho_t) = -\int_\Gamma \rho_t \ln\rho_t d\mu \tag{1}$$

does not depend on time. But it is known that entropy can increase even for isolated systems.

There are several ways to explain the connection between, dynamical (reversible) evolution and thermodynamical (irreversible) evolution. The most popular view is that irreversibility is a result of a "coarse--grained" reduction of the deterministic dynamics. Recently in a series of papers, written by I. Prigogine, B. Misra, M. Courbage and others (cf. [2], [4], [5]), was proposed another approach to dynamical systems. Note first, that the classical dynamics can be described by the family $\{U_t\}$ of linear operators defined by

$$(U_t\rho)(\omega) = \rho(S_{-t}\omega) , \quad \omega \in \Gamma . \tag{2}$$

If we restrict ourself to square integrable functions, $\{U_t\}$ is an unitary group on $L^2(\Gamma,\mu)$. The main feature of Prigogine et al. theory is that for the given unitary group $\{U_t\}$ it is possible to find the nonunitary similarity transformation Λ converting it into a contraction semigroup of a Markov process. Similarity means that Λ must be a nonnegative and invertible bounded linear operator (cf. Th.1 for details). The construction of Λ was first done for a special dynamical system - baker transformation [5] and then for Bernoulli systems [2] and finally for K-systems [4]. Since this construction uses Hilbert space methods, it is assumed that density functions belong to $L^2(\Gamma,\mu)$.

In paper [7] we proposed a different approach to this problem by using a stochastic integral technique. This gives a simple and general construction, and a possibility of further extensions. One of such extensions is presented in this paper. Under the assumption that considered dynamical system is K-system we construct Λ on the space $L^1(\Gamma,\mu)$ in such a way that for densities with finite entropy (1) the reversible evolution $\{\rho_t\}$ is transformed into the evolution $\{\tilde{\rho}_t\}$ with increasing entropy.

2. STOCHASTIC DYNAMICS.

Let $(\Omega, A, \mu, \{S_t\}_{t\in \mathbb{R}})$ be a dynamical system. In the sequel we assume that it is also K-flow. This means that there exists an increasing and right continuous family $\{A_t\}$ of sub-σ-fields of A satisfying:

(i) $S_t(A_t) = A_{s+t}$

(ii) $\bigcup_{-\infty}^{\infty} A_t$ generates the σ-field A

(iii) $\bigcap_{-\infty}^{\infty} A_t = \{\emptyset, \Gamma\} = A_{-\infty}$.

Let $\{M_t\}$ be an operator valued martingale on our dynamical system , i.e. a family of bounded linear operators on $L^1(\Gamma,\mu)$ such that for each $\rho \in L^1(\Gamma,\mu)$ $\{M_t\rho\}$ is a real right continuous martingale with respect to $\{A_t\}$ which is uniformly integrable. Let $\{\{\bar{M}_t\}$ denotes the left "transposed" martingale to $\{M_t\}$, i.e. $M_t\bar{M}_t = E^{A_t}$. Observe that

$$(M_{s_2} - M_{s_1})(\bar{M}_{t_2} - \bar{M}_{t_1}) = 0 \quad \text{for} \quad s_1 \le s_2 < t_1 \le t_2 \, . \qquad (3)$$

Assume that M_∞ is a positive one-to-one operator satisfying

$$M_\infty 1 = 1 \, , \quad M_\infty U_t = U_t M_\infty \quad \text{for each} \quad t \in \mathbb{R}, \qquad (4)$$

where U_t is given by (2).

For example, the conditional expectations

$$E_t = E^{A_t}, \quad t \in \mathbb{R} \tag{5}$$

form an operator valued martingale satisfying the above assumptions.

We use H^1 to denote the space of all martingales $\{m_t\}$ for which $\|\sup_t |m_t|\|_{L^1} < \infty$. Identyfying a uniformly integrable martingale with its limit at infinity we have $H^1 \subset L^1$.

In the papers [7] and [8] we wiedely exploited the concept of stochastic integral $\int f(t)dM_t$ of a deterministic function with respect to the operator valued martingale on $L^2(\Gamma,\mu)$. We will also use this technics in this paper. However, for L^1-martingales we can not define the integral in the same way as in [7]. The theory of integrals $\int f(t)dx_t$ where $\{x_t\}$ is a p-integrable stochastic process, $0 \le p < \infty$, can be found in [1] (cf. also [3]). For f - a simple function (or a simple process) $\int f(t)dx_t$ is defined as usual. Then it is said that x_t is an L^p-integrator, $0 \le p < \infty$, if the linear operator $f \to \int f dx_t$ from the space of simple Borel measurable functions into $L^p(\Gamma,\mu)$ has an extension satisfying the Dominated Convergence Theorem. Since every right continuous martingale is L^0 integrator (cf. [1]), the integral $\int f(t)dM_t\rho$ can be reasonably defined for every $\rho \in L^1(\Gamma,\mu)$. Let f be a Borel measurable function on $\mathbb{R}$. Denote by D_f the set of all $\rho \in L^1(\Gamma,\mu)$ for which $\int f(t)dM_t\rho$ exists. In other words, $\rho \in D_f$ if and only if there exists a sequence $\{f_n\}$ of simple Borel measurable functions on $\mathbb{R}$ such that

$$\begin{aligned} &1^\circ \quad f_n \to f \ \text{a.s.} \\ &2^\circ \quad \int f_n dM\rho \ \text{converges in } L^0(\Gamma,\mu)\,. \end{aligned} \tag{6}$$

Now, we can formulate the main results of this paper.

<u>THEOREM 1</u>. For any decreasing positive and differentiable function f

on $\mathbb{R}$ with properties : $f(-\infty) = 1$, $f(+\infty) = 0$ the linear operator Λ on $L^1(\Gamma,\mu)$ defined by the formula

$$\Lambda\rho = \int f dM\rho + M_{-\infty}\rho$$

satisfies:

(i) Λ is positive, i.e. $\Lambda\rho \geq 0$ if $\rho \geq 0$

(ii) $\Lambda 1 = 1$

(iii) Λ restricted to H^1 is bounded with densely defined right-inverse Λ^{-1} .

Proof. Let $\rho \in L^1(\Gamma,\mu)$ be given. Since f is bounded, the integral $\int f(s)dM_s\rho$ exists as an element of $L^0(\Gamma,\mu)$ (cf. [1]). Moreover, approximating f by step functions it is easy to check that

$$\Lambda\rho = -\int_{-\infty}^{\infty} M_s\rho df(s) \qquad \mu\text{-a.e.} \tag{7}$$

Because M_∞ is positive, so is Λ. This proves (i). Also (ii) is obvious because of (4).

(iii) If $\{M_t\rho\}$ belongs to H^1 then (cf. [1], Th. 7.2) $\{M_t\rho\}$ is L^1-integrator and we get the inequality

$$\|\Lambda\rho\|_{L^1} \leq \|M_{-\infty}\rho\|_{L^1} . \tag{8}$$

Finally, we shall construct the inverse Λ^{-1} . Consider the positive increasing function $g = 1/f$ and its domain D_g . It is easy to show that D_g is a linear space. Let ρ be a simple function measurable with respect to $\bigcup_{-\infty}^{\infty} A_t$. We show that $\rho \in D_g$. Note that there exists t_o such that ρ is A_{t_o} measurable. By the orthogonality of increments of $\{M_t\}$, we have

$$(M_t - M_{t_o})\rho = (M_t - M_{t_o})[(M_{t_o} - M_{-\infty})M_\infty^{-1}\rho + \int_\Gamma \rho d\mu] = 0 .$$

Therefore, the integrability of g with respect to $\{M_t\rho\}$ is equivalent to the integrability of $g1_{(-\infty,t_o]}$. Since the last function is bounded, $\rho \in D_g$. Put

$$\Lambda^{-1}\rho = \int 1/f(s)d\bar{M}_s\rho + \bar{M}_{-\infty}\rho \quad .$$

Using the property

$$\int f dM\rho \int g d\bar{M}\rho = \int f g dE\rho$$

(cf. [7]) we obtain that $\Lambda\Lambda^{-1}\rho = \rho$ on a dense subspace of $L^1(\Gamma,\mu)$.

□

Now, we can use just defined Λ as a similarity transformation converting $\{U_t\}$ into a Markov semigroup. Putting

$$W_t = \Lambda U_t \Lambda^{-1} \tag{9}$$

we obtain a family of linear operators densely defined on $L^1(\Gamma,\mu)$.

<u>THEOREM 2</u>. Let f be as in Theorem 1 and additionally assume that $\ln f$ is concave. Then

$$W_t = (\int f(s)/f(s-t)dE_s)U_t + E_{-\infty} \tag{10}$$

and $\{W_t\}_{t\geq 0}$ restricted to H^1 forms a Markov semigroup such that

(i) $W_t 1 = 1$, $t \geq 0$,

(ii) $\|W_t\rho - 1\|_{L^1} \to 0$ as $t \to +\infty$.

<u>Proof</u>. The crucial point in the proof of (10) is the following commutation relation

$$U_t(\int f(s)dM_s) = (\int f(s-t)dM_s)U_t \tag{11}$$

(cf. [7] Prop. 2) which remains true for uniformly integrable martingales $\{M_t\rho\}$. Since $\ln f$ is concave, the ratio $f(s)/f(s-t)$ is for each $t > 0$ decreasing function of s . Therefore applying Theorem 1 to the operator $\int f(s)/f(s-t)dE_s$ we get that W_t are positive. Since $E_{-\infty}\rho = \int_\Gamma \rho d\mu$ and $E_\infty \rho = \rho$ for every $\rho \in L^1(\Gamma,\mu)$, we obtain that $W_t 1 = 1$ and $\|W_t\rho\|_{L^1} = \|\rho\|_{L^1}$ for $\rho \geq 0$. Then W_t is a contraction for every $t \geq 0$. Finally, because

$$\|W_t\rho - 1\|_{L^1} = \| \int f(s)/f(s-t)dE_s U_t\rho\|_{L^1} = \|U_t \int f(s+t)/f(s)dE_s\rho\|_{L^1}$$
$$\leq \|U_t\|\cdot\| \int f(s+t)/f(s)dE_s\rho\|_{L^1}$$

and $f(s+t)$ tends monotonically to 0 as $t \to +\infty$ we get (ii). □

The next theorem crowns our effort to convert the reversible dynamics into irreversible. We will give a sketch of the proof of Theorem 3. More details one can found in [8], where L^2-version of this theorem was considered.

<u>THEOREM 3</u>. If the measure μ is non-degenerated then for every probability density ρ with $\Omega(\rho) < \infty$, the Markov evolution $\tilde{\rho}_t = W_t\rho$, $t \geq 0$, has strictly increasing entropy.

<u>Proof</u>. Since the function $\rho \ln \rho$ is integrable, the martingale $\{E_t\rho\}$ belongs to H^1 (cf. [3], p. 261). Hence $\tilde{\rho}_t = W_t\rho \in L^1(\Gamma,\mu)$ by Theorem 2. Denote

$$\varphi(x) = x\ln x \ , \quad x > 0 \ .$$

First we show the inequality

$$\int_\Gamma \varphi(W_t\rho)d\mu \leq \int_\Gamma \varphi(\rho)d\mu \ . \tag{12}$$

Indeed, let ρ be a simple function, $\rho = \sum_1^n a_i 1_{A_i}$ such that $\bigcup_1^n A_i = \Gamma$ and $A_i \in \bigcup_t A_t$. Then $W_t\rho \in L^1(\Gamma,\mu)$ and

$$\varphi(W_t\rho) = \varphi(\sum_1^n a_i W_t 1_{A_i}) \leq \sum_1^n \varphi(a_i) W_t 1_{A_i} = W_t\varphi(\rho)$$

because φ is convex. Hence

$$\int_\Gamma \varphi(W_t\rho)d\mu \le \int_\Gamma W_t\varphi(\rho)d\mu = \int_\Gamma \varphi(\rho)d\mu .$$

For a non-simple ρ it is enough to choose a sequence $\rho_n \to \rho$ in L^1 and such that $\varphi(\rho_n) \to \varphi(\rho)$ in L^1 and apply Theorem 2. Now putting $W_{t_2-t_1}$ instead of W_t and $W_{t_1}\rho$ instead of ρ in (12) we have for $t_1 < t_2$

$$\Omega(\tilde{\rho}_{t_1}) \le \Omega(\tilde{\rho}_{t_2}) .$$

Since μ is non-degenerate and φ is strictly convex the above inequality is strict. □

REFERENCES

[1] K. Bichteler, Stochastic integration and L^p theory of semimartingales, Ann. Prob. 9, (1981), 49-89.

[2] M. Courbage, B. Misra, On the equivalence between Bernoulli dynamical systems and stochastic Markov processes, Physica 104A, (1980), 359-377.

[3] C. Dellacherie, P.A. Mayer, Probabilities and Potential B, North--Holland 1982.

[4] S. Goldstein, B. Misra, M. Courbage, On intrinsic randomness of dynamical systems, J. Stat. Phys. 25 (1981), 111-139.

[5] B. Misra, I. Prigogine, M. Courbage, From deterministic dynamics to probabilistic descriptions, Physica 98A (1979), 1-26.

[6] M. Rybaczuk, K. Weron, Z. Suchanecki, A. Weron, Kinetic equation in the Prigogine theory of irreversibility, submitted for publicatio

[7] Z. Suchanecki, A. Weron, Applications of an operator stochastic integral in Prigogine's theory of irreversible dynamical systems, Exp. Math. (to appear).

[8] Z. Suchanecki, A. Weron, M. Rybaczuk, K. Weron, An explicit approach to the Λ operator and the H-theorem in Prigogine's theory of irreversibility, Physica A (to appear).

Probability Theory on Vector Spaces IV
Lancut, June'87, Springer's LNM 1391

VECTOR VALUED FOURIER SERIES AND SAMPLE CONTINUITY OF RANDOM PROCESSES

BY

V.I. TARIELADZE

Academy of Sciences of Georgian SSR

Muskhelishvili Institute of Computer Mathematics

Tbilisi 380093 USSR

1. INTRODUCTION

The following two theorems are well-known.

<u>Theorem 1</u>. (A.Kolmogorov). Let $\xi(t)$, $t \in [0,2\pi]$ be a second order random process and suppose that the inequality

$$(E|\xi(t_1) - \xi(t_2)|^2)^{1/2} \leq |t_1-t_2|^\alpha \ , \quad t_1,t_2 \in [0,2\pi] \qquad (1)$$

holds for some $\alpha > \frac{1}{2}$. Then ξ has a sample continuous modification.

<u>Theorem 2</u>. (S.Bernstein). Let f be a 2π-periodic continuous scalar function on $\mathbb{R}$ and suppose that the inequality

$$|f(t_1) - f(t_2)| \leq |t_1 - t_2|^\alpha \ , \quad t_1,t_2 \in [0,2\pi]$$

holds for some $\alpha > \frac{1}{2}$, then f possesses the absolutely convergent Fourier series.

There are different generalizations and methods of proof of the Kolmogorov's theorem (cf. [4], [5], [7]). We show that the Bernstein's theorem can be easily generalized for Hilbert space valued functions and that this generalization implies the Kolmogorov's theorem. Our method gives also the following refinement: if a 2π-periodic ξ satisfies (1) for some $\alpha > \frac{1}{2}$, then ξ has a modification, all trajectories of which possesss absolutely convergent Fourier series. The same method works also for p-th order $(1 < p < 2)$ processes.

2. THE CASE OF SECOND ORDER PROCESSES

Let H be a Hilbert space over the field C of complex numbers and let $f : \mathbb{R} \to H$ be a 2π-periodic continuous function. Define the Fourier coefficients of f by the equality

$$\hat{f}(n) = \frac{1}{2\pi} \int_0^{2\pi} f(t)e^{-int}dt \ , \quad n = 0,\pm 1,\dots \ .$$

The integral is understood in the Pettis sense.

A generalization of the Bernstein's theorem is formulated as follows.

<u>Theorem 2'</u>. Suppose

$$\|f(t_1) - f(t_2)\| \le \varphi(|t_1 - t_2|) \ , \quad t_1,t_2 \in [0,2\pi] \ ,$$

where $\varphi : [0,2\pi] \to \mathbb{R}^+ = [0,\infty)$ is an increasing continuous at 0 function with the property

$$\int_0^{2\pi} \frac{\varphi(u)}{u^{3/2}} \, du < \infty \ . \tag{2}$$

Then $\sum_n \|\hat{f}(n)\| < \infty$.

Proof is similar to proof in the case of scalar function (see [9], vol. 1, Ch.VI, Theorem 3.1). The main tool is the Bessel inequality

$$\sum_n \|\hat{f}(n)\|^2 \le \frac{1}{2\pi} \int_0^{2\pi} \|f(t)\|^2 dt$$

which follows easily from the scalar version (see also in the next section the proof of Theorem 2" which contains Theorem 2').

The second step is the following proposition.

<u>Proposition 1</u>. Let $\xi(t)$, $t \in \mathbb{R}$ be a measurable stochastically

continuous 2π-periodic random process, the trajectories of which are integrable on $[0,2\pi]$. Define

$$\xi^{\wedge}(n) = \frac{1}{2\pi}\int_0^{2\pi} \xi(t)e^{-int}dt \ , \quad n = 0,\pm 1,\dots \ .$$

If

$$\sum_n |\xi^{\wedge}(n)| < \infty \qquad \text{a.s.,}$$

then ξ has a modification η, all trajectories of which possess absolutely convergent Fourier series; in particular, η is sample continuous.

Proof. Let for $t \in \mathbb{R}$

$$\eta(t) = \sum_n \xi^{\wedge}(n)e^{int}$$

if $\sum_n |\xi^{\wedge}(n)| < \infty$, and $\eta(t) = 0$ otherwise. It is evident that $\eta(t)$, $t \in \mathbb{R}$ possesses desired properties.

Now we can prove our theorem.

Theorem 3. Let $\xi(t)$, $t \in \mathbb{R}$ be a second order mean continuous 2π-periodic random process. Suppose

$$(E|\xi(t_1) - (t_2)|^2)^{1/2} \le \varphi(|t_1 - t_2|), \quad t_1,t_2 \in [0,2\pi] \quad (3)$$

where $\varphi : [0,2\pi] \to \mathbb{R}^+$ is an increasing continuos at 0 function with property (2). Then ξ has a modification η , all trajectories of which possess absolutely convergent Fourier series, in particular, η is sample continuous.

Proof. Let (Ω,A,P) be the probability space on which ξ is defined and let $H = L_2(\Omega,A,P)$; ξ maps $\mathbb{R}$ into H . The assumptions of Theorem 3, according to Theorem 2' imply $\sum_n \|\hat{\xi}(n)\| < \infty$. Thus

$$\sum_n |\hat{\xi}(n)| < \infty \qquad \text{a.s.}$$

Since $\xi^{\wedge}(n) = \hat{\xi}(n)$ a.s. for all n , we obtain

$$\sum_n |\hat{\xi}(n)| < \infty \quad \text{a.s.}$$

and it remains to apply Proposition 1.

Corollary. (M.Hahn [4]). Let $\xi(t)$, $t \in [0,2\pi]$ be a second order random process which satisfies (3), where $\varphi : [0,2\pi] \to \mathbb{R}^+$ is an increasing, continuous at 0 function with property (2). Then ξ possesses a sample continuous modification.

Proof. Consider a new process ξ_1 defined by the equality

$$\xi_1(t) = \xi(t) - \frac{t}{2\pi}(\xi(2\pi) - \xi(0)) , \quad t \in [0,2\pi] .$$

We have $\xi_1(0) = \xi_1(2\pi)$ and it is easy to see that the 2π-periodic continuation of ξ_1 satisfies the assumptions of Theorem 3. Thus, ξ_1 possesses a sample continuous modification η_1 and hence $\eta_1(t) + \frac{t}{2\pi}(\xi(2\pi) - \xi(0))$, $t \in [0,2\pi]$ is the desired sample continuous modification of ξ.

Remark. In [7] it is shown that Theorem 3 implies the Bernstein theorem.

3. THE CASE OF p-th ORDER PROCESSES

Let now X be a complex Banach space and let $f : \mathbb{R} \to X$ be a 2π-periodic continuous function. The Fourier coefficients $\hat{f}(n)$, $n \in Z$ of f are defined in the same way. It can be shown that the analogue of Theorem 2' is not valid for Banach space valued functions (see the end of this section). Nevertheless the following is true.

Theorem 2". Let $X = L_r(\Lambda,\Sigma,\nu)$, where $1 < r < \infty$ and (Λ,Σ,ν) is a σ-finite measure space, $r' = \frac{r}{r-1}$, $p = \min(r,r')$. Suppose

$$\|f(t_1) - f(t_2)\| \le \varphi(|t_1 - t_2|) , \quad t_1,t_2 \in [0,2\pi] ,$$

where $\varphi : [0,2\pi] \to \mathbb{R}^+$ is an increasing continuous at 0 function with the property

$$\int_0^{2\pi} \frac{\varphi(u)}{u^{1+\frac{1}{p}}}\, du < \infty . \tag{4}$$

Then $\sum_n \|\hat{f}(n)\| < \infty$.

From this theorem by means of Proposition 1 we obtain

Theorem 3'. Let $\xi(t)$, $t \in \mathbb{R}$ be p-th order, $1 < p \leq 2$ p-mean continuous 2π-periodic random process. Suppose

$$(E|\xi(t_1) - \xi(t_2)|^p)^{1/p} \leq \varphi(|t_1 - t_2|), \quad t_1,t_2 \in [0,2\pi] ,$$

where $\varphi : [0,2\pi] \to \mathbb{R}^+$ is an increasing continuous at 0 function with property (4). Then ξ has a modification η , all trajectories of which possess absolutely convergent Fourier series, in particular, η is sample continuous.

For the proof of Theorem 2" we need the following generalization of the Hausdorff-Young inequality.

Proposition 2. let $1 < r < \infty$, $1 < p \leq \min(r,r')$ and $f \in L_p([0,2\pi];L_r)$. Then

$$\left(\sum_n \|\hat{f}(n)\|^{p'}\right)^{1/p'} \leq \left(\frac{1}{2\pi} \int_0^{2\pi} \|f(t)\|^p dt\right)^{1/p} .$$

The more general F.Riesz's type assertion is also valid.

Proposition 2'. Let (φ_k) $k \in N$ be an orthonormal sequence in $L_2[0,1]$ with the property $\sup_k \|\varphi_k\|_\infty = \beta < \infty$, and let $X = L_r(\Lambda,\Sigma,\nu)$, $1 < r < \infty$, $1 < p \leq \min(r,r')$. Denote for $f \in L_1([0,1];X)$

$$C_k(f) = \int_0^1 f(t)\bar{\varphi}_k(t)dt , \quad k \in N .$$

Then

$$\left(\sum_k \|C_k(f)\|^{p'}\right)^{1/p'} \leq \beta^{\frac{2}{p}-1} \left(\int_0^1 \|f(t)\|^p dt\right)^{1/p} .$$

Proof. For the simplicity suppose that $f \in L_p([0,1];\ell_r)$, i.e., $f = (f_j)$ $j \in N$ and

$$\left(\int_0^1 \|f(t)\|_r^p dt\right)^{1/p} = \left(\int_0^1 \left(\sum_j |f_j(t)|^r\right)^{p/r} dt\right)^{1/p} < \infty .$$

Then $C_k(f) = (C_k(f_j))$ $j \in N$ and we have by F.Riesz's theorem (see [9], vol. II, Ch. Xii, Theorem 2.8)

$$\left(\sum_k |C_k(f_j)|^{p'}\right)^{1/p'} \leq \beta^{\frac{2}{p}-1} \left(\int_0^1 |f_j(t)|^p dt\right)^{1/p}, \quad j \in N$$

hence

$$\sum_j \left(\sum_k |C_k(f_j)|^{p'}\right)^{r/p'} \leq \beta^{r(\frac{2}{p}-1)} \sum_j \left(\int_0^1 |f_j(t)|^p dt\right)^{r/p} .$$

The inequality

$$\left\|\int_0^1 (|f_j(t)|^p) dt\right\|_{r/p} \leq \int_0^1 \|(|f_j(t)|^p)\|_{r/p} dt$$

gives

$$\left(\sum_j \left(\int_0^1 |f_j(t)|^p dt\right)^{r/p}\right)^{p/r} \leq \int_0^1 \left(\sum_j |f_j(t)|^r\right)^{p/r} dt = \int_0^1 \|f(t)\|_r^p dt .$$

The inequality of the same kind gives also

$$\left(\sum_k \|C_k(f)\|^{p'}\right)^{r/p'} = \left(\sum_k \left(\sum_j |C_k(f_j)|^r\right)^{p'/r}\right)^{r/p'} \leq \sum_j \left(\sum_k |C_k(f_j)|^{p'}\right)^{r/p'} .$$

Therefore

$$\left(\sum_k \|C_k(f)\|^{p'}\right)^{1/p'} \leq \sum_j \left(\sum_k |C_k(f_j)|^{p'}\right)^{r/p'} \leq \beta^{\frac{2}{p}-1} \left(\int_0^1 \|f(t)\|_r^p dt\right)^{1/p} .$$

Remark. If X is a Banach space and for all $f \in L_2([0,1];X)$ we have

$$\sum_n \|\hat{f}(n)\|^2 < \infty ,$$

then X is isomorphic to a Hilbert space (see [6]).

Proof of Theorem 2". Denote

$$q(\tau) = \left(\int_0^{2\pi} \| f(t+\tau) - f(t-\tau)\|^p (2\pi)^{-1} dt\right)^{1/p}, \quad \tau > 0.$$

From Proposition 2 we obtain

$$\sum_n \|\hat{f}(n)|\sin n\tau\|^{p'} \le q(\tau)^{p'} \cdot 2^{-p'} .$$

Hölder's inequality gives for $j \in N$

$$\sum_{2^j \le |n| < 2^{j+1}} \|\hat{f}(n)\| \le 2^{\frac{j+1}{p}} \Big(\sum_{2^j \le |n| < 2^{j+1}} \|\hat{f}(n)\|^{p'} \Big)^{1/p'} . \tag{5}$$

We have also

$$2^{-p'} q\left(\frac{\pi}{3\cdot 2^j}\right)^{p'} \ge \sum_{2^j \le |n| < 2^{j+1}} \|\hat{f}(n)\|^{p'} \left|\sin \frac{n\pi}{3\cdot 2^j}\right|^{p'} .$$

If $\frac{\pi}{3} \le |t| < \frac{2}{3}\pi$, then $|\sin t| \ge \frac{\sqrt{3}}{2}$. Therefore

$$2^{-p'} q\left(\frac{\pi}{3\cdot 2^j}\right)^{p'} \ge \left(\frac{\sqrt{3}}{2}\right)^{p'} \sum_{2^j \le |n| < 2^{j+1}} \|\hat{f}(n)\|^{p'} . \tag{6}$$

We obtain from (6) and (5)

$$\sum_n \|\hat{f}(n)\| = \sum_j \sum_{2^j \le |n| < 2^{j+1}} \|\hat{f}(n)\| \le \sum_j 2^{\frac{j+1}{p}} \Big(\sum_{2^j \le |n| < 2^{j+1}} \|\hat{f}(n)\|^{p'} \Big)^{1/p'}$$

$$\le \frac{4}{\sqrt{3}} \sum_j 2^{\frac{j+1}{p}} q\left(\frac{\pi}{3\cdot 2^j}\right) . \tag{7}$$

Assumptions of Theorem 2" give

$$\sum_n \|\hat{f}(n)\| \le \frac{4}{\sqrt{3}} \sum_j 2^{\frac{j+1}{p}} q\left(\frac{\pi}{3\cdot 2^j}\right) = \frac{4}{\sqrt{3}} \cdot 2^{\frac{1}{p}} \sum_j 2^{\frac{j}{p}} \varphi\left(\frac{2\pi}{3\cdot 2^j}\right) ,$$

and it remains to remark that (4) implies $\sum_j 2^{j/p} \varphi\left(\frac{1}{2^j}\right) < \infty$.

<u>Remark</u>. The conclusion of Theorem 3' about sample continuity remains valid for all p , $0 < p < \infty$ (see [4]), however the conclusion on absolute convergence is no longer valid in case $2 < p < \infty$ (it is enough to consider non-random $f \in Lip_\alpha$ where $\frac{1}{p} < \alpha \le \frac{1}{2}$, such that $\sum_n |\hat{f}(n)| = \infty$).

The following example shows that the Bernstein's theorem is not valid for all Banach spaces.

Example. Let $X = c_o$, $f : \mathbb{R} \to X$ define by the equality

$$f(t) = \sum_k \frac{\lambda_k}{k} \sin kte_k \ , \quad t \in \mathbb{R} \ ,$$

where (c_k) $k \in N$ is the natural bases of c_o , (λ_k) $k \in N$ is a sequence of positive numbers such that $\lambda_k \to 0$ and $\sum_k \frac{\lambda_k}{k} = \infty$. It is easy to see that $f \in Lip_1([0,2\pi];c_o)$, but $\sum_n \|\hat{f}(n)\| = \infty$.

4. ON ZYGMUND'S THEOREMS AND THEIR CONSEQUENCES

The Zygmund's theorem on absolutely convergent Fourier series admits the following generalization.

Theorem 4. Let $X = L_r(\Lambda,\Sigma,\nu)$, where $1 < r < \infty$ and (Λ,Σ,ν) is a σ-finite measure space, $p = \min(r,r')$. Suppose $f : \mathbb{R} \to X$ is a continuous 2π-periodic function which has finite variation $V(f)$ on $[0,2\pi]$ satisfying

$$\|f(t_1) - f(t_2)\| \le \varphi(|t_1 - t_2|) \ , \quad t_1,t_2 \in [0,2\pi] \ ,$$

where $\varphi : [0,2\pi] \to \mathbb{R}^+$ is an increasing continuous at 0 function with the property

$$\int_0^{2\pi} \frac{\varphi(u)^{1/p'}}{u} du < \infty \ . \tag{8}$$

Then $\sum_n \|\hat{f}(n)\| < \infty$.

Proof. In the notations of the proof of Theorem 2" it can be verified for $k \in N$ that

$$kq^p(\tfrac{\pi}{k}) = \int_0^{2\pi} \Big(\sum_{j=1}^{k} \|f(t + \tfrac{(2j-1)\pi}{k}) - f(t + \tfrac{(2j+1)\pi}{k})\|^p\Big)(2\pi)^{-1}dt$$
$$\le V(f)\varphi^{p-1}(\tfrac{2\pi}{k}) \ .$$

This inequality gives for $k = 3 \cdot 2^j$

$$q(\frac{\pi}{3\cdot 2^j}) \le V(f)^{1/p}(\frac{1}{3\cdot 2^j})^{1/p}\ \varphi^{1/p'}(\frac{2\pi}{3\cdot 2^j}) .$$

From this and from inequality (7) of section 3 we obtain

$$\sum_n \|\hat{f}(n)\| \le \frac{4}{\sqrt{3}} \sum_j 2^{\frac{j+1}{p}} V(f)^{1/p}(\frac{1}{3\cdot 2^j})^{1/p}\ \varphi^{1/p'}(\frac{2\pi}{3\cdot 2^j})$$

$$\le \frac{4}{\sqrt{3}} \cdot 3^{-1/p} \cdot 2^{1/p} \cdot V(f)^{1/p} \sum_j \varphi^{1/p'}(\frac{2\pi}{3\cdot 2^j}) ,$$

and it remains to note that (8) implies $\sum_j \varphi^{1/p'}(\frac{1}{2^j}) < \infty$.

This theorem by virtue of Proposition 1 implies

Theorem 5. Let $\xi(t)$, $t \in \mathbb{R}$ be an an r-th order r-mean continuous 2π-periodic process, $p = \min(r,r')$. Suppose ξ as a map from $[0,2\pi]$ into the Banach space $L_r(\Omega,A,P)$ has finite variation and

$$(E|\xi(t_1) - \xi(t_2)|^r)^{1/r} \le \varphi(|t_1 - t_2|), \quad t_1,t_2 \in [0,2\pi] , \tag{9}$$

where $\varphi : [0,2\pi] \to \mathbb{R}^+$ is an increasing continuous at 0 function with property (8). Then ξ has a modification η , all trajectories of which possess absolutely convergent Fourier series, in particular, η is sample continuous.

Remark. Even sample continuity conclusion of the formulated theorem seems to be new. As Professor S.Kwapień remarked, this conclusion can be proved directly by use of the Radon-Nikodym property of L_r . Thus the following is true:

Theorem 6. Let $\xi(t)$, $t \in [0,1]$ be an r-th order, $1 < r < \infty$, r-mean continuous random process. If ξ as a map from $[0,1]$ into $L_r(\Omega,A,P)$ has finite variation, then ξ has a sample continuous modification.

Now we shall generalize another theorem from [9] (Ch. VI, Theorem 3.9).

Theorem 7. Let $X = L_r(\Lambda,\Sigma,\nu)$, where $1 < r < \infty$ and (Λ,Σ,ν) is a σ-finite measure space. Suppose $f : \mathbb{R} \to X$ be a 2π-periodic continuous function which is absolutely continuous on $[0,2\pi]$ and its derivative f' (which does exist, see [9], p. 107) satisfies the condition

$$\int_0^{2\pi} \|f'(t)\| \ln^+ \|f'(t)\| dt < \infty ,$$

then $\sum_n \|\hat{f}(n)\| < \infty$ $(\ln^+ t = \ln(\max(t,1)), \ t \geq 0)$.

For the proof we need the following

Proposition 3. Let (φ_k) $k \in N$ be an orthonormal sequence in $L_2[0,1]$ with the property $\sup_k \|\varphi_k\|_\infty = \beta < \infty$, $X = L_r$, $1 < r < \infty$ and let $g : [0,1] \to X$ be a measurable function satisfying the condition

$$\int_0^1 \|g(t)\| \ln^+ \|g(t)\| dt < \infty . \tag{10}$$

Then

$$\sum_k \frac{1}{k} \left\| \int_0^1 g(t) \overline{\varphi}_k(t) dt \right\| < \infty .$$

Proof. Denote

$$C_k(g) = \int_0^1 g(t) \overline{\varphi}_k(t) dt$$

and

$$Q(g;y) = \sum_{k^{p-1} \|C_k(g)\| > y} \frac{1}{k^p} , \quad y \in \mathbb{R}^+$$

where $p = \min(r,r')$.

Let us verify that

$$\sum_k \frac{\|C_k(g)\|}{k} = \int_0^\infty Q(g;y) dy ; \tag{11}$$

$$yQ(g;y) \leq b \int_0^1 \|g(t)\| dt , \quad y > 0 \tag{12}$$

for a $b > 0$ independent of g; and

$$y^{p'}Q(g;y) \le \beta^{\frac{2-p}{p-1}} \left(\int_0^1 \|g(t)\|^p dt\right)^{p'/p} \quad , \qquad y > 0 . \qquad (13)$$

Introduce a measure μ on $\mathbb{R}^+$ by the equality

$$\mu = \sum_k \frac{1}{k^p} \delta_{k^{p-1}\|C_k(g)\|} \ ,$$

where δ_y means the Dirac measure concentrated at $y \in \mathbb{R}^+$.

It is evident that

$$Q(y;q) = \mu\{x \in \mathbb{R}^+ : x > y\} \quad , \quad y \in \mathbb{R}^+ . \qquad (14)$$

So we have by change variable formula

$$\int_{\mathbb{R}^+} Q(g;y)dy = \int_{\mathbb{R}^+} x d\mu(x) = \sum_k \frac{1}{k^p} \cdot k^{p-1}\|C_k(g)\| = \sum_k \frac{1}{k} \|C_k(g)\| .$$

(12) is checked directly if one note that $\|C_k(g)\| \le \beta\|g\|_1$,

$$Q(g;y) \le \sum_{\beta k^{p-1}\|g\|_1 > y} k^{-p} = \sum_{k > \left(\frac{y}{\beta\|g\|_1}\right)^{\frac{1}{p-1}}} k^{-p} \le b \frac{\|g\|_1}{y} .$$

For the verification of (13) we use Proposition 2'

$$Q(g;y) \le t^{-p'} \int_{\mathbb{R}^+} x^{p'} d\mu(x) = t^{-p'} \sum_k \frac{1}{k^p} (k^{p-1}\|C_k(g)\|)^{p'}$$

$$= t^{-p'} \sum_k \|C_k(g)\|^{p'} \le \beta^{\frac{2-p}{p-1}} t^{-p'} \|g\|_p^{p'} .$$

Using (12) and (13) we obtain $(y > 0)$

$$Q(g;y) \le Q(g1_{[\|g\| \le y]}; y/2) + Q(g1_{[\|g\| > y]}; y/2)$$

$$\le \beta^{\frac{2-p}{p-1}} \cdot 2^{p'} \cdot y^{-p'} \|g1_{[\|g\| \le y]}\|_p^{p'} + 2b \cdot y^{-1} \|g1_{[\|g\| > y]}\|_1 .$$

Since $p' \ge p$, we can write

$$\left(\int_{\mathbb{R}^+} y^{-p'} \|g1_{[\|g\| \le y]}\|_p^{p'} dy\right)^{p/p'} \le \int_0^1 \|g(t)\|^{p'} \left(\int_{\mathbb{R}^+} y^{-p'} 1_{[\|g\| \le y]}(t) dy\right)^{p/p'} dt$$

$$= \int_0^1 \|g(t)\|^{p'} \left(\int_{\|g(t)\|}^{\infty} y^{-p'} dy\right)^{p/p'} dt$$

$$= \int_0^1 \|g(t)\| dt < \infty .$$

Application of Fubini's theorem gives

$$\int_1^\infty y^{-1}\|g1_{[\|g\|>y]}\|_1 dy = \int_0^1 \|g(t)\|(\int_1^\infty y^{-1}1_{[\|g\|>y]}(t)dy)dt$$

$$= \int_{[\|g\|\geq 1]} \|g(t)\|(\int_1^{g(t)} y^{-1}dy)dt$$

$$= \int_0^1 \|g(t)\|\ln^+\|g(t)\|dt < \infty .$$

Finally we derive

$$\int_{\mathbb{R}^+} Q(g;y)dy \leq \int_{\mathbb{R}^+} Q(g1_{[\|g\|\leq y]}; \tfrac{y}{2})dy + \int_{\mathbb{R}^+} Q(g1_{[\|g\|>y]}; \tfrac{y}{2})dy$$

$$\leq 2^{-p'} \int_0^1 \|g(t)\|dt + \int_0^1 Q(g1_{[\|g\|>y]}; \tfrac{y}{2}) +$$

$$+ 2b \int_0^1 \|g(t)\|\ln^+\|g(t)\|dt < \infty ,$$

which, according to (11), gives the desired conclusion.

Remark. The presented proof of Proposition 3 is a slight modification of proof of an analogous assertion for scalar functions from [2].

Proof of Theorem 7. Denote $f' = g$. We have $\hat{g}(n) = n\hat{f}(n)$, $n \in Z$. We have by the assumption

$$\int_0^1 \|g(t)\|\ln^+\|g(t)\|dt < \infty .$$

From Proposition 3 we get

$$\sum_{n\neq 0} \frac{\|\hat{g}(n)\|}{|n|} < \infty \quad , \quad \text{i.e.,} \quad \sum_n \|\hat{f}(n)\| < \infty .$$

For random processes Theorem 7 gives the following result:

Theorem 8. Let $\xi(t)$, $t \in \mathbb{R}$ be an r-th order, $1 < r < \infty$, r-mean continuous 2π-periodic random process such that ξ as a map from $[0,2\pi]$ into $L_r(\Omega,A,P)$ is absolutely continuous and its r-mean derivative satisfies the condition

$$\int_0^{2\pi} \|\xi'(t)\|\ln^+\|\xi'(t)\|dt < \infty . \qquad (15)$$

Then ξ has a modification η, all trajectories of which possess absolutely convergent Fourier series, in particular, η is sample continuous.

Remark. Conclusion of this theorem on sample continuity is known and it is valid without assumption (15) (see [1]).

Finally, let us note that the assumptions of the above formulated Theorem 3, Theorem 6 (the case of $r = 2$) and Theorem 8 (the case of $r = 2$) depend only on the covariance of the considered process ξ, so the result of [8] automatically implies that in the considered situations the distribution of ξ in $C[0,2\pi]$ satisfies the central limit theorem.

It can be shown also that in the same manner the conditions of absolute convergence of multiple Fourier series imply sample continuity and sample absolute convergence conditions for corresponding random fields.

REFERENCES

[1] S. Cambanis and G. Miller, Some path properties of p-th order and symmetric stable processes, Ann. Prob., v. 8 (1980), p. 1148-1156.

[2] O.D. Cereteli, On the interpolation of operators by the truncation method, Proc. Razmadze Inst. Math. Acad. Sci. Georgian SSR, v. 36 (1969), p. 111-122.

[3] J. Diestel and J.J. Uhl. Jr., Vector Mesures, Providence, 1977.

[4] M.G. Hahn, Conditions for sample-continuity and the central limit theorem, Ann. Prob., v. 5 (1977), p. 351-360.

[5] I.A. Ibragimov, On the conditions for the smoothness of trajectories of random functions, Theor. Prob. and its appl., v. 28 (1983), p. 229-249.

[6] S. Kwapień, Isomorphic characterizaitons of inner product spaces by orthogonal series with vector valued coefficients, Studia Math. v. 44 (1972), p. 583-595.

[7] G. Pisier, Conditions d'entropie assurant la continuite de certains processus et applications a l'analyse harmonique, Seminaire d analyse functionelle 1979-1980, Exp. XIII-XIV.

[8] V.I. Tarieladze, Characterization of covariance operators which guarantee the CLT, Lecture Notes in Statistics, v.2 (1980), p. 348-359.

[9] A. Zygmund, Trigonometric Series, v. 1,2, Cambridge, 1959.

Probability Theory on Vector Spaces IV
Lancut, June'87, Springer's LNM 1391

Singular integrals and second-order stationary sequences

by Janusz Woś

1. Introduction. The aim of this paper is to apply some ideas of the theory of vector valued singular integrals to study the behaviour of second-order stationary sequences, i.e. sequences of the form $(U^n f)_{n=-\infty}^{\infty}$, where U is a unitary operator on $L_2(\mu)$ and $f \in L_2(\mu)$.

A basic theorem in the theory of singular integrals states that if k is a measurable function from R^n into the space $B = B(H_1, H_2)$ of bounded linear operators from a Hilbert space H_1 into a Hilbert space H_2, satisfying

$$(1.1) \qquad \|k(x)\|_B < C_1 \|x\|^{-n} \quad \text{for } \|x\| > 0,$$

$$(1.2) \qquad \int_{\|x\|>2\|y\|} \|k(x-y) - k(x)\|_B \, dx < C_2 \quad \text{for } \|y\| > 0,$$

$$(1.3) \qquad \int_{R_1<\|x\|<R_2} k(x)\, dx = 0 \quad \text{for } 0<R_1<R_2<\infty ,$$

and if

$$(T_\varepsilon f)(x) = \int_{\|y\| \geq \varepsilon} k(y) f(x-y)\, dy ,$$

then for each $f \in L_p(dx; H_1)$ and $1<p<\infty$ we have

$$(1.4) \qquad \|T_\varepsilon f\|_{L_p(dx;\, H_2)} \leq A_p \|f\|_{L_p(dx;\, H_1)} ,$$

where A_p does not depend on f and ε. Moreover, for each $f \in L_p(dx; H_1)$

the limit $\lim_{\varepsilon\to 0} T_\varepsilon f = Tf$ exists in the norm of $L_p(dx; H_2)$ and the operator T defined in this way also satisfies (1.4) (see [1], [7], or [13], Theorems II.2 and II.5).

Similar results are also true if $\mathbb{R}^n$ is replaced by $\mathbb{Z}^n$ or some other groups (see [11]). In case of the group of the integers $\mathbb{Z}$ the above theorem (in case of p=2) is concerned with a unitary operator U in $l_2(\mathbb{Z})$, which is induced by the shift transformation on $\mathbb{Z}$.

In the present paper we prove similar results for arbitrary unitary operator U in $L_2(\mu) = L_2(\mu; \mathbb{C})$, where (Ω, Σ, μ) is some measure space. If $k \in B(H_1, H_2)$, we define an operator $U\otimes k$: $L_2(\mu; H_1) \to L_2(\mu; H_2)$ as a composition $U\otimes k = (id\otimes k)(U\otimes id_{H_1})$, where $U\otimes id_{H_1}: L_2(\mu; H_1) \to L_2(\mu; H_1)$ is a natural extension of U from $L_2(\mu)$ to $L_2(\mu; H_1)$, and $id\otimes k$: $L_2(\mu; H_1) \to L_2(\mu; H_2)$ is given by the formula $(id\otimes k)f(\omega) = k(f(\omega))$. In Section 3 of the present paper it is shown that if a sequence $k_n \in B(H_1, H_2)$ satisfies the following conditions (similar to (1.1) and (1.2)):

(K1) $\qquad \|k_n\|_B < C_1 n^{-1}$ for $n = 1, 2, \ldots$,

(K2) $\qquad \Sigma_{n>\alpha m} \|k_{n-m} - k_n\|_B < C_2$ for $m = 1, 2, \ldots$,

where C_1, C_2, and $\alpha \geq 1$ are some constants, then the series

(1.5) $$S = \Sigma_{n=1}^{\infty} (U^n \otimes k_n - U^{-n} \otimes k_n)$$

converges in the strong operator topology and defines a bounded linear operator $S: L_2(\mu; H_1) \to L_2(\mu; H_2)$ (see Theorem 2).

In Section 4 we give some examples of the operator S defined by (1.5). Let $A_n = n^{-1}\Sigma_{i=1}^n U^i$ denote the ergodic averages of the operator U. It is shown that if (u_n) is a sequence of non-negative numbers satisfying

(1.6) $$n^{-1}\Sigma_{i=1}^n u_i < C \quad (n = 1, 2, \ldots)$$

for some constant C, then for each $f \in L_2(\mu)$ the series

(1.7) $$Sf = \Sigma_{n=1}^{\infty} n^{-1}(A_n f - A_n^* f)u_n$$

converges in the norm of $L_2(\mu)$ and defines a bounded linear operator S in $L_2(\mu)$. This result generalizes a theorem of Campbell [2] concerning the case where $u_n = 1$. It is also shown that if $u_n \geq 0$ satisfy (1.6) then the square function

(1.8) $$Gf = (\Sigma_{n=1}^{\infty} n^{-1}|A_n f - A_n^* f|^2 u_n)^{1/2}$$

is of strong type (2,2), i.e. $\|Gf\|_{L_2(\mu)} \leq A\|f\|_{L_2(\mu)}$ for some constant A independent of f (see Proposition 2). By taking a suitable sequence (u_n) we obtain from (1.8) an ergodic Marcinkiewicz integral

$$Mf = (\Sigma_{n=1}^{\infty} n^{-1}|A_n f - A_n^* f|^2)^{1/2}$$

as well as an ergodic dyadic square function

$$Qf = (\Sigma_{n=0}^{\infty} |A_{2^n} f - A_{2^n}^* f|^2)^{1/2} .$$

In Section 5 we apply the above results to the questions of almost everywhere behaviour of the sequence $A_n f$ of ergodic averages.

The author would like to express his deep gratitude to Krzysztof Samotij for his invaluable comments concerning the contents of this paper.

2. Bounded convergence of the sine series. Let $(k_n)_{n=1}^{\infty}$ be a sequence of elements of some comlex Banach space B. We are interested in the bounded convergence of the sequence $s_m(t) = \Sigma_{n=1}^{m} (\sin nt)k_n$.

LEMMA 1. Assume that the sequence $(k_n)_{n=1}^{\infty}$ satisfies (K1) and (K2). Then $\lim_{m\to\infty} s_m(t)$ exists for each $t \in [0, 2\pi)$, and there exists a constant A such that

(2.1) $$\|s_m(t)\| \leq A$$

for each $t \in [0, 2\pi)$ and each $m = 1, 2, \ldots$.

Proof. The proof of the convergence of $s_m(t)$ is the same as in the scalar case (cf. [3], Section 7.2.1). To prove the uniform boundedness of $s_m(t)$ let $t \in (0, \pi]$ be fixed and let $m = [\pi/t]$ be the integral part of π/t. Then for each positive integer N we have

$$(2.2) \qquad \|s_N(t)\| \le \Sigma_{n=1}^{N} nt\|k_n\| < C_1 tN \le \pi C_1 N m^{-1}$$

by (K1).

Now, we shall estimate the sum

$$\sum_{\alpha m< n\le N} (\sin nt)k_n = (2i)^{-1} \sum_{\alpha m<|n|\le N} e^{int}k_n ,$$

where $k_{-n} = -k_n$ by definition and $N>(\alpha+1)m$. In this case

$$\Big(\sum_{\alpha m<|n|\le N} e^{int}k_n\Big)(1 - e^{imt}) =$$

$$\sum_{\alpha m<|n|\le N} e^{int}k_n - \sum_{\alpha m<|n-m|\le N} e^{int}k_{n-m} =$$

$$\sum_{\substack{(\alpha+1)m<n\le N\\ -N+m\le n<-\alpha m}} e^{int}(k_n - k_{n-m}) + \sum_{\substack{-N\le n<-N+m\\ \alpha m<n\le(\alpha+1)m}} e^{int}k_n$$

$$- \sum_{\substack{-\alpha m\le n<(1-\alpha)m\\ N<n\le N+m}} e^{int}k_{n-m} = I_1 + I_2 + I_3.$$

By (K2) we have

$$\|I_1\| \le 2\,\Sigma_{n>\alpha m} \|k_{n-m} - k_n\| < 2C_2 .$$

Next, using (K1) we obtain

$$\|I_2\| \le \sum_{\substack{\alpha m<n\le(\alpha+1)m\\ -N\le n<-N+m}} \|k_n\| < C_1 m(\alpha m)^{-1} + C_1 m(N-m)^{-1} < 2C_1\alpha^{-1},$$

since $N>(\alpha+1)m$. Similarly, $\|I_3\| < 2C_1\alpha^{-1}$.

Since $0<t\le\pi$ and $m = [\pi/t]$, it is easy to see that $\pi/2\le mt\le\pi$. Consequently, $|e^{imt} - 1| \ge \sqrt{2}$. Using the above estimates we obtain

$$(2.3) \qquad \Big\|\sum_{\alpha m<n\le N} (\sin nt)k_n\Big\| < (2C_1\alpha^{-1} + C_2)/\sqrt{2}$$

for $N>(\alpha+1)m$. Finally, using (2.2) and (2.3) we get

$$\|s_N(t)\| < \pi C_1\alpha + \max(\pi C_1,\ \sqrt{2}C_1\alpha^{-1} + C_2/\sqrt{2}) = A$$

for arbitrary $N\ge1$, which completes the proof.

Remark 1. If (k_n) satisfies the condition

(K2') $$\Sigma_{n>\beta m} \|k_{n-1} - k_n\| < C_3 m^{-1} \quad \text{for } m = 1, 2, \ldots ,$$

where C_3 and $\beta>0$ are some constants (cf. [3], Section 7.2.3), then we have (K2). Observe that in a particular case, when $B = \mathbb{C}$ and k_n is a nonincreasing sequence of real numbers such that $nk_n = O(1)$, both the conditions (K1) and (K2') (and hence (K2)) are satisfied.

3. Ergodic analogues of vector valued singular integrals. Let (Ω, Σ, μ) be a measure space and let U be a unitary operator in the Hilbert space $L_2(\mu) = L_2(\mu;\mathbb{C})$ of all square-integrable complex valued functions defined on Ω. If H_1 is some complex Hilbert space, and if a function $f:\Omega \to H_1$ is of the form

(3.1) $$f(\omega) = \Sigma_{j=1}^{l} f_j(\omega)x_j$$

for some $f_j \varepsilon L_2(\mu)$ and $x_j \varepsilon H_1$, we set

$$(U\otimes id_{H_1})f = \Sigma_{j=1}^{l} (Uf_j)x_j .$$

It is clear that $U\otimes id_{H_1}$ extends to a unitary operator, on the Hilbert space $L_2(\mu; H_1)$ of all square-integrable H_1-valued functions, with spectral measure $\tilde{E}(dt) = E(dt)\otimes id_{H_1}$, where $E(dt)$ denotes the spectral measure of U.

On the other hand, if k is a bounded linear operator from H_1 into another Hilbert space H_2, then we put

$$(id\otimes k)f = \Sigma_{j=1}^{l} f_j(kx_j)$$

if f is of the form (3.1). Observe that $(id\otimes k)f(\omega) = k(f(\omega))$. Consequently, as is easy to see, we have

$$\|id\otimes k\|_{B(L_2(\mu;\, H_1),\, L_2(\mu;\, H_2))} = \|k\|_{B(H_1,\, H_2)} .$$

Finally, if U and k are as above, we define a bounded linear operator $U\otimes k\colon L_2(\mu; H_1) \to L_2(\mu; H_2)$ as a composition

(3.2) $$U\otimes k = (id\otimes k)(U\otimes id_{H_1}) = (U\otimes id_{H_2})(id\otimes k).$$

THEOREM 1. Let U be a unitary operator on $L_2(\mu)$ and let $(k_n)_{n=-\infty}^{+\infty}$ be a sequence of bounded linear operators from a Hilbert space H_1 into another Hilbert space H_2. Then for each $f \in L_2(\mu; H_1)$ and for each positive integers n_1 and n_2 we have

$$\| \sum_{n_1<|n|<n_2} (U^n \otimes k_n) f \|^2_{L_2(\mu;\ H_2)} \leq \tag{3.3}$$

$$\leq \int_0^{2\pi} \| \sum_{n_1<|n|<n_2} e^{int} k_n \|^2_{B(H_1,H_2)} \langle \tilde{E}(dt) f, f \rangle_{L_2(\mu;\ H_1)} ,$$

where $\tilde{E}(dt)$ denotes the spectral measure of $U \otimes id_{H_1}$.

Proof. It suffices to prove (3.3) for the functions f having the form (3.1). Using the spectral theorem it is easy to see that in this case

$$\| \sum_{n_1<|n|<n_2} (U^n \otimes k_n) f \|^2_{L_2(\mu;\ H_2)} = \tag{3.4}$$

$$\Sigma_{i=1}^{l} \Sigma_{j=1}^{l} \int_0^{2\pi} \langle s(t)x_i, s(t)x_j \rangle_{H_2} \langle E(dt) f_i, f_j \rangle_{L_2(\mu)} ,$$

where for each $t \in [0, 2\pi)$ the operator $s(t) \in B(H_1,H_2)$ is defined by

$$s(t) = s_{n_1,n_2}(t) = \sum_{n_1<|n|<n_2} e^{int} k_n .$$

The right hand side of the equality (3.4) may be approximated by the sums

$$\Sigma_{m=1}^{r} \Sigma_{i=1}^{l} \Sigma_{j=1}^{l} \langle s(t_m)x_i, s(t_m)x_j \rangle_{H_2} \langle E(\Delta_m) f_i, f_j \rangle_{L_2(\mu)} ,$$

where $\{\Delta_1, \Delta_2, \ldots, \Delta_r\}$ is some partition of the interval $[0, 2\pi)$, and $t_m \in \Delta_m$ for each $m = 1, 2, \ldots, r$. Since $\langle E(\Delta_m) f_i, f_j \rangle = \langle E(\Delta_m) f_i, E(\Delta_m) f_j \rangle$, the above sum is equal to

$$\Sigma_{m=1}^{r} \| \Sigma_{j=1}^{l} (E(\Delta_m) f_j)(s(t_m)x_j) \|^2_{L_2(\mu;\ H_2)} =$$

$$\Sigma_{m=1}^{r} \| (id \otimes s(t_m))(\Sigma_{j=1}^{l} (E(\Delta_m) f_j) x_j) \|^2_{L_2(\mu;\ H_2)} .$$

Since $\| id \otimes s(t_m) \|_{B(L_2(\mu;\ H_1), L_2(\mu;\ H_2))} = \| s(t_m) \|_{B(H_1,H_2)}$, the last sum is not less than

$$\Sigma_{m=1}^{r} \| s(t_m) \|^2_{B(H_1,H_2)} \| \Sigma_{j=1}^{l} (E(\Delta_m) f_j) x_j \|^2_{L_2(\mu;\ H_1)} ,$$

which approximates the integral

$$\int_0^{2\pi} \|s(t)\|^2_{B(H_1,H_2)} \langle \tilde{E}(dt)f,f\rangle_{L_2(\mu;\, H_1)} .$$

This completes the proof of the theorem.

Since $\langle \tilde{E}(dt)f,f\rangle$ is a finite non-negative Borel measure on $[0, 2\pi)$ with total mass equal to $\|f\|^2_{L_2(\mu;\, H_1)}$, Lebesgue dominated convergence theorem, Theorem 1, and Lemma 1 yield the following result.

THEOREM 2. Assume that U is a unitary operator in $L_2(\mu)$, and let $(k_n)_{n=1}^{\infty}$ be a sequence of linear operators from a Hilbert space H_1 into a Hilbert space H_2 satisfying (K1) and (K2). Then the series

$$S = \Sigma_{n=1}^{\infty} (U^n \otimes k_n - U^{-n} \otimes k_n)$$

converges in the strong operator topology and defines a bounded linear operator $S\colon L_2(\mu;\, H_1) \to L_2(\mu;\, H_2)$ with the norm

$$\|S\| \leq \sup_{0 \leq t < 2\pi} \|\Sigma_{n=1}^{\infty} (\sin nt) k_n\|_{B(H_1,H_2)} \leq A$$

(see (2.1)).

Remark 2. Observe that if $H_1 = H_2$ in Theorem 2, then by (3.2) the operators $K_n = id \otimes k_n$ commute with the unitary operator $\tilde{U} = U \otimes id_{H_1}$ defined in the Hilbert space $H = L_2(\mu;\, H_1)$. This particular case of Theorem 2 may be generalized as follows: if $\tilde{U}$ is a unitary operator in a Hilbert space H, and if $(K_n)_{n=1}^{\infty}$ is a sequence of operators from $B(H,H)$, which commute with $\tilde{U}$ and satisfy (K1) and (K2), then the series $S = \Sigma_{n=1}^{\infty} K_n(\tilde{U}^n - \tilde{U}^{-n})$ converges in the strong operator topology. The proof of this result will be given in a forthcoming paper by the author. Let us note that in [12] K. Samotij has shown that, except for some trivial cases, the commutativity assumption in that result cannot be deleted.

Now, we shall consider two particular cases of Theorem 2.

COROLLARY 1. Let U be a unitary operator in $L_2(\mu)$ and let $(k_n)_{n=1}^{\infty}$ be a sequence of elements of a Hilbert space H satisfying (K1) and (K2). Then for each function $f \in L_2(\mu)$ the series

$$Sf(\omega) = \Sigma_{n=1}^{\infty} (U^n f(\omega) - U^{-n} f(\omega)) k_n$$

converges in the norm of $L_2(\mu; H)$ and defines a bounded linear operator $S: L_2(\mu) \to L_2(\mu; H)$.

Proof. It suffices to take $H_1 = \mathbb{C}$ and $H_2 = H$ in Theorem 2. In this case $k_n \varepsilon B(H_1, H_2)$ are determined by a sequence of vectors $k_n \varepsilon H$, and $\|k_n\|_{B(H_1,H_2)} = \|k_n\|_H$.

Let us note that in this case the inequality (3.3) becomes an equality. The next corollary is connected with vector valued strictly stationary sequences.

COROLLARY 2. Let $T: \Omega \to \Omega$ be an invertible bimeasurable transformation which preserves the measure μ, and let $(k_n)_{n=1}^{\infty}$ be a sequence of linear operators from a Hilbert space H_1 into a Hilbert space H_2 satisfying (K1) and (K2). Then for each function $f \varepsilon L_2(\mu; H_1)$ the series

$$Sf(\omega) = \Sigma_{n=1}^{\infty} k_n (f(T^n\omega) - f(T^{-n}\omega))$$

converges in the norm of $L_2(\mu; H_2)$ and defines a bounded linear operator $S: L_2(\mu; H_1) \to L_2(\mu; H_2)$.

Proof. We define a unitary operator $U: L_2(\mu) \to L_2(\mu)$ by the formula $Uf(\omega) = f(T\omega)$. In this case $U \otimes id_{H_1}$ is still given by the formula $(U \otimes id_{H_1}) f(\omega) = f(T\omega)$ and our corollary follows by Theorem 2.

In the sequel we shall be interested in the behaviour of the averages $A_n = n^{-1} \Sigma_{i=1}^{n} U^i$ of a unitary operator U in $L_2(\mu)$. Clearly, $A_n^* = n^{-1} \Sigma_{i=1}^{n} U^{-i}$, and

$$\Sigma_{n=1}^{N} (U^n - U^{-n}) \otimes k_n = \Sigma_{n=1}^{N} (A_n - A_n^*) \otimes n(k_n - k_{n+1}) + (A_N - A_N^*) \otimes N k_N.$$

Since we always have $\lim_{N\to\infty} A_N - A_N^* = 0$ in the strong operator topology so, if (K1) holds, we obtain

$$S = \Sigma_{n=1}^{\infty} (A_n - A_n^*) \otimes n(k_n - k_{n+1}) \tag{3.5}$$

in the strong operator topology.

4. Examples. In this section we shall give some examples of the operator S appearing in Corollary 1. The results of this section are analogous to some results of Ostrow and Stein [10]. Throughout this and the next section we fix, unless stated otherwise, a unitary operator U in $L_2(\mu)$.

We shall need the following lemma on numerical sequences.

LEMMA 2. Let (u_n) be a sequence of non-negative real numbers and let $\lambda>0$. Then (u_n) satisfies (1.6) if and only if

(4.1) $$\Sigma_{i=n}^{\infty} i^{-\lambda-1} u_i = O(n^{-\lambda}) \qquad (n = 1, 2, \ldots).$$

4A. Ergodic Hilbert transform with weights. In this section we shall generalize a theorem of Campbell ([2], Corollary 11), which states that the series $\Sigma_{n=1}^{\infty} n^{-1}(U^n - U^{-n}) = \Sigma_{n=1}^{\infty} (n+1)^{-1}(A_n - A_n^*)$ converges in the strong operator topology.

PROPOSITION 1. Let $(k_n)_{n=1}^{\infty}$ be a nonincreasing sequence of real numbers such that $nk_n = O(1)$. Then for each function $f\varepsilon L_2(\mu)$ the series

$$Sf = \Sigma_{n=1}^{\infty} k_n(U^n f - U^{-n} f)$$

converges in the norm of $L_2(\mu)$ and defines a bounded linear operator in $L_2(\mu)$.

Proof. We take $H = \mathbb{C}$ in Corollary 1. Then (k_n) satisfies (K1) and (K2) by Remark 1.

Observe that a nonincreasing sequence (k_n) of real numbers satisfies $nk_n = O(1)$ if and only if the sequence $u_n = n^2(k_n - k_{n+1})$ of non-negative numbers satisfies (4.1) with $\lambda = 1$. Consequently, using Lemma 2 and (3.5) we may reformulate Proposition 1 as follows.

PROPOSITION 1'. Let (u_n) be a sequence of non-negative numbers satisfying (1.6). Then for each function $f\varepsilon L_2(\mu)$ the series

(4.2) $$Sf = \Sigma_{n=1}^{\infty} n^{-1}(A_n f - A_n^* f)u_n$$

converges in the norm of $L_2(\mu)$ and defines a bounded linear operator S in $L_2(\mu)$.

Observe that by taking $u_n = 1$ in Proposition 1' we obtain the series

$$S = \Sigma_{n=1}^{\infty} n^{-1}(A_n - A_n^*) ,$$

while in case, where $u_n = n$ if $n = 2^k$ for some $k = 0, 1, 2, \ldots$, and $u_n = 0$ otherwise, we get

$$S = \Sigma_{n=0}^{\infty} (A_{2^n} - A_{2^n}^*) .$$

Also, if (δ_n) is an arbitrary sequence taking values 1 and -1 (or, more generally, arbitrary bounded sequence of real numbers) then, by distracting two suitable operators of the form (4.2), we infer that the series $S = \Sigma_{n=1}^{\infty} n^{-1}(A_n - A_n^*)\delta_n$ converges in the strong operator topology (cf. [10], Section 5).

Finally, let us note that in case where the operator U is given by $Uf(t) = e^{it}f(t)$ for $f \varepsilon L_2(\mu)$, where μ denotes the Lebesgue measure on $[0, 2\pi)$, then using ([3], Section 7.2.2) it is easy to see that the condition (1.6) is also necessary for Proposition 1' to hold.

4B. Ergodic Marcinkiewicz integral with weights. In this section we shall define a square function Gf which is analogous to the classical Marcinkiewicz integral with weights (cf. [10]).

Let $H = l_2$ be the Hilbert space of sequences $a = (a(i))_{i=1}^{\infty}$ with the norm $||a|| = (\Sigma_{i=1}^{\infty} |a(i)|^2)^{1/2}$, and let $k_n \varepsilon l_2$ be given by

$$k_n(i) = i^{-3/2} u_i^{1/2} 1_{[n,+\infty)}(i)$$

for $i,n = 1, 2, \ldots$, where (u_i) is a sequence of non-negative numbers satisfying (1.6).

Applying Lemma 2 with $\lambda = 2$ we obtain

$$|| k_n||_{l_2}^2 = \Sigma_{i=n}^{\infty} i^{-3} u_i = O(n^{-2}) ,$$

so that (K1) is satisfied. Next, we have

$$\|k_{n-m} - k_n\|_{1_2}^2 = \Sigma_{i=n-m}^{n-1} i^{-3} u_i \leq (n-m)^{-3} \Sigma_{i=n-m}^{n-1} u_i$$

for n>m. Using Schwarz inequality and Lemma 2 with $\lambda = 1/2$ we obtain

$$\Sigma_{n>2m} \|k_{n-m} - k_n\|_{1_2} \leq \Sigma_{l=m+1}^{\infty} l^{-3/2} (\Sigma_{i=1}^{l+m-1} u_i)^{1/2}$$

$$\leq (\Sigma_{l=m+1}^{\infty} l^{-3/2})^{1/2} (\Sigma_{j=0}^{m-1} \Sigma_{l=m+1}^{\infty} ((l+j)/l)^{3/2} u_{l+j} (l+j)^{-3/2})^{1/2}$$

$$\leq C(m^{-1/2})^{1/2} (m\ m^{-1/2})^{1/2} = C ,$$

since $(l+j)/l \leq 2$ for $j<m<l$. Consequently, the condition (K2) is satisfied. Now, for arbitrary $f \in L_2(\mu)$ we have

$$Sf(\omega)(i) = \Sigma_{n=1}^{\infty} (U^n f(\omega) - U^{-n} f(\omega)) k_n(i)$$

$$= \Sigma_{n=1}^{\infty} (U^n f(\omega) - U^{-n} f(\omega)) i^{-3/2} u_i^{1/2} \mathbf{1}_{[1,i]}(n)$$

$$= (i^{-1} u_i)^{1/2} (A_i f(\omega) - A_i^* f(\omega)) .$$

Let us consider the square function

$$Gf(\omega) = \|Sf(\omega)\|_{1_2} = (\Sigma_{n=1}^{\infty} n^{-1} |A_n f(\omega) - A_n^* f(\omega)|^2 u_n)^{1/2} .$$

Applying Corollary 1 we obtain

PROPOSITION 2. Assume that (u_n) is a sequence of non-negative numbers satisfying (1.6). Then there exists a constant A such that for each $f \in L_2(\mu)$ we have $\|Gf\|_{L_2(\mu)} \leq A \|f\|_{L_2(\mu)}$. In other words, the function G is of strong type (2,2).

Similarly as in Section 4A, by taking a suitable sequence (u_n) we infer that the ergodic Marcinkiewicz integral

$$Mf = (\Sigma_{n=1}^{\infty} n^{-1} |A_n f - A_n^* f|^2)^{1/2} ,$$

as well as the dyadic square function

$$Qf = (\Sigma_{n=0}^{\infty} | A_{2^n} f - A_{2^n}^* f|^2)^{1/2}$$

are of strong type (2,2). In particular, they are finite a.e. for each $f \in L_2(\mu)$.

Remark 3. Let us note that the sequence (k_n) corresponding to Q satisfies (K2'), while in case of M the condition (K2') is not satisfied.

5. Some applications. In this section we shall apply the results of the preceding section to the questions of a.e. behaviour of second-order stationary sequences.

It is well known that in general the pointwise ergodic theorem does not hold for unitary operators in $L_2(\mu)$ (see [9], Section 5.2.3). In fact, it was shown by Gaposhkin that if $f \in L_2(\mu)$ then $\lim_{n\to\infty} A_n f$ exists a.e. if and only if

$$\lim_{n\to\infty} \int_{0<|t|<2^{-n}} E(dt)f = 0 \quad \text{a.e.},$$

where E(dt) denotes the spectral measure of U (see [5] and [6]). The following corollary shows that we always have the a.e. convergence of symmetrized ergodic averages.

COROLLARY 3. For each $f \in L_2(\mu)$ we have $\lim_{n\to\infty} A_n f - A_n^* f = 0$ a.e.

Proof. It was shown by Gaposhkin [5] that the function

$$Rf = (\Sigma_{n=0}^{\infty} \sup_{2^n<k<2^{n+1}} |A_{2^n} f - A_k f|^2)^{1/2}$$

is of strong type (2,2). Consequently, it is finite a.e. Combining this with the a.e. finiteness of Qf (see Section 4B) we get the desired result.

It should be noted that the above corollary follows also directly from some general results of Gaposhkin ([4], Theorem 1), and the corresponding proof also goes through the a.e. finiteness of Rf and Qf.

It is known that in general the ergodic Hilbert transform

$$Hf = \Sigma_{n=1}^{\infty} (n+1)^{-1}(A_n f - A_n^* f) \tag{5.1}$$

does not exist a.e. (see [8], Remark 3.2 and Theorem 3.3). Thus the following corollary shows that in general the series (5.1) is not even Abel summable a.e.

COROLLARY 4. Assume that (u_n) is a bounded sequence of real numbers. If $f \in L_2(\mu)$ and if the series

$$\Sigma_{n=1}^{\infty} n^{-1}(A_n f - A_n^* f)u_n \tag{5.2}$$

is Abel summable a.e. then it is convergent a.e.

Proof. By Proposition 2 we have

$$\Sigma_{n=1}^{\infty} n^{-1} |A_n f - A_n^* f|^2 u_n^2 < \infty \quad \text{a.e.} \tag{5.3}$$

On the other hand, (5.3) is a Tauberian condition for Abel summability of (5.2).

Another application of the a.e. finiteness of Mf, to the questions of a.e. strong Cesàro limitability of the sequence of ergodic averages of a strictly stationary sequence, may be found in [14].

REFERENCES

[1] A. Benedek, A. P. Calderón, R. Panzone, Convolution operators on Banach space valued functions, Proc. Nat. Acad. Sci. U. S. A. 48 (1962), 356-365.

[2] J. T. Campbell, Spectral analysis of the ergodic Hilbert transform, Indiana Univ. Math. J. 35 (1986), 379-390.

[3] R. E. Edwards, Fourier Series. A Modern Introduction, Springer-Verlag, New York - Heidelberg - Berlin, 1979.

[4] V. F. Gaposhkin, A theorem on almost everywhere convergence of a sequence of measurable functions and its application to sequences of stochastic integrals, Mat. Sb. 104 (1977), 3-21 (in Russian).

[5] V. F. Gaposhkin, Criteria for the strong law of large numbers for some classes of second-order stationary processes and homogenous random fields, Teor. Veroyatn. Primen. 22 (1977), 295-319 (in Russian).

[6] V. F. Gaposhkin, On the individual ergodic theorem for normal operators in L_2, Funkts. Analiz Prilozh. 15 (1981), 18-22 (in Russian).

[7] J. García-Cuerva, J. L. Rubio de Francia, Weighted Norm Inequalities and Related Topics, North-Holland, Amsterdam 1985.

[8] E. Hensz, R. Jajte, Pointwise convergence theorems in L_2 over a von Neumann algebra, Math. Z. 193 (1986), 413-429.

[9] U. Krengel, Ergodic Theorems, de Gruyter, Berlin-New York, 1985.

[10] E. Ostrow, E. M. Stein, A generalization of lemmas of Marcinkiewicz and Fein with applications to singular integrals, Ann. Scuola Norm. Sup. Pisa 11 (1957), 117-135.

[11] N. Riviere, Singular integrals and multiplier operators, Ark. Mat. 9 (1971), 243-278.

[12] K. Samotij, On the convergence of the series $\Sigma K_n(U^n - U^{-n})$, These Proceedings.

[13] E. M. Stein, Singular Integrals and Differentiability Properties of Functions, Princeton Univ. Press, Princeton, N. J., 1970.

[14] J. Woś, On Marcinkiewicz-Zygmund law of large numbers and Cesàro limitability of ergodic averages (to appear).

Institute of Mathematics
Technical University
Wybrzeże Wyspiańskiego 27
50-370 Wroclaw
Poland

Probability Theory on Vector Spaces IV
Lancut, June'87, Springer's LNM 1391

On the difference of Gaussian measure of two balls in Hilbert spaces.

by

Tomasz Zak

The aim of this paper is to give a short proof of a theorem proved in |3|. Let γ be a symmetric Gaussian measure on a separable Hilbert space H. Denote $K_t = \{x \in H: \|x\| \le t\}$ and let $\lambda_1 \ge \lambda_2 \ge \dots$ be the eigenvalues of the covariance operator of γ. Then, as it was proved in (|3|, Theorem 1), there exists a constant c_t, depending only on t, such that for every $r \in H$ we have the following estimate:

$$(1) \quad \gamma(K_t) - \gamma(K_t + r) \le \frac{c_t}{(\bar{\lambda}_1 \bar{\lambda}_2)^{5/2}} \|r\|^2 , \text{ where } \bar{\lambda}_i = \min(1, \lambda_i).$$

The proof was long and complicated. Here we give a simpler proof which, we hope, can be generalized at least to l_p spaces, $p \ge 2$. This proof is based on ideas belonging to Rhee and Talagrand |1|, developed by Ryznar in |2|. We use only the Taylor formula and simple facts concerning Gaussian measures in Hilbert spaces. We get constants in the estimate (1) slightly different than in the paper |3| but they may be even better for some purposes.

Theorem. *Let* γ *be a symmetric Gaussian measure on a separable Hilbert space* H *and let* $\lambda_1 \ge \lambda_2 \ge \dots$ *be the eigenvalues of its covariance operator. Then there exists a constant* C_t*, which may be taken* $C_t = \max(1, t^4)$, *such that for every* $r \in H$ *we have:*

$$(2) \quad \gamma(K_t) - \gamma(K_t + r) \le \frac{C_t}{(\lambda_1 \lambda_2 \lambda_3 \bar{\lambda}_4^3)^{1/2}} \|r\|^2, \text{ where } \bar{\lambda}_4 = \min(1, \lambda_4).$$

Proof. It is well-known that we may also assume that γ is a product measure. Let $\lambda_1 \geq \lambda_2 \geq \lambda_3 \geq \lambda_4$ be the greatest eigenvalues of the covariance operator of γ and let H_1 be a four-dimensional subspace generated by their eigenvectors. Then $H = H_1 + H_1^{\perp}$ and $\gamma = \gamma_1 \times \gamma_2$, where γ_1 is concentrated on H_1 and has the density (with respect to the Lebesgue measure on H_1):

$$(3) \qquad f(x_1,x_2,x_3,x_4) = \frac{c_1}{(\lambda_1\lambda_2\lambda_3\lambda_4)^{1/2}} \exp\left(-\tfrac{1}{2}\left(\frac{x_1^2}{\lambda_1} + \frac{x_2^2}{\lambda_2} + \frac{x_3^2}{\lambda_3} + \frac{x_4^2}{\lambda_4}\right)\right) .$$

Here and in the sequel by $c_1, c_2, \ldots$ we denote absolute constants. Observe, that

$$(4) \qquad \sup_x f(x) = \frac{c_1}{(\lambda_1\lambda_2\lambda_3\lambda_4)^{1/2}} , \qquad \sup_i \sup_x \left|\frac{\partial f(x)}{\partial x_i}\right| \leq \frac{c_2}{(\lambda_1\lambda_2\lambda_3\lambda_4^2)^{1/2}}$$

and

$$(5) \qquad \sup_{i,j} \sup_x \left|\frac{\partial^2 f(x)}{\partial x_i \partial x_j}\right| \leq \frac{c_3}{(\lambda_1\lambda_2\lambda_3\lambda_4^3)^{1/2}} .$$

For every measurable set $A \subset H$ we have:

$$(6) \qquad \gamma(A) = \int \gamma_1(A^y)\, d\gamma_2(y) , \text{ where } A^y = \{x \in H_1 : x+y \in A\} .$$

By the symmetry of both γ and K_t we have the equality: $\gamma(K_t) - \gamma(K_t+r) = \gamma(K_t) - \gamma(K_t-r)$. Hence in order to estimate $\gamma(K_t) - \gamma(K_t+r)$ we may estimate the quantity:

$$(7) \qquad \gamma(K_t) - \gamma(K_t+r) = \tfrac{1}{2}\{ 2\gamma(K_t) - \gamma(K_t+r) - \gamma(K_t-r)\} =$$

$$= \tfrac{1}{2} \int_{H_1^{\perp}} \{2\gamma_1(K_t^y) - \gamma_1((K_t+r)^y) - \gamma_1((K_t-r)^y)\}\, d\gamma_2(y), \text{ by (6).}$$

We will show that for $\|r\| \leq \min(1,t/10)$ we have the following estimate:

$$(8) \qquad I = 2\gamma_1(K_t^y) - \gamma_1((K_t+r)^y) - \gamma_1((K_t-r)^y) \leq \frac{C_t}{(\lambda_1\lambda_2\lambda_3\lambda_4)^{1/2}} \|r\|^2 .$$

It is enough to show (8) because γ_2 is a probability measure and if $\|r\| \geq \min(1,t/10)$ then $\|r\|^2$ is bigger than a constant so in this case we may change C_t to get (7). Observe, that I is always less than 2.

Let us fix $y \in H_1$ and let $r = r_1 + r_2$, where $r_1 \in H_1$ and $r_2 \in H_1^{\perp}$. Of course $\|r\|^2 = \|r_1\|^2 + \|r_2\|^2$. Assume that $\|r\| \leq \min(1,t/10)$. We consider two cases.

Case 1. $\|y\| \geq t - 2\|r\|$. In this case all three four-dimensional balls K_t^y, $(K_t+r)^y$ and $(K_t-r)^y$ have radii less than $(t^2 - (t-3\|r\|)^2)^{\frac{1}{2}} \leq$ $\leq (6t\|r\|)^{\frac{1}{2}}$ (they may be empty), hence their Lebesgue measures are less than an absolute constant multiply of $36t^2 \|r\|^2$. In this case, using (4), we get:

$$(9) \qquad I \le \frac{c_4 t^2}{(\lambda_1\lambda_2\lambda_3\lambda_4)^{1/2}} \|r\|^2 .$$

Case 2. $\|y\| \le t - 2\|r\|$. In this case all three balls are non-empty. To estimate I we use the polar coordinates in H_1. Let u_y, u_{y-r} and u_{y+r} denote the radii of balls K_t^y, $(K_t-r)^y$ and $(K_t+r)^y$, respectively. Observe, that these radii depend on r_2 only (when t and y are fixed). Simple computations give:

$$(10) \quad u_y = (t^2 - \|y\|^2)^{\frac{1}{2}} , \quad u_{y-r} = (t^2 - \|y-r_2\|^2)^{\frac{1}{2}}, \quad u_{y+r} = (t^2 - \|y+r_2\|^2)^{\frac{1}{2}}$$

and, by our assumptions on r and y, we have the following estimates:

$$(11) \qquad |u_y^2 - u_{y-r}^2| \le c_5 t\|r\| , \qquad |u_y^2 - u_{y+r}^2| \le c_5 t\|r\| ,$$

and

$$(12) \qquad |u_y - u_{y-r}| \le \frac{\|y\| + \|r\|}{(t^2 - (\|y\| + \|r\|)^2)^{\frac{1}{2}}} \|r\| .$$

We need one more estimate of this type; using two terms of the Taylor formula we get (in our assumptions) :

$$(13) \qquad |2u_y^4 - u_{y+r}^4 - u_{y-r}^4| \le c_6 t^2\|r\|^2 .$$

Denote by σ the surface measure on the unit sphere S_1 in H_1 and by u_m the minimum of the values u_y, u_{y-r} and u_{y+r}. By the convexity arguments $u_m = u_{y-r}$ or $u_m = u_{y+r}$. The roles of r and -r are the same so we may assume that $u_m = u_{y-r}$. Recall that in polar coordinates (s,α) in H_1 we have:

$$(14) \quad I = \int_{S_1} \{ \int_0^{u_y} 2f(\alpha s)\alpha^3 d\alpha - \int_0^{u_{y-r}} f(\alpha s - r_1)\alpha^3 d\alpha - \int_0^{u_{y+r}} f(\alpha s + r_1)\alpha^3 d\alpha \} \sigma ds.$$

Now we can estimate the first part of I. Using two terms of the Taylor formula once again, the estimate (5) and the fact that $u_m \le t$ we get:

$$(15) \quad |2f(x) - f(x-r_1) - f(x+r_1)| \le \frac{c_7}{(\lambda_1\lambda_2\lambda_3\lambda_4^3)^{1/2}} \|r_1\|^2 ,$$

which gives the estimate:

$$(16) \quad I_1 = \int_{S_1} \int_0^{u_m} \{2f(\alpha s) - f(\alpha s - r_1) - f(\alpha s + r_1)\} \alpha^3 d\alpha \sigma ds \le \frac{c_8 t^4}{(\lambda_1\lambda_2\lambda_3\lambda_4)^{\frac{1}{2}}} \|r_1\|^2$$

In order to end the calculations we must estimate $I_2 = I - I_1$:

$$I_2 = \int_{S_1}\{\int_0^{u_y} 2f(\alpha s)\alpha^3 d\alpha - \int_{u_{y-r}}^{u_{y+r}} f(\alpha s+r_1)\alpha^3 d\alpha\}\sigma(ds) =$$

$$= \int_{S_1}\int_{u_{y-r}}^{u_y}\{f(\alpha s) - f(\alpha s+r_1)\}\alpha^3 d\alpha\sigma(ds) + \int_{S_1}\int_{u_{y-r}}^{u_y} f(\alpha s)\alpha^3 d\alpha\sigma(ds) -$$

$$- \int_{S_1}\int_{u_y}^{u_{y+r}} f(\alpha s+r_1)\alpha^3 d\alpha\sigma(ds) = A + B - C .$$

Because σ is a finite measure it is enough to estimate only terms under the first integral sign. From the estimate $|f(x) - f(x+r_1)| \leq \leq \frac{c_2}{(\lambda_1\lambda_2\lambda_3\lambda_4^2)^{1/2}}\|r_1\|$, using (11) we get:

$$\int_{u_{y-r}}^{u_y}\{f(\alpha s) - f(\alpha s+r_1)\}\alpha^3 d\alpha \leq \frac{c_2}{(\lambda_1\lambda_2\lambda_3\lambda_4^2)^{1/2}}\|r_1\|\tfrac{1}{4}(u_y^2+u_{y-r}^2)|u_y^2-u_{y-r}^2| \leq$$

$$\leq \frac{c_9 t^3}{(\lambda_1\lambda_2\lambda_3\lambda_4^2)^{1/2}}\|r\|^2 ,$$ which ends the estimate for A.

In the similar way we estimate B - C :

$$(17) \quad \int_{u_{y-r}}^{u_y}\{f(\alpha s) - f(u_y s)\}\alpha^3 d\alpha + \int_{u_{y-r}}^{u_y} f(u_y s)\alpha^3 d\alpha -$$

$$- \int_{u_y}^{u_{y+r}}\{f(\alpha s+r_1) - f(u_y s)\}\alpha^3 d\alpha - \int_{u_y}^{u_{y+r}} f(u_y s)\alpha^3 d\alpha .$$

The second and fourth terms of the above we estimate as follows:

$$\left|\int_{u_{y-r}}^{u_y} f(u_y s)\alpha^3 d\alpha - \int_{u_y}^{u_{y+r}} f(u_y s)\alpha^3 d\alpha\right| = |\tfrac{1}{4}f(u_y s)(2u_y^4-u_{y-r}^4-u_{y+r}^4)| \leq$$

$$\leq \frac{c_{10}t^2}{(\lambda_1\lambda_2\lambda_3\lambda_4)^{1/2}}\|r\|^2 ,$$ by (4) and (13) .

The first and the third terms can be estimated in the same way so we show the estimate only for one of them. Using (4) and (11) we get

$$\int_{u_y}^{u_{y+r}}\{f(\alpha s+r_1) - f(u_y s)\}\alpha^3 d\alpha \leq \frac{c_2}{(\lambda_1\lambda_2\lambda_3\lambda_4^2)^{1/2}}(|u_{y+r}-u_y| + \|r_1\|)\tfrac{1}{4}(u_{y+r}^2+u_y^2)\times$$

$$\times|u_{y+r}^2-u_y^2| \leq$$

$$\leq \frac{c_{11}t^4}{(\lambda_1\lambda_2\lambda_3\lambda_4^2)^{1/2}}||r||^2 + \frac{c_{12}}{(\lambda_1\lambda_2\lambda_3\lambda_4^2)^{1/2}} |u_{y+r} - u_y| c_{13}t||r|| \max(u^2_{y+r}, u^2_y).$$

But, using (12) and taking supremum over y, where $\|y\| \leq t - 2\|r\|$, we have the following inequality:

$$\sup \max (u^2_y, u^2_{y+r}) \times |u_{y+r} - u_y| \leq \sup \frac{\max(t^2 - \|y\|^2, t^2 - \|y+r\|^2)}{(t^2 - (\|y\| + \|r\|)^2)^{\frac{1}{2}}} t\|r\| \leq$$

$\leq c_{14}t^2\|r\|$, which ends the estimate for (17).

Now, denoting $\bar{\lambda}_4 = \min(1, \lambda_4)$, we can take the greatest value in all previous estimates:

$$\gamma(K_t) - \gamma(K_t + r) \leq \frac{c_{15}\max(1, t^4)}{(\lambda_1\lambda_2\lambda_3\bar{\lambda}_4^3)^{1/2}} \| r \|^2 ,$$

which ends the proof of the theorem.

Remark. In the proof of the above theorem we used only three properties of γ:

1) there exists at least four dimensional subspace of H and a measure γ_1 on it such that $\gamma = \gamma_1 \times \gamma_2$.
2) γ_1 has the density (with respect to the Lebesgue measure) which is twice differentiable
3) this density and its derivatives are bounded by a constant c_γ .

If γ is a measure with the above three properties then similar theorem is valid for γ with a constant depending on c_γ .

References.

[1] Rhee W.S., Talagrand M.(1986) Uniform convexity and the distribution of the norm for a Gaussian measure. Probab. Theory Rel. Fields 71, 59-67

[2] Ryznar M. (1988) On density of a stable uniformly convex norm.(submitted)

[3] Zak T. (1987) A formula for the density of the norm of stable random vectors in Hilbert spaces. (to appear in Probability and Math. Stat. 1989)

Institute of Mathematics
Technical University
Wybrzeze Wyspianskiego 27
50-370 Wroclaw
Poland

Probability Theory on Vector Spaces IV
Lancut, June'87, Springer's LNM 1391

Brownian Sheets in a Locally Pseudoconvex Metric Linear Space

A. M. Zapała

Institute of Mathematics
Maria Curie-Skłodowska University, Lublin, Poland

1. Introduction and preliminaries.

Let M be a real locally pseudoconvex metric linear space. Recall that the metric linear space is called locally pseudoconvex if it possesses a basis of neighbourhoods of zero consisted of starlike sets A characterized by the property $bA \subset A$ for all $b \in (0,1]$, with a finite modulus of concavity $c(A) = \inf\ (b>0 : A + A \subset bA)$. Under these assumptions a topology of M is generated by the F-norm

$$\|x\| = \sum_{L=1}^{\infty} \|x\|_L/(1+\|x\|_L)2^L \quad ,$$

where each $\|.\|_L$ is a p_L-homogeneous seminorm in M (see Rolewicz (1984), Th. 3.1.4). Denote by $\mathfrak{M}$ the smallest σ-field in M making Borel measurable all the seminorms $\|.\|_L : M \to R$ and suppose that $(M,\mathfrak{M})$ is a measurable vector space. The last condition is always true if M is separable, but we need not restrict considerations to such a case. Formulating some further results we will assume however that M is complete. It is worth to mention here that there are nonseparable spaces satisfying all the above requirements.

Let now ξ be a symmetric Gaussian random element taking values in the space $(M,\mathfrak{M})$; in other words if ξ_1 , ξ_2 are independent copies of ξ and t , s are real numbers such that $t^2 + s^2 = 1$, then $t\xi_1 + s\xi_2$ and $s\xi_1 - t\xi_2$ are independent random elements distributed according to ξ . Denoting by μ_t the distribution of $t^{1/2}\xi$, $t \geq 0$ we see that $\mu_t * \mu_s = \mu_{t+s}$, so the family $(\mu_t, t \geq 0)$ forms a convolution semigroup of probability measures. By means of such a convolution semigroup we can construct a continuous M-valued multiparameter Wiener process, for brevity called Brownian sheet in M .

Let $T = (R_+)^r$, $r \geq 1$, where R_+ is the set of nonnegative real numbers, and let ∂T be the boundary of T , i.e. the set of all r-tuples $t = (t_1, \dots t_r) \in T$ such that $t_i = 0$ for some $i = 1, \dots r$. In the sequel for $t, s \in T$ we also use the notation : $|t| = t_1 \cdot \dots \cdot t_r$, $t \vee s = (\max(t_1, s_1), \dots \max(t_r, s_r))$, and similarly $t \wedge s$ with max replaced by min . By Brownian sheet with values in M we mean a stochastic process $X = (X(t), t \in T)$ defined on a complete probability space $(\Omega, \mathcal{F}, P)$, satisfying the following conditions :

(1.1) for each $t \in T$, $X(t) : \Omega \to M$ is a random element in $(M, \mathfrak{M})$,

(1.2) the process X has independent increments

$$\Delta X(V) = \sum_{(1 \leq i \leq r:\ t_i = a_i \text{ or } t_i = b_i)} (-1)^{\operatorname{card}(j \leq r:\ t_j = a_j)} X(t_1, \dots t_r)$$

on disjoint rectangles $V = [a,b) = (t \in T : a_i \leq t_i < b_i , i = 1, \dots r)$,

(1.3) $\Delta X(V)$ has distribution $\mu_{\mathrm{vol}\, V}$ for $V = [a,b) \subset T$, where $\mathrm{vol}\, V = |b-a|$,

(1.4) realizations of the process X are a.s. continuous.

Condition (1.3) implies immediately that $X(t)$ is $\mu_{|t|}$ - distributed whenever $t \in T$, in particular $X(t) = 0$ a.s. iff $t \in T$.

Brownian sheets with values in Banach spaces were investigated by Morrow (1981), and the reader who would like to find some more extensive information about this subject is also referred to the previous paper by Zapała (1987). In Section 2 below we propose a classical and very simple method leading to the construction of continuous Brownian sheets in M . The more modern approach being a modification of Dudley's (1973) metric entropy criterion enables us to obtain continuous Gaussian processes indexed by metric compact spaces with values in metric Abelian groups (see Zapała (1988)) . The main result of the present article is Theorem 3.4 of Section 3, where an expansion of Brownian sheet in M into a series of independent Gaussian processes concentrated on 1-dimensional subspaces is given. Section 4 is devoted to a brief exposition of some further properties of the constructed process, e.g. we describe there the space of all admissible shifts for the distribution of Brownian sheet on the space of its continuous trajectories, moreover the most important properties of covariance operators of Brownian sheets are discussed.

2. Construction of Brownian sheets.

Let $M^T = \prod_{t\in T} M^t$ where $M^t = M$ for $t \in T$, and let $\mathcal{B}M^T$ be the cylindrical σ-field in M^T generated by $\mathcal{M}$. Using the well-known Kolmogorov s consistency theorem it is not difficult to check that on the space $(M^T, \mathcal{B}M^T)$ there exists the unique probability measure P whose projections onto finite products $M^{t(1)}\times \ldots \times M^{t(n)}$ coincide with distributions of $(X(t(1)),\ldots X(t(n)))$. Defining $Y = (Y(t), t\in T)$ on $\Omega = M^T$ by the formula $Y(t,\omega) = \omega(t)$, $t\in T$, $\omega \in \Omega$, we see that the process Y satisfies (1.1) - (1.3), thus it is the canonical representation of the Brownian sheet in M. In general the process Y need not satisfy (1.4), but we shall prove that Y possesses a modification X with a.s. continuous trajectories defined on $(\Omega,\mathcal{F},P)$, where $\mathcal{F}$ is the completion of $\mathcal{B}M^T$ under P.

Let $T_m = (t\in T: 0\le t_i \le m_i$ for $i=1,\ldots r)$, $\partial T_m = T_m \cap \partial T$ and $N = (1,2,\ldots)$. Throughout the paper $m \in N^r$, but this is not a substantial restriction. Consider a family of subsets of the set T_m with binary rational coordinates of the form $Q_m(n) = (k2^{-n} = (k_1 2^{-n},\ldots k_r 2^{-n}) \in T_m : k_i \in (0,1,\ldots m_i 2^n)$, $1\le i\le r)$. Introduce next a sequence of stochastic processes $(Y_n(t), t \in T_m)$ such that $Y_n(t) = Y(t)$ for $t \in Q_m(n)$ and $Y_n(t)$ is an interpolation of $Y|_{Q_m(n)}$ if $t \in T_m - Q_m(n)$. The mentioned interpolation we define as follows. Suppose $V = [k2^{-n},(k+1)2^{-n}]$, where $k2^{-n}, (k+1)2^{-n} \in Q_m(n)$ and $t \in V$. Let $v(1) = v(t,1),\ldots v(2^r) = v(t,2^r)$ denote the vertices of the rectangle V and let $v = v(t) = (2k+1)2^{-n-1}$. Further, let $V_i(t) = [(2v-v(i))\wedge t, (2v-v(i))\vee t]$ for each $v(i)$, $i=1,\ldots r$. Now we put

$$(2.1)\qquad Y_n(t) = (\mathrm{vol}\ V)^{-1} \sum_{i=1}^{2^r} Y(v(t,i))\ \mathrm{vol}\ V_i(t) \qquad \text{if } t \in T_m - Q_m(n).$$

The above definition is unambigious even if t is situated at the boundary of V and any other rectangle determined by the nearest points of $Q_m(n)$. Consequently Y_n are continuous functions of $t \in T_m$ for every $\omega \in \Omega$.

Lemma 2.1. <u>For each</u> $t \in T$

$$(2.2)\qquad \sup_{t\in T_m} [E\,\|Y_n(t) - Y(t)\|^q]^{1/q} \to 0 \qquad \text{as } n \to \infty.$$

Proof. It suffices to prove (2.2) for $m \in N^r$ and a nondecreasing sequence

$q_k = 2^{b_k}$ with $b_k \in N$, because $(E \|.\|^q)^{1/q}$ is a nondecreasing function of $q \in (0,\infty)$. Let b_k be positive integers such that $p_k' - r2^{b_k+1} > 0$, where $p_k' = \min (p_L : L<k)$. Obviously, we have

$$(2.3) \qquad \sum_{L\geq k} 2^{-L} \|.\| / (1 + \|.\|_L) \leq 2^{-k+1} ,$$

and it is readily seen that for $a_i \in R$ and $b,j \in N$,

$$(2.4) \qquad (a_1 + \dots + a_j)^{2^b} \leq j^{2^b-1} (a_1^{2^b} + \dots + a_j^{2^b}) .$$

Denote by $\mathcal{U}_n$ the collection of all the rectangles $V = [k2^{-n},(k+1)2^{-n}]$ with vertices in $Q_m(n)$ and choose $n > \log_2 (r|m|)$. Applying (2.4) we obtain

$$(2.5) \qquad \sup_{t\in T_m} E \| Y_n(t) - Y(t) \|^{2^b} \leq$$

$$\leq \max_{V\in\mathcal{U}_n} k^{2^b-1} \sum_{L<k} 2^{-L2^b} \sup_{t\in V} E \| Y_n(t) - Y(t) \|_L^{2^b} + k^{-1} (k/2^{k-1})^{2^b} .$$

Moreover, basing on a theorem proved by Byczkowski and Żak (1980), for $q \in (0,\infty)$ and $L \in N$ we have

$$(2.6) \qquad \| \mu_1 \|_{L,q} := (\int_M \| x \|_L^q \, d\mu_1(x))^{1/q} < \infty ,$$

because in fact each $\|.\|_L$ is exponentially integrable w.r.t. μ_1 . Therefore

$$(2.7) \qquad \sup_{t\in V} E \| Y_n(t) - Y(t) \|_L^{2^b} \leq$$

$$\leq 2^{r(2^b-1)+p_L nr2^b} \sup_{t\in V} \sum_{i=1}^{2^r} (\mathrm{vol}\, V_i(t))^{p_L 2^b} E \| Y(v(t,i)) - Y(t) \|_L^{2^b} .$$

Observe now that $Y(s) - Y(t)$ may be represented as a finite sum of increments $\Delta Y(U_j)$ with signs +,- on disjoint rectangles $U_j = U_j(s,t)$, $1\leq j\leq j_0\leq 2^r$, having vertices composed of coordinates of points s,t and 0 , such that the sum of U_j's is contained in the set $[0,s\vee t) - [0,s\wedge t)$. It follows that $Y(s) - Y(t)$ has distribution μ_A , where

$$A = \sum_{j=1}^{j_0} \mathrm{vol}\, U_j \leq \mathrm{vol}\, [0,s\vee t) - \mathrm{vol}\, [0,s\wedge t) \leq r\,|m|\, 2^{-n} , \quad s = v(t,i) .$$

Hence

$$(2.8) \qquad E \| Y(v(t,i)) - Y(t) \|_L^{2^b} \leq \| \mu_1 \|_{L,2^b}^{2^b} (r|m|)^{p_L 2^{b-1}} 2^{-np_L 2^{b-1}} .$$

Since $\text{vol}\, V_i(t) \le 2^{-nr}$, substituting (2.8) to (2.7) we conclude that

$$(2.9) \qquad \sup_{t \in V} E \, \| \, Y_n(t) - Y(t) \, \|_L^{2^b} \le 2^{r2^b} \, \|\mu_1\|_{L,2^b}^{2^b} \, (r|m|2^{-n})^{p_L 2^{b-1}} \quad ,$$

thus

$$(2.10) \qquad \sup_{t \in T_m} (E \, \| \, Y_n(t) - Y(t) \, \|^{2^{b_k}})^{1/2^{b_k}} \le$$

$$\le C(r,m,b,k) \, 2^{-n(p_k' - r/2^{b_k})} + k^{-1/2^{b_k}} \cdot k/2^{k-1} \quad ,$$

where $C(r,m,b,k) \le k2^r (r|m|)^{1/2} |m|^{1/2^{b_k}} \max_{L<k} \|\mu_1\|_{L,2^{b_k}} < \infty$. Given $\varepsilon > 0$ we can select sufficiently large $k \in N$ such that $k^{-1/2^{b_k}} \cdot k/2^{k-1} < \varepsilon/2$, and next large enough n such that $C(r,m,b,k) \, 2^{-n(p_k' - r/2^{b_k})} < \varepsilon/2$. Then $\sup_{t \in T_m} (E \, \| \, Y_n(t) - Y(t) \, \|^{2^{b_k}})^{1/2^{b_k}} < \varepsilon$, which completes the proof. #

Remark. It is worth to mention that Y may be the process with discontinuous trajectories, but it has to be continuous in $L^q(\Omega,\mathcal{F},P;M)$ (and hence stochastically) uniformly on each set T_m . It follows easily from the fact that Y on T_m is the uniform limit in $L^q(\Omega,\mathcal{F},P;M)$ of continuous stochastic processes $Y_n(t)$.

Lemma 2.2. <u>Under the assumption that</u> M <u>is a complete locally pseudo-convex metric linear space the sequences of continuous functions</u> $(Y_n(t,\omega), n \ge 1)$ <u>are convergent P - a.e. in the F-norm of</u> M <u>uniformly on</u> T_m <u>to some limits</u> $X_m(t,\omega)$ <u>being continuous functions of</u> $t \in T_m$ <u>with probability 1. The processes</u> $X_m = (X_m(t) \, , \, t \in T_m)$ <u>satisfy also conditions (1.1) - (1.3) restricted to</u> T_m.

Proof. As above without loss of generality we consider only the case $m \in N^r$. Repeating finitely many times if necessary the same seminorms in the sequence $(\|.\|_L)$ we obtain an F-norm equivalent to the original one, but coefficients of homogeneity are such that $p_n' 2^{b_n-1} - r \ge 1$, $2^{b_n} \le \ln n$ and $\max_{L<n} \| \mu_1 \|_{L,2^{b_n}} \le n$ for all sufficiently large n (if $2^{b_n} \in N$, then for the last inequality it is enough that $E_\mu \exp (\|x\|_n) (2^{b_n})! / n^{2^{b_n}} \le 1$) .

Let $d_n = \sup_{t \in T_m} \| \, Y_{n+1}(t) - Y_n(t) \, \|$ and $D_n = [d_n > \varepsilon_n]$. Arguing as in (2.3) - (2.9) we infer that

$$(2.11)\qquad P[D_n] \le \varepsilon_n^{-2^{b_n}} n^{2^{b_n}-1} \sum_{V\in U_n} \sum_{t\in(Q_m(n+1)-Q_m(n))\cap V} 2^{r(2^{b_n}-1)} \sum_{L<n} (2^{-L}2^{nrp_L})^{2^{b_n}} \cdot$$

$$\cdot \sum_{i=1}^{2^r} (\operatorname{vol} V_i(t))^{p_L 2^{b_n}} E\,\| Y(t) - Y(v(t,i)) \|_L^{2^{b_n}} + \varepsilon_n^{-2^{b_n}} n^{2^{b_n}-1} 2^{(-n+1)2^{b_n}} \le$$

$$\le \varepsilon_n^{-2^{b_n}} n^{2^{b_n}-1} |m|\, 2^{nr} (3^r-2^r)\, 2^{r2^{b_n+1}} (r|m|)^{2^{b_n}} \max_{L<n} \|\mu_1\|_{L,2^{b_n}}^{2^{b_n}} 2^{-np_n' 2^{b_n-1}} + (2n/\varepsilon_n 2^n)^{2^{b_n}} .$$

Let $\varepsilon_n = n^{-2}$. Then the last expression can be estimated from above by $|m|\,(3^r-2^r)\,(n^4 2^{2r}\sqrt{r|m|})^{2^{b_n}}\, 2^{-n(p_n' 2^{b_n-1}-r)} + (2n^3/2^n)^{2^{b_n}}$. Denoting by C a finite positive constant dependent only on r and m , for sufficiently large n we have $(n^4 2^{2r}\sqrt{r|m|})^{2^{b_n}} = \exp[\,2^{b_n} \ln(n^4 2^{2r}\sqrt{r|m|})\,] \le \exp[\,C(\ln n)^2\,]$. Consequently $\sum_n \varepsilon_n < \infty$ and $\sum_n P[D_n] < \infty$. Hence with probability 1 occur only finitely many events D_n . However, if $k<n$, then $\sup_{t\in T_m} \| Y_n(t) - Y_k(t) \| \le \sum_{j=k}^{n-1} d_j$ and $d_j \le \varepsilon_j$ for $j \ge j(\omega)$ a.s., thus the sequence Y_n converges uniformly on T_m in the F-norm of M with probability 1 . Denote by Ω_1 the convergence set of Y_n and put $X_m(t) = \lim_n Y_n(t)$ for $\omega \in \Omega_1$ and $X_m(t) = 0$ otherwise. Since X_m is P - a.e. the uniform limit of continuous functions, the process X_m is continuous. Lemma 2.1 implies that $P[\,X_m(t) = Y_m(t)\,] = 1$ for each $t \in T_m$, thus X_m satisfies (1.1) - (1.4) on T_m .

Theorem 2.3. <u>The stochastic process</u> Y <u>has a modification</u> X <u>with a.s. continuous trajectories.</u>

Proof. For each $m \in N^r$ we extend the definition of X_m setting $\overline{X}_m(t) = X_m(t)$ if $t \in T_m$ and $\overline{X}_m(t) = X_m(t_m)$ otherwise, where t_m is determined uniquely as a point of intersection of the line segment joining 0 to t with the boundary of T_m . Now $(\,\overline{X}_m(t)\,,\ t \in T_m\,)$ are continuous stochastic processes and $\overline{X}_{m+m'}(t) = \overline{X}_m(t)$ for $t \in T_m$ and $m,m' \in N^r$. Therefore there exists an a.s. limit

$$(2.12)\qquad X(t) = \lim_{m\to\infty} \overline{X}_m(t) \quad , \quad \text{where } m \to \infty \text{ means that } m_1,\dots m_r \to \infty .$$

Let $X(t) = 0$ on the excepcional set of all those $\omega \in \Omega$ for which the limit in (2.12) does not exist. The process X satisfies then conditions (1.1) - (1.4). #

3. Expansion of Brownian sheets in metric linear spaces into a series of 1-dimensional processes.

Another way leading to the construction of a multiparameter Wiener process in a metric linear space is an expansion of the process into a series of Brownian sheets concentrated on 1 - dimensional subspaces. To derive a.s. continuity of such a process the mentioned series should converge strongly uniformly on each set T_m with probability 1 .

Let M be a metric linear space. We say that a Gaussian random element ξ in $(M, \mathfrak{M})$ possesses an orthogonal expansion if there exist a sequence $(x_k) \subset M$ and a sequence $(g^{(k)})$ of independent standard normal random variables defined on the same probability space $(\Omega, \mathcal{F}, P)$ such that $\xi = \sum_{k=1}^{\infty} g^{(k)} x_k$ a.s. and the series converges with probability 1 in the F-norm of M . Series expansions for Gaussian random elements in separable Banach spaces were described by several authors, e.g. Jain and Kallianpur (1970), Kuelbs (1971) and LePage (1972). Later Byczkowski (1979) has proved the existence of orthogonal expansions for Gaussian random elements in separable Orlicz spaces L_ϕ , where ϕ is the Young function satisfying (Δ_2) condition, and recently (1987) he extended this result to Gaussian random elements with values in spaces that are isomorphic to separable metric linear spaces satisfying some additional assumptions. Roughly speaking he has shown that a symmetric Gaussian random vector possesses an orthogonal expansion if the space of additive functionals is sufficiently rich and separates the points of the basic space. Moreover Jurlewicz (1984) obtained a series expansion for the 1 - parameter Wiener process in a separable Frechét space with a p - homogeneous F - norm . We derive below an analogous expansion for the multiparameter Wiener process taking values in a pseudoconvex metric linear space. For our purposes the space M need not be separable if we already know that ξ has an orthogonal expansion. Unfortunately series expansions for Gaussian random elements were obtained only in a proper subclass of separable metric linear spaces, so if we want to have a constructive definition of Brownian sheets the method proposed in this section is substantiated in some narrow scope.

Let $(h_{n,j}^{(i)})$ be the system of Haar functions on $L^2([0,m_i])$, where $m_i \in N$ and $L^2(S)$ is the space of square-integrable with respect to Lebesgue measure Λ mappings on S equipped with the scalar product $(f,g) = \int_S f g \, d\Lambda$

The Haar functions are defined as follows:

$$h_{0,j_i}^{(i)}(u) = \begin{cases} 1 & \text{for } j_i \leq u < j_i + 1 , \\ 0 & \text{otherwise ,} \end{cases}$$

$j_i = 0,1,\ldots (m_i-1)$, and

$$h_{n_i,j_i}^{(i)}(u) = \begin{cases} 2^{(n_i-1)/2} & \text{if } j_i/2^{n_i-1} \leq u < (2j_i+1)/2^{n_i} , \\ -2^{(n_i-1)/2} & \text{if } (2j_i+1)/2^{n_i} \leq u < (j_i+1)/2^{n_i-1} , \\ 0 & \text{otherwise ,} \end{cases}$$

for $j_i = 0,1,\ldots (m_i 2^{n_i-1} - 1)$ and $n_i = 1,2,\ldots$. Denote $N_o = (0,1,\ldots)$, $W_n^{(z)} = (j=(j_1,\ldots j_r): (z_i-1)(2^{n_i-1} \vee 1) \leq j_i < z_i(2^{n_i-1} \vee 1))$ for $z_i = 1,\ldots m_i$, $i = 1,\ldots r$, and $W_n = \bigcup_{z\in[1,m]} W_n^{(z)}$ for a fixed $n \in N_o^r$. Introduce the functions

$$b_{n,j}(t) = \prod_{i=1}^{r} h_{n_i,j_i}^{(i)}(t_i) , \quad j \in W_n , \quad n \in N_o^r .$$

It can be easily seen that the family $(b_{n,j} : j \in W_n , n \in N_o^r)$ forms a complete orthonormal system in the space $L^2(T_m)$. In our further considerations we shall use the functions

$$H_{n,j}(t) = (\mathbf{1}_t , b_{n,j}) , \quad t \in T_m ,$$

where $\mathbf{1}_t$ is the indicator of the set $[0,t]$. Obviously $H_{n,j}(t)$ are continuous functions of t on T_m.

Suppose $\xi = \sum_{k=1}^{\infty} g^{(k)} x_k$ is an orthogonal expansion of the Gaussian random element ξ in $(M , \mathcal{M})$, where $(x_k) \subset M$ is a sequence of vectors and $g^{(k)}$ are standard normal random variables defined on the probability space $(\Omega,\mathcal{F},P)$. Let us observe that the transformation $g = (g^{(k)}) : \Omega \to R^{\infty}$ induces a distribution P' of g on the space R^{∞} with the cylindrical σ- field $\mathcal{R}^{\infty}$, namely P' is the product probability measure of standard normal distributions. Denoting $A_c = (g \in R^{\infty} : \sum_k g^{(k)} x_k$ converges $)$, and for each fixed arbitrarily sequence of numbers $\theta_k = \pm 1$, $B_c = (g \in R^{\infty} : \sum_k \theta_k g^{(k)} x_k$ converges) , by symmetry we have $P'[A_c] = P'[B_c]$. However $P'[A_c] = 1$, therefore $\sum_k \theta_k g^{(k)} x_k$ converges in $(M, \|.\|)$ with probability 1 , for $g^{-1}(B_c)$ is the convergence set of the last series and $P[g^{-1}(B_c)] = P'[B_c] = 1$. Similar arguments (via the space $(R^{\infty}, \mathcal{R}^{\infty})$) show that if $(f^{(k)})$ is a sequence of independent normal random variables with mean zero and variance $t > 0$ on an arbitrary probability space then the series $\zeta_t = \sum_k f^{(k)} x_k$ converges with probability 1 and $\mathcal{L}(\zeta_t) = \mathcal{L}(t^{1/2}\zeta)$, where $\zeta = \sum_k g^{(k)} x_k$ may be defined on another probability space.

Let $(\Omega^{\infty}, \mathcal{F}^{\infty}, P^{\infty})$ be the product probability space such that $\Omega^{\infty} = \prod_{n \in N_0^{\Gamma}} \prod_{j \in W_n} \Omega_{n,j}$ and $\Omega_{n,j} = \Omega$. Consider sequences $(g_{n,j}^{(k)})$ defined on $(\Omega^{\infty}, \mathcal{F}^{\infty}, P^{\infty})$ being independent copies of $(g^{(k)})$, i.e. $g_{n,j}^{(k)}(\omega) = g^{(k)}(\omega_{n,j})$ for $\omega = (\omega_{n,j} : j \in W_n , n \in N_0^{\Gamma})$. Then

$$\zeta_{n,j} = \sum_k \theta_{n,j}^{(k)} g_{n,j}^{(k)} x_k \quad P^{\infty}\text{- a.s.} , \quad \theta_{n,j}^{(k)} = \pm 1 ,$$

are independent random elements defined on $(\Omega^{\infty}, \mathcal{F}^{\infty}, P^{\infty})$ distributed according to ζ on $(\Omega, \mathcal{F}, P)$.

Lemma 3.1. _Let_ $\zeta_{n,j}$ _be random elements defined as above taking values in a complete pseudoconvex metric linear space_ M . _Then the series_

$$\sum_{n \in N_0^{\Gamma}} \sum_{j \in W_n} \zeta_{n,j} H_{n,j}(t)$$

converges with probability 1 in the F-norm $\|.\|$ _of_ M _uniformly in_ $t \in T_m$.

Proof. Since M is complete we only have to prove that partial sums of the considered series satisfy Cauchy's condition. First we will show that for a fixed seminorm $\|.\|_L$ the functions $\| \sum_{j \in W_n} \zeta_{n,j}(\omega) H_{n,j}(t) \|_L$, $n \in N^{\Gamma}$, are bounded uniformly in $t \in T_m$ by sequences of real numbers $K_n^{(L)}(\omega) > 0$ such that $\sum_{n \in N^{\Gamma}} K_n^{(L)}(\omega) < \infty$ P^{∞} - a.s. .

Observe that $H_{n,j}$ for $j \in W_n$, $n \in N^{\Gamma}$, are non-overlapping , so

$$(3.1) \qquad \sup_{t \in T_m} \| \sum_{j \in W_n} \zeta_{n,j}(\omega) H_{n,j}(t) \|_L \le$$

$$\leq \sup_{t \in T_m} \max_{j \in W_n} \| \zeta_{n,j}(\omega) \prod_{i=1}^{r} 2^{-(n_i+1)/2} H_{n,j}(t) \prod_{i=1}^{r} 2^{(n_i+1)/2} \|_L \leq$$

$$\leq \max_{j \in W_n} \| \zeta_{n,j}(\omega) \prod_{i=1}^{r} 2^{-(n_i+1)/2} \|_L .$$

On the basis of exponential integrability of a Gaussian random vector (see e.g. Byczkowski and Żak (1980)) for every $L \in N$ there exist finite positive constants A_L and C_L such that

$$(3.2) \qquad P^{\infty} [\| \zeta_{n,j} \|_L > \prod_{i=1}^{r} (m_i 2^{n_i - 1} + j_i)^{p_L/4}] \leq$$

$$\leq C_L \exp[-A_L \prod_{i=1}^{r} (m_i 2^{n_i-1} + j_i)^{1/2}] .$$

Moreover

$$(3.3) \qquad \sum_{n \in N^r} \sum_{j \in W_n} \exp [-A_L \prod_{i=1}^{r} (m_i 2^{n_i-1} + j_i)^{1/2}] \leq$$

$$\leq \sum_{n \in N^r} \sum_{j \in W_n} \exp [-A_L \sum_{i=1}^{r} (m_i 2^{n_i-1} + j_i)^{1/2} / r] \leq$$

$$\leq \prod_{i=1}^{r} \left[\sum_{n_i \in N} \exp [-A_L (m_i 2^{n_i-1})^{1/2} / r] m_i 2^{n_i-1} \right] < \infty .$$

Thus, according to the Borel-Cantelli lemma, for P^{∞} - a.a. $\omega \in \Omega^{\infty}$ we can find an r-tuple $n(\omega) \in N^r$ such that for all $n \in N^r$, $n \not\leq n(\omega)$, and $j \in W_n$ the inequality

$$\| \zeta_{n,j}(\omega) \|_L \leq \prod_{i=1}^{r} (m_i 2^{n_i-1} + j_i)^{p_L/4}$$

holds. Since $\|.\|_L$ is a p_L-homogeneous seminorm, on account of (3.1) we conclude that

$$(3.4) \qquad \sum_{n \in N^r} \sup_{t \in T_m} \| \sum_{j \in W_n} \zeta_{n,j}(\omega) H_{n,j}(t) \|_L \leq$$

$$\leq \sum_{\substack{n \leq n(\omega) \\ n \in N^r}} \max_{j \in W_n} \| \zeta_{n,j}(\omega) \|_L \prod_{i=1}^{r} 2^{-p_L(n_i+1)/2} + \sum_{\substack{n \not\leq n(\omega) \\ n \in N^r}} |m|^{p_L/4} \prod_{i=1}^{r} 2^{-p_L n_i/4} ,$$

and the last expression is finite because

$$\sum_{n \in N^r} \prod_{i=1}^{r} 2^{-p_L n_i/4} = \prod_{i=1}^{r} \left[\sum_{n_i \in N} 2^{-p_L n_i/4} \right] \leq (1 - 2^{p_L/4})^{-r} .$$

The right-hand side of (3.4) determines the required numbers $K_n^{(L)}(\omega)$.

The conclusion is now quite obvious. Given $\varepsilon > 0$ we select a number $h \in N$ such that $2^{-h+1} < \varepsilon/2$, and next we choose $n(\omega)$ such that for each $L \le h$ the inequality $\sum_{n \not\le n(\omega),\, n \in N^r} K_n^{(L)}(\omega) < \varepsilon/2$ holds. Then from (3.4) it follows that for $I, J > n(\omega)$,

$$\sup_{t \in T_m} \Big\| \sum_{\substack{n \le I \\ n \in N^r}} \sum_{j \in W_n'} \xi_{n,j}(\omega)\, H_{n,j}(t) - \sum_{\substack{n \le J \\ n \in N^r}} \sum_{j \in W_n''} \xi_{n,j}(\omega)\, H_{n,j}(t) \Big\| < \varepsilon ,$$

where $W_n' = W_n$ if $n < I$, $W_n'' = W_n$ if $n < J$, and $W_n', W_n'' \subset W_n$ otherwise. Hence we infer immediately that the series $\sum_{n \in N^r} \sum_{j \in W_n} \xi_{n,j} H_{n,j}(t)$ converges in the F-norm $\|.\|$ uniformly in $t \in T_m$. To estimate the sum $\sum_{n \in N_0^r - N^r} \sum_{j \in W_n} \xi_{n,j} H_{n,j}(t)$ we proceed as above replacing r by a smaller number $r' = r - s$, $1 \le s < r$, for fixed $n_{i_1} = \ldots = n_{i_s} = 0$.

In our further considerations we need some results which allow us to make various rearrangements of indices in a random series. Introduce the denotation $\mathcal{W} = [\, U = ((i_1,j_1),\ldots(i_n,j_n)) \subset N \times N : n = 1,2,\ldots \,]$. We say that a double series $\sum_{i,j} x_{i,j}$ in a Frechét space $(\mathcal{X}, |.|)$ converges with respect to (w.r.t.) the family $\mathcal{W}$ and write $\lim_{U \in \mathcal{W}} \sum_{(i,j) \in U} x_{i,j} = x$, if for each $\varepsilon > 0$ there exists a set $U \in \mathcal{W}$ such that for every $U' \in \mathcal{W}$, $U \subset U'$, we have $\| \sum_{(i,j) \in U'} x_{i,j} - x \| < \varepsilon$. Convergence of a single series $\sum_i x_i$ w.r.t. the family W of finite subsets of N is defined in an analogous manner. By a permutation p of the set N we mean a sequence $(p(n))$ (but not an array of sequences) such that the set $(p(1), p(2), \ldots)$ equals to N and $p(n) \ne p(n')$ for $n \ne n'$. A rearrangement (r_1, r_2) of the product $N \times N$ is defined to be a sequence of pairs $[\, (r_1(n), r_2(n)),\ n \in N \,]$ such that for each $i,j \in N$, $\sum_{n \in N,\, r_1(n) = i} [\, r_2(n) \,] = N$ and $\sum_{n \in N,\, r_2(n) = j} [\, r_1(n) \,] = N$, and furthermore $(r_1(n), r_2(n)) \ne (r_1(n'), r_2(n'))$ for $n \ne n'$. The next result is true as well for usual series, however having in mind direct applications there is more convenient to formulate it for double series.

Lemma 3.2. Let $(x_{i,j},\ (i,j) \in N \times N)$ be a double sequence in a Frechét space $\mathcal{X}$. Then the following conditions are equivalent:

1^0. For each array of numbers $(\theta_{i,j})$ with $\theta_{i,j} = \pm 1$ the series $\sum_{i\in N}\sum_{j\in N} \theta_{i,j} x_{i,j}$ converges,

2^0. For each array of numbers $(\delta_{i,j})$ with $\delta_{i,j} = 0$ or 1 the series $\sum_{i\in N}\sum_{j\in N} \delta_{i,j} x_{i,j}$ converges,

3^0. The limit $\lim_{U\in \mathcal{U}} \sum_{(i,j)\in U} x_{i,j}$ exists,

4^0. For each array $(\theta_{i,j})$ with $\theta_{i,j} = \pm 1$ the series $\sum_{i,j} \theta_{i,j} x_{i,j}$ converges w.r.t. the family $\mathcal{U}$,

5^0. For every permutations p, q of the set N and every rearrangement (r_1,r_2) of the set $N\times N$ three series

$$\sum_{i\in N}\sum_{j\in N} x_{p(i),q(j)} \ , \quad \sum_{j\in N}\sum_{i\in N} x_{p(i),q(j)} \quad \text{and} \quad \sum_{n\in N} x_{r_1(n),r_2(n)}$$

converge.

<u>Moreover, if one of the conditions</u> $1^0 - 5^0$ <u>is satisfied, then for all permutations</u> p, q <u>of the set</u> N <u>and each rearrangement</u> (r_1,r_2) <u>of</u> $N\times N$ <u>we have</u>

$$\sum_{i\in N}\sum_{j\in N} x_{p(i),q(j)} = \sum_{j\in N}\sum_{i\in N} x_{p(i),q(j)} = \sum_{n\in N} x_{r_1(n),r_2(n)} = x \ ,$$

<u>where</u> x <u>is the limit appearing in</u> 3^0.

We do not want to present here a detailed proof of this lemma because an analogous assertion for single series is rather known, e.g. the proof of implications $1^0 \Rightarrow 2^0 \Rightarrow 5^0 \Rightarrow 3^0$ given by Singer (1970) in Lemma 16.1 for a Banach space works too in a Frechét space, and the fact that 5^0 implies 2^0 is ensured by certain Orlicz's result (see Rolewicz (1984), Th. 3.8.2). Equivalence of conditions $1^0 - 4^0$ for double series in a Frechét space was proved by Jurlewicz (1984), Lemma 2.5, and under $1^0 - 4^0$ the additional statement without permutations, namely the equality

$$\sum_{i\in N}\sum_{j\in N} x_{i,j} = \sum_{j\in N}\sum_{i\in N} x_{i,j} = x \ ,$$

was also established by Jurlewicz (1984), Lemma 2.6. The proof of our Lemma 3.2 can be obtained by a modification of arguments known for single series and used by Jurlewicz, furthermore some other conditions equivalent to $1^0 - 5^0$ can be given.

Let Z be a random element in a Frechét space $(\mathcal{X}, |.|)$ with a σ-field $\mathfrak{X}$ such that $(\mathcal{X}, \mathfrak{X})$ is a measurable vector space and $|.|$ is $\mathfrak{X}$-measurable, and let $|Z|_0 = E\,|Z|/(1+|Z|)$. Recall that the F-norm $|.|_0$ metrizes convergence in probability in the Frechét space $L_0(\mathcal{X})$ of equivalence classes of random elements on a probability space $(\Omega,\mathcal{F},P)$ with values in $\mathcal{X}$.

Lemma 3.3. *Let $(Z_{i,j}, (i,j)\in N\times N)$ be a double sequence of independent separably valued random elements in a Frechét space $(\mathcal{X}, |.|)$. If for each array $(\theta_{i,j})$ with $\theta_{i,j} = \pm 1$ the series $\sum_{i\in N}\sum_{j\in N} \theta_{i,j} Z_{i,j}$ converges in probability, then the limit*

$$\lim_{U\in\mathcal{U}} \sum_{(i,j)\in U} Z_{i,j} = Z$$

exists in $(L_0(\mathcal{X}), |.|_0)$, and moreover for every permutations p, q of the set N and each rearrangement (r_1,r_2) of $N\times N$,

$$\sum_{i\in N}\sum_{j\in N} Z_{p(i),q(j)} = \sum_{j\in N}\sum_{i\in N} Z_{p(i),q(j)} = \sum_{n\in N} Z_{r_1(n),r_2(n)} = Z \quad \text{a.s.}.$$

Proof. This is a straightforward consequence of Lemma 3.2 applied to the Frechét space $(L_0(\mathcal{X}), |.|)$ and the fact that for a series of independent random elements taking values in a separable Frechét space $\mathcal{X}_0$ generated by $(Z_{i,j})$ convergence in probability and a.s. are equivalent (cf. Tortrat (1965) and Vahania, Tarieladze, Chobanian (1985), Th. 5.2.2, p. 212). The random sums of considered series are equal a.s. because they are equal in $L_0(\mathcal{X})$. #

Theorem 3.4. *Let M be a complete pseudoconvex metric linear space and let ζ be a symmetric Gaussian random element in M having an orthogonal expansion $\sum_{k\in N} g^{(k)} x_k = \zeta$ a.s., where $(x_k)\subset M$ and $(g^{(k)})$ are independent standard normal random variables defined on $(\Omega,\mathcal{F},P)$. Then for each $m\in N^r$ there exists a sequence $(B_k(t) : t\in T_m)$ of independent real-valued Brownian sheets defined on the probability space $(\Omega^\infty, \mathcal{F}^\infty, P^\infty)$ such that the series*

$$X(t) = \sum_{k\in N} B_k(t)\, x_k$$

converges in the F-norm $\|.\|$ of M uniformly in $t\in T_m$ with P^∞-probability 1 and its sum $(X(t), t\in T_m)$ is a Brownian sheet with values in M.

Proof. Let $C(T_m,M)$ be the space of continuous mappings $x : T_m \to M$ vanishing at the boundary ∂T_m endowed with the F-norm $\|x\|^{T_m} = \sup_{t\in T_m} \| x(t) \|$. Then $(C(T_m,M), \|.\|^{T_m})$ is a complete metric linear space.

Let $C_1(T_m,M)$ be a (separable) subspace generated by $(x_k H_{n,j}(t),\ k\in N, j\in W_n, n\in N_o^r)$. By Lemma 3.1 for each array $(\theta_{n,j}^{(k)})$, $\theta_{n,j}^{(k)} = \pm 1$, the series

$$\sum_{n\in N_o^r} \sum_{j\in W_n} \sum_{k\in N} \theta_{n,j}^{(k)} g_{n,j}^{(k)}(\omega)\, x_k\, H_{n,j}(t)$$

converges P^∞- a.s. in $C_1(T_m,M)$ and defines a random element in that space. Basing on Lemma 3.3 we conclude that with P^∞- probability 1 ,

$$X(t) := \sum_{n\in N_o^r} \sum_{j\in W_n} \Big(\sum_{k\in N} g_{n,j}^{(k)} x_k \Big) H_{n,j}(t) =$$

$$= \sum_{k\in N} \Big[\sum_{z\in[1,m]} \Big(\sum_{p=0}^{\infty} \sum_{n_1+\ldots+n_r=p} \sum_{j\in W_n^{(z)}} g_{n,j}^{(k)} H_{n,j}(t)\, x_k \Big) \Big] .$$

In view of Theorem 1, Corollary 1 and Theorem 2 by Park (1970), for all $k \in N$ the sums in brackets express Brownian sheets $(B_k(t)\ ,\ t\in T_m)$ and these processes are independent, therefore

$$X(t) = \sum_{k\in N} B_k(t)\, x_k \qquad P^\infty \text{ - a.s.} \quad .$$

Hence it follows easily that X satisfies (1.1) - (1.4) , because B_k satisfy analogous conditions in the real case. In particular, (1.3) is a consequence of the equality $\mathcal{L}(\Delta X(V)) = \mathcal{L}(\Delta(\sum_{k\in N} B_k x_k)(V)) = \mathcal{L}(\sum_{k\in N}(\Delta B_k(V))\, x_k) =$ $= \mathcal{L}((\mathrm{vol}\, V)^{1/2} \gamma)$, for $\Delta B_k(V)$ are independent zero mean normal random variables with variance $\mathrm{vol}\, V$. #

Further steps towards the extension of the process X on the whole set T may be done as in Theorem 2.3, but it is not necessary and we can obtain a direct series expansion for the Brownian sheet in M indexed by T .

Consider the functions $h_{n_i,j_i}^{(i)}$ for $j_i \geq 1$, $W_n^{(z)}$ without inequalities $z_i \leq m_i$, and $W_n = \bigcup_{z\in N^r} W_n^{(z)}$. Let $C(T,M)$ be the space of continuous functions $x : T \to M$ such that $x|_{\partial T} = 0$ viewed with a family of F-pseudonorms $(\|.\|^{T_m}\ ,\ m\in N^r)$. Investigating convergence in $C(T,M)$ we can restrict ourselves to a fixed set T_m , thus we have the following result.

C o r o l l a r y 3 . 5 . Under the assumptions of Theorem 3.4 there exists a sequence of independent real Brownian sheets $(B_k(t)\ ,\ t\in T)$ defined on

$(\Omega^{\infty}, \mathcal{F}^{\infty}, P^{\infty})$ such that the series

$$X(t) = \sum_{k \in N} B_k(t)\, x_k$$

converges in the space $C(T,M)$ with P^{∞}- probability 1 and its sum $(X(t) , t \in T)$ is a Brownian sheet with values in M .

4. Distributions of Brownian sheets.

In this section we describe more precisely properties of distributions of Brownian sheets with parameter sets T_m and T .

Let $(C(T_m,M) , \|.\|^{T_m})$ and $(C(T,M) , \|.\|^{T_m} , m \in N^r)$ be the spaces of M - valued continuous functions on T_m and T respectively vanishing on boundaries of these sets defined as previously. Denote $\| x \|_L^{T_m} = \sup_{t \in T_m} \| x(t) \|_L$. Obviously the topology in $C(T_m,M)$ is equivalent to the one determined by the family of seminorms $(\|.\|_L^{T_m} , L \in N)$, therefore $C(T_m,M)$ is a locally pseudoconvex metric linear space. The same is true for the space $C(T,M)$, because its topology is generated by the family $(\|.\|_L^{T_m} , L \in N , m \in N^r)$. We shall use besides the following denotations:

$p_t : C(T_m,M)$ (or $C(T,M)$) $\rightarrow M$ are projections given by $p_t(x) = x(t)$,

$\mathcal{C}_m$ = σ- field induced by the family $(p_t : C(T_m,M) \rightarrow (M,\mathcal{M}) , t \in T_m)$

$\mathcal{D}_m = (D = A \cap C(T_m,M) : A \in \mathcal{C}M^{T_m})$,

$\mathcal{E}_m$ = the smallest σ- field making Borel measurable all the mappings

$\| p_t(.) \|_L : C(T_m,M) \rightarrow R$, $L \in N$, $t \in T_m$.

We define next by analogy $\mathcal{C}$, $\mathcal{D}$ and $\mathcal{E}$ replacing T_m by T .

It can be easily seen that $\mathcal{C}_m = \mathcal{D}_m = \mathcal{E}_m$ and $\mathcal{C} = \mathcal{D} = \mathcal{E}$, furthermore all the seminorms $\|.\|_L^{T_m}$, $L \in N$, as well as the F-norm $\|.\|^{T_m}$ are $\mathcal{C}_m$ - measurable, while $\|.\|^{T_m}$ and $\|.\|_L^{T_m}$ for $L \in N$ and $m \in N^r$ are $\mathcal{C}$ - measurable. Since for cylindrical sets $C_m \in \mathcal{C}_m$ and $C \in \mathcal{C}$, $[X_m \in C_m]$ and $[X \in C]$ are cylinders in $M^T = \Omega$, X_m and X are random elements in $C(T_m,M)$ and $C(T,M)$ respectively. Thus we can define distributions of X_m and X :

$$P_{X_m}(D) = P [X_m^{-1}(D)] , \quad D \in \mathcal{C}_m ,$$

and

$$P_X(D) = P [X^{-1}(D)] , \quad D \in \mathcal{C} .$$

Note that $P_{X_m}(D) = P_X[h_m^{-1}(D)]$ for $D \in \mathcal{C}_m$, where h_m is given by the restriction of domain, $h_m(x) = x|_{T_m}$.

Proposition 4.1. <u>For an arbitrary</u> $m \in N^r$, P_{X_m} <u>is a symmetric Gaussian probability measure on</u> $(C(T_m,M), \mathcal{C}_m)$.

Proof. For the proof it is enough to check that distributions of "finite dimensional" vectors $(\Delta X(V_1),\ldots \Delta X(V_n)) : C(T_m,M) \to (M^n, \mathcal{M}^n)$, $n \geq 1$, where $V_1,\ldots V_n$ are disjoint rectangles contained in T_m, are symmetric Gaussian probability measures in the sense of definition employed in Section 1. #

By analogy to Proposition 4.1 we have the following result.

Proposition 4.2. P_X <u>is a symmetric Gaussian probability measure on</u> $(C(T,M), \mathcal{C})$.

The direct proof of Proposition 4.2 is essentially the same as that of Proposition 4.1.

Assume now that ξ is a symmetric Gaussian M-valued random element with distribution μ having an orthogonal expansion

$$(4.1) \qquad \xi = \sum_{k \in N} g^{(k)} x_k ,$$

where $(g^{(k)})$ is a sequence of independent standard normal random variables on $(\Omega,\mathcal{F},P)$, $(x_k) \subset M$ - a complete metric linear space, and the series on the right-hand side of (4.1) converges in the F-norm $\|.\|$ P - a.s. . In the next theorem we describe the space of all admissible shifts for μ, i.e. the set of all elements $y \in M$ such that $\mu * y$ and μ are mutually absolutely continuous.

Theorem 4.3. <u>Under the above assumptions the set of all admissible shifts for</u> μ <u>is equal to the Hilbert space</u> $H = \overline{\text{span}}(x_k, k \in N)$ <u>considered with the scalar product induced by equalities</u> $(x_k,x_j)_H = \delta_{k,j}$, <u>where the closure is taken in the topology of</u> H.

Proof. The proof may be obtained by a modification of arguments used by Byczkowski (1987), Theorem 2. #

Theorem 4.3 applies in particular to the distributions P_{X_m} and P_X, thus it allows us to formulate the next result.

Corollary 4.4. Let M be a locally pseudoconvex complete metric linear space and let ξ be a symmetric Gaussian random element in $(M, \mathfrak{M})$ with distribution μ having an orthogonal expansion (4.1). Then the set of all admissible shifts for the measure P_{X_m} is equal to the Hilbert space

$$\mathbf{H}_m = C'(T_m) \otimes_2 H ,$$

where $C'(T_m)$ consists of such functions $f : T_m \to R$, $f|_{\partial T_m} = 0$ that are absolutely continuous with respect to Lebesgue measure on T_m and satisfy the condition

$$| f |^2_{C'(T_m)} := \int_{T_m} (Df(t))^2 \, dt < \infty ,$$

$$Df(t) = \lim_{\substack{\mathrm{vol}\, K \to 0 \\ t \in K}} \Delta f(K) / \mathrm{vol}\, K , \quad K = r\text{-dimensional cubs} .$$

The Hilbert space of all admissible shifts for the measure P_X is equal to

$$\mathbf{H} = C'(T) \otimes_2 H ,$$

where $C'(T)$ is the Hilbert space defined analogously as $C'(T_m)$ with T_m replaced by T.

Considering Corollary 4.4 it is not difficult to establish now the full analogue of Lemma 4 given by Kuelbs and LePage (1973) for the Banach space case. We do not formulate this result here, but refer the reader to the mentioned paper.

Observe that if ξ has an orthogonal expansion (4.1), then by Lemma 3.1 X_m and X can be expanded into orthogonal series too. Therefore it is possible to define covariance operators Γ_ξ, Γ_m and Γ of ξ, X_m and X resp. . Since a separable Hilbert space is isomorphic to its topological dual, we define these operators as follows:

$$(\Gamma_\xi u) (v) = \int_M u(x) \, v(x) \, d\mu(x) = (u , v)_H ,$$

for $u, v \in H^*$, $\Gamma_\xi : H^* \to M \supset H$,

$$(\Gamma_m F)(G) = \int_{C(T_m,M)} F(x)\, G(x)\, dP_{X_m}(x) = (F , G)_{\mathbf{H}_m} ,$$

for $F, G \in \mathbf{H}_m^*$, $\Gamma_m : \mathbf{H}_m^* \to C(T_m,M) \supset \mathbf{H}_m$, and Γ is defined analogously as Γ_m . Clearly Γ_ξ , Γ_m and Γ possess similar properties as the usual covariance operators of Gaussian measures, in particular they are symmetric , positive definite and compact. The proof of their compactness will be given elsewhere. Note here only that as a consequence of this property the unit ball in H is a compact subset of M , and the same remark regards to the pairs of spaces $\mathbf{H}_m$, $C(T_m,M)$ and $\mathbf{H}$, $C(T,M)$.

Suppose now that ξ is a symmetric Gaussian random element in (M , $\mathfrak{M}$) in the sense of Bernstein, i.e. if ξ' is an independent copy of ξ , then random elements $\xi + \xi'$ and $\xi - \xi'$ are independent. In such a case all the above results are true as well, but for their proof instead of the quoted theorem by Byczkowski and Żak (1980) (see Lemma 2.1 , (2.6) and Lemma 3.1 , (3.2)) a result obtained recently by the author (1987) for symmetric Gaussian measures on groups must be applied.

Acknowledgements. This work was done while the author was visiting the Technical University at Wrocław. He is grateful to Prof. C. Ryll-Nardzewski and Prof. A. Weron for their hospitality during his stay and helpful suggestions leading to an improvment of the paper. The author also thanks sincerely Prof. T. Byczkowski for valuable discussions and for making accessible the typescript of a new article (1987) prepared just for publication.

References

1. Billingsley P. (1968) Convergence of Probability Measures. Wiley, New York.

2. Byczkowski T. (1979) Norm convergent expansion for L_ϕ-valued Gaussian random elements. Studia Math. 64, 87-95.

3. Byczkowski T. (1987) RKHS for Gaussian measures on metric spaces. Bull. Acad. Pol. Sci. Math. 35, 93-103.

4. Byczkowski T., Żak T. (1980) On the integrability of Gaussian random vectors. Lect. Notes in Math. 828, 21-29.

5. Dudley R.M. (1973) Sample functions of the Gaussian process. Ann. Probab. 1 , 66-103.

6. Jain N.C., Kallianpur G. (1970) Norm convergent expansions for Gaussian processes in Banach spaces. Proc. Amer. Math. Soc. 25, 890-895.

7. Jurlewicz T. (1984) Wiener process with values in an Orlicz space. Ph.D. thesis, Wrocław Technical Univ. Prep. No 16 (in Polish).

8. Kuelbs J. (1971) Expansions of vectors in a Banach space related to Gaussian measures. Proc. Amer. Math. Soc. 27, 364-370.

9. Kuelbs J., LePage R.D. (1973) The law of the iterated logarithm for Brownian motion in a Banach space. Trans. Amer. Math. Soc. 185, 253-264.

10. LePage R.D. (1972) Note relating Bochner integrals and reproducing kernels to series expansions on a Gaussian Banach space. Proc. Amer. Math. Soc. 32, 285-288.

11. Morrow G.J. (1981) Approximation of rectangular sums of B-valued random variables. Z. Wahr. verw. Geb. 57, 265-291.

12. Park W.J. (1970) A multi-parameter Gaussian process. Ann. Math. Stat. 41 , 1582-1595.

13. Rolewicz S. (1984) Metric Linear Spaces. PWN Warsaw, D. Reidel, Dordrecht.

14. Singer I. (1970) Bases in Banach Spaces I. Springer-Verlag, Berlin.

15. Tortrat A. (1965) Lois de probabilité sur un espace topologique complètement régulier et produits infinis â termes indépendants dans un group topologique. Ann. Inst. H. Poincaré 1, 217-237.

16. Vahania N.N., Tarieladze V.I., Chobanian S.A. (1985) Probability Distributions in Banach Spaces. Nauka, Moscow (in Russian).

17. Zapała A.M. (1987) Brownian sheets with values in a Banach space. Prep. No 79, Wrocław University Press.

18. Zapała A.M. (1987) On exponential integrability of a transferable pseudometric with respect to a Gaussian measure on a group. Bull. Acad. Pol. Sci. Math. 35, 597-600.

19. Zapała A.M. (1988) Continuity of Gaussian processes in groups (to appear).

Institute of Mathematics
Maria Curie-Skłodowska University
ul. Nowotki 10
20-031 Lublin , Poland .

LECTURE NOTES IN MATHEMATICS

Edited by A. Dold and B. Eckmann

Some general remarks on the publication of proceedings of congresses and symposia

Lecture Notes aim to report new developments - quickly, informally and at a high level. The following describes criteria and procedures which apply to proceedings volumes. The editors of a volume are strongly advised to inform contributors about these points at an early stage.

§1. One (or more) expert participant(s) of the meeting should act as the responsible editor(s) of the proceedings. They select the papers which are suitable (cf. §§ 2, 3) for inclusion in the proceedings, and have them individually refereed (as for a journal). It should not be assumed that the published proceedings must reflect conference events faithfully and in their entirety. Contributions to the meeting which are not included in the proceedings can be listed by title. The series editors will normally not interfere with the editing of a particular proceedings volume - except in fairly obvious cases, or on technical matters, such as described in §§ 2, 3. The names of the responsible editors appear on the title page of the volume.

§2. The proceedings should be reasonably homogeneous (concerned with a limited area). For instance, the proceedings of a congress on "Analysis" or "Mathematics in Wonderland" would normally not be sufficiently homogeneous.

One or two longer survey articles on recent developments in the field are often very useful additions to such proceedings - even if they do not correspond to actual lectures at the congress. An extensive introduction on the subject of the congress would be desirable.

§3. The contributions should be of a high mathematical standard and of current interest. Research articles should present new material and not duplicate other papers already published or due to be published. They should contain sufficient information and motivation and they should present proofs, or at least outlines of such, in sufficient detail to enable an expert to complete them. Thus resumes and mere announcements of papers appearing elsewhere cannot be included, although more detailed versions of a contribution may well be published in other places later.

Surveys, if included, should cover a sufficiently broad topic, and should in general not simply review the author's own recent research. In the case of surveys, exceptionally, proofs of results may not be necessary.

"Mathematical Reviews" and "Zentralblatt für Mathematik" require that papers in proceedings volumes carry an explicit statement that they are in final form and that no similar paper has been or is being submitted elsewhere, if these papers are to be considered for a review. Normally, papers that satisfy the criteria of the Lecture Notes in Mathematics series also satisfy this

.../...

requirement, but we would strongly recommend that the contributing authors be asked to give this guarantee explicitly at the beginning or end of their paper. There will occasionally be cases where this does not apply but where, for special reasons, the paper is still acceptable for LNM.

§4. Proceedings should appear soon after the meeeting. The publisher should, therefore, receive the complete manuscript within nine months of the date of the meeting at the latest.

§5. Plans or proposals for proceedings volumes should be sent to one of the editors of the series or to Springer-Verlag Heidelberg. They should give sufficient information on the conference or symposium, and on the proposed proceedings. In particular, they should contain a list of the expected contributions with their prospective length. Abstracts or early versions (drafts) of some of the contributions are very helpful.

§6. Lecture Notes are printed by photo-offset from camera-ready typed copy provided by the editors. For this purpose Springer-Verlag provides editors with technical instructions for the preparation of manuscripts and these should be distributed to all contributing authors. Springer-Verlag can also, on request, supply stationery on which the prescribed typing area is outlined. Some homogeneity in the presentation of the contributions is desirable.

Careful preparation of manuscripts will help keep production time short and ensure a satisfactory appearance of the finished book. The actual production of a Lecture Notes volume normally takes 6 -8 weeks.

Manuscripts should be at least 100 pages long. The final version should include a table of contents and as far as applicable a subject index.

§7. Editors receive a total of 50 free copies of their volume for distribution to the contributing authors, but no royalties. (Unfortunately, no reprints of individual contributions can be supplied.) They are entitled to purchase further copies of their book for their personal use at a discount of 33.3 %, other Springer mathematics books at a discount of 20 % directly from Springer-Verlag. Contributing authors may purchase the volume in which their article appears at a discount of 33.3 %.

Commitment to publish is made by letter of intent rather than by signing a formal contract. Springer-Verlag secures the copyright for each volume.

Vol. 1232: P.C. Schuur, Asymptotic Analysis of Soliton Problems. VIII, 180 pages. 1986.

Vol. 1233: Stability Problems for Stochastic Models. Proceedings, 1985. Edited by V.V. Kalashnikov, B. Penkov and V.M. Zolotarev. VI, 223 pages. 1986.

Vol. 1234: Combinatoire énumérative. Proceedings, 1985. Edité par G. Labelle et P. Leroux. XIV, 387 pages. 1986.

Vol. 1235: Séminaire de Théorie du Potentiel, Paris, No. 8. Directeurs: M. Brelot, G. Choquet et J. Deny. Rédacteurs: F. Hirsch et G. Mokobodzki. III, 209 pages. 1987.

Vol. 1236: Stochastic Partial Differential Equations and Applications. Proceedings, 1985. Edited by G. Da Prato and L. Tubaro. V, 257 pages. 1987.

Vol. 1237: Rational Approximation and its Applications in Mathematics and Physics. Proceedings, 1985. Edited by J. Gilewicz, M. Pindor and W. Siemaszko. XII, 350 pages. 1987.

Vol. 1238: M. Holz, K.-P. Podewski and K. Steffens, Injective Choice Functions. VI, 183 pages. 1987.

Vol. 1239: P. Vojta, Diophantine Approximations and Value Distribution Theory. X, 132 pages. 1987.

Vol. 1240: Number Theory, New York 1984–85. Seminar. Edited by D.V. Chudnovsky, G.V. Chudnovsky, H. Cohn and M.B. Nathanson. V, 324 pages. 1987.

Vol. 1241: L. Gårding, Singularities in Linear Wave Propagation. III, 125 pages. 1987.

Vol. 1242: Functional Analysis II, with Contributions by J. Hoffmann-Jørgensen et al. Edited by S. Kurepa, H. Kraljević and D. Butković. VII, 432 pages. 1987.

Vol. 1243: Non Commutative Harmonic Analysis and Lie Groups. Proceedings, 1985. Edited by J. Carmona, P. Delorme and M. Vergne. V, 309 pages. 1987.

Vol. 1244: W. Müller, Manifolds with Cusps of Rank One. XI, 158 pages. 1987.

Vol. 1245: S. Rallis, L-Functions and the Oscillator Representation. XVI, 239 pages. 1987.

Vol. 1246: Hodge Theory. Proceedings, 1985. Edited by E. Cattani, F. Guillén, A. Kaplan and F. Puerta. VII, 175 pages. 1987.

Vol. 1247: Séminaire de Probabilités XXI. Proceedings. Edité par J. Azéma, P.A. Meyer et M. Yor. IV, 579 pages. 1987.

Vol. 1248: Nonlinear Semigroups, Partial Differential Equations and Attractors. Proceedings, 1985. Edited by T.L. Gill and W.W. Zachary. IX, 185 pages. 1987.

Vol. 1249: I. van den Berg, Nonstandard Asymptotic Analysis. IX, 187 pages. 1987.

Vol. 1250: Stochastic Processes – Mathematics and Physics II. Proceedings 1985. Edited by S. Albeverio, Ph. Blanchard and L. Streit. VI, 359 pages. 1987.

Vol. 1251: Differential Geometric Methods in Mathematical Physics. Proceedings, 1985. Edited by P.L. García and A. Pérez-Rendón. VII, 300 pages. 1987.

Vol. 1252: T. Kaise, Représentations de Weil et GL_2 Algèbres de division et GL_n. VII, 203 pages. 1987.

Vol. 1253: J. Fischer, An Approach to the Selberg Trace Formula via the Selberg Zeta-Function. III, 184 pages. 1987.

Vol. 1254: S. Gelbart, I. Piatetski-Shapiro, S. Rallis. Explicit Constructions of Automorphic L-Functions. VI, 152 pages. 1987.

Vol. 1255: Differential Geometry and Differential Equations. Proceedings, 1985. Edited by C. Gu, M. Berger and R.L. Bryant. XII, 243 pages. 1987.

Vol. 1256: Pseudo-Differential Operators. Proceedings, 1986. Edited by H.O. Cordes, B. Gramsch and H. Widom. X, 479 pages. 1987.

Vol. 1257: X. Wang, On the C*-Algebras of Foliations in the Plane. V, 165 pages. 1987.

Vol. 1258: J. Weidmann, Spectral Theory of Ordinary Differential Operators. VI, 303 pages. 1987.

Vol. 1259: F. Cano Torres, Desingularization Strategies for Three-Dimensional Vector Fields. IX, 189 pages. 1987.

Vol. 1260: N.H. Pavel, Nonlinear Evolution Operators and Semigroups. VI, 285 pages. 1987.

Vol. 1261: H. Abels, Finite Presentability of S-Arithmetic Groups. Compact Presentability of Solvable Groups. VI, 178 pages. 1987.

Vol. 1262: E. Hlawka (Hrsg.), Zahlentheoretische Analysis II. Seminar, 1984–86. V, 158 Seiten. 1987.

Vol. 1263: V.L. Hansen (Ed.), Differential Geometry. Proceedings, 1985. XI, 288 pages. 1987.

Vol. 1264: Wu Wen-tsün, Rational Homotopy Type. VIII, 219 pages. 1987.

Vol. 1265: W. Van Assche, Asymptotics for Orthogonal Polynomials. VI, 201 pages. 1987.

Vol. 1266: F. Ghione, C. Peskine, E. Sernesi (Eds.), Space Curves. Proceedings, 1985. VI, 272 pages. 1987.

Vol. 1267: J. Lindenstrauss, V.D. Milman (Eds.), Geometrical Aspects of Functional Analysis. Seminar. VII, 212 pages. 1987.

Vol. 1268: S.G. Krantz (Ed.), Complex Analysis. Seminar, 1986. VII, 195 pages. 1987.

Vol. 1269: M. Shiota, Nash Manifolds. VI, 223 pages. 1987.

Vol. 1270: C. Carasso, P.-A. Raviart, D. Serre (Eds.), Nonlinear Hyperbolic Problems. Proceedings, 1986. XV, 341 pages. 1987.

Vol. 1271: A.M. Cohen, W.H. Hesselink, W.L.J. van der Kallen, J.R. Strooker (Eds.), Algebraic Groups Utrecht 1986. Proceedings. XII, 284 pages. 1987.

Vol. 1272: M.S. Livšic, L.L. Waksman, Commuting Nonselfadjoint Operators in Hilbert Space. III, 115 pages. 1987.

Vol. 1273: G.-M. Greuel, G. Trautmann (Eds.), Singularities, Representation of Algebras, and Vector Bundles. Proceedings, 1985. XIV, 383 pages. 1987.

Vol. 1274: N. C. Phillips, Equivariant K-Theory and Freeness of Group Actions on C*-Algebras. VIII, 371 pages. 1987.

Vol. 1275: C.A. Berenstein (Ed.), Complex Analysis I. Proceedings, 1985–86. XV, 331 pages. 1987.

Vol. 1276: C.A. Berenstein (Ed.), Complex Analysis II. Proceedings, 1985–86. IX, 320 pages. 1987.

Vol. 1277: C.A. Berenstein (Ed.), Complex Analysis III. Proceedings, 1985–86. X, 350 pages. 1987.

Vol. 1278: S.S. Koh (Ed.), Invariant Theory. Proceedings, 1985. V, 102 pages. 1987.

Vol. 1279: D. Ieşan, Saint-Venant's Problem. VIII, 162 Seiten. 1987.

Vol. 1280: E. Neher, Jordan Triple Systems by the Grid Approach. XII, 193 pages. 1987.

Vol. 1281: O.H. Kegel, F. Menegazzo, G. Zacher (Eds.), Group Theory. Proceedings, 1986. VII, 179 pages. 1987.

Vol. 1282: D.E. Handelman, Positive Polynomials, Convex Integral Polytopes, and a Random Walk Problem. XI, 136 pages. 1987.

Vol. 1283: S. Mardešić, J. Segal (Eds.), Geometric Topology and Shape Theory. Proceedings, 1986. V, 261 pages. 1987.

Vol. 1284: B.H. Matzat, Konstruktive Galoistheorie. X, 286 pages. 1987.

Vol. 1285: I.W. Knowles, Y. Saitō (Eds.), Differential Equations and Mathematical Physics. Proceedings, 1986. XVI, 499 pages. 1987.

Vol. 1286: H.R. Miller, D.C. Ravenel (Eds.), Algebraic Topology. Proceedings, 1986. VII, 341 pages. 1987.

Vol. 1287: E.B. Saff (Ed.), Approximation Theory, Tampa. Proceedings, 1985–1986. V, 228 pages. 1987.

Vol. 1288: Yu. L. Rodin, Generalized Analytic Functions on Riemann Surfaces. V, 128 pages, 1987.

Vol. 1289: Yu. I. Manin (Ed.), K-Theory, Arithmetic and Geometry. Seminar, 1984–1986. V, 399 pages. 1987.

Vol. 1290: G. Wüstholz (Ed.), Diophantine Approximation and Transcendence Theory. Seminar, 1985. V, 243 pages. 1987.

Vol. 1291: C. Mœglin, M.-F. Vignéras, J.-L. Waldspurger, Correspondances de Howe sur un Corps p-adique. VII, 163 pages. 1987

Vol. 1292: J.T. Baldwin (Ed.), Classification Theory. Proceedings, 1985. VI, 500 pages. 1987.

Vol. 1293: W. Ebeling, The Monodromy Groups of Isolated Singularities of Complete Intersections. XIV, 153 pages. 1987.

Vol. 1294: M. Queffélec, Substitution Dynamical Systems – Spectral Analysis. XIII, 240 pages. 1987.

Vol. 1295: P. Lelong, P. Dolbeault, H. Skoda (Réd.), Séminaire d'Analyse P. Lelong – P. Dolbeault – H. Skoda. Seminar, 1985/1986. VII, 283 pages. 1987.

Vol. 1296: M.-P. Malliavin (Ed.), Séminaire d'Algèbre Paul Dubreil et Marie-Paule Malliavin. Proceedings, 1986. IV, 324 pages. 1987.

Vol. 1297: Zhu Y.-l., Guo B.-y. (Eds.), Numerical Methods for Partial Differential Equations. Proceedings. XI, 244 pages. 1987.

Vol. 1298: J. Aguadé, R. Kane (Eds.), Algebraic Topology, Barcelona 1986. Proceedings. X, 255 pages. 1987.

Vol. 1299: S. Watanabe, Yu.V. Prokhorov (Eds.), Probability Theory and Mathematical Statistics. Proceedings, 1986. VIII, 589 pages. 1988.

Vol. 1300: G.B. Seligman, Constructions of Lie Algebras and their Modules. VI, 190 pages. 1988.

Vol. 1301: N. Schappacher, Periods of Hecke Characters. XV, 160 pages. 1988.

Vol. 1302: M. Cwikel, J. Peetre, Y. Sagher, H. Wallin (Eds.), Function Spaces and Applications. Proceedings, 1986. VI, 445 pages. 1988.

Vol. 1303: L. Accardi, W. von Waldenfels (Eds.), Quantum Probability and Applications III. Proceedings, 1987. VI, 373 pages. 1988.

Vol. 1304: F.Q. Gouvêa, Arithmetic of p-adic Modular Forms. VIII, 121 pages. 1988.

Vol. 1305: D.S. Lubinsky, E.B. Saff, Strong Asymptotics for Extremal Polynomials Associated with Weights on $\mathbb{R}$. VII, 153 pages. 1988.

Vol. 1306: S.S. Chern (Ed.), Partial Differential Equations. Proceedings, 1986. VI, 294 pages. 1988.

Vol. 1307: T. Murai, A Real Variable Method for the Cauchy Transform, and Analytic Capacity. VIII, 133 pages. 1988.

Vol. 1308: P. Imkeller, Two-Parameter Martingales and Their Quadratic Variation. IV, 177 pages. 1988.

Vol. 1309: B. Fiedler, Global Bifurcation of Periodic Solutions with Symmetry. VIII, 144 pages. 1988.

Vol. 1310: O.A. Laudal, G. Pfister, Local Moduli and Singularities. V, 117 pages. 1988.

Vol. 1311: A. Holme, R. Speiser (Eds.), Algebraic Geometry, Sundance 1986. Proceedings. VI, 320 pages. 1988.

Vol. 1312: N.A. Shirokov, Analytic Functions Smooth up to the Boundary. III, 213 pages. 1988.

Vol. 1313: F. Colonius, Optimal Periodic Control. VI, 177 pages. 1988.

Vol. 1314: A. Futaki, Kähler-Einstein Metrics and Integral Invariants. IV, 140 pages. 1988.

Vol. 1315: R.A. McCoy, I. Ntantu, Topological Properties of Spaces of Continuous Functions. IV, 124 pages. 1988.

Vol. 1316: H. Korezlioglu, A.S. Ustunel (Eds.), Stochastic Analysis and Related Topics. Proceedings, 1986. V, 371 pages. 1988.

Vol. 1317: J. Lindenstrauss, V.D. Milman (Eds.), Geometric Aspects of Functional Analysis. Seminar, 1986–87. VII, 289 pages. 1988.

Vol. 1318: Y. Felix (Ed.), Algebraic Topology – Rational Homotopy. Proceedings, 1986. VIII, 245 pages. 1988

Vol. 1319: M. Vuorinen, Conformal Geometry and Quasiregular Mappings. XIX, 209 pages. 1988.

Vol. 1320: H. Jürgensen, G. Lallement, H.J. Weinert (Eds.), Semigroups, Theory and Applications. Proceedings, 1986. X, 416 pages. 1988.

Vol. 1321: J. Azéma, P.A. Meyer, M. Yor (Eds.), Séminaire de Probabilités XXII. Proceedings. IV, 600 pages. 1988.

Vol. 1322: M. Métivier, S. Watanabe (Eds.), Stochastic Analysis. Proceedings, 1987. VII, 197 pages. 1988.

Vol. 1323: D.R. Anderson, H.J. Munkholm, Boundedly Controlled Topology. XII, 309 pages. 1988.

Vol. 1324: F. Cardoso, D.G. de Figueiredo, R. Iório, O. Lopes (Eds.), Partial Differential Equations. Proceedings, 1986. VIII, 433 pages. 1988.

Vol. 1325: A. Truman, I.M. Davies (Eds.), Stochastic Mechanics and Stochastic Processes. Proceedings, 1986. V, 220 pages. 1988.

Vol. 1326: P.S. Landweber (Ed.), Elliptic Curves and Modular Forms in Algebraic Topology. Proceedings, 1986. V, 224 pages. 1988.

Vol. 1327: W. Bruns, U. Vetter, Determinantal Rings. VII,236 pages. 1988.

Vol. 1328: J.L. Bueso, P. Jara, B. Torrecillas (Eds.), Ring Theory. Proceedings, 1986. IX, 331 pages. 1988.

Vol. 1329: M. Alfaro, J.S. Dehesa, F.J. Marcellan, J.L. Rubio de Francia, J. Vinuesa (Eds.): Orthogonal Polynomials and their Applications. Proceedings, 1986. XV, 334 pages. 1988.

Vol. 1330: A. Ambrosetti, F. Gori, R. Lucchetti (Eds.), Mathematical Economics. Montecatini Terme 1986. Seminar. VII, 137 pages. 1988.

Vol. 1331: R. Bamón, R. Labarca, J. Palis Jr. (Eds.), Dynamical Systems, Valparaiso 1986. Proceedings. VI, 250 pages. 1988.

Vol. 1332: E. Odell, H. Rosenthal (Eds.), Functional Analysis. Proceedings, 1986–87. V, 202 pages. 1988.

Vol. 1333: A.S. Kechris, D.A. Martin, J.R. Steel (Eds.), Cabal Seminar 81–85. Proceedings, 1981–85. V, 224 pages. 1988.

Vol. 1334: Yu.G. Borisovich, Yu. E. Gliklikh (Eds.), Global Analysis – Studies and Applications III. V, 331 pages. 1988.

Vol. 1335: F. Guillén, V. Navarro Aznar, P. Pascual-Gainza, F. Puerta, Hyperrésolutions cubiques et descente cohomologique. XII, 192 pages. 1988.

Vol. 1336: B. Helffer, Semi-Classical Analysis for the Schrödinger Operator and Applications. V, 107 pages. 1988.

Vol. 1337: E. Sernesi (Ed.), Theory of Moduli. Seminar, 1985. VIII, 232 pages. 1988.

Vol. 1338: A.B. Mingarelli, S.G. Halvorsen, Non-Oscillation Domains of Differential Equations with Two Parameters. XI, 109 pages. 1988.

Vol. 1339: T. Sunada (Ed.), Geometry and Analysis of Manifolds. Procedings, 1987. IX, 277 pages. 1988.

Vol. 1340: S. Hildebrandt, D.S. Kinderlehrer, M. Miranda (Eds.), Calculus of Variations and Partial Differential Equations. Proceedings, 1986. IX, 301 pages. 1988.

Vol. 1341: M. Dauge, Elliptic Boundary Value Problems on Corner Domains. VIII, 259 pages. 1988.

Vol. 1342: J.C. Alexander (Ed.), Dynamical Systems. Proceedings, 1986–87. VIII, 726 pages. 1988.

Vol. 1343: H. Ulrich, Fixed Point Theory of Parametrized Equivariant Maps. VII, 147 pages. 1988.

Vol. 1344: J. Král, J. Lukeš, J. Netuka, J. Veselý (Eds.), Potential Theory – Surveys and Problems. Proceedings, 1987. VIII, 271 pages. 1988.

Vol. 1345: X. Gomez-Mont, J. Seade, A. Verjovski (Eds.), Holomorphic Dynamics. Proceedings, 1986. VII, 321 pages. 1988.

Vol. 1346: O. Ya. Viro (Ed.), Topology and Geometry – Rohlin Seminar. XI, 581 pages. 1988.

Vol. 1347: C. Preston, Iterates of Piecewise Monotone Mappings on an Interval. V, 166 pages. 1988.

Vol. 1348: F. Borceux (Ed.), Categorical Algebra and its Applications. Proceedings, 1987. VIII, 375 pages. 1988.

Vol. 1349: E. Novak, Deterministic and Stochastic Error Bounds in Numerical Analysis. V, 113 pages. 1988.